IET ENERGY ENGINEERING SERIES 92

Power Quality in Future Electrical Power Systems

Other volumes in this series:

Volume 1 **Power Circuit Breaker Theory and Design** C.H. Flurscheim (Editor)
Volume 4 **Industrial Microwave Heating** A.C. Metaxas and R.J. Meredith
Volume 7 **Insulators for High Voltages** J.S.T. Looms
Volume 8 **Variable Frequency AC Motor Drive Systems** D. Finney
Volume 10 **SF_6 Switchgear** H.M. Ryan and G.R. Jones
Volume 11 **Conduction and Induction Heating** E.J. Davies
Volume 13 **Statistical Techniques for High Voltage Engineering** W. Hauschild and W. Mosch
Volume 14 **Uninterruptible Power Supplies** J. Platts and J.D. St Aubyn (Editors)
Volume 15 **Digital Protection for Power Systems** A.T. Johns and S.K. Salman
Volume 16 **Electricity Economics and Planning** T.W. Berrie
Volume 18 **Vacuum Switchgear** A. Greenwood
Volume 19 **Electrical Safety: A guide to causes and prevention of hazards** J. Maxwell Adams
Volume 21 **Electricity Distribution Network Design, 2nd Edition** E. Lakervi and E.J. Holmes
Volume 22 **Artificial Intelligence Techniques in Power Systems** K. Warwick, A.O. Ekwue and R. Aggarwal (Editors)
Volume 24 **Power System Commissioning and Maintenance Practice** K. Harker
Volume 25 **Engineers' Handbook of Industrial Microwave Heating** R.J. Meredith
Volume 26 **Small Electric Motors** H. Moczala *et al.*
Volume 27 **AC-DC Power System Analysis** J. Arrillaga and B.C. Smith
Volume 29 **High Voltage Direct Current Transmission, 2nd Edition** J. Arrillaga
Volume 30 **Flexible AC Transmission Systems (FACTS)** Y.-H. Song (Editor)
Volume 31 **Embedded Generation** N. Jenkins *et al.*
Volume 32 **High Voltage Engineering and Testing, 2nd Edition** H.M. Ryan (Editor)
Volume 33 **Overvoltage Protection of Low-Voltage Systems, Revised Edition** P. Hasse
Volume 36 **Voltage Quality in Electrical Power Systems** J. Schlabbach *et al.*
Volume 37 **Electrical Steels for Rotating Machines** P. Beckley
Volume 38 **The Electric Car: Development and future of battery, hybrid and fuel-cell cars** M. Westbrook
Volume 39 **Power Systems Electromagnetic Transients Simulation** J. Arrillaga and N. Watson
Volume 40 **Advances in High Voltage Engineering** M. Haddad and D. Warne
Volume 41 **Electrical Operation of Electrostatic Precipitators** K. Parker
Volume 43 **Thermal Power Plant Simulation and Control** D. Flynn
Volume 44 **Economic Evaluation of Projects in the Electricity Supply Industry** H. Khatib
Volume 45 **Propulsion Systems for Hybrid Vehicles** J. Miller
Volume 46 **Distribution Switchgear** S. Stewart
Volume 47 **Protection of Electricity Distribution Networks, 2nd Edition** J. Gers and E. Holmes
Volume 48 **Wood Pole Overhead Lines** B. Wareing
Volume 49 **Electric Fuses, 3rd Edition** A. Wright and G. Newbery
Volume 50 **Wind Power Integration: Connection and system operational aspects** B. Fox *et al.*
Volume 51 **Short Circuit Currents** J. Schlabbach
Volume 52 **Nuclear Power** J. Wood
Volume 53 **Condition Assessment of High Voltage Insulation in Power System Equipment** R.E. James and Q. Su
Volume 55 **Local Energy: Distributed generation of heat and power** J. Wood
Volume 56 **Condition Monitoring of Rotating Electrical Machines** P. Tavner, L. Ran, J. Penman and H. Sedding
Volume 57 **The Control Techniques Drives and Controls Handbook, 2nd Edition** B. Drury
Volume 58 **Lightning Protection** V. Cooray (Editor)
Volume 59 **Ultracapacitor Applications** J.M. Miller
Volume 62 **Lightning Electromagnetics** V. Cooray
Volume 63 **Energy Storage for Power Systems, 2nd Edition** A. Ter-Gazarian
Volume 65 **Protection of Electricity Distribution Networks, 3rd Edition** J. Gers
Volume 66 **High Voltage Engineering Testing, 3rd Edition** H. Ryan (Editor)
Volume 67 **Multicore Simulation of Power System Transients** F.M. Uriate
Volume 68 **Distribution System Analysis and Automation** J. Gers
Volume 69 **The Lightening Flash, 2nd Edition** V. Cooray (Editor)
Volume 70 **Economic Evaluation of Projects in the Electricity Supply Industry, 3rd Edition** H. Khatib
Volume 72 **Control Circuits in Power Electronics: Practical issues in design and implementation** M. Castilla (Editor)
Volume 73 **Wide Area Monitoring, Protection and Control Systems: The enabler for smarter grids** A. Vaccaro and A. Zobaa (Editors)
Volume 74 **Power Electronic Converters and Systems: Frontiers and applications** A.M. Trzynadlowski (Editor)
Volume 75 **Power Distribution Automation** B. Das (Editor)
Volume 76 **Power System Stability: Modelling, analysis and control** B. Om P. Malik
Volume 78 **Numerical Analysis of Power System Transients and Dynamics** A. Ametani (Editor)
Volume 79 **Vehicle-to-Grid: Linking electric vehicles to the smart grid** J. Lu and J. Hossain (Editors)
Volume 81 **Cyber-Physical-Social Systems and Constructs in Electric Power Engineering** Siddharth Suryanarayanan, Robin Roche and Timothy M. Hansen (Editors)
Volume 82 **Periodic Control of Power Electronic Converters** F. Blaabjerg, K. Zhou, D. Wang and Y. Yang
Volume 86 **Advances in Power System Modelling, Control and Stability Analysis** F. Milano (Editor)
Volume 88 **Smarter Energy: From Smart Metering to the Smart Grid** H. Sun, N. Hatziargyriou, H.V. Poor, L. Carpanini and M.A. Sánchez Fornié (Editors)
Volume 89 **Hydrogen Production, Separation and Purification for Energy** A. Basile, F. Dalena, J. Tong, and T.N. Veziroğlu (Editors)
Volume 93 **Cogeneration and District Energy Systems: Modelling, Analysis and Optimization** M.A. Rosen and S. Koohi-Fayegh
Volume 90 **Clean Energy Microgrids** S. Obara and J. Morel (Editors)
Volume 95 **Communication, Control and Security Challenges for the Smart Grid** S.M. Muyeen and S. Rahman (Editors)
Volume 97 **Synchronized Phasor Measurements for Smart Grids** M.J.B. Reddy and D.K. Mohanta (Editors)
Volume 100 **Modeling and Dynamic Behaviour of Hydropower Plants** N. Kishor and J. Fraile-Ardanuy (Editors)
Volume 101 **Methane and Hydrogen for Energy Storage** R. Carriveau and David S.-K. Ting
Volume 905 **Power System Protection, 4 volumes**

Power Quality in Future Electrical Power Systems

Edited by
Ahmed F. Zobaa and Shady H. E. Abdel Aleem

The Institution of Engineering and Technology

Published by The Institution of Engineering and Technology, London, United Kingdom

The Institution of Engineering and Technology is registered as a Charity in England & Wales (no. 211014) and Scotland (no. SC038698).

First published 2017

The Institution of Engineering and Technology
Michael Faraday House
Six Hills Way, Stevenage
Herts, SG1 2AY, United Kingdom

www.theiet.org

British Library Cataloguing in Publication Data
A catalogue record for this product is available from the British Library

ISBN 978-1-78561-123-0 (hardback)
ISBN 978-1-78561-124-7 (PDF)

Typeset in India by MPS Limited
Printed in the UK by CPI Group (UK) Ltd, Croydon

Contents

Preface

Power quality is an important subject of electric power systems for the customers, equipment manufacturers and service suppliers. In general, power quality is an arrangement of requirements or constraints that allow electrical systems to operate with reliable, secure and continuous good quality of power without any additional loss, deterioration of performance, or ageing of any particular equipment due to abnormalities, that is, variations from the standard case.

Because of the on-going complication of the electrical power systems structure and the developments in the smart grid domain around the world, new challenges and opportunities in the power quality area have emerged. Consequently, much attention on power system quality should be paid to meet the future electrical power quality requirements. Besides, extensive analysis of the common and recent power quality problems, improved indices, enhanced standards, better passive/active conditioners and more powerful capabilities for monitoring equipment are required for better readiness of the future electrical power systems.

This book highlights the recent developments in the power systems that have led to the new challenges in the quality of power domain such as the large-scale renewable energy-based generation technologies, and the optimal utilisation of the existing power grid with the growing penetration of renewable-based generation technologies as a necessity for service operators and customers alike. In addition, the increased electricity demands and economic operation of power systems in a deregulated environment causes interconnected power grids to operate closer to their stability margins. Another challenge is the advance of nonlinear loads which may cause some problems such as power system harmonics distortion that may reduce the voltage quality and increase the transmission and distribution losses if they exceed their permissible levels. The book highlights this issue, causes and effects in detail whereas presenting the modern facilities of power conditioners that can solve the problem efficiently.

In addition, some bright opportunities of the future electricity grids with the increased potential of power quality monitoring devices and the enhanced capabilities of the next-generation power grid, namely smart grid, with their improved communication infrastructures are presented and discussed. As these smart networks need to be built in a better hierarchical design approach with enhanced quality, improved controllability and higher reliability and security.

Briefly, this book aimed at introducing novel research outcomes, programmes and ideas that join the past, present and future of the electrical power grids from a quality of power perspective. It is a tool for the planners, designers, operators and practising engineers of electrical power systems that are concerned with the power networks quality, reliability and security. Likewise, it is a key resource for advanced students, postgraduates, academics and researchers who had some background in electrical power systems. The book is principally focused on applications, but each of the book's chapters begins with the fundamental structure of the problem required for a rudimentary understanding of the methods described. The book is sorted out and organised in 12 chapters.

Chapter 1 reviews power-quality definitions, issues and impacts of poor power quality on customers, equipment manufacturers and service suppliers. The primary power quality disturbances are interpreted in this chapter. In addition, power quality indices and an overview of the international standards cover the various power quality perspectives are presented. The cost of poor power quality, investment analysis to mitigate these power quality costs and economic mechanisms for improving power quality aspects and their discussion are described in the last section of this chapter.

Chapter 2 first presents the widely known apparent power definitions and their resolutions for nonsinusoidal and unbalanced systems. They are summarised and qualitatively analysed, and the measurement results are presented to interpret main properties of the apparent powers and their resolutions included in IEEE and DIN standards. Second, the methods and indices, which are proposed for detection of the harmonic producing loads and quantifying the harmonic pollution responsibility of the loads in the literature, are summarised and qualitatively analysed, and the statistical analysis of the measurement results are presented to illustrate their response for different load cases under several supply voltage cases.

Chapter 3 presents the passive harmonic filters commonly used in industry, their classifications, configurations, advantages and disadvantages, theoretical as well as practical design considerations and the data required for their installation. The goal of this chapter is to present the frequency characteristics variation and consequently the filtration properties of different passive filters.

Chapter 4 addresses the industrial load models and characteristics, active power filter topologies and design considerations, configurations and control strategies. In addition, two case studies for the development and optimisation of an industrial active power filter and hybrid active power filters for AC–DC system are developed.

Chapter 5 discusses the methodology to perform harmonic performance studies for commercial shunt FACTS and all devices regulated with the aid of power electronics switches, such as wind energy plants to assess that the harmonic impact of the devices on the transmission system is below the acceptable limits and to specify equipment ratings for the manufacturers.

Chapter 6 develops simple modelling for the series FACTS into load flow method that can be used to analysis and improve the power quality. In these developed series controllers, the real and/or reactive power flow in single or multi-line transmission system can be controlled, and the problem that arises when the series FACTS device is the only link between two areas has been solved.

Consequentially, the complexity of load flow algorithm with series FACTS has been reduced.

Chapter 7 discusses the power quality issues with the application of distributed generation (DG) technologies. Different modes of DG operation are considered such as the islanded mode, standby mode and grid-connected mode of operation. Control techniques and power quality in DG systems are addressed and discussed in detail. In addition, harmonics issues in DG, power filter configurations with their design principles are proposed and simulated.

Chapter 8 presents a novel stochastic strategy to split an entire power network into several stable subsystems. The proposed stochastic scenario generation algorithm can evaluate the steady-state stability of the created islands in each generated solution. The impact of distributed generators on the splitting problem has comprehensively been studied. It is proved that the proposed optimisation procedure can efficiently reduce the total active power losses and improve voltage profiles in the presence of DGs, which guarantee the power quality of the created islands.

Chapter 9 presents a methodology to use the voltage profile as power quality index in smart microgrids. It addresses the decentralised voltage control in smart grids, the decentralised and distributed control systems and the centralised hierarchical control of the distributed energy resources (DERs). In addition, the reactive power dispatch, the DER integration concealment and the distributed voltage control schemes are presented in detail.

Chapter 10 focuses on the tariff-related issues that are correlated with the quality of power, the different approaches for finding out power quality impact on the tariff, the design of modules to associate disturbance and economic loss, and the cost–benefit investigation indices. It addresses the need of revenues for improving the current behaviour towards power quality through updating or changing the current policies to include more incentives to enhance the quality of power. Two case studies on power quality investment in developing countries are presented.

Chapter 11 introduces the field of energy management systems in the operation of microgrid applications. It proposes a novel formulation for robust optimisation-based EMS for microgrid applications. The novelty recast in the introduction of transmission constraints, and in the avoidance of the use of prediction models for the uncertain variables.

Finally, *Chapter 12* addresses the contracts of PQ in a reconfigured electric power industry, power quality monitoring, measurement, performance indices, equipment requirements, emerging issues and power quality impacts as well as power quality trends, future interface problems with the smart grids. A case study exhibiting three different low-cost FACTS-based modulated/switched filter-compensation devices and smart dynamic control strategies using dynamic fast controlled PWM switching strategies are proposed. In addition, future problems and challenges in measurements, monitoring and analysis tools are presented.

Chapter 1
Power quality definitions

Ramani Kannan[1] *and Jagabar Sathik Mohd. Ali*[2]

1.1 Introduction to various power quality indices

The power quality (PQ) can take into the different behavior of the electrical signals in power systems, which supply to the loads economically with uninterrupted power supply services, and various influences arise to change the waveform characteristics in electrical signals.

PQ was not considered as an important issue in the field of electrical engineering until the entry of electronic equipment. This electronic equipment is more sensitive than electrical equipment. *In a human body, if there is any problem in the blood or blood circulation, it will collapse the entire organ system. Likewise, if there are any defects in the PQ, it collapses the entire high sensitive equipment.* In a power system, all the equipment are connected linearly. If a problem occurs at one end, it can be felt at the other end of the power system. If a short circuit occurs, it results in the opening of the circuit breaker to run the generator on no-load condition, and after fault clearance, the generator is shifted to on load, which gives a serious impact on all the equipment connected to the power system.

Due to the numerous growths of industries and commercial sectors, the electric power demand is getting increased day by day. In order to overcome this situation, it is mandatory not only to satisfy the power demand but also to maintain the PQ. The electrical systems are defined as high PQ systems, the PQ is necessary for low-to-high-power applications. In order to maintain the quality of power, various standards are presented, depending on the load characteristics and nature of the disturbance; various FACTS devices are used in low to high power applications.

1.1.1 Why are we concerned about power quality?

Economically, if the PQ is not maintained, huge loss will occur. Both current and voltage should be in sinusoidal shape with specified magnitude at a constant frequency in order to obtain a good PQ as stated in [1]. It allows the electric system to

[1]Electrical and Electronic Engineering Department, Universiti Teknologi Petronas, Seri Iskandar, Perak, Malaysia
[2]SRM University, Kattankullathur, Tamilnadu, India

operate as expected with high performance, thereby reducing the loss and increasing the life of the device. If any deviation takes place in these four parameters, the PQ decreases and thereby damaging the equipment. The modern society is mostly depending on electronic equipment like laptop, personal computers, mobile phones, and many more; these are highly sensitive to the electrical system and PQ. In the case of any small variations in power supply, it has a great economic impact on the utility and industrial consumers.

1.1.2 Definition of power quality

PQ is defined based on four important measurements of electrical parameters, that is, *voltage*, *current*, *frequency*, and *phase*; therefore, both the voltage and current should be in sinusoidal shape with specified magnitude at a constant frequency without any change in phase. An ideal voltage sine wave can be provided by a generator, but the current passing through the impedance of the system can cause several disturbances to the ideal sinusoidal voltage waveform. With any deviation from these parameters, the system is said to be low PQ.

PQ impacts on customers, equipment manufacturers, and service suppliers.

The PQ issues are a major concern for equipment manufacturers. The following list can describe the various PQ issues:

Home appliances: Low-power applications, such as TV, laptop, air conditioner and refrigerator, and others, will damage due to variation in electrical power, and the cost of the damage is high [2,3].

Electrical systems in aircraft: Uninterruptible and reliable power supply are required for flight load, but various problems arise and affect the PQ by switching different load and AC power supply disturbed due to other transient conditions such as DC source supply to flight control actuators, electrical machines like fuel-transfer pump, and compressor or discontinue AC load variations [4].

AC contractors, relays, and power switches: Contactors and relays are operating a range of voltages and ratings of devices. These devices are mostly used as either open or close but no control power, and this is widely used in the substation and high-voltage applications. Nowadays, the contractors function has been replaced by power electronics devices such as gate turn off (GTOs); insulated-gate bipolar transistor (IGBTs) with more advanced control techniques and relay functions are replaced by programmable logic controller [5,6].

The high power quality is maintained by high-cost devices to save high-cost equipment.

The above-stated issues will reduce the PQ of the electrical system; it will further lead to the reduction in the lifetime of equipment, which will result in increasing of cost. The PQ may also define as the ability of electric power that enables the equipment to work properly and increase the life of the equipment. PQ is also defined as the distortion that takes place from its ideal sinusoidal waveform. If the distortion of the waveform consumed is more, PQ is said to be poor.

Factors affecting the PQ are as follows:

PQ resembles the reliability of power supplied to the consumer. It depends on various external and internal factors. External factors include the following:

- Lightning
- Switching effects
- Nonlinear load
- High-power switched-mode converters.

Internal factors include the following:

- Electromagnetic interference
- Electrostatic discharges
- Environmental factors (i.e., excessive temperature, excessive vibration, etc.).

Other factors include the following:

- Misoperation of equipment
- Equipment creates a disturbance at overloading conditions
- Long-time running equipment
- Not a proper maintenance of equipment
- High-quality materials are not used
- Other problems are related to grounding and earthing.

1.2 Various conventional power quality indices

The following factors are considered as basic PQ indices:

1. Harmonics and interharmonic
2. Voltage fluctuations and flicker
3. Voltage unbalance
4. Power frequency variation
5. Transients
6. Sag, swell, interruption, under voltage, and overvoltage.

These can be classified into three major aspects: (i) voltage stability, (ii) uninterrupted power supply, and (iii) voltage waveform, which is shown in Figure 1.1, and various PQ issues and also their effects are listed in Table 1.1.

1.2.1 Harmonics and interharmonic

1.2.1.1 Harmonics

Definition: "*As per the electrical, the harmonic may define as multiple integer frequencies of the fundamental frequency (50 or 60 Hz) presented in electrical signal either in voltage or current waveform.*" The fundamental voltage waveform with harmonic waveform is illustrated in Figure 1.2.

1.2.1.2 Interharmonic

The interharmonic is generated between the frequency of both voltage and current, and its frequency is a noninteger multiple of the fundamental frequency.

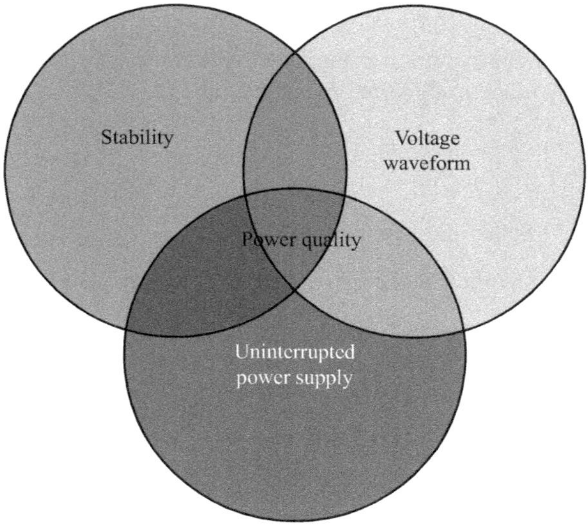

Figure 1.1 Three major aspects of power quality

Table 1.1 Effects of power quality indices in various aspects

Power quality indices	**Voltage stability**	**Power supply continuity**	**Voltage waveform**
Harmonics	✓	×	✓
Interharmonics	×	×	✓
Voltage unbalance	✓	×	✓
Voltage fluctuations and flicker	✓	×	✓
Frequency	✓	×	✓
Transients	✓	×	✓
Sag, swell, under voltage, and over voltage	✓	×	✓
Interruption	×	✓	×

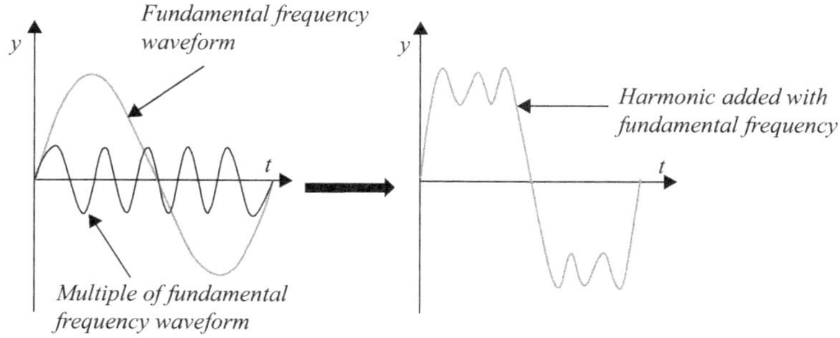

Figure 1.2 Prescribed as harmonics with fundamental waveform

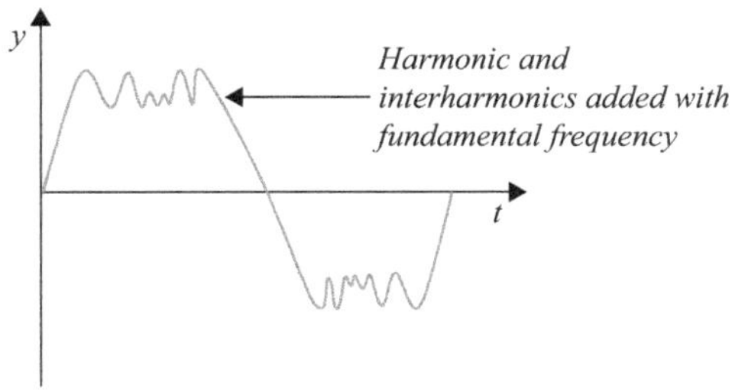

Figure 1.3 Fundamental waveform with harmonics and interharmonics

Definition: "*Noninteger of the fundamental frequency is called interharmonics, and this can appear as discrete frequencies.*" Both the harmonics and interharmonics added together in the fundamental waveform are shown in Figure 1.3.

1.2.1.3 Causes

Harmonics

The energy losses and distortion in the main supply voltage is generated by unwanted current flows in the network. The improper usage of defective devices such as fluorescent lamps, home appliances, or other equipment and also the malfunction of ripple control, protective relays, and mains signaling. Additional losses in passive elements and rotating machines will affect the efficiency of motors and create a harmonics in line voltage or current.

Interharmonics

Cycloconverter: Unbalance load and asymmetric phase voltages and firing angle create interharmonics.

Arc welding and furnace: These loads create a low-voltage flicker, resultant low-frequency voltage variations, and it may introduce the interharmonics. Doubly fed configurations of induction motor also are the sources of interharmonics.

Others: Ripple control, heating applications, and induction furnace are sources of interharmonics.

1.2.1.4 Impact

Impact of current harmonics: Neutral overloading, transformer overheating, anonymous tripping of circuit breaker, skin effect, and capacitors used in power factor correction.

Impact of voltage harmonics: Voltage waveform distortions, induction motors, and increasing current harmonics.

Impact of voltage interharmonics: Overheating of the devices or equipment, effects of interharmonics are the impact on light, flicker, overload of series tuned filters, and saturation of current transformer.

Reducing effects of the harmonics and interharmonics for various loads:

There are several techniques employing to control the harmonics which are: shunt-active filter, phase multiplication, harmonic injection, modulation technique, and passive filters.

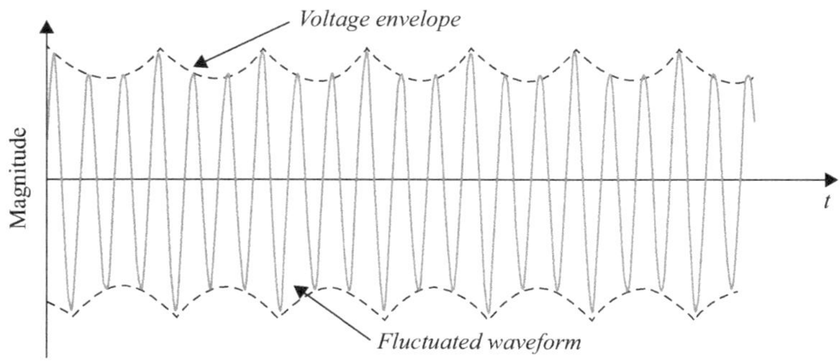

Figure 1.4 Voltage fluctuating at the load [7]

1.2.2 Voltage fluctuations and flicker

Definition: "*The voltage fluctuations can be defined as the periodic or nonperiodic variation in the voltage envelope due to uncertain changes in the real and reactive power by the load with allowable limits of* $\pm 5\%$ *variation in nominal voltage.*"

Figure 1.4 illustrates the voltage fluctuation and nonuniformly current drawn at the load as illustrated in [7].

1.2.2.1 Causes

Electric arc furnaces: An electric arc furnace load is considered as current harmonics producer during the meltdown period, and also it affects three-phase quantities such as voltage, real and reactive power, frequency, and total harmonic distortion in respective phases.

Another cause of voltage fluctuations is the frequent start of home appliances like air-conditioner units, fans, and pumping motor and mine hoists, rolling mills, woodchippers, and car shredders. The main cause of the flicker and voltage fluctuation on power system in utility grid is due to arc welding, and it creates electromagnetic interferences (e.g., resistance welder or an electric arc furnace), continues tripping the relay, interferes with communication equipment and may damage the electronics devices, and nonstop fluctuation of voltage frequencies may damage the high-sensitive loads.

The flickers are caused by voltage fluctuations under any load variations. This flicker should not exceed the perception of short-term flicker (P_{st}), which limits to 1.0, and perception of long-term flicker (P_{lt}), which limits to 0.8 for both low and medium-voltage levels. Less than 10% of voltage fluctuation does not create a major problem in low-power devices (electronic equipment). The voltage fluctuation is classified based on the nature of disturbance with respect to time

- Step-voltage changes which may be regular or irregular.
- Cyclic voltage fluctuation.
- Random voltage fluctuation.

1.2.2.2 Effects

The flickers mainly occur due to lighting systems, and it changes the magnitude of voltage and current which leads to the instability in electronics equipment. This effect damages the high-sensitive electronics equipment like TV and monitoring equipment.

1.2.2.3 Reducing effects of the voltage fluctuations for various loads

Increasing the fault level at the point of connection, by using the FACTS devices to compensate the reactive power flow through the network, inclusive of star-delta starter for smooth start of the three-phase motor, by adding cooling system, reduces the heat of system; installations of filters, static volt-ampere reactive (VAR) systems, or distribution static compensators are used to mitigate the voltage fluctuations. Isolation transformer is used to separate the load and main supply and replace the equipment.

1.2.3 Voltage unbalance

Definition: "*In three-phase system, the magnitudes of each phase voltage or line voltage are different, and phase angles also differ from a balanced system.*" Figure 1.5 illustrates the unbalanced voltage waveform of three-phase system.

1.2.3.1 Causes

The unequal distribution of single-phase load: Single-phase load is connected to the three-phase supply will create a voltage unbalance in any one of the phases. During unbalance impedance of transmission and distribution network, three-phase AC/DC power electronic converter (three-phase diode rectifier) is the most common unbalance system which draws nonsinusoidal current from the main AC supply. Most of the industrial and domestic loads like lighting loads are single phase which is supplied from the three-phase supply. The single-phase loads on the three-phase system causes less than 2% of unbalance loads, and greater than 5% can result from single phasing conditions.

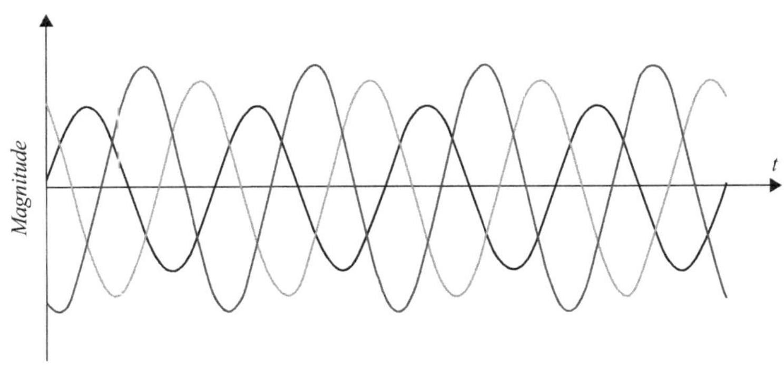

Figure 1.5 Voltage unbalance in three-phase system

1.2.3.2 Effects

Most of the power electronics converters are designed for closed-loop applications in which the firing pulses are derived from the output voltages, in both DC and AC side source the unbalances, and can cause the presence of noncharacteristic (produced by power semiconductor devices in normal operating conditions) harmonics.

Other than this, the power system stability is reduced due to voltage unbalancing on the network and draw more of reactive power; improper operation of instruments leads to reduction in the life span of equipment. In electrical machines, voltage unbalances lead to increased losses by drawing unbalanced current, and it produces uneven heating and creates oscillating torque in motors.

1.2.3.3 Reduction of the unbalance voltages

Impossible to reduce unbalance voltage to zero due to following reasons:

Uneven connection and disconnection of single-phase load on three-phase supply, even though there are some mitigating techniques available in utility and industrial level systems, which equally distributed the single-phase loads. Single-phase regulators, by using passive systems or active systems, load balancing in industrial systems; high-sensitive loads are not connected in a system which supplies the single-phase loads, using relay or tripper to protect the loads. Reexamine or recheck the distribution of the single-phase loads on the three-phase system.

1.2.4 Power frequency variations

The fundamental frequency deviates from its specified actual value (i.e., 50 or 60 Hz), this frequency variation does not often occur in stable interconnected power grid systems. The frequency often occurs in very poor power system network and standby generators, and generators are heavily loaded due to sudden changes in the load.

1.2.4.1 Causes

In addition, separations of cascading system may provide slight deviations in frequency because electric systems are closely connected and depended on the synchronous operation. The imbalance between load and generations can cause power frequency variations. As the frequency variation exceeds, the accepted limits ($\pm$0.5 Hz) faults will occur in steady-state power system, and maximum loads are disconnected, resulting to a large source of generation going off-line.

1.2.4.2 Effects

It results in damage to sensitive electronic equipment, such as hard drive crash, keyboard lockup, erratic operations, data corruptions and program failure, and even component damage. Small variations in frequency produce high torque, and it leads to severe damage in generators and turbine shafts, and also in load side, the motor to run inefficiently, which lead to added heat to increase current.

1.2.4.3 Reduction of the power frequency variations

The frequency variation is controlled by advanced power electronics converter with the proper balance between prime mover of the generator and active power demand

Table 1.2 Various events with frequency range and magnitude of voltage [1]

S. no.	Event type	Spectral content	Event duration	Magnitudes (or) event voltage magnitude in per unit
1	Voltage imbalance		Steady state	0.5%–2%
2	Waveform distortion			
	DC offset		Steady state	0%–0.1%
	Harmonics	0–100 Hz	Steady state	0%–20%
	Interharmonic	0–6 kHz	Steady state	0%–2%
	Notching		Steady state	
	Noise	Broadband	Steady state	0%–1%
3	Voltage fluctuations	<25 Hz	Intermittent	0.1%–7%
4	Frequency variations		<10 seg	

at the load. By using power electronics, circuits can operate at close to the fundamental frequency; dynamic load control method can be employed and can disconnect the heavy load. The different events are occurring with various frequencies and voltage are detailed in Table 1.2.

1.2.5 Transients

The PQ indices and transients are considered as uncertain event because it appears for a very short duration with a large magnitude of voltage, current, or even both. The more familiar term of transients is "spikes." The transients can be classified into two types: (i) impulsive transients and (ii) oscillatory transients.

1.2.5.1 Impulsive transients

The magnitude of voltage, current, or even both suddenly may rise or lower the level for a very short period of fewer than 50 ns. As per *IEEE 1159* definition, as a sudden, non-power frequency change in the steady-state condition of voltage, current, or both—that is unidirectional in polarity—is either primarily positive or negative. It is normally a single, very high impulse like lightning.

1.2.5.2 Causes

The impulse transients may occur due to lightning, poor grounding system, the switching of inductive loads, utility fault clearing, electrostatic discharge, or other issues. In addition to this, impulsive transients corrupt the data and physical damage to equipment. The change of the voltage magnitude during lighting is depicted in Figure 1.6.

1.2.5.3 Oscillatory transients

The instantaneous value of either voltage or current varies either from positive to negative or vice versa quickly.

As per IEEE 1159 definition, as a sudden, non-power frequency change in the steady-state condition of voltage, current, or both has both positive and negative effect. This is subdivided into three types as listed in Table 1.3.

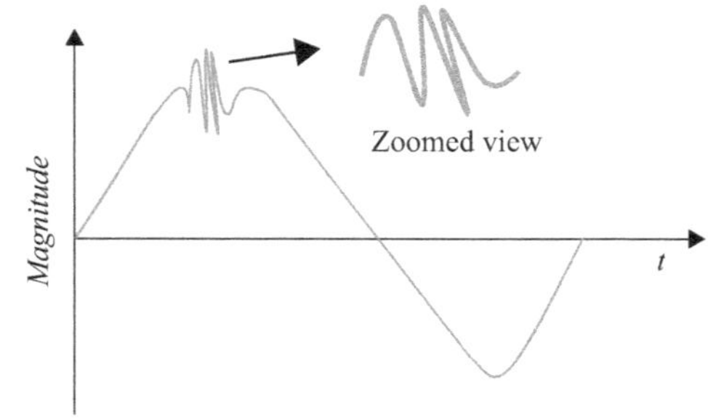

Figure 1.6 Current impulsive transient caused by lightning

Table 1.3 Classification oscillatory transients based on frequency [8]

Oscillatory transients	Spectral content	Typical duration	Voltage magnitude
Low frequency	<5 kHz	0.3–50 ms	0–4 pu
Medium frequency	5–500 kHz	20 μs	0–8 pu
Frequency	0.5–5 MHz	5 μs	0–4 pu

1.2.5.4 Cause of transients

Lightning strikes, electrostatic discharge, malfunction due to high-voltage stress (dv/dt) and improper switching of power electronics switches, load switching activities, and switching of bulk capacitor bank.

Circuit breaker reclosing operations, by using tap changing on transformers, improper connections in the distribution system and human errors, accidents, or broken machines or line.

1.2.5.5 Effects

Electronic equipment

Malfunction of electronics equipment and produces corrupted results, destroys computer chips, and TV regulators; multiple zero crossing will damage the low-power electronics devices like timer, disks, and others.

Lights

Fluorescent bulb and ballast failure, black rings appear at the fluorescent tube ends and premature filament damage of incandescent light.

Electrical equipment

Circuit breakers and switches contacting surface are affected by transients, false activation of breaker, reduce transformer efficiency (increase hysteresis losses), motor vibration, excessive heat, high losses (hysteresis), increasing operating temperature, and noise, and motor winding performance will degrade, resulting in equipment failure.

Solution for transients: Surge arresters mounted on transmission and distributed system, uninterrupted power supply system to protect the sensitive load from surge and loss of power, filters, and isolation transformers.

1.2.6 Short duration voltage variations

1.2.6.1 Voltage sag

Sag is sometimes called as voltage dip. Voltage dips (sags) are PQ disturbances. The reduction of root mean square (RMS) voltage of fundamental magnitude for a short duration to an extent between 0.1 and 0.9 pu. It lasts for durations of 0.5 cycles to 1 min. Sag is classified based on the time duration and voltage magnitude as follows:

- Instantaneous sag, momentary sag, and temporary sag.

1.2.6.2 Voltage sag effects

Most of the voltage sag occurs due to malfunctions on industrial drives power converters, electronic equipment, unexpected drop out of relays, and contractors; the induction motor can draw a higher current, loss of production in manufacturing industries, and computer and other sensitive electronics system damage.

Voltage sags are usually caused by, short circuit in the electrical power system, a fault in the power system network, excessively large inrush current in the network, energization of heavy inductive loads like arc furnace, initially, a large amount of current drawn by induction motors, the line to ground fault, transferring load from one source to another, and line–line symmetrical faults [9].

1.2.6.3 Voltage swell

The voltage swell is an inversion of voltage sag, an increase in RMS voltage at the frequency for durations from 0.5 cycles to 1 min. Swells are caused by single-line-to-ground failures, upstream failures, and switching off a heavy load or switching to a large capacitor bank. The voltage swell is classified based on the time duration and voltage magnitude as follows: instantaneous swell, momentary swell, and temporary swell. Voltage swells are usually caused by sudden load decreases and switching off large capacitor bank.

1.2.6.4 Effects of swell

Malfunctions of uninterruptible power supply, reduction of grid, or bus voltage for a short duration lead to voltage instability, malfunction of low-voltage devices, malfunction of measuring and control equipment, grid-tie inverter trip out in generation side, and at load side such as variable speed drives tripping out.

1.2.6.5 Preventive measures

Uninterruptible power supply or voltage conditioners are best choice to prevent voltage sags and swell; constant voltage transformer can be used to maintain the voltage. Energy storage, like battery, SMES, flywheels, and others, can be used as electronic equipment such as laptop, FACTS Devices like DVR, UPQC, and others, for improving the immunity of equipment.

Interruption (short-duration and long-duration voltage variations):

An interruption is an event which occurs randomly with zero magnitude voltage or current for a time period. In other words, the magnitude of voltage or current is less than 0.1 pu.

The reduction voltage is classified based on the duration and standards as follows:

- *EN 50160-2000* classifies interruptions as:
  - Pre-arranged interruptions: customers are informed during maintenance in advance.
  - Accidental interruptions: the unexpected failure of equipment or transients fault, and it may take a long time to rectify the failure. This may be long or short interruption depending on the fault.

Based on the standard IEEE 1250-1995

These standards are especially striking for interruptions,

1. An instantaneous interruption—between 0.5 and 30 cycles (half a second)
2. A momentary interruption—between 30 cycles and 2 s
3. A temporary interruption—between 2 s and 2 min
4. A sustained interruption longer than 2 min.

Interruptions are caused by

External events are a major cause for permanent or transients faults, an unexpected disconnection of contactor or relay during electromagnetic interference, control malfunctions of equipment; maintenance of network can require voltage interruption for a particular area, and blown fuse or breaker opening.

1.2.6.6 Effects

Short circuit fault or overloading transformer leads to long interruption, and this affects the large group of customer, as well as serious damage to electrical equipment. Due to voltage loss, the equipment is not able to operate in a normal condition which leads to degrading the performance of the system.

Induction motor speed slows down and adjustable speed drives (ASDs) are more sensitive to short interruption, and it will cause the loss of the load.

Electronics equipment, such as video recorder, micro ovens, and others, will blink when the supply is interrupted; IT equipment are more sensitive to voltage variation and the common effect of interruption is a loss of signal transmission and control.

These interrupt the industrial process and performance of equipment.

Solutions:

1. Minimizing the number of faults: entire short circuits faults cannot be eliminated, but it can be minimized by regular cutting of trees, replacing overheads lines with cables, periodic maintenance, increased insulation, and others. Minimize the fault clearance time: after the fault occurred, the operating time of switches, fuses, and others and different power supply are connected to the same bus bar for sensitive customers.
2. Power converters: Using stored energy like UPS, SEMS, and others, and the solid state devices DVR, STATCOM, UPQC, and others.

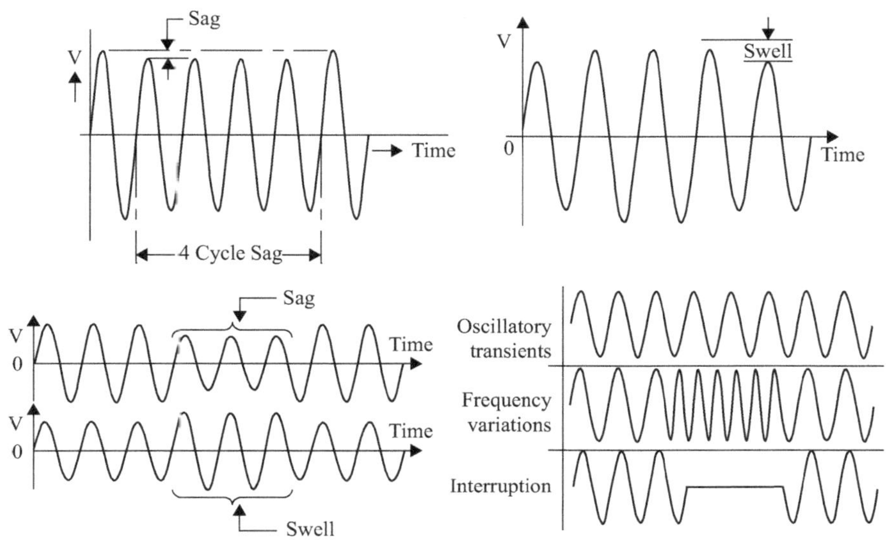

Figure 1.7 Disturbed voltage waveform of sag, swell, and interruptions, frequency variations and transients

The voltage magnitude variations and effects different events are shown in Figure 1.7.

1.2.6.7 Over voltage

The over voltage is an increase of RMS voltage greater than required value (i.e., 1.1–1.2 pu) for the period of more than 1 min.

Reasons for over voltage to occur in power system:

Abrupt variation of system conditions, sudden or unexpected reduction of load, resonance phenomena, poor voltage regulation of system, an insulation fault, alternator regulators faults, tap changer transformer, or overcompensation and lightning.

Switching of the capacitive bank or opening of protective devices.

1.2.6.8 Effect of over voltage

Tripping of protective devices due to dielectric failure, equipment damage: over voltage is a risk for computer data and healthcare centers (safety), huge loss in industrial consumers (production loss), and the risk of network destabilization.

1.2.6.9 Under voltage

RMS voltage of the system is lesser than required value, that is, 0.8–0.9 pu for more than 1 min is known as under voltage.

Causes of under voltage:

- Less supply capability, under voltages, may occur due to switching on a large load that is, high inductor, overloaded circuits can also lead to under voltages and not proper wiring connection, or corroded connections can also cause under voltages.

Effects of under voltage:

- Voltage instability, the excessive current drawn by the motor, high reactive power demand, and others.

1.2.6.10 Preventive measures

1. By reducing the ferromagnetic resonance, neutral earthing is a solution for phase/earth resonance, inserting resistance in circuit breaker to avoid more energizing in transmission network and using good dielectric materials and increasing the contact separation speed.
2. Quality earthing connections to withstand the lighting stroke, surge arresters increase the insulation levels for equipment and corresponding over voltage protection system and using protective devices like neutral/earthing, shielding (LPS), repressive protection, spark gap arresters, and others.

The magnitude of voltage varies for a different event and the frequency also differs. In [14], the magnitude variations, different frequency ranges, and duration of effect are listed in Table 1.4.

Table 1.4 Power quality variations according to IEEE 1159 [10,11]

Categories (or) event type	Spectral content	Event duration	Event voltage magnitude in per unit
Transients			
Impulsive			
○ Nanoseconds	5 ns rise	<50 ns	
○ Microseconds	1 μs rise	50 ns–1 ms	
○ Milliseconds	0.1 ms rise	>1 ms	
Oscillatory			
○ Low frequency	<5 kHz	0.3–50 ms	0–4 pu
○ Medium frequency	5–500 kHz	20 μs	0–8 pu
○ High frequency	0.5–5 MHz	5 μs	0–4 pu
Short duration variations			
Instantaneous			
○ Sag		0.5–30 cycles	0.1–0.9 pu
○ Swell		0.5–30 cycles	1.1–1.8 pu
Momentary			
○ Interruption		0.5 cycles–3 seg	<0.1 pu
○ Sag		30 cycles–3 seg	0.1–0.9 pu
○ Swell		30 cycles–3 seg	1.1–1.4 pu
Temporary			
○ Interruption		3 seg–1 min	<0.1 pu
○ Sag		3 seg–1 min	0.1–0.9 pu
○ Swell		3 seg–1 min	1.1–1.2 pu
Long duration variations			
○ Interruption, sustained		>1 min	0.0 pu
○ Under voltage		>1 min	0.8–0.9 pu
○ Over voltage		>1 min	1.1–1.2 pu

1.3 International standards

The various international societies have their own standards to testing and measurement, requirement and installation procedure.

1.3.1 The Institute of Electrical and Electronic Engineering (IEEE) Standards

1. IEEE 1048-2003—Protective grounding of power lines
2. IEEE 1100-2005—Powering and grounding electronic equipment
3. IEEE 1122-1998—Digital recorders for measurements in high-voltage impulse tests
4. IEEE 1124-2003—IEEE guide for analysis and definition of DC Side harmonic performance of HVDC transmission systems
5. IEEE 1159-2009—Monitoring electric PQ
6. IEEE 1250-2011—Identifying and improving voltage quality in power systems
7. IEEE 1366-2012—Electric power distribution reliability indices
8. IEEE 1409-2012—Application of power electronics for PQ improvement on distribution systems rated 1 through 38 kV
9. IEEE 1410-2010—Improving the lightning performance of electric power overhead distribution lines
10. IEEE 142-2007—Grounding of industrial and commercial power systems
11. IEEE 1453.1-2012—(Adoption of IEC/TR 61000-3-7:2008, Electromagnetic compatibility)—limits—assessment of emission limits for the connection of fluctuating installations to MV, HV, and EHV power systems
12. IEEE 1453-2015—Analysis of fluctuating installations on power systems
13. IEEE 1531-2003—Application and specification of harmonic filters
14. IEEE 1564-2014—Voltage sag indices
15. IEEE 1585-2002—Functional specification of medium voltage (1–35 kV) electronic series devices for compensation of voltage fluctuations
16. IEEE 1668-2014—(Trial-Use) Voltage sag and short interruption ride-through testing for end-use electrical equipment rated less than 1,000 V
17. IEEE 1775-2010—Power-line communication equipment-electromagnetic compatibility (EMC) requirements testing and measurement methods
18. IEEE 1860-2014—Voltage regulation and reactive power compensation at 1,000 kV AC and above
19. IEEE 1894-2015—Online monitoring and recording systems for transient overvoltages in electric power systems
20. IEEE 3004.5-2014—Application of low-voltage circuit breakers in industrial and commercial power systems
21. IEEE 367-2012—Determining the electric power station ground potential rise and induced voltage from a power fault
22. IEEE 430—Measurement of radio noise from overhead power lines and substations
23. IEEE 433-2009—Insulation testing of AC electric machinery with high voltage at very low frequency

24. IEEE 446-1995—Emergency and standby power systems for industrial and commercial applications
25. IEEE 449-1998—Ferro resonant voltage regulators
26. IEEE 495-2007—Guide for testing faulted circuit indicators
27. IEEE 515.1-2012—The testing, design, installation, and maintenance of electrical resistance trace heating for commercial applications
28. IEEE 519-2014—Practice and requirements for harmonic control in electric power systems
29. IEEE 539-2005—Definitions of terms relating to corona and field effects of overhead power lines
30. IEEE 656-1992—The measurement of audible noise from overhead transmission lines
31. IEEE 81-2012—Guide for measuring earth resistivity, ground impedance, and earth surface potentials of a grounding system
32. IEEE 82-2002—Standard test procedure for impulse voltage tests on insulated conductors
33. IEEE 998-2012—Direct lightning stroke shielding of substations
34. IEEE C37.90.1-2012—Surge withstand capability (SWC) tests for relays and relay systems associated with electric power apparatus
35. IEEE C62.31-2006—Low-voltage gas-tube surge-protective device components

1.3.2 American National Standards Institute (IEEE/ANSI)

1. ANSI C84.1—Electric power systems and equipment—voltage ratings (60 Hz)
2. ANSI 70—National electric code
3. ANSI C62.61-1993—Gas-tube surge arresters on wire line telephone circuits
4. ANSI C63.12-2015—Electromagnetic compatibility limits and test levels
5. ANSI C63.22-2004—Automated electromagnetic interference measurements
6. ANSI C78.377-2015—For electric lamps—specifications for the chromaticity of solid state lighting (SSL) products
7. ANSI C82.77-10-2014—Lighting equipment–harmonic emission limits–related PQ requirements
8. ANSI C82.77-5-2015—Lighting equipment—voltage surge requirements
9. ANSI C84.1-2011—Electric power systems and equipment—voltage ratings (60 Hz)
10. IEEE C37.90.1-2012—Surge withstand capability (SWC) tests for relays and relay systems associated with electric power apparatus
11. IEEE/ANSI 15C62.48-1995—Interactions between power system disturbances and surge-protective devices
12. IEEE/ANSI 446-1995—Emergency and standby power systems for industrial and commercial applications
13. IEEE/ANSI 644-1994—IEEE standard procedures for measurement of power frequency electric and magnetic fields from AC power lines
14. IEEE/ANSI C37.011-2011—Application of transient recovery voltage for AC high-voltage circuit breakers

15. IEEE/ANSI C37.04b-2008—Rating structure for AC high-voltage circuit breakers rated on a symmetrical current basis amendment 2: to change the description of transient recovery voltage for harmonization with IEC 62271-100
16. IEEE/ANSI C37.117-2007—Protective relays used for abnormal frequency load shedding and restoration
17. IEEE/ANSI C37.16—Preferred ratings, related requirements, and application recommendations for low-voltage AC (635 V and below) and DC (3,200 V and below) power circuit breakers
18. IEEE/ANSI C37.17-2012—Trip systems for low-voltage (1,000 V and below) AC and general purpose (1,500 V and below) DC power circuit breakers
19. IEEE/ANSI C37.233-2009—Power system protection testing
20. IEEE/ANSI C37.26-2014—Methods of power-factor measurement for low-voltage (1,000 V AC or lower) inductive test circuits
21. IEEE/ANSI C37.91-2008—Protecting power transformers
22. IEEE/ANSI C57.110-2008—Establishing liquid-filled and dry-type power and distribution transformer capability when supplying nonsinusoidal load currents
23. IEEE/ANSI C57.142-2010—Describe the occurrence and mitigation of switching transients induced by transformers, switching device, and system interaction
24. IEEE/ANSI C57.32-2015—Standard for requirements, terminology, and test procedures for neutral grounding devices
25. IEEE/ANSI C62.36-2014—Test methods for surge protectors used in low-voltage data, communications, and signaling circuits
26. IEEE/ANSI C62.39-2012—Test methods and preferred values for self-restoring current-limiter components used in telecommunication surge protection
27. IEEE/ANSI C62.41.2-2002—Characterization of surges in low-voltage (1,000 V and less) AC power circuits
28. IEEE/ANSI C62.48-2005—Interactions between power system disturbances and surge-protective devices
29. IEEE/ANSI C62.72-2016—Application of surge-protective devices for use on the load side of the service equipment in low-voltage (1,000 V or less, 50 or 60 Hz) AC power circuits
30. NEMA ICS 61800-1-2002 (R2007)—Adjustable speed electrical power drive systems, Part 1: general requirements—rating specifications for low voltage adjustable-speed DC power drive systems
31. NEMA LSD 8-2014—PQ implications of self-ballasted lamps in residences

1.3.3 British Standards (BS) with IEC Standards

General

1. BS EN 50157-2-1:1999—Domestic and similar electronic equipment interconnection requirements (Part1 = AV.link)
2. BS EN 55014-1:2006—Electromagnetic compatibility—requirements for household appliances, electric tools, and similar apparatus

3. EN 61000-1-4—Historical rationale for the limitation of power-frequency conducted harmonic current emissions from equipment, in the frequency range up to 2 kHz
4. EN 61000-1-6—Guide to the assessment of measurement uncertainty.

Environment:

1. EN 61000-2-2—Compatibility levels for low-frequency conducted disturbances and signaling in public low-voltage power supply systems
2. EN 61000-2-4—Compatibility levels in industrial plants for low-frequency conducted disturbances
3. EN 61000-2-9—Description of HEMP environment. Radiated disturbance. Basic EMC publication
4. EN 61000-2-10—Description of HEMP environment. Conducted disturbance
5. EN 61000-2-12—Compatibility levels for low-frequency conducted disturbances and signaling in public medium-voltage power supply systems.

Limits:

1. EN 61000-3-2—Limits for harmonic current emissions (equipment input current up to and including 16 A per phase)
2. EN 61000-3-3—Limitation of voltage changes, voltage fluctuations, and flicker in public low-voltage supply systems, for equipment with rated current ≤ 16 A per phase and not subject to conditional connection
3. EN 61000-3-11—Limitation of voltage changes, voltage fluctuations, and flicker in public low-voltage supply systems. Equipment with rated voltage current ≤ 75 A and subject to conditional connection
4. EN 61000-3-12—Electromagnetic compatibility (EMC).

Testing and measurement techniques:

1. EN 61000-4-1—Overview of IEC 61000-4 series
2. EN 61000-4-2—Electrostatic discharge immunity test. Basic EMC publication
3. EN 61000-4-3—Radiated, radio-frequency, electromagnetic field immunity test
4. EN 61000-4-4—Electrical fast transient/burst immunity test
5. EN 61000-4-5—Surge immunity test
6. EN 61000-4-6—Immunity to conducted disturbances, induced by radio-frequency fields
7. EN 61000-4-7—General guide on harmonics and interharmonics measurements and instrumentation, for power supply systems and equipment connected thereto
8. EN 61000-4-8—Power frequency magnetic field immunity test. Basic EMC publication
9. EN 61000-4-11—Voltage dips, short interruptions, and voltage variations immunity tests
10. EN 61000-4-12—Oscillatory wave's immunity test. Basic EMC publication
11. EN 61000-4-13—Harmonics and interharmonics including mains signaling at a.c. power port, low-frequency immunity tests

12. EN 61000-4-14—Voltage fluctuation immunity test for equipment with input current not exceeding 16 A per phase
13. EN 61000-4-15—Flicker meter. Functional and design specifications. Basic EMC publication
14. EN 61000-4-16—Test for immunity to conducted, common mode disturbances in the frequency range of 0 Hz to 150 kHz
15. EN 61000-4-17—Ripple on d.c. input power port immunity test
16. EN 61000-4-18—Damped oscillatory wave immunity test
17. EN 61000-4-19—Test for immunity to conducted, differential mode disturbances, and signaling in the frequency range 2 kHz to 150 kHz at a.c. power ports
18. EN 61000-4-23—Test methods for protective devices for HEMP and other radiated disturbances
19. EN 61000-4-24—Test methods for protective devices for HEMP conducted disturbance. Basic EMC publication
20. EN 61000-4-28—Variation of power frequency, immunity test for equipment with input current not exceeding 16 A per phase
21. EN 61000-4-29—Voltage dips, short interruptions, and voltage variations on d.c. input power ports. Immunity tests. Basic EMC Publication
22. EN 61000-4-30—Testing and measurement techniques. Power quality measurement methods
23. EN 61000-4-34—Voltage dips, short interruptions, and voltage variations immunity tests for equipment with mains current more than 16 A per phase
24. IEC 61000-5-1—General considerations. Basic EMC publication.

Generic standards:

1. EN 61000-6-1—Immunity for residential, commercial and light-industrial environments
2. EN 61000-6-2—Immunity for industrial environments
3. EN 61000-6-3—Emission standard for residential, commercial, and light-industrial environments
4. EN 61000-6-4—Emission standard for industrial environments

1. EN 50121—Electromagnetic compatibility
2. EN 50122-1,2,3—Fixed installations—electrical safety, earthing, and the return circuit—Part 1: protective provisions against electric shock for AC and DC traction systems
3. EN 50178—Electronic equipment for use in power installations
4. EN 50522—Earthing of power installations exceeding 1 kV a.c.
5. EN 55014—Electromagnetic compatibility—requirements for household appliances, electric tools and similar apparatus.

1.3.4 International Electrotechnical Commission (IEC) Standards

1. IEC 60051-1:2016—Direct acting indicating analogue electrical measuring instruments and their accessories—Part 1: definitions and general requirements common to all parts

2. IEC 60255-127—Measuring relays and protection equipment—functional requirements for over/under voltage protection
3. IEC 60469:2013—Transitions, pulses and related waveforms—terms, definitions, and algorithms
4. IEC 61000—Testing and measurement techniques—PQ measurement methods
5. IEC 61000-4-11 and 4.34—Voltage sag immunity—16 amps or less and more than 16 amps
6. IEC 61000-4-30—PQ measurement methods
7. IEC 62586-1:2013—PQ measurement in power supply systems—Part 1: PQ instruments (PQI)
8. IEC 62586-2—PQ measurement in power supply systems—Part 2: functional tests and uncertainty requirements
9. IEC 62586-2:2013—PQ measurement in power supply systems—Part 2: functional tests and uncertainty requirements
10. IEC SC77A—Low-frequency EMC Phenomena—essentially equivalent of "power quality" in American terminology
11. IEC TR 60725:2012—Consideration of reference impedances and public supply network impedances for use in determining the disturbance characteristics of electrical equipment having a rated current ≤ 75 A per phase
12. IEC TR 61869-103—Instrument transformers—the use of instrument transformers for power quality measurement
13. IEC TR 62511:2014—Guidelines for the design of interconnected power systems
14. IEC TS 62749:2015—Assessment of power quality—characteristics of electricity supplied by public networks

Standards development on progress:

1. EC TR 62544—High-voltage direct current (HVDC) systems—application of active filters
2. IEC 62747:2015—Terminology for voltage-sourced converters (VSC) for high-voltage direct current (HVDC) systems
3. SC 77A/PT 61000-3-1—Electromagnetic compatibility (EMC)—Part 3-1: limits—overview of emission standards and guides. Technical report
4. SC 77A/WG 1—Harmonics and other low-frequency disturbances
5. SC 77A/WG 2—Voltage fluctuations and other low-frequency disturbances
6. SC 77A/WG 6—Low-frequency immunity tests
7. SC 77A/WG 8—Electromagnetic interference related to the network frequency
8. SC 77A/WG 9—Power quality measurement methods

1.4 Cost of poor power quality

Different aspects of production and delivery of energy have been done in demand of power, quality is increasing to meet the consumer requirements. The PQ problems are not raised from the source of power generation, but the PQ problems arise between power transmission, distribution, and end users (load) side. The cost reduction is a more effective method while corrective action is taken at designing the equipment [12,13].

- *The various sources of poor PQ:*

The various poor PQ sources basically started from installation to maintenance of power networks, which includes not proper grounding and not undersized distribution, the equipment or devices operated in beyond the designed value, power factor due to heavy load and cable insulation, and improper grounding.

The economic impacts are high due to poor PQ. In order to improve the PQ, various compensation techniques are used which resulting in high cost. The cost reduction in PQ is a major challenge. The cost reduction analysis of PQ should do as like other capital investments, and it is necessary to perform in different solutions. The precise amount of losses estimating in PQ is very difficult, but approximate calculation can be done by proper field survey, studies, and customer interaction.

The economic impact of PQ can be dividing into three categories:

Direct economic impacts:

- Long or unrecoverable downtime
  Production loss, wages during nonproductive hours, loss of resources for example, raw materials and product damage—semi-finished production.
- Restart cost of process—After-fault clearance or new unit.
- Equipment damage, human health and safety, environmental financial penalties, and utility cost associated with the interruption.

Indirect economic impacts:

- Product sales or unavailability of product in the market
- Premature equipment damage
- The financial market shares loss
- The cost of restoring brand equity.

Social-economic impacts:

- Uncomfortable building temperatures which may reduce efficiency, health, or safety
- Personal injury or fear
- Evacuating nearby residential buildings as a result of a failure of industrial safety.

The graphical representation of the cost of PQ based on various PQ indices is shown in Figure 1.8.

The process of investments can be evaluating in four basic steps:

1. PQ performance of present systems
2. Estimate the cost of poor PQ
3. Characterize the different solutions in terms of their cost and effectiveness
4. A comparative economic analysis of the different solutions.

1.4.1 Investment analysis to mitigate costs of power quality

The economic decision-making tools used to compare the different possible solutions and performing comparative investment analysis [9].

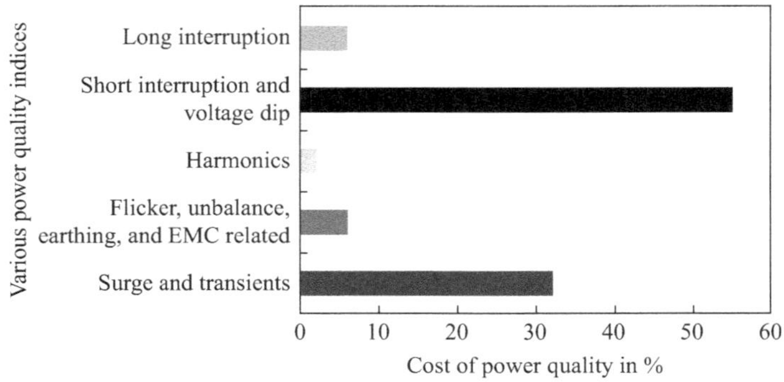

Figure 1.8 Various power quality indices versus cost of power quality

1.4.1.1 The various investment analysis of power quality

Industries investment analysis: investment of money for development of firm either investing on a project, research and development, sales, and marketing to follow the rules and regulations as per local government, international standards, and others.

The investment analysis is further classified into two groups which are:

- Return of investment: investing in projects and development of industries like more funding for research and development, improve market share, wages of employee, and advanced machines and others.
- Nonreturn of investment: this is a nonprofit-based investment; the firm should obey the rules and regulations of their government and international standards for pollution control, the safety of equipment, machines, and employees' health and safety policies.

Capital budgeting of industries: the capital budgeting is a decision process; the PQ investment is conducted in a similar way like other capital investment, and considering all the options is on an equal basis. The capital investment relies on the precise cash flows estimation, and possible accurate forecast result should be provided by decision makers. The decision makers should identify the various factors affecting the project, determine the key factors for the success of the project, and reliable forecasting of accuracy is also required. These factors are considered in decision makers, and this run in both deterministic and stochastic method to define the cash flow of precise and degree of uncertainty, respectively.

Individual or mutually exclusive project: the project can be classified as either independent or mutually exclusive.

Independent project: the independent project cash flow should not be affected by any other projects and should meet the company's capital budgeting criterion.

Mutually exclusive: a group or set of projects, more than one project, may satisfy the company norms and capital budgeting, but among a set of the project which only one will be accepted.

The cost of capital: the required rate of return on company securities which is a collection of financial assets such as cash stocks, loan, share, and bonds. Weighted average cost of capital is fund raised from various sources like the cost of debt (long and short term) and cost of equity (common and preferred stock).

Time value of money: The amount of money received today is more beneficial rather than later. The generalized formula to calculate the present net value is as follows:

$$\text{Present net value} = \frac{\text{Future value}}{1 + \text{rate of interest}}$$

Future value is specifying that today's investment will grow in future based on the present assets or cash value.

Present value is estimating the worth of money received in future is worth in today's money. Future value of a single cash flow: future value of a single cash flow refers to the growth of investment today, how much a single cash flow to over a period of time if it is invested to earn a compound rate of interest.

Present value of a single cash flow and of a cash flow stream: estimating the future value of money by evaluating cash flow of today's money or determining what percentage of cash flow to be received in future is worth in today's money.

1.4.1.2 Deterministic approach to PQ investment analysis

The various fundamental steps are followed in the decision process in which one of the major processes is economic analysis and investigation of investment against PQ because cost reduction is the main target in PQ investment. As said earlier, the deterministic method defines the precise cash flow in cost budgeting. The main factors of an investment are as follows:

- Capital investment
- Reduction of cost
- Operating and maintenance expenses for the investment
- Investment economic life.

Several evaluation methods are used to evaluate internal evaluation criteria, and depending on the importance of the investment, they select more or less sophisticated methods. An evaluation method provides the equivalent present value by estimating the net present value and internal rate of return.

1.4.1.3 Discounted cash flow methods

Net present value and internal rate of return: the general decision process rules and regulations are applicable for both NPV and IIR with consideration of all the projects' cash flows and time value of money, but these two have separate consideration in reinvestment, that NPV assumes that project cash flow is reinvested at the company capital cost, but in the case of IRR, it is assumed that the cash flows can be reinvested at the project IRR. As per engineering economic analysis, the NPV is an appropriate method to provide the base investment decision. The IIR focuses on particular problems but does not provide a unique solution.

1.4.1.4 Nondiscounted cash flow methods

Payback time: The payback time estimates the amount of time needed to recover the initial cost of the project. The payback time differs from each project if independent projects within the payback time are accepted when the number of years is less than specified. In mutually exclusive projects, shortest payback is preferred.

Break-even analysis: This is applicable for a complex project and rare for PQ investment; the break-even analysis is used for a gradual buildup of the project and returns their benefits over a time. The investment in companies will need several years to build the facilities, labor, training and service, and maintenance. After a while, the plant returns the money gradually as experience grows.

Stochastic approach to PQ investment analysis: The actual loss or negative flow is the main cause of PQ problem, the losses occur due to either continuous or discrete event such as harmonics generated in a transformer or induction motor or unmitigated voltage sags. The decision makers further classified the PQ problems into three categories based on the events that occur which are certainty, risk, and uncertainty.

1.4.2 Economic impact of power quality disturbances

The PQ disturbances have a direct impact on the end user, the electrical energy is converted into another form of energy like thermal or mechanical energy. The PQ disturbance affects the load or equipment. Modernized industries are automated plant with very sensitive equipment and highly complex processes. The PQ disturbance causes the long or short downtime, and this is correlated with loss of production which reduces the profit and revenue of the industries [14].

The economic and investment analysis of PQ disturbance is estimated by using deterministic and stochastic methods. The deterministic method is suitable when all the items or variable are definitely known, but in the case of stochastic method, variables are uncertain which cannot predict the time of event occur. The various impact of PQ disturbance is listed as follows:

- Cost of sag in deterministic and stochastic analysis method.

1.4.2.1 Deterministic analysis method

The three evaluation steps are followed to estimate the voltage sags at a particular site in the network [7]:

1. Estimating the performance of system.
2. Evaluating the components and equipment failure during a voltage sag.
3. Analysis the economic impact.

The performance of power supply system is estimated based on the critical distance (CD) and fault position (FP) of voltage sags. These methods based on the simulation of the system in short-circuited conditions for assigned FP.

Critical distance: CD method allows the computation voltage sags when the particular node is experienced by a short circuit, and this method is more suitable for the radial system (distributed network).

Fault position: This method is applied to meshed systems; it obtains the voltage sags in all the nodes of the system due to faults in every bus system. This method offers a global response to faults, even for radial network, and this method provides following the detailed information about the electric power system network.

The above information helps the customer to observe the knowledge about voltage sags and requires tolerance rate of their components and equipment. Different manufacturer and equipment have various tolerance limits for voltage variation. The economic impact of voltage sags is directed to the cost of shutting down the process of plant, fault clearance, restarting the process, loss of production, and replacing the component or equipment if any damage during shut down.

Stochastic analysis method: The estimation of economic consequence of voltage sags in various uncertainty scenarios is taken into consideration and related to equipment and process sensitivity.

Pohjanheimo and Milanovic have proposed two different methods to assess the economic value by using stochastic analysis method.

- Pohjanheimo proposed the sensitivity using discrete state on and off and stochastic values;
- A voltage sag characteristic depends on fault occurrence, and assigning a different cost value for voltage sag varies based on their fault and its characteristics.
- The probability of interconnections between equipment is taken into account and includes with the effect of mitigation devices.

Milanovic analyzed the stochastic method, the calculation of economic value takes account the uncertainty associated with voltage sag, different equipment interconnected in the plant, the type of customer, and the location of the process.

This method differs from Pohjanheimo method by treating the equipment based on their sensitivity level. Typically, the same brand type of equipment has different sensitivity, and each equipment respond to voltage sags in a different way. The sensitivity of equipment is classified into four major equipment types that is computers (PCs and Laptops), automation controller (PLC), adjustable speed drives, and AC contactors.

1.4.2.2 Cost of harmonics in deterministic and stochastic analysis method

Any electrical equipment will experience the harmonics presented in either voltage or current. The effect of distortion on voltage or current is producing the additional losses, misoperation, and premature aging of components and equipment.

To estimate or compute the economic value of additional losses (fundamental losses) during operating condition, the following information are required:

1. The study acquires deep knowledge about system operating condition
2. Knowledge of type and configuration of the system
3. Assignment of the variation rate of the electric energy unit cost and the discount rate.

The uncertain changes in the system will affect the real-time systems for changes in nonlinear load demands and operate modes, and due to changes in network configuration, the economic model should be in a random basis. This will introduce the random variables and stochastic analysis method.

Misoperation: The performance of equipment is degraded compared to normal condition. The cost estimation of improper is a complex subject and some considerations for deterministic methods are used to estimate the cost of improper operation.

To estimate the cost of misoperation, the following data are required:

- Due to harmonics presence in the system, the equipment or components are malfunctioning;
- The type of equipment is used in process or activity;
- The economic values of all the items contributing to lower productivity.

Premature aging: The aging level of equipment is decreased because of increasing the stress level compared normal condition. Due to premature, the equipment life will decrease result error in measurement, to overcome this replacing damaged equipment. To estimate the economic value of premature aging (replacing the equipment), the following data is required:

- Deep study of network structure and duration configurations
- Knowledge of type and configuration of the system
- Knowledge equipment and components life model.

1.4.3 Economic mechanisms for improving power quality levels

The various mechanisms for improving PQ levels are system-side solutions, customer service entrance solutions, power-electronics-based conditioner, improved specifications, and equipment design.

The value of mitigation of PQ problems is based on the base nominal financial loss per process trip, location of the plant in the network system, and the immunity and interdependence levels of the processes. The economic value of mitigation is achieved by reduction of financial loss by direct installing of mitigation devices.

- Need variety of network planning and access to reduce the investment
- Incentives are provided to utilities, which maintain the good power factor, and this can continuously monitor the variation of power factor. The quality of service is provided by the power distribution companies, if they failed to impose a penalty for poor service and quality.
- By improving the voltage quality and continuity power supply.
- The customer is penalized for violation of exceeding reference level of disturbance or event; if reference level exceeds, the customer does not reduce the emission level. The customer side reliability/PQ concerns are to be evaluated and different performance indices to be used and fix the expected level of performance.
- Incentives and penalties given are based on the performance either above or below the expected level. Each parties' responsibility is to achieve the desired performance or resolving the PQ problems.

Contractual method:

- Contract between transmission and distribution companies
- Contracts between distribution and end-user
- Contracts between retail marketers and the end-use customer
- Contracts between distribution companies and independent power producers.

References

[1] Dugan R. C., McGranaghan M. F., Santosa S., Beaty H. W. (eds). *Electrical Power Systems Quality*, 2nd edn, McGraw-Hill; 2003, pp. 528.

[2] Djokic A., Vanalme G., Milanovic J. V., Stockman K. 'Sensitivity of Personal Computers to Voltage Sags and Short Interruptions'. *IEEE Transactions on Power Delivery*. 2005, vol. 20 (1), pp. 375–383.

[3] Koval D. O., Carter C. 'Power Quality Characteristics of Computer Loads'. *IEEE Transactions on Industry Applications*. 1997, vol. 33 (3), pp. 613–621.

[4] Woods, E. J. 'Aircraft Electrical Systems—Coping with Harmonics for Changing Power Demands,' *Proceeding of the Energy Conversion Engineering Conference*. 1990. pp. 84–89.

[5] Djokic S. Z., Milanovic J. V., Kischen D. S. 'Sensitivity of AC Coil Contactors to Voltage Sags, Short Interruptions, and Undervoltage Transients'. *IEEE Transactions on Power Delivery*. 2004, vol. 19 (3), pp. 1299–1307.

[6] Gomez J. C., Morcos M. M., Reineri C. A., Campetelli G. N. 'Behavior of Induction Motor Due to Voltage Sags and Short Interruptions'. *IEEE Transactions on Power Delivery*. 2002, vol. 17 (2), pp. 434–440.

[7] Sankaran C., *Power Quality*. Boca Raton, FL: CRC Press; 2002.

[8] *IEEE-1159 Recommended Practice for Monitoring Electric Power Quality-IEEE Power & Energy Society*.

[9] Caramia P., Carpinelli G., Verde P. *Power Quality Indices in Liberalized Market*. West Sussex, UK: Wiley; 2009.

[10] Moreno-Muñoz, A. Ed., *Power Quality*: Mitigation Technologies in a Distributed Environment, Springer Science & Business Media; 2007.

[11] Fuchs E., Masoum M. A. *Power Quality in Power Systems and Electrical Machines*, Academic Press; 2011.

[12] Fuchs E. F., Mausoum M. A. S. *Power Quality in Power Systems and Electrical Machines*. London: Elsevier; 2008.

[13] Baggini A. *Handbook of Power Quality*. New York: John Wiley & Sons; 2008.

[14] Reid W. E. 'Power Quality Issues—Standards and Guidelines.' *IEEE Transactions on Industry Applications*. 1996, vol. 32.3, pp. 625–632.

Chapter 2

Frequency-domain power theory and metering of harmonic-pollution responsibility

Murat Erhan Balci[1] and Mehmet Hakan Hocaoglu[2]

2.1 Introduction

Historically, for single-phase systems, apparent power (S) is defined as the product of voltage and current rms values, and it is mainly used to size the power system equipment [1]. In addition, apparent power is decomposed to two power components as active (P) and reactive (Q) powers and is expressed as their vector sum. Active power is the average value of the instantaneous power, which is the product of instantaneous voltage and current, and it transports the net energy from source to load. However, this is not the case for reactive power. Due to this, reactive power is also called useless power in electrical engineering community. Accordingly, to measure the system how effectively be used, the power factor is defined as the ratio of active and apparent powers. The capacitors are conventionally used to compensate inductive reactive powers drawn by the loads [2]. It is well known for sinusoidal systems that the power of the compensation capacitors, which improve power factor to its maximum (near unity), is equal to the inductive reactive power measured at the load terminal when the variation of the terminal voltage is negligible with the compensation.

On the other hand, for three-phase systems, the oldest apparent power definitions are arithmetic and vector apparent powers [3,4]. They are defined by treating each phase individually of the three-phase system. First definition is calculated as the arithmetic sum of each phase's apparent power, and it has no power decomposition. Vector apparent power is expressed as the vector sum of total active and total reactive powers of the system.

In the last 20 years, the power-electronic-circuits based non-linear devices, which draw distorted (or non-sinusoidal) current wave shapes, have largely been employed in the power systems. However, classical apparent power and its

[1]Faculty of Engineering, Department of Electrical and Electronics Engineering, Balikesir University, Balikesir, Turkey
[2]Faculty of Engineering, Department of Electronics Engineering, Gebze Technical University, Gebze, Kocaeli, Turkey

decomposition, which are briefly summarized above, are defined for the systems with sinusoidal voltage and current conditions. Therefore, for the non-sinusoidal systems, several apparent power definitions and their resolutions are proposed in the literature [5–16]. These definitions and resolutions are analysed and interpreted in many studies [17–23]. It can be concluded from [17–21] that the power resolutions generally aimed to provide a tool for sizing and operating of the non-active power compensators and measuring non-linearity degree level of the loads. Balci and co-workers [22,23] also pointed out that the power factor calculated with respect to arithmetic and vector apparent powers do not provide true information on the system efficiency for non-sinusoidal and unbalanced three-phase systems. Due to this, for three-phase systems not only harmonic distortion but also unbalance conditions are considered to redefine the apparent power and its decomposition. However, the concept of apparent power or frequency-domain power theory is still a topic that remained controversial in electrical engineering community. Thus, the different apparent power definitions and their decompositions are encouraged by the international standards such as IEEE standard 1459 [11] and DIN standard 40110 [12].

One of the most significant adverse effects of the harmonic pollution can be mentioned as the increased losses and decreased life expectancy of the power system equipment [24]. Accordingly, the costs related with harmonics and any required mitigation device should be recovered from the consumers that generate harmonic distortion. However, the guidelines and standards on the limitation of harmonic pollution do not include any method to determine the harmonic-producing consumers [25–31]. In addition, the conventional harmonic-pollution metering index as current total harmonic distortion (THDI) is not capable of metering consumer's harmonic-pollution responsibility due to the fact that it depends on the distortion of utility side's supply voltage. Therefore, to fulfil this gap, for the harmonic source detection and sharing harmonic responsibility between utility and consumer sides, several methods and indices are proposed, and their validities are evaluated by simulation and experimental based analysis in the literature [32–49].

In this chapter, first, for non-sinusoidal and unbalanced systems, the widely known apparent power definitions and their resolutions will be summarized and qualitatively analysed, and the measurement results will be presented to interpret main properties of the apparent powers and their resolutions included in IEEE and DIN standards. Second, the methods and indices, which are proposed for detection of the harmonic-producing loads and quantifying the harmonic-pollution responsibility of the loads in the literature, are summarized and qualitatively analysed, and the statistical analysis of the measurement results will be presented to illustrate the response of the practical ones for the different load cases under several supply voltage cases.

2.2 Power resolutions for non-sinusoidal single-phase systems

For single-phase systems, there are three distinct definitions of the apparent power, namely classical apparent power [1], Wilczynski's apparent power [13] and Ghassemi's apparent power [14]. As mentioned in Section 2.1, the classical one is defined as the product of rms voltage (V) and rms current (I) values. Accordingly, for time- and harmonic-domain, it can be calculated as in the following equation:

$$S = \sqrt{\frac{1}{T}\int_0^T v(t)^2 dt}\sqrt{\frac{1}{T}\int_0^T i(t)^2 dt} = \sqrt{\sum_n V_n^2}\sqrt{\sum_n I_n^2} \tag{2.1}$$

where $v(t)$ and $i(t)$ denote time–domain expressions of voltage and current,

$$v(t) = \sum_n \sqrt{2}V_n \sin(\omega_n t + \theta_n) \tag{2.2}$$

$$i(t) = \sum_n \sqrt{2}I_n \sin(\omega_n t + \delta_n) \tag{2.3}$$

and V_n and I_n are voltage and current rms values for nth harmonic, θ_n and δ_n denote phase angles of nth harmonic voltages and currents, and ω_n is the nth harmonic angular frequency.

Classical apparent power is conventionally used for sizing the system equipment and measuring the energy transfer efficiency. On the other side, the definitions of Wilczynski (S_W) and Ghassemi (S_G) have the expressions given in (2.4) and (2.5), respectively:

$$S_W = \sqrt{\int_0^T p(t)^2 dt} \tag{2.4}$$

$$S_G = \sqrt{\int_0^T |\tilde{p}(t)|^2 dt} \tag{2.5}$$

where $p(t)$ is the instantaneous power

$$p(t) = v(t)i(t) \tag{2.6}$$

and $\tilde{p}(t)$ is the complex instantaneous power,

$$\tilde{p}(t) = \tilde{v}(t)i(t), \quad \tilde{v}(t) = \sum_n \sqrt{2}V_n e^{j(\omega_n t + \theta_n)} \tag{2.7}$$

Since both definitions consider the oscillations of instantaneous and complex instantaneous powers, they are more relevant for the sizing of the system equipment when compared to the classical apparent power. However, the losses are more significant than the sizing for the proper utilization of the system. Due to this case, classical one is still recommended and enforced by the international standards. Accordingly, the widely known resolutions proposed for the classical apparent power will be summarized and qualitatively analysed in this section.

2.2.1 Budeanu's power resolution

Budeanu postulated that apparent power has two orthogonal components, namely active and non-active powers [5,17–19]. The first one is average power that can easily be calculated by averaging the product of voltage and current over one period (T) in time domain or by convolution in harmonic domain:

$$P = \frac{1}{T}\int_0^T p(t)dt = \sum_n P_n, \quad P_n = V_n I_n \cos(\theta_n - \delta_n) \tag{2.8}$$

This component was considered by Budeanu since it is an actual power converted to the physical work. The second one, non-active power, is separated into two components as reactive and distortion powers. Budeanu's reactive power is expressed as arithmetic sum of the individual harmonic reactive powers:

$$Q_b = \sum_n Q_n = \sum_n V_n I_n \sin(\theta_n - \delta_n) \tag{2.9}$$

Distortion power is expressed as

$$D_b = \sqrt{S^2 - P^2 - Q_b^2} \tag{2.10}$$

It is seen from the analysis results presented in [18] that power factor does not attain its maximum when Budeanu's reactive power is completely compensated with a basic capacitor under non-sinusoidal conditions. The same study also shows that Budeanu's distortion power has the parts compensable with a basic capacitor. Thus, it is clear that Budeanu's reactive power does not give true information on the sizing of the optimal compensation capacitor, which maximizes the power factor.

2.2.2 *Fryze's power resolution*

Fryze divided the load current into two orthogonal components as active and non-active (or reactive) currents to provide his power resolution [5,17–19]. The first one is calculated by using active power:

$$i_{ac}(t) = \frac{P}{V^2} v(t) \tag{2.11}$$

and the second one is the remaining current component when the active part is subtracted from the total current:

$$i_{fr}(t) = i(t) - i_{ac}(t) \tag{2.12}$$

Since both active and reactive current parts are orthogonal, the rms value of the total current is vector sum of their rms values:

$$I = \sqrt{I_{ac}^2 + I_{fr}^2} \tag{2.13}$$

By multiplying both sides of the expression given in (2.13) with the rms voltage (V), he defined two power components:

active power

$$P = VI_{ac} \tag{2.14}$$

and reactive power,

$$Q_f = VI_{fr} \tag{2.15}$$

Therefore, he suggested that the classical apparent power should be the vector sum of active and reactive powers:

$$S = \sqrt{P^2 + Q_f^2} \tag{2.16}$$

Balci and Hocacglu [18] show that Fryze's reactive power attains its minimum when the power factor is maximized by simple capacitor under non-sinusoidal conditions. However, it could not be used for direct provision of the power of optimum compensation capacitor. On the other hand, Fryze's current resolution leads to the idea of the active compensators, which injects his reactive current component into the system, for providing unity power factor compensation under non-sinusoidal conditions.

2.2.3 Shepherd and Zakikhani's power resolution

Shepherd and Zakikhani [6,7,17–19] divide the current into three orthogonal components, which are:

active current,

$$I_R = \sqrt{\sum_{n \in n} I_n^2 \cos^2(\theta_n - \delta_n)} \tag{2.17}$$

reactive current

$$I_X = \sqrt{\sum_{n \in n} I_n^2 \sin^2(\theta_n - \delta_n)} \tag{2.18}$$

and distortion current,

$$I_D = \sqrt{I^2 - I_R^2 - I_X^2} \tag{2.19}$$

where n_1 is defined as the common harmonics of voltage and current.

They proposed a power resolution related with the rms currents given in (2.17)–(2.19):

$$S = \sqrt{S_R^2 + S_X^2 + S_D^2} \tag{2.20}$$

with active apparent power,

$$S_R = \sqrt{\sum_{n \in n_1 \cup n_2} V_n^2 I_R} \tag{2.21}$$

reactive apparent power

$$S_X = \sqrt{\sum_{n \in n_1 \cup n_2} V_n^2 I_X} \tag{2.22}$$

and distortion apparent power,

$$S_D = \sqrt{\sum_{n \in n_1 \cup n_2} V_n^2} I_D = \sqrt{\sum_{n \in n_1} V_n^2 \sum_{n \in n_3} I_n^2 + \sum_{n \in n_2} V_n^2 \left(\sum_{n \in n_1} I_n^2 + \sum_{n \in n_3} I_n^2 \right)} \tag{2.23}$$

where n_2 and n_3 denote the harmonic numbers of voltage in which there are no current harmonics and the harmonic numbers of current in which there are no voltage harmonics.

This power resolution has the reactive power, which attains the minimum value when the power factor is maximized by a simple capacitor [18]. However, it could not be used for direct provision of the power of optimum compensation capacitor. In addition to that, Shepherd and Zakikhani's resolution does not provide any information on power factor since active apparent power (S_R) is different from active power (P). Lastly, it should be mentioned that for the practical systems, Shepherd and Zakikhani's distortion apparent power is nil since voltage and current have the same harmonic orders for those systems.

2.2.4 Sharon's power resolution

Sharon's power resolution [8,18,19] has three components active power (P), Shepherd and Zakikkani's reactive apparent power (S_X) and complementary apparent power (S_C), which are expressed as follows:

$$S_C = \sqrt{S^2 - P^2 - S_X^2} \tag{2.24}$$

Sharon's power resolution inherits the same problem, which is the lack of direct provision of optimum compensation capacitor's power, from Shepherd and Zakikhani's resolution [18].

2.2.5 Kusters and Moore's power resolution

The power resolution of Kusters and Moore aims to determine the value of the shunt compensation capacitor or inductor for maximization of power factor in the non-sinusoidal systems where the variation of the load voltage is negligible with the compensation [9,17–19]. Here, their power resolution is summarized for the inductive load case, since in the power systems, the loads generally have inductive characteristic.

Kusters and Moore separated current into three components to propose their power resolution. These three current components are:

active current given in (2.11), reactive current, which can be compensated by shunt capacitor in the case of inductive load

$$i_{kr}(t) = \frac{1/T \int_0^T \mathring{v}(t) i(t) dt}{\mathring{V}^2} \mathring{v}(t) \tag{2.25}$$

and residual reactive current;

$$i_{krr}(t) = i(t) - i_{ac}(t) - i_{kr}(t) \tag{2.26}$$

where $\overset{\circ}{v}(t)$ and $\overset{\circ}{V}$ are the instantaneous voltage's derivative and the rms value of instantaneous voltage's derivative, respectively.

Due to the fact that four current components are orthogonal, vector sum of their rms values gives the rms value of total current:

$$I = \sqrt{I_{ac}^2 + I_{kr}^2 + I_{krr}^2} \tag{2.27}$$

By multiplying both sides of (2.27) with the rms value of the load voltage, they found three power components. These power components are: P is the active power, Q_{kr} is the reactive power

$$Q_{kr} = VI_{kr} = V\frac{\sum_n nV_nI_n \sin(\theta_n - \delta_n)}{\sqrt{\sum_n n^2V_n^2}} \tag{2.28}$$

and Q_{krr} is the residual reactive power,

$$Q_{krr} = VI_{krr} \tag{2.29}$$

Therefore, they suggested the power resolution as in (2.30),

$$S = \sqrt{P^2 + Q_{kr}^2 + Q_{krr}^2} \tag{2.30}$$

The main advantage of Kusters and Moore's resolution is that its reactive power component gives the power of optimum compensation capacitor, which maximizes the power factor under non-sinusoidal conditions, if the variation of the load voltage is negligible before and after compensation [18].

2.2.6 Czarnecki's power resolution

The need of identification of the reason on the decreasing source efficiency under non-sinusoidal conditions is the main motivation of Czarnecki's resolution [10,17–19]. For this aim, he distinguished current into four parts. These current parts are

active current,

$$i_{ac}(t) = G_e \sum_{n\in l} \sqrt{2}V_n \sin(\omega_n t + \theta_n) \tag{2.31}$$

reactive current,

$$i_{cr}(t) = \sum_{n\in l} \sqrt{2}B_nV_n \sin\left(\omega_n t + \theta_n - \frac{\pi}{2}\right) \tag{2.32}$$

scattered current

$$i_s(t) = \sum_{n \in l} \sqrt{2}(G_n - G_e)V_n \sin(\omega_n t + \theta_n) \tag{2.33}$$

and generated harmonic current,

$$i_g(t) = \sum_{n \in k} \sqrt{2} I_n \sin(\omega_n t + \delta_n) \tag{2.34}$$

where G_e denotes the load's equivalent conductance,

$$G_e = \frac{P}{V^2} \tag{2.35}$$

G_n and B_n are the nth harmonic conductance and susceptance parameters, which are real and imaginary parts of nth harmonic-load admittance,

$$Y_n = G_n - jB_n = \frac{P_n}{V_n^2} - j\frac{Q_n}{V_n^2} \tag{2.36}$$

and k represents current harmonic numbers do not present in the set of voltage harmonic numbers, and l is the voltage harmonic numbers.

Thus, he provided the current resolution expressed as follows:

$$i(t) = i_{ac}(t) + i_{cr}(t) + i_s(t) + i_g(t) \tag{2.37}$$

Since four current components are orthogonal, vector sum of their rms values gives the rms value of total current:

$$I = \sqrt{I_{ac}^2 + I_{cr}^2 + I_s^2 + I_g^2} \tag{2.38}$$

The rms values of the current components can be written in harmonic domain as follows:

$$I_{ac} = G_e V \tag{2.39}$$

reactive current,

$$I_{cr} = \sqrt{\sum_{n \in l} B_n^2 V_n^2} \tag{2.40}$$

scattered current

$$I_s = \sqrt{\sum_{n \in l} (G_n - G_e)^2 V_n^2} \tag{2.41}$$

and generated harmonic current,

$$I_g = \sqrt{\sum_{n \in k} I_n^2} \tag{2.42}$$

Consequently, he suggested a power resolution based on the current resolution expressed in (2.38) as follows:

$$S = \sqrt{P^2 + Q_{cr}^2 + D_s^2 + D_g^2} \tag{2.43}$$

where P denotes active power, and other three powers are as follows:
reactive power,

$$Q_{cr} = VI_{cr} \tag{2.44}$$

scattered power

$$D_s = VI_s \tag{2.45}$$

and generated harmonic power,

$$D_g = VI_g \tag{2.46}$$

The main property of Czarnecki's power resolution is identification of the physical phenomena responsible for the source current increase. It can be mentioned from the analysis given in [18] that the reactive powers of the resolutions proposed by Sharon and Czarnecki have the same values. In addition to that, generated harmonic power is nil in the practical systems since voltage and current have the same harmonic orders for those systems. As a result, for the practical systems, Czarnecki's scattered power and Sharon's complementary apparent power give the same results.

2.2.7 IEEE standard power resolution

The IEEE standard power resolution [11] has two main power components: one of them is fundamental effective apparent power

$$S_1 = V_1 I_1 = \sqrt{P_1^2 + Q_1^2} \tag{2.47}$$

and other is non-fundamental apparent power (S_{NF}),

$$S_{NF}^2 = S^2 - S_1^2 \tag{2.48}$$

In (2.47), V_1 and I_1 denote the fundamental rms voltage and current.

Non-fundamental apparent power (S_{NF}) is decomposed into the power components:

$$S_{NF}^2 = D_I^2 + D_V^2 + S_H^2 \tag{2.49}$$

where the current distortion power, the voltage distortion power and the harmonic apparent power are defined as in (2.50)–(2.52), respectively:

$$D_I = V_1 I_H \tag{2.50}$$

$$D_V = V_H I_1 \tag{2.51}$$

$$S_H = V_H I_H \tag{2.52}$$

In (2.50)–(2.52), V_H and I_H denote the non-fundamental rms voltage and current values calculated as

$$V_H = \sqrt{V^2 - V_1^2} \tag{2.53}$$

$$I_H = \sqrt{I^2 - I_1^2} \tag{2.54}$$

2.2.8 Balci and Hocaoglu's power resolution

Balci and Hocaoglu aim to provide a power resolution that can be used as a tool for two goals as (i) collective design and operation of the several kinds of passive and active compensators and (ii) detection of the non-linear or harmonic producing loads. To achieve both goals, first, for non-sinusoidal single-phase systems, they separated the current into four components as active, reactive, scattered conductance and scattered susceptance currents, which are all expressed in terms of the load impedance parameters [15,21]. The load impedance parameters, which are used to establish the current resolution, are equivalent admittance (G_e) given in (2.35), the nth harmonic conductance (G_n) and nth harmonic susceptance (B_n) given in (2.36), and the nth harmonic equivalent susceptance (B_{en}) defined in the following equation:

$$B_{en} = nB_{e1} = \frac{n\sum_n nQ_n}{\sum_n n^2 V_n^2} \tag{2.55}$$

Accordingly, they proposed current resolution as in the following equation:

$$i(t) = i_{ac}(t) + i_r(t) + i_{sc}(t) + i_{ss}(t) \tag{2.56}$$

where the current components are expressed as
active current,

$$i_{ac}(t) = \sum_n \sqrt{2} G_e V_n \sin(\omega_n t + \theta_n) \tag{2.57}$$

reactive current,

$$i_r(t) = \sum_n \sqrt{2} B_{en} V_n \sin\left(\omega_n t + \theta_n - \frac{\pi}{2}\right) \tag{2.58}$$

scattered conductance current

$$i_{sc}(t) = \sum_n \sqrt{2}(G_n - G_e) V_n \sin(\omega_n t + \theta_n) \tag{2.59}$$

and scattered susceptance current,

$$i_{ss}(t) = \sum_n \sqrt{2}(B_n - B_{en}) V_n \sin\left(\omega_n t + \theta_n - \frac{\pi}{2}\right) \tag{2.60}$$

The rms values of these four current components written as in harmonic domain are as follows:

active current's rms value as given in (2.39),

reactive current's rms value,

$$I_r = \sqrt{\sum_n B_{en}^2 V_n^2} \tag{2.61}$$

scattered conductance current's rms value

$$I_{sc} = \sqrt{\sum_n (G_n - G_e)^2 V_n^2} \tag{2.62}$$

and scattered susceptance current's rms value,

$$I_{ss} = \sqrt{\sum_n (B_n - B_{en})^2 V_n^2} \tag{2.63}$$

Since the current components are orthogonal, the rms value of total current can be calculated as follows:

$$I = \sqrt{I_{ac}^2 + I_r^2 + I_{sc}^2 + I_{ss}^2} \tag{2.64}$$

Consequently, they obtained their apparent power resolution by multiplying both sides of (2.64) with the square of voltage's rms value:

$$\begin{aligned} S^2 &= (VI_c)^2 + (VI_r)^2 + (VI_{sc})^2 + (VI_{ss})^2 \\ &= P^2 - Q_r^2 + D_{sc}^2 + D_{ss}^2 \end{aligned} \tag{2.65}$$

In (2.65), the power components are named as active (P), reactive (Q_r), scattered conductance (D_{sc}) and scattered susceptance (D_{ss}) powers. Q_r gives the power of the optimal capacitor, which maximizes the power factor under non-sinusoidal conditions. The nth harmonic reactance value of the optimal capacitor can be written as

$$X_{Cn} = \frac{1}{B_{en}} = \frac{\sum_n n^2 V_n^2}{n \sum_n n Q_n} \tag{2.66}$$

In addition, D_{sc} is useful for identification of the load's non-linearity or harmonic source detection.

When the resolution of Balci and Hocaoglu is compared with the similar resolutions, some features should be mentioned as follows:

- The residual reactive power component of Kusters and Moore's resolution is separated into two power components, namely, scattered conductance and scattered susceptance powers in the resolution of Balci and Hocaoglu.
- The reactive power component of Czarnecki's resolution is separated into two power components, namely, reactive and scattered susceptance powers in the resolution of Balci and Hocaoglu.

2.3 Power resolutions for non-sinusoidal and unbalanced three-phase systems

It can be seen from the previous section that for single-phase systems, there is a consensus on the classical apparent definition, and the controversial issue is its resolution under non-sinusoidal conditions. However, for non-sinusoidal and unbalanced three-phase systems, there is no consensus on both definition and resolution of the apparent power, and several apparent power definitions and their resolutions are encouraged by the international standards such as IEEE standard 1459 and DIN standard 40110.

Accordingly, the widely known apparent power definitions and their resolutions are summarized and qualitatively analysed in the following sections.

2.3.1 Vector apparent power and its resolution

Vector apparent power [3,4,22,23] is the vector sum of the active, Budeanu's reactive and Budeanu's distortion powers, which are calculated individually for each phase ($m = a$, b and c):

active powers,

$$P_m = \sum_n P_{mn} = \sum_n V_{mn} I_{mn} \cos(\theta_{mn} - \delta_{mn}) \tag{2.67}$$

Budeanu's reactive powers

$$Q_{bm} = \sum_n Q_{mn} = \sum_n V_{mn} I_{mn} \sin(\theta_{mn} - \delta_{mn}) \tag{2.68}$$

and Budeanu's distortion powers,

$$D_{bm} = \sqrt{S_m^2 - P_m^2 - Q_{bm}^2} \tag{2.69}$$

where apparent powers,

$$S_m = \sqrt{\sum_n V_{mn}^2}\sqrt{\sum_n I_{mn}^2} \tag{2.70}$$

Thus, vector apparent power is calculated as follows:

$$\begin{aligned} S_V &= \sqrt{\left(\sum_{m=a,b,c} P_m\right)^2 + \left(\sum_{m=a,b,c} Q_{bm}\right)^2 + \left(\sum_{m=a,b,c} D_{bm}\right)^2} \\ &= \sqrt{P^2 + Q^2 + D^2} \end{aligned} \tag{2.71}$$

It can be seen from the expressions given in (2.67)–(2.71) that vector apparent power threats each phase as an independent single-phase system. Thus, according to this apparent power definition, the unity power factor case can be achieved when each phase is individually compensated. This practically means that for unity power factor case of the vector apparent power, the unbalanced and distorted current will exist under the unbalanced and distorted terminal voltage condition.

2.3.2 *Arithmetic apparent power*

Arithmetic apparent power [3,4,22,23] is defined as the arithmetic sum of each phase's apparent power, and it handles three-phase systems by treating each phase individually:

$$S_{Ar} = \sum_{m=a,b,c} S_m \tag{2.72}$$

It can be mentioned from (2.70) and (2.72) that arithmetic apparent power can be measured by means of few computational efforts. This case is its advantage for practical evaluation of the system sizing. However, its main limitation is that it is not decomposed into the power components.

2.3.3 *Bucholz's apparent power and its resolutions*

For three-phase systems (*a*, *b*, *c* phases) with neutral line (*N*), Bucholz's apparent power definition can be written as [12,22,23],

$$S_\Sigma = V_\Sigma I_\Sigma = \sqrt{V_{a0}^2 + V_{b0}^2 + V_{c0}^2 + V_{N0}^2}\sqrt{I_a^2 + I_b^2 + I_c^2 + I_N^2} \tag{2.73}$$

where the instantaneous value of the *m* line-to-virtual star point (0) voltage is

$$v_{m0}(t) = v_{mN}(t) - \frac{1}{4}\left[v_{aN}(t) + v_{bN}(t) + v_{cN}(t)\right] \tag{2.74}$$

and the line-to-virtual star point rms voltages (V_{a0}, V_{b0}, V_{c0}, V_{N0}) can be calculated in terms of the positive-, negative- and zero-sequence voltage harmonics of order $n(V_n^+, V_n^-, V_n^0)$;

$$V_{m0} = \sqrt{\sum_n \left(V_n^+\right)^2 + \sum_n \left(V_n^-\right)^2 + \frac{1}{16}\sum_n \left(V_n^0\right)^2} \quad \text{for } m = a, b, c \text{ phases} \tag{2.75}$$

$$V_{N0} = \sqrt{\frac{9}{16}\sum_n \left(V_n^0\right)^2} \tag{2.76}$$

The Bucholz's apparent power have three resolutions in the literature. They are summarized below.

2.3.3.1 Resolution contained in DIN standard 40110

DIN standard 40110 resolution of Bucholz's apparent power has two main power components as active and non-active powers [12,22]:

$$S_\Sigma = \sqrt{P^2 + Q_{tot\Sigma}^2} \tag{2.77}$$

Active power is expressed as follows:

$$P = \frac{1}{T}\int_0^T \sum_{m=a,b,c,N} v_{m0}(t) i_m(t) dt \tag{2.78}$$

In addition, the non-active power is divided into two parts:

$$Q_{tot\Sigma} = \sqrt{Q^2_{tot\Sigma\perp} + Q^2_{tot\Sigma||u}} \tag{2.79}$$

where $Q_{tot\Sigma||u}$ is caused by the unbalanced part of the load conductance, and $Q_{tot\Sigma\perp}$ is caused by the current component orthogonal with the voltage.

The DIN standard gives the expressions for the non-active components as in the following equations:

$$Q_{tot\Sigma||u} = V_\Sigma \sqrt{\sum_m (G_m - G)^2 V^2_{m0}} \tag{2.80}$$

$$Q_{tot\Sigma\perp} = V_\Sigma \sqrt{\sum_{m=a,b,c,N} \left[I^2_m - G^2_m V^2_{m0}\right]} \tag{2.81}$$

where the conductance parameters placed in (2.80) and (2.81) are as follows:

$$G_m = \frac{P_m}{V^2_{m0}} \tag{2.82}$$

$$G = \frac{P}{V^2_\Sigma} \tag{2.83}$$

According to DIN standard 40110 resolution, the fully compensated load is equivalent to four identical resistances connected in star and producing a virtual neutral point 0. Due to this, each line current waveform is a perfect replica of its respective line-to-virtual neutral voltage, and the neutral wire is considered as a fourth phase.

Consequently, for non-sinusoidal and unbalanced terminal voltages, if the compensation is achieved by regarding DIN standard power resolution, some imbalance will exist in the line currents, the neutral current will not be nil and the line currents will still have non-sinusoidal or distorted wave shapes.

2.3.3.2 Resolution proposed by Czarnecki

Czarnecki [10,17,19] proposed a resolution of Buchollz's apparent power for three-phase and three-line systems with non-sinusoidal and balanced phase voltages as follows:

$$\vec{v}(t) = \begin{bmatrix} \sum_n \sqrt{2} V_n \sin(\omega_n t + \theta_{an}) \\ \sum_n \sqrt{2} V_n \sin(\omega_n t + \theta_{bn}) \\ \sum_n \sqrt{2} V_n \sin(\omega_n t + \theta_{cn}) \end{bmatrix} \tag{2.84}$$

and non-sinusoidal and unbalanced phase currents as follows:

$$\vec{i}(t) = \begin{bmatrix} \sum_n \sqrt{2} I_{an} \sin(\omega_n t + \delta_{an}) \\ \sum_n \sqrt{2} I_{bn} \sin(\omega_n t + \delta_{bn}) \\ \sum_n \sqrt{2} I_{cn} \sin(\omega_n t + \delta_{cn}) \end{bmatrix} \tag{2.85}$$

For the case of the system considered by Czarnecki, Buchollz's apparent power can be rewritten as in the following equation:

$$S_\Sigma = V_\Sigma I_\Sigma = \sqrt{\sum_n V_{\Sigma n}^2} \sqrt{\sum_n I_{\Sigma n}^2} \tag{2.86}$$

where $V_{\Sigma n}$ and $I_{\Sigma n}$ are expressed as follows:

$$V_{\Sigma n} = \sqrt{V_{an}^2 + V_{bn}^2 + V_{cn}^2} = \sqrt{3 V_n^2} \tag{2.87}$$

$$I_{\Sigma n} = \sqrt{I_{cn}^2 + I_{bn}^2 + I_{cn}^2} \tag{2.88}$$

Czarnecki's three-phase power resolution has the same motivation with his single-phase power resolution. Thus, similar to his single-phase power resolution, he employed a current-based decomposition in which current is divided into five orthogonal components, that is active, reactive, scattered, generated harmonic and unbalanced currents. These current components are as follows:

active current,

$$\vec{i}_{ac}(t) = \begin{bmatrix} \sum_{n \in l} \sqrt{2} G_e V_n \sin(\omega_n t + \theta_{an}) \\ \sum_{n \in l} \sqrt{2} G_e V_n \sin(\omega_n t + \theta_{bn}) \\ \sum_{n \in l} \sqrt{2} G_e V_n \sin(\omega_n t + \theta_{cn}) \end{bmatrix} \tag{2.89}$$

reactive current,

$$\vec{i}_{cr}(t) = \begin{bmatrix} \sum_{n \in l} \sqrt{2} B_{en} V_n \sin\left(\omega_n t + \theta_{an} - \frac{\pi}{2}\right) \\ \sum_{n \in l} \sqrt{2} B_{en} V_n \sin\left(\omega_n t + \theta_{bn} - \frac{\pi}{2}\right) \\ \sum_{n \in l} \sqrt{2} B_{en} V_n \sin\left(\omega_n t + \theta_{cn} - \frac{\pi}{2}\right) \end{bmatrix} \tag{2.90}$$

scattered current,

$$\vec{i}_s(t) = \begin{bmatrix} \sum_{n \in l} \sqrt{2}(G_{en} - G_e)V_n \sin(\omega_n t + \theta_{an}) \\ \sum_{n \in l} \sqrt{2}(G_{en} - G_e)V_n \sin(\omega_n t + \theta_{bn}) \\ \sum_{n \in l} \sqrt{2}(G_{en} - G_e)V_n \sin(\omega_n t + \theta_{cn}) \end{bmatrix} \tag{2.91}$$

generated harmonic current

$$\vec{i}_g(t) = \begin{bmatrix} \sum_{n \in k} \sqrt{2}I_{an} \sin(\omega_n t + \delta_{an}) \\ \sum_{n \in k} \sqrt{2}I_{bn} \sin(\omega_n t + \delta_{bn}) \\ \sum_{n \in k} \sqrt{2}I_{cn} \sin(\omega_n t + \delta_{cn}) \end{bmatrix} \tag{2.92}$$

and unbalanced current,

$$\vec{i}_u(t) = \vec{i}(t) - \vec{i}_{ac}(t) - \vec{i}_{cr}(t) - \vec{i}_s(t) - \vec{i}_g(t) \tag{2.93}$$

where the conductance and susceptance parameters of the load are as follows:
equivalent conductance,

$$G_e = \frac{\sum_{m=a,b,c} P_m}{V_\Sigma^2} \tag{2.94}$$

nth harmonic equivalent conductance

$$G_{en} = \frac{\sum_{m=a,b,c} P_{mn}}{V_{\Sigma n}^2} \tag{2.95}$$

and nth harmonic equivalent susceptance,

$$B_{en} = \frac{\sum_{m=a,b,c} Q_{mn}}{V_{\Sigma n}^2} \tag{2.96}$$

The collective rms values of the current components defined by Czarnecki can be written in harmonic domain as
active current,

$$I_{\Sigma ac} = G_e V_\Sigma \tag{2.97}$$

reactive current,

$$I_{\Sigma cr} = \sqrt{\sum_{n \in l} B_{en}^2 V_{\Sigma n}^2} \tag{2.98}$$

scattered current,

$$I_{\Sigma s} = \sqrt{\sum_{n \in l} (G_{en} - G_e)^2 V_{\Sigma n}^2} \tag{2.99}$$

generated harmonic current

$$I_{\Sigma g} = \sqrt{\sum_{n \in k} I_{\Sigma n}^2} \tag{2.100}$$

and unbalanced current,

$$I_{\Sigma u} = \sqrt{I_{\Sigma}^2 - I_{\Sigma ac}^2 - I_{\Sigma cr}^2 - I_{\Sigma s}^2 - I_{\Sigma g}^2} \tag{2.101}$$

Consequently, he suggested that power resolution related with his current components as follows:

$$S_{\Sigma} = \sqrt{P^2 + Q_{cr}^2 + D_s^2 + D_g^2 + D_u^2} \tag{2.102}$$

where P denotes active power and other four powers are as follows:
reactive power,

$$Q_{cr} = V_{\Sigma} I_{\Sigma cr} \tag{2.103}$$

scattered power,

$$D_s = V_{\Sigma} I_{\Sigma s} \tag{2.104}$$

generated harmonic power

$$D_g = V_{\Sigma} I_{\Sigma g} \tag{2.105}$$

and unbalanced power,

$$D_u = V_{\Sigma} I_{\Sigma u} \tag{2.106}$$

2.3.3.3 Resolution proposed by Balci and Hocaoglu

Balci and Hocaoglu proposed a resolution of Bucholłz's apparent power for three-phase and three-line systems with non-sinusoidal and balanced voltages and non-sinusoidal and unbalanced current conditions [16]. It has the same motivation with their single-phase power resolution [15,16]. Accordingly, similar to their single-phase resolution, they separated three-phase currents into six components as active, reactive, scattered conductance, scattered susceptance, unbalanced in-phase and unbalanced quadrature currents, which are all expressed in terms of the load-impedance parameters. The load-impedance parameters, which are used to establish the current resolution, are:

*n*th harmonic balanced conductance, which draws *n*th harmonic balanced active power (P_{Bn}) from each phase,

$$G_{Bn} = \frac{P_{Bn}}{V_n^2} = \frac{\sum_{m=a,b,c} P_{mn}}{3V_n^2} \tag{2.107}$$

equivalent conductance, which draws balanced active power from each phase,

$$G_e = \frac{\sum_n P_{Bn}}{\sum_n V_n^2} = \frac{P}{3\sum_n V_n^2} \tag{2.108}$$

*n*th harmonic-balanced susceptance, which draws *n*th harmonic-balanced reactive power (Q_{Bn}) from each phase,

$$B_{Bn} = \frac{Q_{Bn}}{V_n^2} = \frac{\sum_{m=a,b,c} Q_{mn}}{3V_n^2} \tag{2.109}$$

*n*th harmonic-equivalent susceptance,

$$B_{en} = nB_{e1} = \frac{n\sum_n nQ_{Bn}}{\sum_n n^2 V_n^2} \tag{2.110}$$

*n*th harmonic conductance, which draws *n*th harmonic in-phase power from $m = a$, b, c phases,

$$G_{mn} = \frac{P_{mn}}{V_n^2} \tag{2.111}$$

*n*th harmonic susceptance, which draws *n*th harmonic quadrature power from $m = a$, b, c phases,

$$B_{mn} = \frac{Q_{mn}}{V_n^2} \tag{2.112}$$

*n*th harmonic unbalanced conductance, which draws *n*th harmonic unbalanced in-phase power from $m = a$, b, c phases,

$$G_{mn}^u = G_{mn} - G_{Bn} \tag{2.113}$$

nth harmonic unbalanced susceptance, which draws nth harmonic unbalanced quadrature power from $m = a, b, c$ phases,

$$B^u_{mn} = B_{mn} - B_{Bn} \tag{2.114}$$

Accordingly, they defined six orthogonal current components:
active current,

$$\vec{i}_{ac}(t) = G_e \vec{V}(t) \tag{2.115}$$

reactive current,

$$\vec{i}_r(t) = \begin{bmatrix} \sum_n \sqrt{2} B_{en} V_n \sin\left(\omega_n t + \theta_{an} - \frac{\pi}{2}\right) \\ \sum_n \sqrt{2} B_{en} V_n \sin\left(\omega_n t + \theta_{bn} - \frac{\pi}{2}\right) \\ \sum_n \sqrt{2} B_{en} V_n \sin\left(\omega_n t + \theta_{cn} - \frac{\pi}{2}\right) \end{bmatrix} \tag{2.116}$$

scattered conductance current,

$$\vec{i}_{sc}(t) = \begin{bmatrix} \sum_n \sqrt{2}(G_{Bn} - G_e) V_n \sin(\omega_n t + \theta_{an}) \\ \sum_n \sqrt{2}(G_{Bn} - G_e) V_n \sin(\omega_n t + \theta_{bn}) \\ \sum_n \sqrt{2}(G_{Bn} - G_e) V_n \sin(\omega_n t + \theta_{cn}) \end{bmatrix} \tag{2.117}$$

scattered susceptance current,

$$\vec{i}_{ss}(t) = \begin{bmatrix} \sum_n \sqrt{2}(B_{Bn} - B_{en}) V_n \sin\left(\omega_n t + \theta_{an} - \frac{\pi}{2}\right) \\ \sum_n \sqrt{2}(B_{Bn} - B_{en}) V_n \sin\left(\omega_n t + \theta_{bn} - \frac{\pi}{2}\right) \\ \sum_n \sqrt{2}(B_{Bn} - B_{en}) V_n \sin\left(\omega_n t + \theta_{cn} - \frac{\pi}{2}\right) \end{bmatrix} \tag{2.118}$$

unbalanced in-phase current

$$\vec{i}_{up}(t) = \begin{bmatrix} \sum_n \sqrt{2} G^u_{an} V_n \sin(\omega_n t + \theta_{an}) \\ \sum_n \sqrt{2} G^u_{bn} V_n \sin(\omega_n t + \theta_{bn}) \\ \sum_n \sqrt{2} G^u_{cn} V_n \sin(\omega_n t + \theta_{cn}) \end{bmatrix} \tag{2.119}$$

and unbalanced quadrature current,

$$\vec{i}_{uq}(t) = \begin{bmatrix} \sum_n \sqrt{2}B^u_{an}V_n \sin\left(\omega_n t + \theta_{an} - \frac{\pi}{2}\right) \\ \sum_n \sqrt{2}B^u_{bn}V_n \sin\left(\omega_n t + \theta_{bn} - \frac{\pi}{2}\right) \\ \sum_n \sqrt{2}B^u_{cn}V_n \sin\left(\omega_n t + \theta_{cn} - \frac{\pi}{2}\right) \end{bmatrix} \tag{2.120}$$

giving the current resolution as follows:

$$\vec{i}(t) = \vec{i}_{ac}(t) + \vec{i}_r(t) + \vec{i}_{sc}(t) + \vec{i}_{ss}(t) + \vec{i}_{up}(t) + \vec{i}_{uq}(t) \tag{2.121}$$

By considering the collective rms current expression of the Buchollz's apparent power definition, the rms values of the currents placed in (2.121) can be calculated as follows:

active current's rms value,

$$I_{\Sigma ac} = \sqrt{\frac{1}{T}\int_0^T \left[\vec{i}_{ac}(t)\right]^{Tr} \cdot \vec{i}_{ac}(t)dt} = \sqrt{3G_e^2\sum_n V_n^2} \tag{2.122}$$

reactive current's rms value,

$$I_{\Sigma r} = \sqrt{\frac{1}{T}\int_0^T \left[\vec{i}_r(t)\right]^{Tr} \cdot \vec{i}_r(t)dt} = \sqrt{3\sum_n B_{en}^2 V_n^2} \tag{2.123}$$

scattered conductance current's rms value,

$$I_{\Sigma sc} = \sqrt{\frac{1}{T}\int_0^T \left[\vec{i}_{sc}(t)\right]^{Tr} \cdot \vec{i}_{sc}(t)dt} = \sqrt{3\sum_n (G_{Bn} - G_e)^2 V_n^2} \tag{2.124}$$

scattered susceptance current's rms value,

$$I_{\Sigma ss} = \sqrt{\frac{1}{T}\int_0^T \left[\vec{i}_{ss}(t)\right]^{Tr} \cdot \vec{i}_{ss}(t)dt} = \sqrt{3\sum_n (B_{Bn} - B_{en})^2 V_n^2} \tag{2.125}$$

unbalanced in-phase current's rms value

$$I_{\Sigma up} = \sqrt{\frac{1}{T}\int_0^T \left[\vec{i}_{up}(t)\right]^{Tr} \cdot \vec{i}_{up}(t)dt} = \sqrt{\sum_n \left[\left(G^u_{an}\right)^2 + \left(G^u_{bn}\right)^2 + \left(G^u_{cn}\right)^2\right] V_n^2} \tag{2.126}$$

and unbalanced quadrature current's rms value,

$$I_{\Sigma uq} = \sqrt{\frac{1}{T}\int_0^T \left[\vec{i}_{uq}(t)\right]^{Tr} \cdot \vec{i}_{uq}(t)dt} = \sqrt{\sum_n \left[\left(B^u_{an}\right)^2 + \left(B^u_{bn}\right)^2 + \left(B^u_{cn}\right)^2\right] V_n^2} \tag{2.127}$$

In (2.122)–(2.127), the superscript '*Tr*' denotes transpose of the respected three-phase current vector. Thus, the collective rms value of the total current can be expressed as the vector sum of the collective rms values of the current components:

$$I_\Sigma = \sqrt{I_{\Sigma ac}^2 + I_{\Sigma r}^2 + I_{\Sigma sc}^2 + I_{\Sigma ss}^2 + I_{\Sigma up}^2 + I_{\Sigma uq}^2} \tag{2.128}$$

Consequently, they obtained their apparent power resolution by multiplying both sides of (2.128) with the square of collective rms voltage value:

$$\begin{aligned} S_\Sigma &= \sqrt{(V_\Sigma I_{\Sigma a})^2 + (V_\Sigma I_{\Sigma r})^2 + (V_\Sigma I_{\Sigma sc})^2 + (V_\Sigma I_{\Sigma ss})^2 + \left(V_\Sigma I_{\Sigma up}\right)^2 + \left(V_\Sigma I_{\Sigma uq}\right)^2} \\ &= \sqrt{P^2 + Q_r^2 + D_{sc}^2 + D_{ss}^2 + D_{up}^2 + D_{uq}^2} \end{aligned} \tag{2.129}$$

In (2.129), the power components are named as active (P), reactive (Q_r), scattered conductance (D_{sc}), scattered susceptance (D_{ss}), unbalanced in-phase (D_{up}) and unbalanced quadrature (D_{uq}) powers. Q_r gives three-phase power of the star connected three identical capacitors, which maximizes the power factor under non-sinusoidal and unbalanced conditions. The nth harmonic reactance value of one capacitor placed in this three-phase capacitors bank can be written as follows:

$$X_{Cn} = \frac{1}{B_{en}} = \frac{\sum_n n^2 V_n^2}{n \sum_n n Q_{Bn}} \tag{2.130}$$

When the resolution of Balci and Hocaoglu is compared with Czarnecki's resolution, some features should be mentioned as follows:

- The reactive power component of Czarnecki's resolution is divided into two power components, namely reactive power and scattered susceptance power in the resolution proposed by Balci and Hocaoglu. The reactive power component of Balci and Hocaoglu is completely compensated when the power factor is maximized by the balanced three-phase capacitors bank. However, this is not the case for the reactive power component of Czarnecki's resolution.
- And also, the unbalanced power component of Czarnecki's power resolution is decomposed into unbalanced in-phase and unbalanced quadrature powers in the power resolution of Balci and Hocaoglu. As a result, in the proposed resolution, all power components are defined in terms of the conductance and susceptance parameters of the load. This should be useful for cooperative design and control of different types of harmonic and reactive power compensators.

In [20], they show that their resolution can be employed to design cost-effective unity power factor compensator including balanced capacitors bank and an active filter.

2.3.4 IEEE standard apparent power and its resolution

By IEEE-power-definitions working group, for three-phase systems with neutral line, apparent power [11,22] is proposed as follows:

$$S_e = 3V_eI_e \tag{2.131}$$

where the equivalent rms voltage is as follows:

$$V_e = \sqrt{\frac{1}{18}\left[3\left(V_a^2 + V_b^2 + V_c^2\right) + V_{ab}^2 + V_{bc}^2 + V_{ca}^2\right]} \tag{2.132}$$

and the equivalent rms current is as follows:

$$I_e = \sqrt{\frac{I_a^2 + I_b^2 + I_c^2 + I_N^2}{3}} \tag{2.133}$$

It has two main power components: one of them is fundamental effective apparent power

$$S_{e1} = 3\,V_{e1}I_{e1} \tag{2.134}$$

and other is non-fundamental apparent power

$$S_{eNF}^2 = S_e^2 - S_{e1}^2 \tag{2.135}$$

In (2.134), V_{e1} and I_{e1} denote the fundamental equivalent rms voltage and current, which are calculated by substituting only fundamental harmonic voltages and currents in (2.132) and (2.133).

S_{e1} is separated into fundamental harmonic positive-sequence apparent power (S_1^+) and fundamental harmonic unbalanced apparent power (S_{U1}):

$$S_{e1} = \sqrt{\left(S_1^+\right)^2 + S_{U1}^2} \tag{2.136}$$

By means of fundamental harmonic-positive sequence active power ($P_1{}^+ = 3V_1{}^+I_1{}^+\cos\varphi_1{}^+$) and fundamental harmonic-positive sequence reactive power ($Q_1{}^+ = 3V_1{}^+I_1{}^+\sin\varphi_1{}^+$), fundamental harmonic-positive sequence apparent power is expressed as follows:

$$S_1^+ = \sqrt{\left(P_1^+\right)^2 + \left(Q_1^+\right)^2} \tag{2.137}$$

Non-fundamental apparent power (S_{eNF}) has three power components:

current distortion power,

$$D_{eI} = 3V_{e1}I_{eH} \tag{2.138}$$

voltage distortion power

$$D_{eV} = 3V_{eH}I_{e1} \tag{2.139}$$

and harmonic apparent power,

$$S_{eH} = 3V_{eH}I_{eH} \tag{2.140}$$

Thus, it is decomposed as follows:

$$S_{eNF}^2 = D_{eI}^2 + D_{eV}^2 + S_{eH}^2 \tag{2.141}$$

V_{eH} and I_{eH} denote the non-fundamental equivalent rms voltage and current, which are calculated as follows:

$$V_{eH} = \sqrt{V_e^2 - V_{e1}^2} \tag{2.142}$$

$$I_{eH} = \sqrt{I_e^2 - I_{e1}^2} \tag{2.143}$$

Consequently, the expression of IEEE apparent power resolution can be written as follows:

$$S_e = \sqrt{\left(P_1^+\right)^2 + \left(Q_1^+\right)^2 + S_{U1}^2 + D_{eI}^2 + D_{eV}^2 + S_{eH}^2} \tag{2.144}$$

It is seen from (2.144) that IEEE apparent power considers the powers other than fundamental harmonic positive-sequence active power as non-active powers. According to this consideration, voltage and current must have in phase sinusoidal and balanced wave shapes for the unity power factor case of the IEEE definition.

2.4 Practical implementation of apparent powers and their power resolutions included in IEEE standard 1459 and DIN standard 40110

In this section, for the practical implementation of arithmetic, vector and effective apparent powers included in IEEE standard 1459 and Buchollz's apparent power encouraged by DIN standard 40110, a power meter is developed via LabView software [50]. In addition to that, for some test cases, the measurement results will be presented to interpret main properties of the apparent powers and their resolutions included in both standards.

2.4.1 LabView blocks of developed power meter

The LabView blocks, which is developed for the practical implementation of the powers, are presented in Figures 2.1–2.4.

Figure 2.1(a) shows the measurement blocks of active, Budeanu's reactive and Budeanu's distortion powers for one phase. The output of the measurement block given in Figure 2.1(b) gives the vector apparent power, which is the vector sum of these powers calculated for three phases.

The block, given in Figure 2.2, measures the apparent powers of each phase and then sums them to find the arithmetic apparent power.

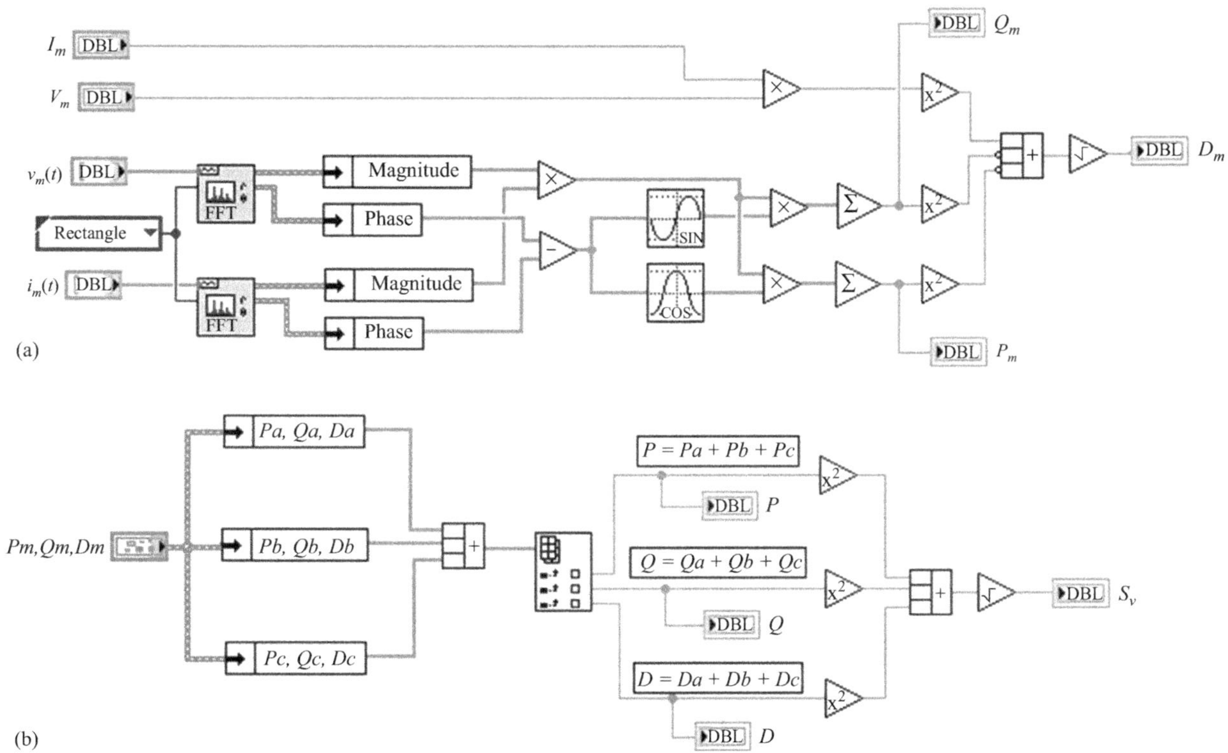

Figure 2.1 LabView blocks for the measurement of vector apparent power and its power components

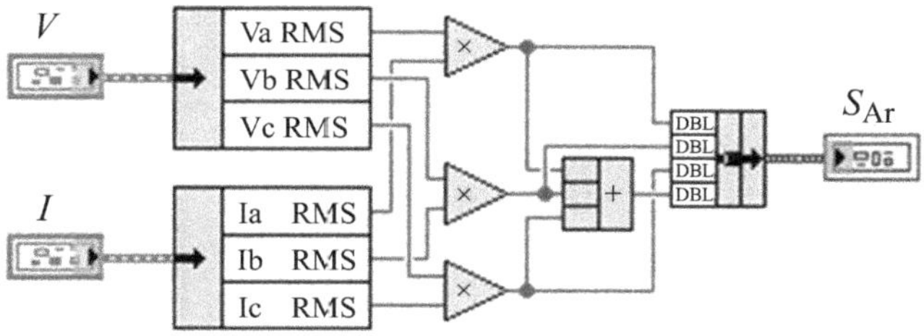

Figure 2.2 LabView blocks for the measurement of arithmetic apparent power

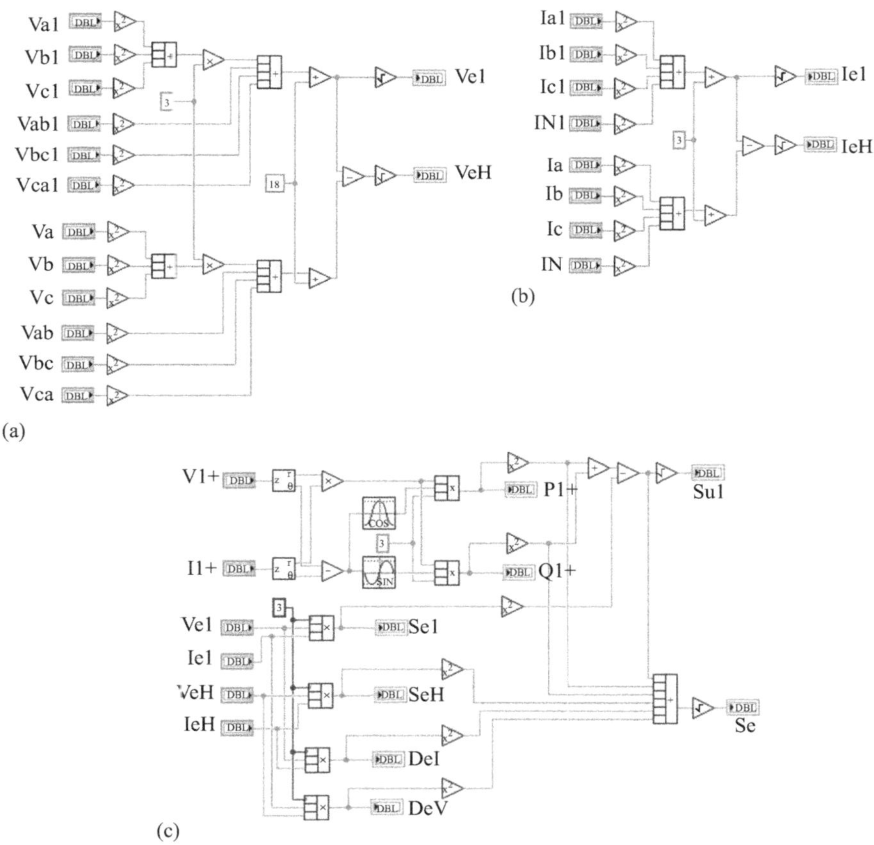

Figure 2.3 LabView blocks for the measurement of effective apparent power and its power components

The block, given in Figure 2.3(a), measures fundamental harmonic and non-fundamental harmonic equivalent rms voltages (V_{e1}, V_{eH}). Figure 2.3(b) presents that the block measures fundamental harmonic and non-fundamental harmonic equivalent rms currents (I_{e1}, I_{eH}). The blocks designed for the measurement of

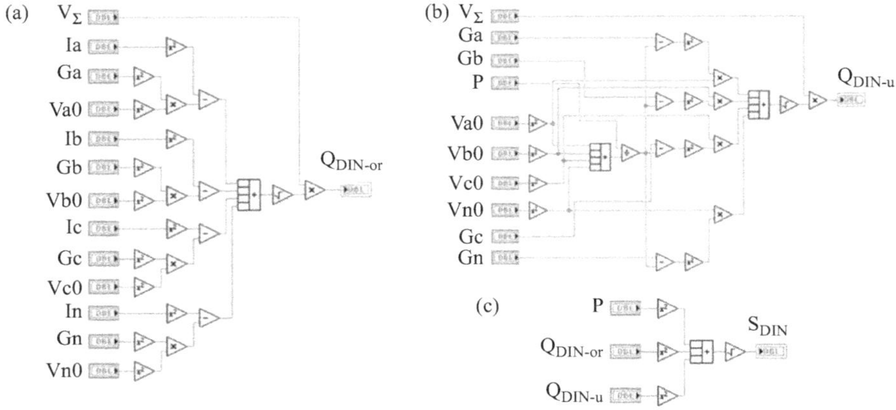

Figure 2.4 LabView blocks for the measurement of Buchollz's apparent power and its DIN standard power components

fundamental harmonic positive-sequence active ($P_1{}^+$), fundamental harmonic positive-sequence reactive ($Q_1{}^+$), fundamental harmonic unbalanced apparent (S_{U1}), fundamental effective apparent (S_{e1}), current distortion (D_{eI}), voltage distortion (D_{eV}) and harmonic apparent (S_{eH}) powers are given in Figure 2.3(c). In addition, the vector sum of the power components as $P_1{}^+$, $Q_1{}^+$, S_{U1}, D_{eI}, D_{eV} and S_{eH} which gives IEEE Std. apparent power (S_e) is presented in the same figure. Note that for simple illustration, the measurement of fundamental harmonic positive sequence voltage and current ($\overline{V}_1^+, \overline{I}_1^+$) are not shown in Figure 2.3.

The blocks, given in Figure 2.4(a) and (b), are used to measure the non-active power caused by the current components orthogonal with the voltages ($Q_{tot\Sigma\perp} = Q_{DIN-or}$), and the non-active power caused by the load's unbalanced conductance ($Q_{tot\Sigma||u} = Q_{DIN-u}$), respectively. As a result, in Figure 2.4(c), Buchollz's apparent power ($S_\Sigma = S_{DIN}$) is calculated as the vector sum of active (P) and two non-active powers.

2.4.2 Measurement results

To illustrate main properties of the apparent powers, they are measured in some cases of a test system, of which schematic given in Figure 2.5, by means of the developed power meter.

Figure 2.5 shows that the system consists of a three-phase programmable power supply, which can generate desired voltage wave shape, a data acquisition card used for sampling voltage and currents, a PC used for running LabView and load. The data acquisition card can sample voltage and current with 120 kHz frequency and 12 bit resolution from eight analogue channels simultaneously.

Accordingly, by using the developed power meter, the powers are measured for four voltages and load cases in the test system. The results are given in following sections.

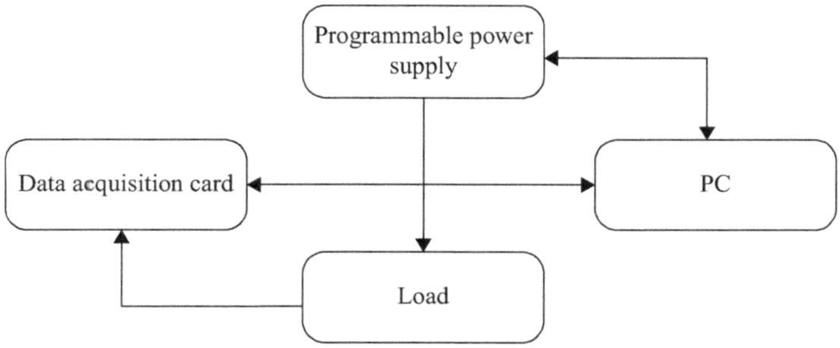

Figure 2.5 The schematic of the test system

2.4.2.1 Case I: distorted-unbalanced voltage and single-phase resistive load

In this case, the voltages, which are given in Figure 2.6, have *THDV* values measured for each phase, V_1^-/V_1^+ and V_1^0/V_1^+ as 8%, 2.04% and 10.35%, respectively. The load side consists of a resistive load connected between a phase and neutral line. For this case, the screen of designed power meter is given in Figure 2.6.

It is seen from Figure 2.6 that P, S_V and S_{Ar} are measured as 1,160 W/VA. This case means that the power factors with respect to both apparent powers are unity. However, S_e is measured as 2,555 VA. The components of S_e are that P_{1+}, S_{u1}, D_{eV} and D_{eI} have the values measured as 1,035 W, 2,322 VA, 163 VA and 192 VA, respectively. In addition, values of S_{eH} and Q_{1+} are negligible for the case. Therefore, the power factor calculated with respect to S_e is 0.454 for this case. S_{DIN}, $Q_{\text{DIN-u}}$ and $Q_{\text{DIN-or}}$ are measured as 2,550 VA, 2,065 VA and 942 VAr, respectively. Accordingly, the power factor calculated regarding the DIN approach is 0.455.

2.4.2.2 Case II: distorted-unbalanced voltage and three identical resistive loads in star connection

The voltage waveforms of Case I are kept in Case II. In addition, the load side consists of three identical resistive loads in star connection. For the case, the screen of power meter is given in Figure 2.7.

Figure 2.7 shows that P, S_V and S_{Ar} have the same value measured as 2,837 W/VA. Thus, the power factors calculated for S_V and S_{Ar} are unity. On the other hand, S_e, P_{1+}, S_{u1}, D_{eV} and D_{eI} are measured as 2,893 VA, 2,785 W, 636 VA, 183 VA and 413 VA, respectively. It is also seen that S_{eH} and Q_{1+} are negligible. Power factor calculated according to S_e is 0.980. In addition to these results, S_{DIN}, $Q_{\text{DIN-u}}$ and $Q_{\text{DIN-or}}$ are measured as 2,887 VA, 489 VA and 218 VAr, respectively, and power factor of S_{DIN} is calculated as 0.982.

2.4.2.3 Case III: sinusoidal-balanced voltage and three identical resistive loads in star connection

For Case 3, the test voltages are sinusoidal and balanced, and the load side consists of three identical resistive loads in star connection. For the same case, the screen of power meter is given in Figure 2.8.

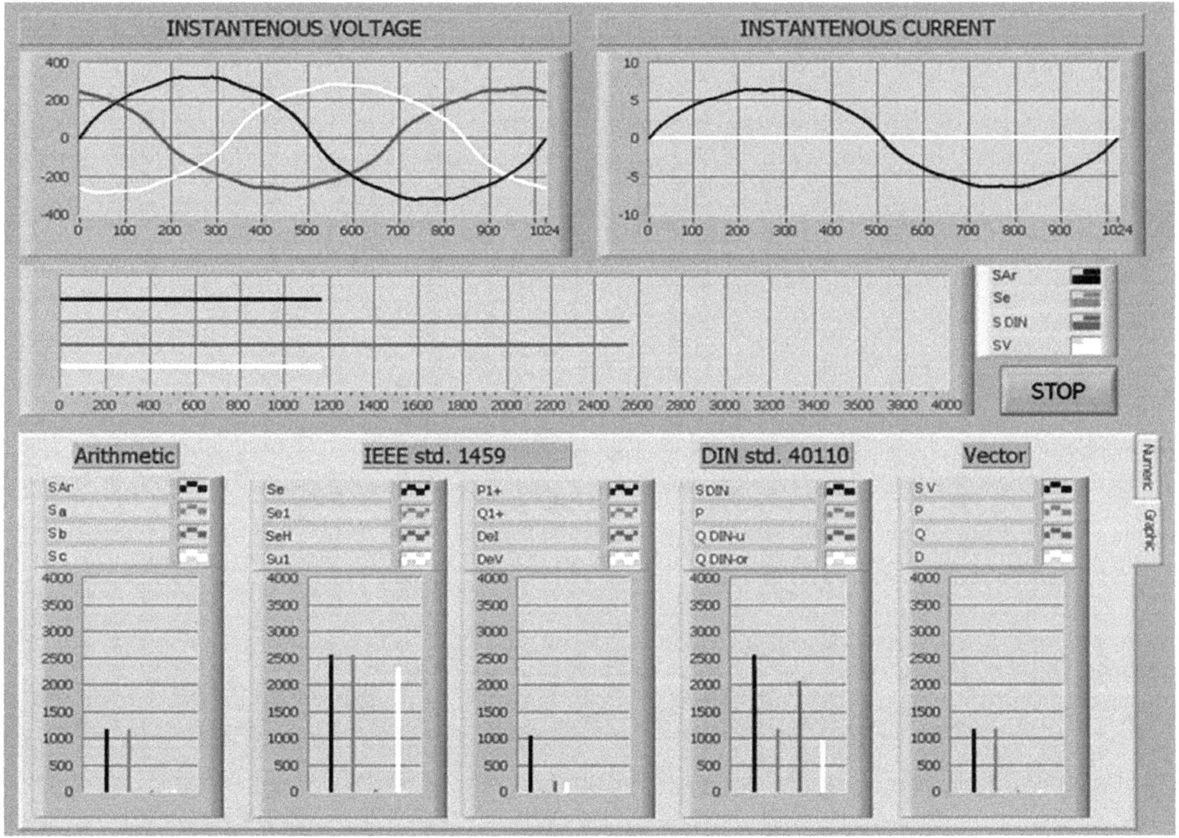

Figure 2.6 The screen of the developed power meter for Case I

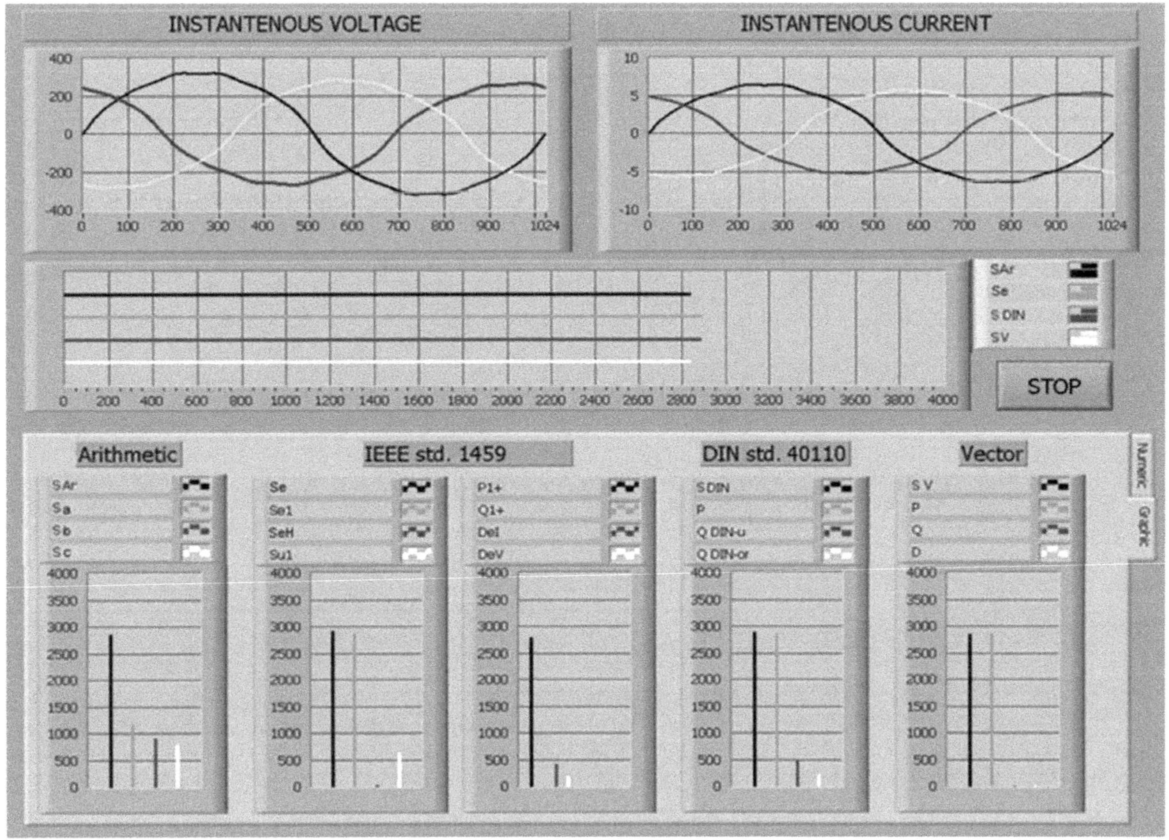

Figure 2.7 The screen of the developed power meter for Case II

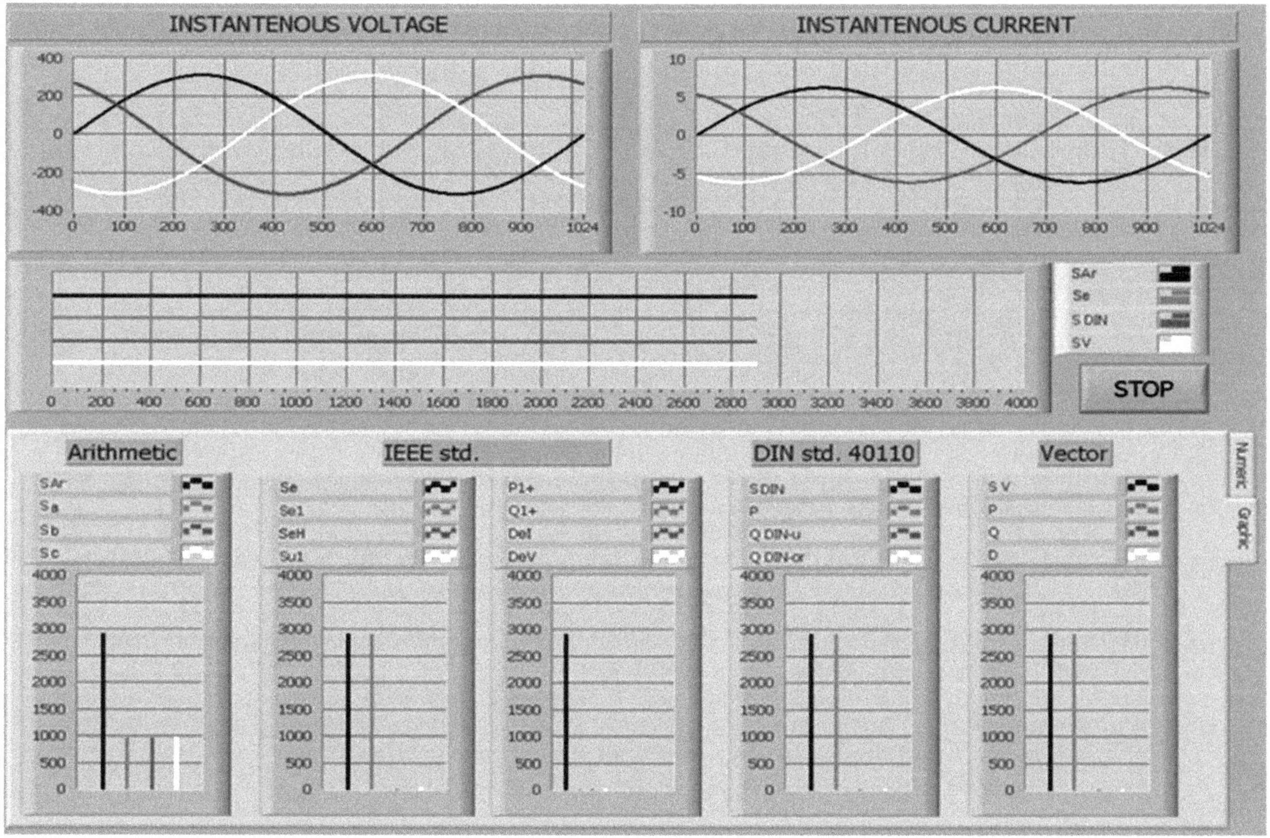

Figure 2.8 The screen of the developed power meter for Case III

The screen shows that four apparent powers as S_{Ar}, S_V, S_e and S_{DIN} are equal to P measured as 2,900 W, and power factors according to these four apparent powers are unity.

2.4.2.4 Case IV: distorted-unbalanced voltage and four identical resistive loads in star connection

For Case IV, the test voltage waveforms of Case I are kept, and the load side consists of four identical resistive loads in star connection. For Case IV, the screen of power meter is given in Figure 2.9.

It can be mentioned from Figure 2.9 that P and S_{DIN} have the same value as 3,012 W/VA. Accordingly, power factor calculated with respect to S_{DIN} is unity for the same case. However, S_e, P_{1+}, S_{u1}, D_{eV} and D_{eI} are measured as 3,018 VA, 2,994 W, 277 VA, 192 VA and 159 VA, respectively. It is also seen that S_{eH} and Q_{1+} are negligible. As a result, power factor of the IEEE definition is calculated as 0.998. Due to the distortion power, D, measured as 161 VA, S_V is larger than P and has the value measured as 3,016 VA. Thus, power factor calculated regarding S_V is calculated as 0.998. In addition to these results, S_{Ar} is measured as 3,022 VA, and the power factor of S_{Ar} is found as 0.996.

2.5 Metering of harmonic-pollution responsibility

The methods and indices proposed for the harmonic-pollution responsibility can be divided into two subgroups according to their measurement strategies: single-point [32] and multi-point (or distributed) [33] ones. Since the multi-point measurements based methods and indices can be implemented by using synchronous measurement devices placed in different sections of the system, they are more reliable than the methods and indices based on single-point measurements strategy [34]. But, they have higher costs and also are hindered by their requirements for the precise time synchronization. Therefore, for the harmonic source detection and sharing harmonic responsibility, single-point measurement-based methods and indices are widely preferred in the literature. Accordingly, in this section, they are taken under consideration.

2.5.1 The indices based on active power direction method

Active power direction (APD) method is the oldest harmonic-source detection method, and it considers the sign of the harmonic active power to determine the source of the respective harmonic order [34–36]. It determines source side as the dominant harmonic source if the respective harmonic active power has a positive sign; otherwise, consumer side is the dominant for the considered harmonic orders.

According to this approach, two indices were suggested according to the same rule: one is that supply load quality index (*SLQI*) [34,35] measures the harmonic generation quantity of the consumer or load side as

$$SLQI = \frac{P}{P_1} = \frac{P_1 + P_H}{P_1} \tag{2.145}$$

where P, P_1 and P_H are the active power, fundamental harmonic active power and sum of the harmonic active powers, respectively. Thus, the load is detected as the

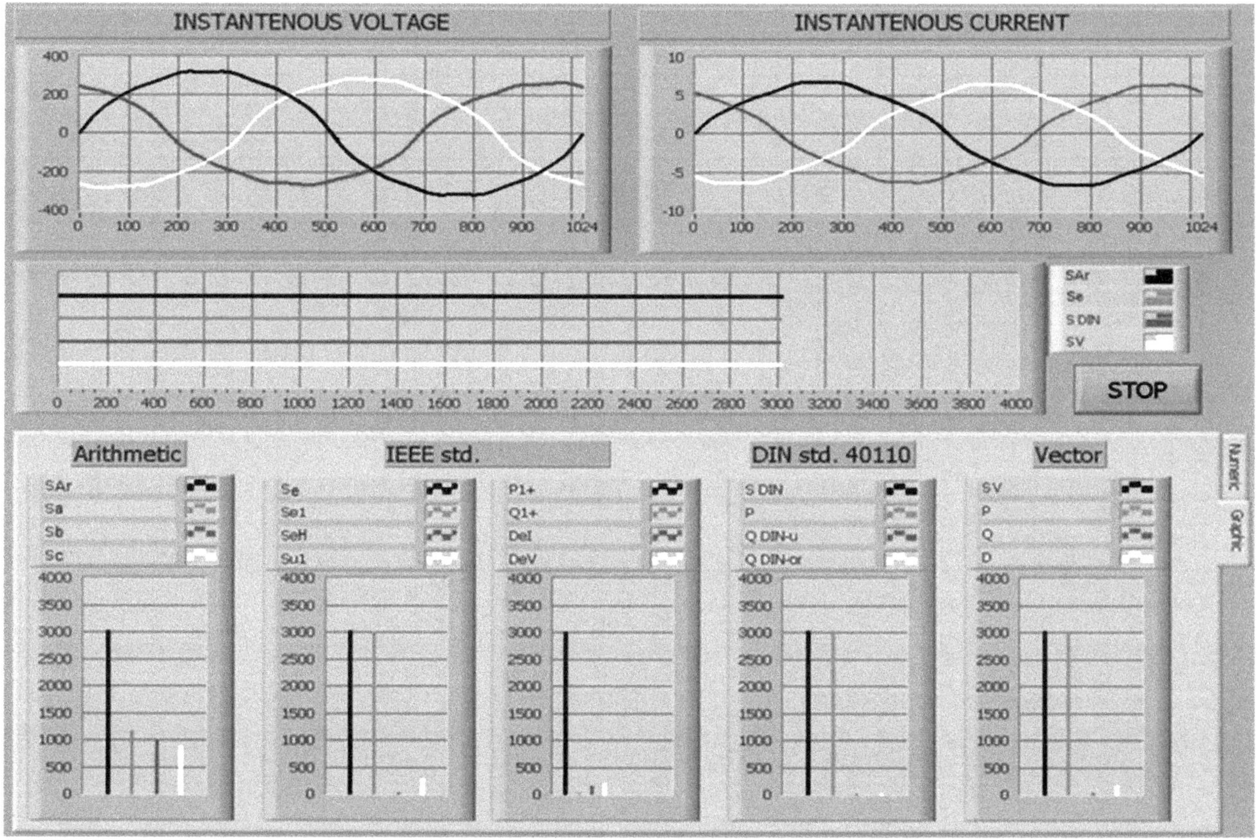

Figure 2.9 The screen of the developed power meter for Case IV

dominant source for the harmonic distortion when the value for *SLQI* is smaller than unity; otherwise, supply side is the dominant source for the harmonic distortion. However, for the distorted utility voltages without exceeding the distortion limits presented in the standards, P_H has very small value when compared to P_1. This means that *SLQI* may not give sensitive information on the harmonic responsibility in the practical applications with the systematic measurement errors.

Second index, harmonic global index [34,36], is defined as

$$HGI = \frac{\sqrt{\sum_{n\in\partial} I_n^2}}{\sqrt{\sum_{n\in\rho} I_n^2}} \tag{2.146}$$

where ∂ is the harmonic orders related to the harmonic active powers, which have the negative sign, and ρ is vice versa. A non-zero value of *HGI* shows that the load causes harmonic distortion.

2.5.2 The methods based on the harmonic analysis of the system

The widely known methods based on the harmonic modelling and analysis of the system are Superposition and projection (SP) [34,37] and critical impedance (CI) [34,38]. Both methods are summarized in the following sections.

2.5.2.1 Superposition and projection method

In SP method, for nth harmonic, utility and consumer sides are modelled as Norton equivalent. Accordingly, for nth harmonic, single-phase equivalent circuit of a typical distribution system is represented as in Figure 2.10.

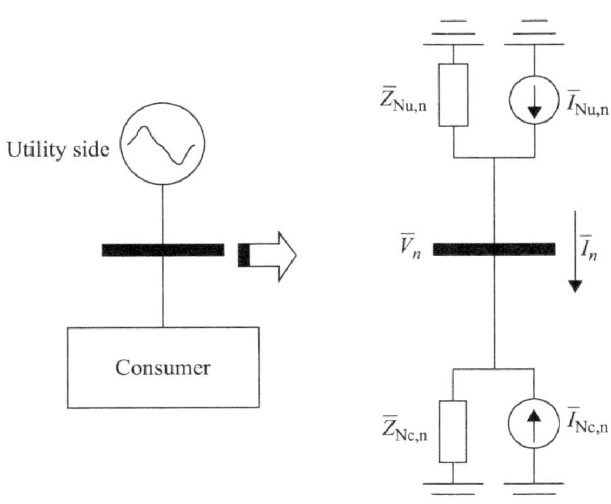

Figure 2.10 The modelling of a typical distribution system for the SP method

In the equivalent circuit, by regarding Superposition theorem, the contributions of utility and consumer sides on the *n*th harmonic line current can be expressed as

the *n*th harmonic line current part contributed by utility side,

$$\overline{I}_{u,n} = \frac{\overline{Z}_{Nu,n}\overline{I}_{Nu,n}}{\overline{Z}_{Nc,n} + \overline{Z}_{Nu,n}} \tag{2.147}$$

the *n*th harmonic line current part contributed by consumer side,

$$\overline{I}_{c,n} = \frac{\overline{Z}_{Nc,n}\overline{I}_{Nc,n}}{\overline{Z}_{Nc,n} + \overline{Z}_{Nu,n}} \tag{2.148}$$

where sum of both current part gives the *n*th harmonic line current:

$$\overline{I}_n = \overline{I}_{u,n} + \left(-\overline{I}_{c,n}\right) \tag{2.149}$$

And then, their projections on the line current are provided:

for utility side's harmonic contribution,

$$I_{Pu,n} = \frac{\mathrm{Re}\left(\overline{I}_{u,n}\overline{I}_n^{\,*}\right)}{I_n} \tag{2.150}$$

for consumer's side harmonic contribution,

$$I_{Pc,n} = \frac{\mathrm{Re}\left(\overline{I}_{c,n}\overline{I}_n^{\,*}\right)}{I_n} \tag{2.151}$$

where the sum of the projections of the current parts is the magnitude of the *n*th harmonic line current:

$$I_n = I_{Pu,n} + I_{Pc,n} \tag{2.152}$$

The obtained projection values show the harmonic contribution of the utility and consumer sides. They are scalars and can have opposite signs. If they have the same signs for *n*th harmonic, both utility and consumer sides contribute the respected line current harmonic. Otherwise, the negative one compensates it.

Some data of the system must be known to implement the SP method. These data are the harmonic Norton equivalent impedances of consumer and utility sides, and they may not be available in most practical cases. The harmonic impedance of utility side is generally regarded as the short circuit impedance of the system. On the other hand, it is not easy to find the harmonic impedance for the consumer side. As a result, for proper modelling of the consumer side, two approaches were proposed in [39,40].

2.5.2.2 Critical impedance method

In CI method, for *n*th harmonic, utility and consumer sides are modelled as Thevenin equivalent.

Thus, for *n*th harmonic, single-phase equivalent circuit of a typical distribution system is represented as in Figure 2.11.

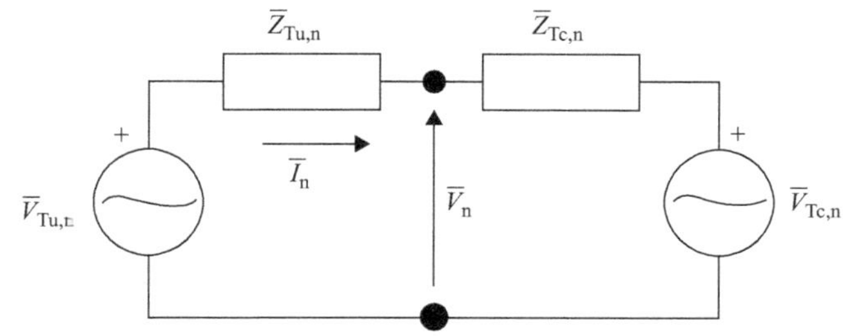

Figure 2.11 The modelling of a typical distribution system for the CI method

By regarding the single-phase equivalent circuit, for nth harmonic, the relations among utility side's Thevenin voltage, consumer side's Thevenin voltage and line current can be written as follows:

$$\overline{V}_{Tu,n} = \overline{I}_n(\overline{Z}_{Tu,n} + \overline{Z}_{Tc,n}) + \overline{V}_{Tc,n} \tag{2.153}$$

where $\overline{V}_{Tu,n}$ and $\overline{V}_{Tc,n}$ are nth harmonic phasor values of Thevenin voltages of the utility and consumer sides, and $\overline{Z}_{Tu,n}$ and $\overline{Z}_{Tc,n}$ are nth harmonic phasor values of Thevenin impedances of the utility and consumer sides.

To simply interpret the CI method, the resistance parts of the Thevenin impedances of the utility and consumer sides as $\overline{Z}_{Tu,n} \approx jX_{Tu,n}$ and $\overline{Z}_{Tc,n} \approx jX_{Tc,n}$ can be neglected, and their sum can be expressed as follows:

$$jX_n = jX_{Tu,n} + jX_{Tc,n} \tag{2.154}$$

For nth harmonic, CI value is defined as the magnitude of the impedance value where the magnitudes of Thevenin voltages of the utility and consumer sides are equal to each other ($V_{Tu,n} = V_{Tc,n}$). Thus, it can be written as follows:

$$X_n' = 2\frac{V_n}{I_n}\sin(\theta_n - \delta_n) \tag{2.155}$$

By comparing X_n and X_n', the relation between $V_{Tu,n}$ and $V_{Tc,n}$ is determined:

- $V_{Tu,n}$ is larger than $V_{Tc,n}$ for $X_n > X_n'$,
- $V_{Tc,n}$ is larger than $V_{Tu,n}$ for $X_n < X_n'$.

Therefore, the side with the larger Thevenin voltage is identified as the main harmonic source. Due to the fact that both SP and CI methods give the similar results [35], the CI method may inherit the same problem about erroneous detection of the linear loads with the compensation capacitors as harmonic source under distorted supply voltage conditions.

2.5.3 The current decomposition based indices

The widely known indices based on the current decomposition are non-conforming current [34,41] and non-linear current indices [34,42]. They are summarized in the following sections.

2.5.3.1 Non-conforming current index

The non-conforming current index identifies the harmonic contributions of utility and consumer sides by dividing the load current into two components, namely conforming current, which has the same replica or same *THD* with the voltage wave shape, the remaining (or non-conforming) current is determined when the conforming current is subtracted from total current in time domain. Both current parts can be written as follows:

conforming current

$$i_{co}(t) = \sum_{n} Y_{L1} V_n \sin[\omega_n t + \theta_n + n(\delta_1 - \theta_1)] \tag{2.156}$$

and non-conforming current,

$$i_{nco}(t) = i(t) - i_{co}(t) \tag{2.157}$$

where Y_{L1} and δ_1 denote the magnitude of the fundamental harmonic-load admittance and the angle of the fundamental harmonic-load current, respectively, and the supply voltage is expressed as follows:

$$v(t) = \sum_{n} \sqrt{2} V_n \sin(\omega_n t + \theta_n) \tag{2.158}$$

Accordingly, in terms of the rms values of the total load current and the non-conforming current part (I and I_{nco}), the index, which quantifies the harmonic responsibility of the load side, is written as

$$\mathrm{NCI} = \frac{I_{nco}}{I} \tag{2.159}$$

The THD levels of the voltage and current do not have the same values for the linear R–L impedances. Thus, this method errenously quantifies some harmonic responsibility for the linear loads under distorted voltage conditions.

2.5.3.2 Non-linear current index

Non-linear current index defines the amount of the harmonic responsibilities for utility and consumer sides by dividing the load current into two parts, namely linear current, which is drawn by the R–L equivalent linear impedance of the load, and the non-linear current is obtained when the linear current is removed from total current. Both current parts can be written as follows:

linear current

$$i_{\ell}(t) = \sum_{n} \sqrt{2} \frac{\mathrm{V}_n}{Z_{Ln}} \sin(\omega_n t + \theta_n - \phi_n) \tag{2.160}$$

and non-linear current,

$$i_{n\ell}(t) = i(t) - i_{\ell}(t) \tag{2.161}$$

where $\overline{Z}_{Ln} = Z_{Ln}\angle\phi_n$ is the nth harmonic linear load impedance, which is equivalent with the serial connection of the resistance, $R_L = \mathrm{Re}(V_1\angle\theta_1/I_1\angle\delta_1)$ and the nth harmonic inductive reactance, $X_{Ln} = n \cdot \mathrm{Im}(V_1\angle\theta_1/I_1\angle\delta_1)$.

Therefore, in terms of the rms values of the total load current and the non-linear current part (I and $I_{n\ell}$), the index, which quantifies the harmonic responsibility of the load side, is written as

$$NLI = \frac{I_{n\ell}}{I} \tag{2.162}$$

The non-linear current index wrongly detects the linear loads with compensation capacitor bank as harmonic-producing loads under the distorted voltage conditions [34,43]. To avoid this wrong detection, in [44], an algorithm is provided to model the linear loads with the capacitor.

2.5.4 The methods based on the evaluation of the non-active powers

It is seen from the accessible literature that there are four harmonic-source detection methods based on the evaluation of the non-active powers. They are (i) the method based on the comparative evaluation of fundamental harmonic-Sharon's–Fryze's reactive powers [45], (ii) the method based on evaluation of the scattered (or scattered conductance) power [20,46,47], (iii) the method based on evaluation Budeanu's distortion power [48] and (iv) the method based on the evaluation of IEEE standard 1459 power components [49]. Their principles are detailed in the following sections.

2.5.4.1 The method based on the comparative evaluation of fundamental harmonic-Sharon's–Fryze's reactive powers

This method compares three different non-active powers, which are fundamental harmonic reactive, Sharon's reactive and Fryze's reactive powers, for the detection of the main harmonic source at point of common coupling (PCC). With respect to this method, under non-sinusoidal supply voltage and a linear load case, Sharon's reactive power should considerably be closer to the fundamental harmonic reactive power than to Fryze's reactive power. On the contrary, under a sinusoidal supply voltage and a non-linear load case, Sharon's reactive power should considerably be closer to Fryze's reactive power than to the fundamental harmonic reactive power. Under a non-sinusoidal supply voltage and a non-linear load case, Sharon's reactive power has as an intermediate value between fundamental harmonic reactive power and Fryze's reactive power. When supply voltage is sinusoidal and load is linear, these reactive powers have the same numerical value.

However, by means of numerical examples, Balci and Hocaoglu [34] figured out that the above summarized method erroneously detect the linear loads with power-factor correction capacitors as harmonic-producing load under non-sinusoidal supply voltages. In addition, since this method is based on the measurement of three reactive powers, which are placed in different power resolutions, it should complicate the software of a power meter.

2.5.4.2 The method based on the evaluation of scattered power

Scattered power (D_s) or scattered conductance power (D_{sc}), which is contained in the power resolutions proposed by Czarnecki and Balci and Hocaoglu, occurs by the difference between *n*th harmonic conductance and equivalent conductance, and it is considered as an indicator of the non-linearity or harmonic generation degree of the load in some studies [20,46,47]. These studies underlined the following:

- **Sinusoidal Voltage (or Voltage with Negligible *THDV*):** For a linear load under negligible voltage distortion, apparent power is very close to the fundamental harmonic apparent power due to the fact that both voltage and current harmonics have negligible magnitudes. As a result, for the same load–voltage case, D_{sc} has negligible value. On the contrary, for a non-linear (or harmonic producing) load under the same voltage case, the load's balanced conductances calculated for the harmonics a part from fundamental one have the considerable values due to the fact that the load injects current harmonics, which is extremely higher than the respective voltage harmonics. Therefore, the rms value of the scattered conductance current (I_{sc}) and D_{sc} have considerably high values for the non-linear load condition.
- **Non-sinusoidal Voltage:** In IEEE standard 519 [25], the maximum permissible value of *THDV* is determined as 5% at the bus voltages lower than 69 kV, 2.5% at the bus voltages between 69 kV and 161 kV, and 1.5% at the bus voltages higher than 161 kV. Thus, according to the same standard, the maximum value of *THDV* measured at PCC can be assumed as 5%. For the maximum permissible voltage distortion level, a linear load draws small amount of harmonic currents if there is no resonance condition in the system. As a result, I_{sc} and D_{sc} of the linear load will be small. For a harmonic-producing load under the same voltage distortion level, it is feasible that its current have some harmonic components, which is extremely larger than respective harmonic components of the voltage, in other words, it behaves as considerably large conductances for the respective harmonic numbers. Consequently, I_{sc} and D_{sc} drawn by the harmonic-producing load will have considerable values under the distorted supply voltage condition.

According to the manners mentioned above, it can be concluded whether the load has a non-harmonic producing characteristic or not. Scattered power method has an important advantage for cost-effective detection of harmonic-producing loads since the energy meters can be used for their implementation.

2.5.4.3 The method based on the evaluation of Budeanu's distortion power

This method advocates that the non-linear or harmonic producing loads draw high-level Budeanu's distortion power. Thus, according to this method, if the distortion power of the load has considerable value, it is detected as the non-linear load.

However, it is seen from [18] that the distortion power of Budeanu's power resolution was significantly affected from the capacitive compensation, and it may have high values for linear loads with power-factor correction capacitors under the distorted voltage conditions. Thus, it can be mentioned that the method based on measurement of distortion power have the same problematic case, which was observed for the method based on comparison of non-active powers [34].

2.5.4.4 The method based on the evaluation of IEEE standard 1459 power components

In this method, by using the harmonic numbers in consumer and utility responsibilities determined according to APD method, IEEE Std. 1459-2010 power components are measured for three frequency intervals: the powers measured at fundamental frequency, frequencies associated with the distribution system distortion and frequencies associated with consumer distortion. Thus, several power factors were resulted from the new proposal by regarding the harmonic-pollution responsibilities of consumer and utility sides. However, it seems that too many power factors should complicate the software of a power meter.

2.6 The statistical evaluation of the *HGI, NLI* and D_s harmonic source detection approaches for different load types under several supply voltage waveforms

It is seen from the summary and qualitative analysis of the methods and indices given up to here that *HGI, NLI* and D_s approaches may practically be used for the harmonic-source detection and sharing harmonic responsibility in the practical systems. Accordingly, in this section, their statistical evaluation is presented for different load types under sinusoidal and non-sinusoidal supply voltage conditions without exceeding the highest permissible *THDV* as 5% determined in the IEEE standard 519. The schematic of the system used for the statistical evaluation is depicted in Figure 2.5. In addition to that, the loads used in the test system are R–L impedance with $X/R = 0.9$, the constant speed and constant torque (under full load) cases of an induction machine, a dimmer-controlled R–L impedance ($X/R = 0.9$ and with the triac conduction angles: 90°–270°), a number of computers and a number of compact fluorescent lamps. For sinusoidal excitation, they have voltage and current pairs plotted in Figure 2.12.

It is shown from Figure 2.12(a) that the current of an R–L impedance has a sinusoidal wave shape for sinusoidal supply voltage. In addition to that, Figure 2.12(b) and (c) shows that induction machine working with constant speed and constant torque draws the current with small amount of *THDI*, which is measured as 5%, for sinusoidal supply voltage. The currents of the dimmer-controlled R–L impedance, computers and compact fluorescent lamps are seen as highly distorted from Figure 2.12(d)–(f). They have *THDI* values that are measured as 50%, 185% and 115%, respectively.

D_s (normalized), *HGI* and *NLI* are measured for each one of these loads supplied with sinusoidal voltage and 100 different voltage wave shapes with 5% value of *THDV*. The results are plotted in Figures 2.13 and 2.14.

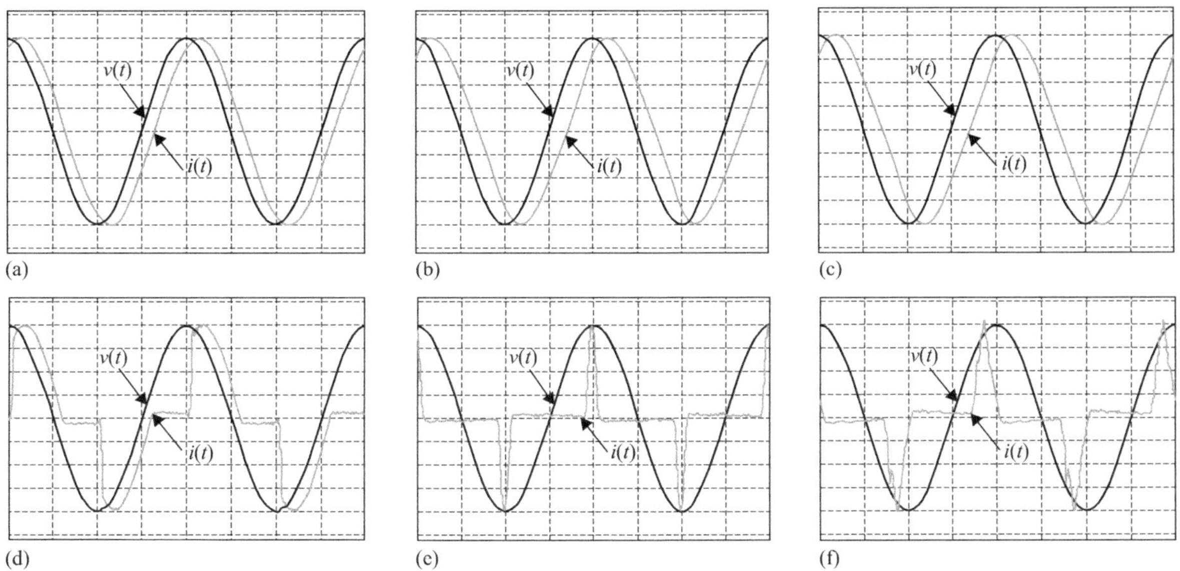

Figure 2.12 *The sinusoidal supply voltage and current pairs of (a) the R–L impedance, (b) the induction machine working with constant speed, (c) the induction machine working with constant torque, (d) the R–L impedance controlled with a dimming circuit, (e) a number of computers and (f) a number of compact fluorescent lamps*

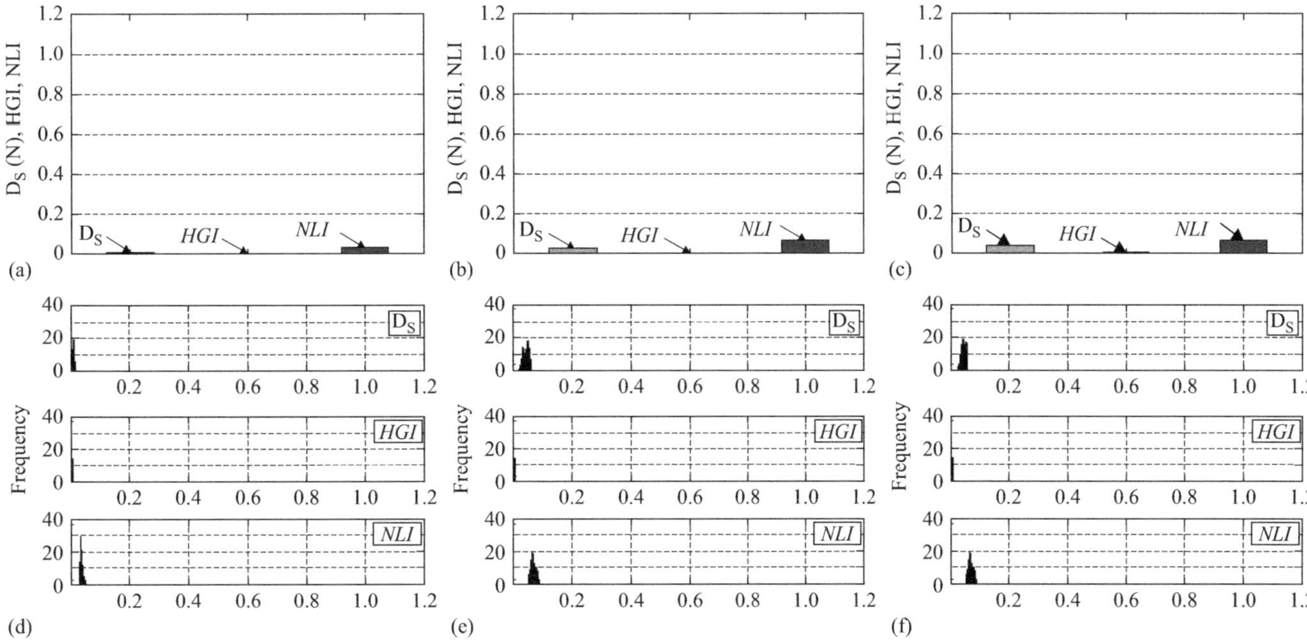

Figure 2.13 (a)–(c) The values of D_s, HGI and NLI measured under sinusoidal supply voltage for an R–L impedance and the constant speed and constant torque cases of an induction machine, respectively, and (d)–(f) the histograms of these quantities measured under 100 different non-sinusoidal supply voltage with 5% of THDV for the respective load cases

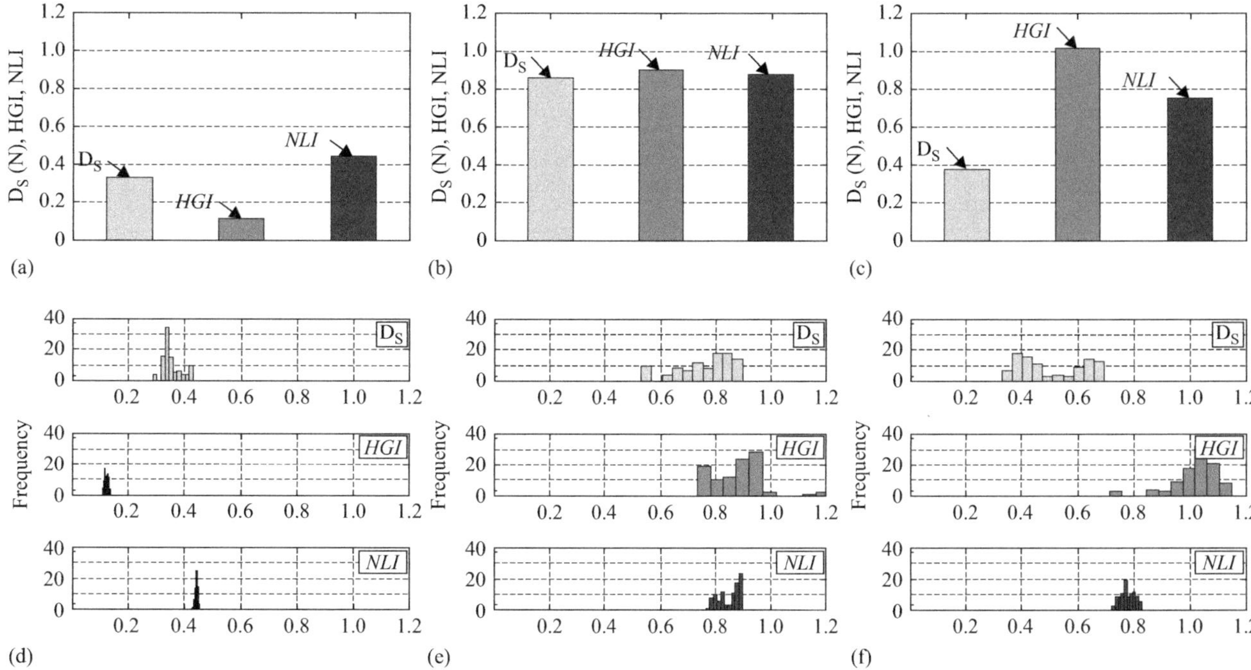

Figure 2.14 *(a)–(c) The values of D_s, HG and NLC measured under sinusoidal supply voltage for an R–L impedance controlled with a dimming circuit, a number of computers and a number of compact fluorescent lamps, respectively, and (d)–(f) the histograms of these quantities measured under 100 different non-sinusoidal supply voltage with 5% of THDV for the respective load cases*

It is seen from Figure 2.13(a) and (d) that the D_S, *HGI* and *NLI* values measured for the R–L impedance under sinusoidal and non-sinusoidal test voltages are very close to zero. Figure 2.13(b), (c), (e) and (f) shows that *HGI* attains zero for the induction machine working with constant speed and constant torque under the same test voltages. For both load cases under sinusoidal and non-sinusoidal test voltages, it is also seen that D_S and *NLI* are around 0.04 and 0.07, respectively. Figure 2.14(a) shows that for the R–L impedance controlled with a dimming circuit under sinusoidal voltage D_S, *HGI* and *NLI* are measured as 0.335, 0.116 and 0.443, respectively. For the same load under non-sinusoidal test voltages, it is observed from Figure 2.14(d) that D_S varies between 0.28 and 0.43, and *HGI* and *NLI* are around 0.12 and 0.44, respectively. From Figure 2.14(b), it can be mentioned that for the computers under sinusoidal voltage D_S, *HGI* and *NLI* are 0.860, 0.903 and 0.878, respectively. In addition, Figure 2.14(e) points out that for the same load under the non-sinusoidal test voltages D_S, *HGI* and *NLI* are in the intervals from 0.53 to 0.9, from 0.73 to 1.21 and from 0.77 to 0.9, respectively. For the compact fluorescent lamps under sinusoidal voltage, it is seen from Figure 2.14(c) that D_S, *HGI* and *NLI* are measured as 0.376, 1.020 and 0.751, respectively. Finally, Figure 2.14(f) reveals that for the same load under the non-sinusoidal test voltages D_S, *HGI* and *NLI* vary in the intervals from 0.33 to 0.69, from 0.72 to 1.15 and from 0.72 to 0.83, respectively.

2.7 Conclusions

In non-sinusoidal and unbalanced systems, the definitions of apparent power and its power resolutions are still a topic that remained controversial in electrical engineering community. Thus, the various definitions were proposed in the literature, and different ones are encouraged by the standards focussed on power measurement.

In addition, the guidelines and standards on the limitation of harmonic pollution do not include any method to determine the harmonic-producing consumers, and the conventional harmonic-pollution metering index as current total harmonic distortion is not capable of metering consumer's harmonic-pollution responsibility due to the fact that it depends on the distortion of utility side's supply voltage. Therefore, to fulfil this gap, for the harmonic-source detection and sharing harmonic responsibility between utility and consumer sides, several methods and indices are proposed in the literature.

In this chapter, both frequency-domain power theory and harmonic-pollution metering approaches are summarized, and they are qualitatively and quantitatively analysed to clearly show their principles, limits, advantages and disadvantages.

References

[1] IEEE Std 100-2000: 'IEEE Standard for the Authoritative Dictionary of IEEE Standards Terms'. Dec. 2000.

[2] Miller, T. J. E.: '*Reactive Power Control in Electric Systems*'. Wiley-Interscience, New York, 1982.

[3] Emanuel, A. E.: 'On the Definition of Power Factor and Apparent Power in Unbalanced Polyphase Circuits with Sinusoidal Voltage and Currents'. *IEEE Transactions on Power Delivery*. Jul. 1993, vol. 8(3), pp. 841–852.

[4] Emanuel, A. E.: 'Apparent Power Definitions for Three-Phase Systems'. *IEEE Transactions on Power Delivery*. Jul. 1999, vol. 14(3), pp. 767–772.

[5] Czarnecki, L. S.: 'Budeanu and Fryze: Two Frameworks for Interpreting Power Properties of Circuits with Nonsinusoidal Voltages and Currents'. *Electrical Engineering (Archiv fur Elektrotechnik)*. Dec. 1997, vol. 80(6), pp. 359–367.

[6] Shepherd, W., Zakikhani, P.: 'Power Factor Correction in Nonsinusoidal Systems by the Use of Capacitance'. *Journal of Physics D: Applied Physics*. 1973, vol. 6, pp. 1850–1861.

[7] Shepherd, W., Zand, P.: 'Energy Flow and Power Factor in Nonsinusoidal Circuits'. Cambridge University Press, New York, 1979.

[8] Sharon, D.: 'Reactive Power Definition and Power-Factor Improvement in Nonlinear Systems'. *Proceedings of the Institution of Electrical Engineers*. 1973, vol. 120(6), pp. 704–706.

[9] Kusters, N. L., Moore, W. J. M.: 'On the Definition of Reactive Power under Nonsinusoidal Conditions'. *IEEE Transactions on Power Apparatus and Systems*. Sep. 1980, vol. 99(5), pp. 1845–1854.

[10] Czarnecki, L. S.: 'Physical Reasons of Currents RMS Value Increase in Power Systems with Nonsinusoidal Voltage'. *IEEE Transactions on Power Delivery*. Jan. 1993, vol. 8(1), pp. 437–447.

[11] IEEE Std. 1459: 'IEEE Standard Definitions for the Measurement of Electric Power Quantities under Sinusoidal, Non-Sinusoidal, Balanced or Unbalanced Conditions'. Mar. 2010.

[12] DIN Std. 40110: 'AC Quantities—Part 2: Polyphase Circuits'. Nov. 2002 (in German).

[13] Wilczynski, E.: 'Total Apparent Power of the Electrical System for Periodic, Deformed Waveforms'. *IEE Proceedings Electric Power Application*. Jul. 2000, vol. 147(4), pp. 281–285.

[14] Ghassemi, F.: 'What Is Wrong with the Electric Power Theory and How It Should Be Modified'. *Ninth International Conference of Metering and Tariffs for Energy Supply-Mates*. pp. 109–114, May 1992.

[15] Balci, M. E., Hocaoglu, M. H.: 'New Power Decomposition for Sinusoidal and Nonsinusoidal Conditions'. *IEEE ICHQP 2006*. Cascais, Portugal, Sep. 2006.

[16] Balci, M. E., Hocaoglu, M. H.: 'A Power Resolution for Nonsinusoidal and Unbalanced Systems—Part II: Theoretical Background'. *ELECO 2011*. Bursa, Turkey, pp. I-173–I-178, Dec. 2011.

[17] Czarnecki, L. S.: 'Comparison of Power Definitions for Circuits with Nonsinusoidal Waveforms'. *IEEE Tutorial Course 90EH0327-7-PWR*. 1990, pp. 43–50.

[18] Balci, M. E., Hocaoglu, M. H.: 'Quantitative Comparison of Power Decomposition'. *Electric Power System Research*. Mar. 2008, vol. 78(3), pp. 318–329.

[19] Balci, M. E., Hocaoglu, M. H.: 'A Power Resolution for Nonsinusoidal and Unbalanced Systems—Part I: Literature Overview and Motivation'. *ELECO 2011*, Bursa, Turkey, pp. I-167–I-172, Dec. 2011.

[20] Balci, M. E., Hocaoglu, M. H.: 'Addendum to a Power Resolution for Nonsinusoidal and Unbalanced Systems: Evaluation Examples'. *ELECO 2011*, Bursa, Turkey, pp. I-179–I-182, Dec. 2011.

[21] Balci, M. E., Hocaoglu, M. H.: 'An Analysis on the Nonactive Powers for Nonsinusoidal Conditions'. *Journal of the Faculty of Engineering and Architecture of Gazi University*. Jun. 2011, vol. 26(2), pp. 307–313 (in Turkish).

[22] Balci, M. E., Emanuel, A. E.: 'Apparent Power Definitions: A Comparison Study'. *International Review of Electrical Engineering*. Nov. 2011, vol. 6(6), pp. 2713–2722.

[23] Canturk, S., Balci, M. E., Hocaoglu, M. H.: 'On the Definition of Apparent Power'. *Electric Power Quality and Utilization Journal*. Dec. 2015, vol. 18(2), pp. 1–6.

[24] Wanger, W. E.: 'Effects of Harmonics on Equipment'. *IEEE Transactions on Power Delivery*. Apr. 1993, vol. 8(2), pp. 672–680.

[25] IEEE Std. 519: 'IEEE Recommended Practices and Requirements for Harmonic Control in Electrical Power Systems'. 2014.

[26] IEC Std. 61000-3-2: 'Limitation of Emission of Harmonic Currents in Low-Voltage Power Supply Systems for Equipment with Rated Current Less Than 16A'. 2000.

[27] IEC Std. 61000-3-4: 'Limitation of Emission of Harmonic Currents in Low-Voltage Power Supply Systems for Equipment with Rated Current Greater Than 16A'. 1998.

[28] IEC Std. 61000-3-6: 'Assessment of Emission Limits for Distorting Loads in MV and HV Power Systems'. 1996.

[29] EN Std. 50160: 'Voltage Characteristics of the Electricity Supplied by Public Distribution Systems'. 2004.

[30] IEC Std. 61000-4-7: 'General Guide on Harmonics and Inter harmonics Measurement and Instrumentation for Power Supply Systems and Equipment Connected Thereto'. 2002.

[31] IEC Std. 61000-4-30: 'Power Quality Measurement Methods'. 2003.

[32] Davis, E. J., Emanuel, A. E., Pileggi, D. J.: 'Evaluation of Single-Point Measurement Method for Harmonic Pollution Cost Allocation'. *IEEE Transactions on Power Delivery*. Jan. 2000, vol. 15(1), pp. 14–18.

[33] Locci, N., Muscas, C., Sulis, S.: 'Multi-Point Measurement Techniques for Harmonic Pollution Monitoring: A Comparative Analysis'. *Sixth International Workshop Power Definitions and Measurements under Non-sinusoidal Conditions*. Milano, Italy, pp. 103–107, Oct. 2003.

[34] Balci, M. E., Hocaoglu, M. H.: 'On the Validity of Harmonic Source Detection Methods and Indices'. *ICHQP 2010*. Bergamo, Italy, pp. 1–5, Sep. 2010.

[35] Ferrero, A., Menchetti, A., Sasdelli, R.: 'Measurement of the Electric Power Quality and Related Problems'. *International Transactions on Electrical Energy Systems*. Nov. 1996, vol. 6(6), pp. 401–406.
[36] Muscas, C.: 'Assessment of Electric Power Quality: Indices for Identifying Disturbing Loads'. *International Transactions on Electrical Energy Systems*. Jul. 1998, vol. 8(4), pp. 287–292.
[37] Xu, W., Liu, Y.: 'A Method for Determining Customer and Utility Harmonic Contributions at the Point of Common Coupling'. *IEEE Transactions on Power Delivery*. Apr. 2000, vol. 15(2), pp. 804–811.
[38] Chen, C., Liu, X., Koval, D., Xu, W., Tayjasanant, T.: 'Critical Impedance Method—A New Detecting Harmonic Sources Method in Distribution Systems'. *IEEE Transactions on Power Delivery*. Jan. 2004. vol. 19(1), pp. 288–297.
[39] de Paula Silva, S. F., de Oliveira, J. C.: 'The Sharing of Responsibility between the Supplier and the Consumer for Harmonic Voltage Distortion: A Case Study'. *Electric Power Systems Research*. Nov. 2008, vol. 78(11), pp. 1959–1964.
[40] Pfajfar, T., Blazic, B., Papic, I.: 'Harmonic Contributions Evaluation with the Harmonic Current Vector Method'. *IEEE Transactions on Power Delivery*. Jan. 2008, vol. 23(1), pp. 425–433.
[41] Srinivasan, K., Jutras, R.: 'Conforming and Nonconforming Current for Attributing Steady State Power Quality Problems'. *IEEE Transactions on Power Delivery*. Jan. 1998, vol. 13(1), pp. 212–217.
[42] Dellapos Aquila, A., Marinelli, M., Monopoli, V. G., Zanchetta, P.: 'New Power-Quality Assessment Criteria for Supply Systems under Unbalanced and Nonsinusoidal Conditions'. *IEEE Transactions on Power Delivery*. Jul. 2004, vol. 19(3), pp. 1284–1290.
[43] Locci, N., Muscas, C., Sulis, S.: 'On the Measurement of Power-Quality Indexes for Harmonic Distortion in the Presence of Capacitors'. *IEEE Transactions on Instrumentation and Measurement*. Oct. 2007, vol. 56(5), pp. 1871–1876.
[44] Herrera, S. R., Salmeron, P.: 'Harmonic Disturbance Identification in Electrical Systems with Capacitor Banks'. *Electric Power Systems Research*. Jan. 2012, vol. 82(1), pp. 18–26.
[45] Barbaro, P. V., Cataliotti, A., Cosentino, V., Nuccio, S.: 'A Novel Approach Based on Nonactive Power for the Identification of Disturbing Loads in Power Systems'. *IEEE Transactions on Power Delivery*. Jul. 2007, vol. 22(3), pp. 1782–1789.
[46] Balci, M. E., Karacasu, O., Hocaoglu, M. H.: 'A Detection Method for Harmonic Producing Loads'. *ELECO 2009*. Bursa, Turkey, pp. I-149–I-153, Dec. 2009.
[47] Bazina, M., Tomisa, T.: 'Comparison of Various Methods for Determining Direction of Harmonic Distortion by Measuring in Point of Common Coupling'. *ENERGYCON 2014*. Cavtat, Croatia, pp. 392–399, May 2014.

[48] Stevanovic, D., Petkovic, P.: 'A Single-Point Method Based on Distortion Power for the Detection of Harmonic Sources in a Power System'. *Metrology and Measurement Systems*. Mar. 2014, vol. 21(1), pp. 3–14.
[49] Ortiz, A., Manana, M., Renedo, C. J., Perez, S., Delgado, F.: 'A New Approach to Frequency Domain Power Measurement Based on Distortion Responsibility'. *Electric Power Systems Research*. Jan. 2011, vol. 81(1), pp. 202–208.
[50] National Instruments Co.: 'Manuals of LabView software'. http://www.ni.com/manuals/, last access date: May 2015.

Chapter 3

Passive harmonic filters

Chamberlin Stéphane Azebaze Mboving[1] *and Zbigniew Hanzelka*[1]

Summary

The current chapter presents the harmonic passive filters commonly used in the industry, their classifications, configurations, advantages and disadvantages, theoretical as well as practical design considerations and the data required for their installation. The goal of this chapter is to present the frequency characteristics variation and consequently the filtration properties of different passive filters.

3.1 Introduction

Nowadays, the increase of non-linear loads, such as power electronic devices, arc furnaces and fluorescent lamps, has become a serious problem for the distribution network because of harmonic generation and often low rate of displacement power factor (DPF). The current harmonics produced by power electronic devices can enter the electrical grid and cause the overheating, overcharging and functionality defect (even damage) of the supply network components (transformers, cables, etc.) and loads connected to that system as well [3,23]. Many devices designed to improve the power quality (e.g. passive filters) are applied to reduce the harmonic disturbances and improve the DPF.

3.2 General concept of passive harmonic filters

The passive harmonic filters (PHFs) are classified into different groups (series, shunt and hybrid) [9,22], and despite their disadvantages (robust construction and resonance phenomena with the grid impedance, etc.), they are widely used in industry and can be applied in high, medium and low voltage of distribution system

[1]Department of Power Electronics and Energy Control Systems, Faculty of Electrical Engineering, Automatics, Computer Science and Biomedical Engineering, AGH University of Science and Technology, Cracow, Poland

because of their low cost, simplicity in construction and ability to compensate the required reactive power (fundamental harmonic) [3].

PHFs are organized in different topologies that vary by their components number, filtration band, immunity on detuning phenomena, power losses and so on. The design and operating cost of each topology depend on the highlighted features and also on the grid and load parameters [9]. Therefore, it is important to know the disadvantages and advantages of each topology while deciding on its choice and in optimal decision taking in terms of technical and economical point of view.

The current chapter is focused on the studies of different PHF topologies presenting the influence of the variation of different parameters on the filter functionality and improvement level of grid power quality as well.

To illustrate the considerations presented in this chapter, the power system example has been modelled in MATLAB® Simulink™ (Figure 3.1(a)). It contains three-phase electrical power supply system with transformer (T), line (l) feeding three-phase controlled thyristor bridge supplying DC adjustable speed drive and

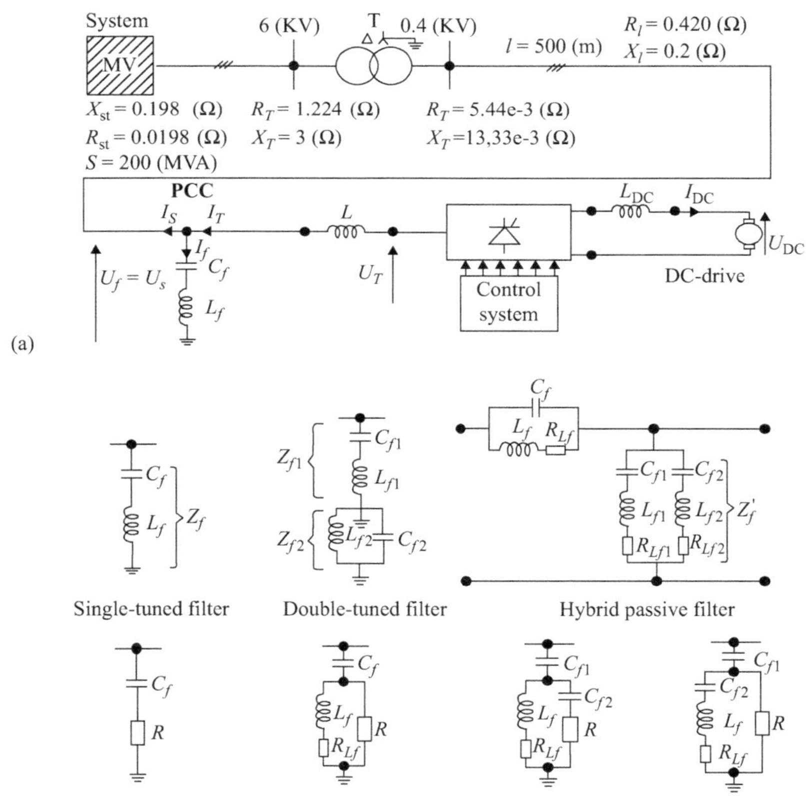

Figure 3.1 (a) Simulated power system, (b) considered passive parallel harmonic filter topologies

input reactor (L) at the AC side. The PHFs are connected at the point of common coupling (PCC).

The filters are tuned to the frequency a bit smaller than the frequency of the first high harmonic characteristic for 6-pulse thyristor rectifier which is the fifth harmonic (4.85th). For each considered filter topology, the capacitor resistance is neglected, the reactors are air-core and the quality factor ($q = 85$) is used to compute their resistance; the reactive power (fundamental harmonic) used to compensate the power electronic converter is constant: $Q_f = -2172.5$ Var. The voltage and current data and characteristics at the AC side of thyristor bridge are considered at the steady state.

3.3 Series passive filters

The series passive filter (SPF) is connected between PCC and the non-linear load. There are different topologies of SPF [22], but in this chapter, the studies are focused on the topology considered in the equivalent circuit of Figure 3.2(a) [8].

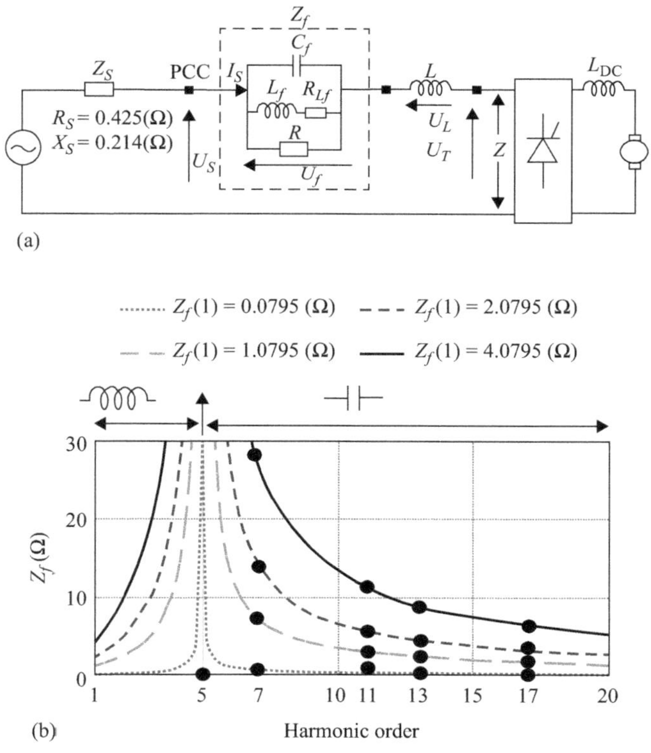

Figure 3.2 (a) Equivalent circuit of simulated power system with example of series passive filter, (b) impedance frequency characteristics for different values of filter impedance at the fundamental harmonic ($Z_{f(1)}$)

The purpose of SPF in power system is to block the harmonic(s) generated by the non-linear load so that it cannot enter the electrical grid. The SPF is much less used than the shunt passive filter [22].

The filter is designed to produce high-impedance Z_f (parallel resonance) for the harmonic to which it is tuned and small impedance for the fundamental harmonic $Z_{f(1)}$ (minimum voltage drop and power losses).

Figure 3.2(b) presents the impedance frequency characteristics of SPF for different values of fundamental harmonic impedance (also different reactive powers in basic harmonic domain). The filter impedance for fifth-order harmonic increases with the increase of $Z_{f(1)}$ as well as the impedance of other harmonics. The SPF impedance is inductive below and capacitive above the resonance frequency (fifth, Figure 3.2(b)). For the small values of $Z_{f(1)}$, the filter capacitor is high, whereas the filter inductor is small along with the filter voltage drop as well.

If the SPF is tuned to the frequency lower than the frequency of harmonic to be blocked (fifth), the filter impedance of harmonic to be blocked will decrease and will be on the capacitive side of the filter characteristic (Figure 3.3(a)), therefore can be amplified.

For the SPF detuned to the frequency higher than the frequency of harmonic to be blocked (fifth), the filter impedance of harmonic to be blocked will also decrease and will be on the inductive part of the filter characteristic (Figure 3.3(a)), and none amplified will be observed. But the filter efficiency on the fifth harmonic reduction will reduce. The impedance of harmonics higher than the fifth is small when the tuning frequency decreases and big when the tuning frequency increases (Figure 3.3(a)).

With the increase of filter capacity or inductor by $+10\%$, the filter characteristic has moved to the left side of fifth harmonic (Figure 3.3(b) and (c)). The filter impedance of harmonic to be blocked has decreased and is on the capacitive side of the filter characteristic. The impedance of high harmonics has decreased as well.

For the decrease of filter capacity or inductor by -10%, the filter characteristic has moved to the right side of fifth harmonic (Figure 3.3(b) and (c)). The filter impedance of high harmonics has increased, whereas the filter impedance of fifth harmonic, which is inductive, has decreased. The filter efficiency is then reduced.

The influence of the reactor resistance on the filter characteristic is presented in Figure 3.3(d). The higher the resistance, the smaller is filter impedance of fifth harmonic. The SPF is less efficient when the reactor is an iron core with small quality factor.

The damping resistance (R) connected in parallel with filter reactor and capacitor (Figure 3.2(a)) is rarely applied because of power losses. Its influence on SPF functionality is showed in Figure 3.4(a). The filter impedance for the fifth-order harmonic is equal to the value of damping resistance. The small damping resistance reduces the efficiency of SPF.

The impedance frequency characteristics of power system (Z) observed from the rectifier input terminals (Figure 3.2(a)) presents the series and parallel

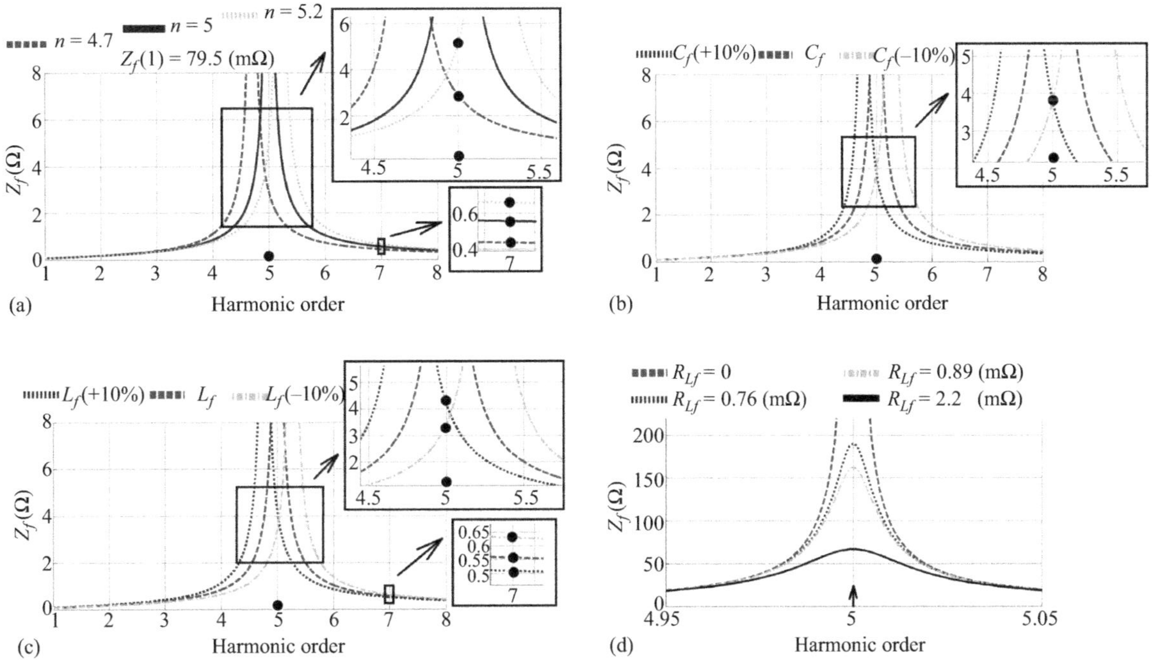

Figure 3.3 Impedance frequency characteristics of series passive filter for different tuning frequency (a), influence of filter capacitor (b), filter reactor (c), and filter reactor resistance (d)

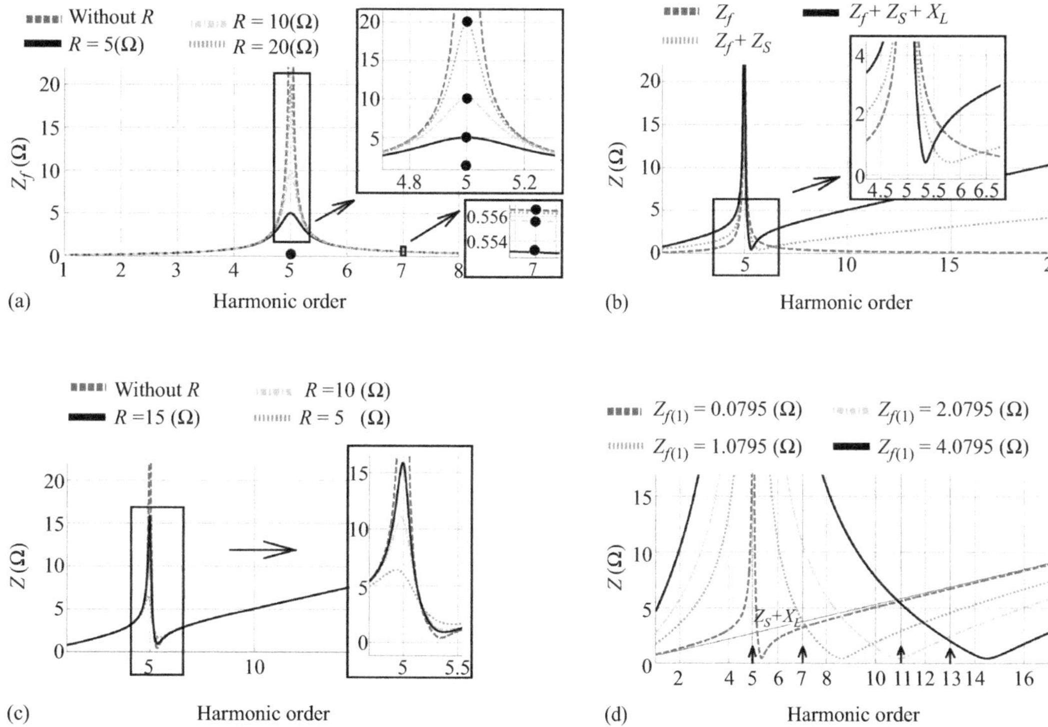

Figure 3.4 *(a) Impedance frequency characteristics of series passive filter for different value of damping resistance R (Figure 3.2(a)), (b) influence of the grid inductance and rectifier input reactor on the power system series resonance, (c) and (d) are, respectively, the power system impedance for different values of filter damping resistance R and filter impedance (fundamental harmonic)*

resonance (Figure 3.4(b)–(d)). The power system series resonance is influenced by the grid inductance and rectifier input reactor as well (Figure 3.4(b)). The influence of filter damping resistance and filter impedance (fundamental harmonic) on the system impedance (Z) are presented in Figure 3.4(c) and (d). The power system parallel and series resonances are damped with the decrease of filter damping resistance (Figure 3.4(c)). The series resonance occurring for the harmonics higher than the fifth (Figure 3.4(d)) moves with the increase of filter impedance ($Z_{f(1)}$).

The comparison between the waveforms and spectrums of the grid current and voltage before and after the SPF connection is presented in Figure 3.5. The SPF is tuned to the frequency of the fifth harmonic. The connection of SPF in the power system has improved the grid voltage waveform by reducing the commutation notches (Figure 3.5(a)) and the THD_{Us} as well (Figure 3.7). The spectrums of Figure 3.5(b) and (d) presents an important reduction of fifth harmonic amplitude at the grid side after the SPF connection and the amplification of the 7th, 11th, 13th, 17th, 19th and 25th as well. The THD_{Is} of PCC current has increased a little after the filter connection (Figure 3.7). The shape of grid current before and after the connection of filter is presented in Figure 3.5(c). The voltage (U_{T}) at the thyristor bridge input terminals is more distorted after the filter connection (Figure 3.6(a) and (b)), and its THD is higher (Figure 3.7).

3.4 Shunt passive filters

3.4.1 Single-tuned filter

Because of the aging, the single filter must be tuned to the frequency of harmonic a bit smaller than the frequency of lowest generated harmonic (ω_r); in the considered example, the fifth order. There are different opinions on how much it should be detuned [6]. According to Muhammad *et al.* [17], the detuning frequency should be in the range of 3%–15% below the frequency of harmonic to be eliminated.

For the tuning frequencies lower or higher than ω_r (Figure 3.8(b) and (c)), the filter impedance of fifth harmonic is high, which reduces the filtration efficiency. In the case when the filter is detuned to the frequencies higher than ω_r, the amplification of filtered harmonic can occur.

The equivalent impedance of filter is capacitive for all harmonic frequencies (also fundamental) lower than the resonance frequency and inductive for the harmonics higher than the resonance component (fifth order in Figure 3.8(b) and (c)).

The waveform of current and voltage at the PCC and their spectrum are presented in Figure 3.9. After the filter connection, a little decrease of commutation notches can be observed (Figure 3.9(a)). The waveform of PCC current is improved (Figure 3.9(c)). With the decrease of tuning frequency (5th to 4.1st), the amplitude of fifth harmonic of PCC voltage and current (Figure 3.9(b) and (d)) has increased, as well as the amplitude of higher harmonics (11th–29th).

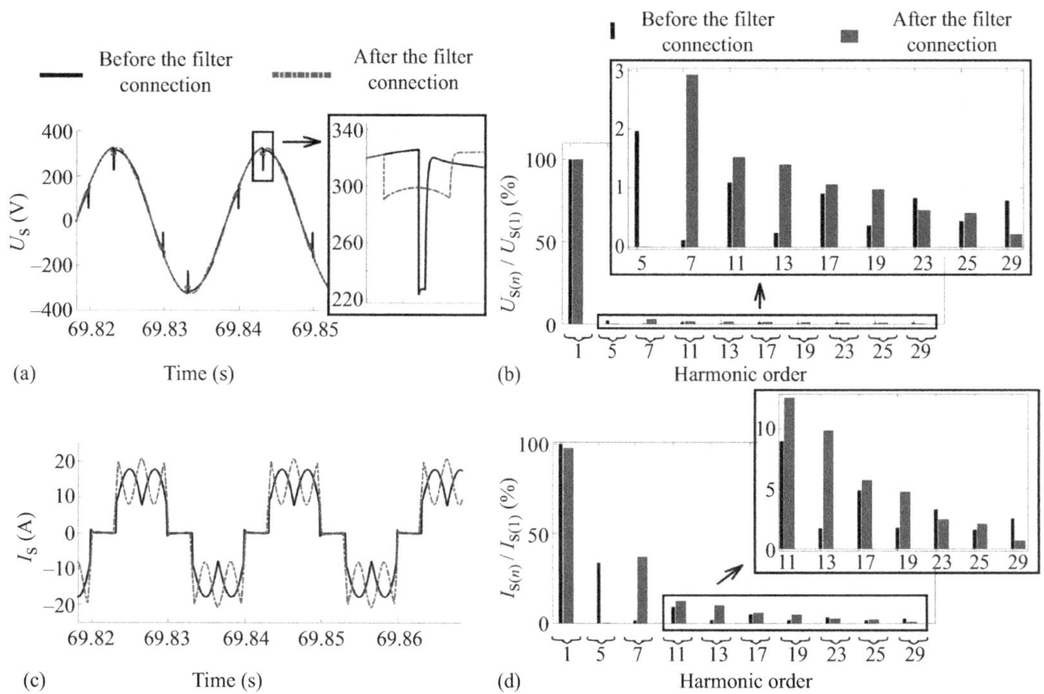

Figure 3.5 (a) Grid voltage and (b) its spectrum, (c) grid current and (d) its spectrum

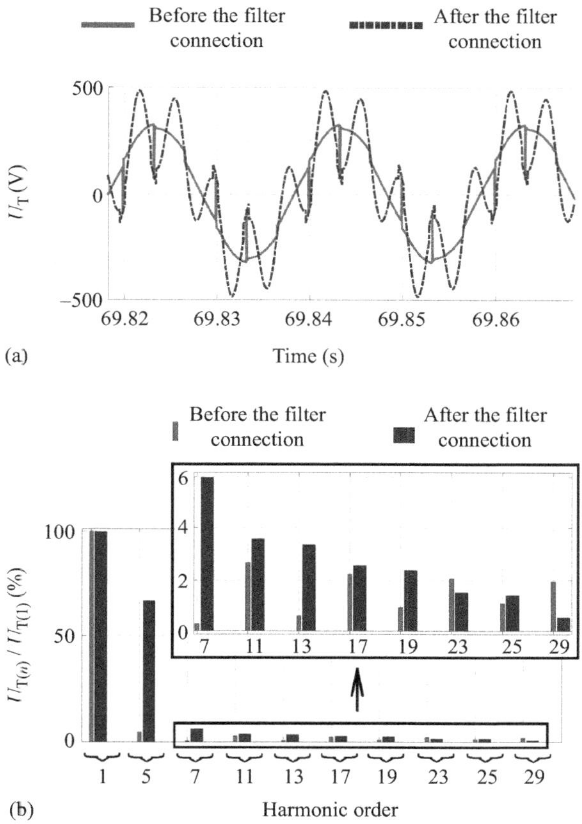

Figure 3.6 (a) Voltage at the input of thyristor bridge and (b) its spectrum

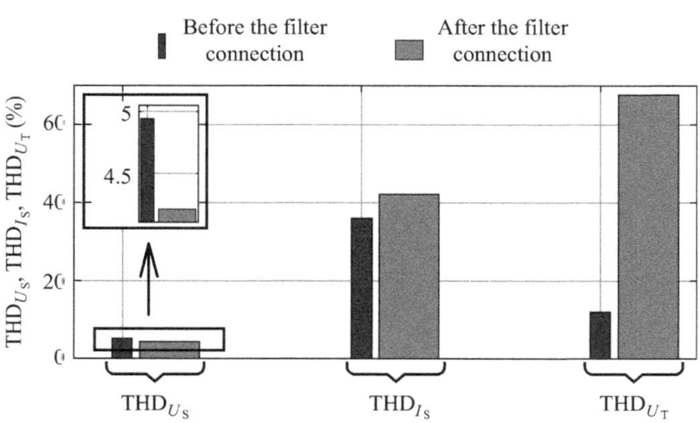

Figure 3.7 THD of grid voltage (U_S), grid current (I_S) and voltage at the input of thyristor bridge (U_T)

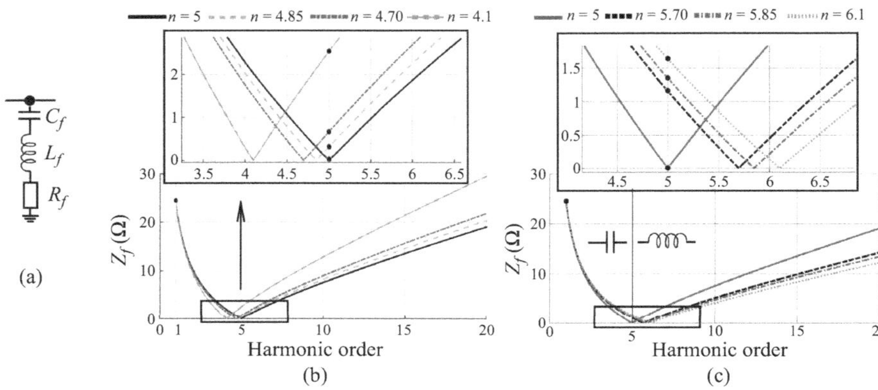

Figure 3.8 Single-branch filter (a) and its impedance frequency characteristics for different values of tuning frequency order, (b) the single filter is tuned to the frequencies of harmonic lower than the frequency of fifth harmonic, and (c) the single filter is tuned to the frequency of harmonic higher than the frequency of fifth harmonic

The grid voltage and current for fundamental harmonic are almost constant during the change of the tuning frequency. But after the filter connection, the current has considerably decreased (from 15.86 to 8.43 A) because of the reactive component compensation. The THD of PCC current and voltage has increased with the decrease of tuning frequency (Table 3.1).

The filter current spectrum is presented in Figure 3.10(b) and the waveform in Figure 3.10(a). The filtered harmonic (fifth) flowing through the filter depends on the tuning frequency. Its value is high for the tuning frequency near to the frequency of fifth harmonic (250 Hz). The single filter is much loaded by the filtered harmonic than the other harmonics (Figure 3.10(b)).

The filter effectiveness is presented in Figure 3.10(c) and (d). The filter is more effective when it is tuned to the frequency smaller but near to the frequency of the filtered harmonic.

The power system impedance measured from the input terminals of thyristor bridge is presented in Figure 3.11. On the zoom of this figure, the top of the characteristics represents higher impedance value (parallel resonance), and the bottom represents lower impedance value (series resonance). With the decrease of tuning frequency, the displacement of parallel and series resonances from the higher to lower frequency can be observed. The amplitude of power system impedance at the parallel resonance has decreased.

To present the influence of filter resistance (R_f) on the filter efficiency in the reduction of filtered harmonic at the PCC, it has been assumed that the analysed single filter tuned to the resonance frequency of 245.5 Hz and the filter resistance is inversely proportional the q-factor. According to James [12], the filter resistance

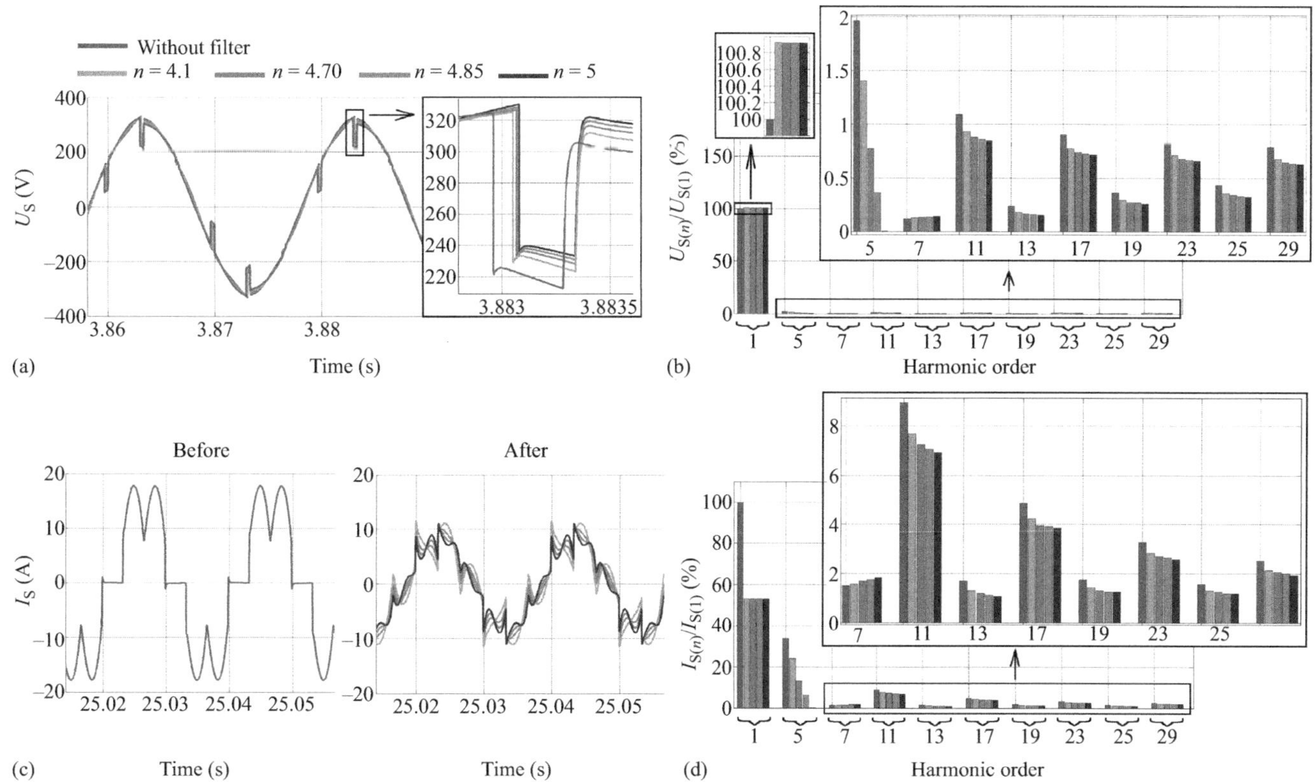

Figure 3.9   (a) Grid voltage and its spectrum (b), (c) grid current and its spectrum (d)

Table 3.1 Grid voltage [U_{Smax} (V)] and current [I_{Smax} (A)] parameters

n	Without filter		$n = 4.1$		$n = 4.70$		$n = 4.85$		$n = 5$	
	Voltage	Current	Voltage	Current	Voltage	Current	Voltage	Current	Voltage	Current
First	319.97	15.86	322.93	8.43	322.92	8.42	322.91	8.42	322.91	8.42
Fifth	6.28	5.34	4.48	3.82	2.46	2.09	1.17	0.99	0.02	0.01
Seventh	0.38	0.24	0.41	0.25	0.42	0.27	0.43	0.28	0.45	0.29
11th	3.48	1.42	2.98	1.22	2.81	1.15	2.75	1.12	2.71	1.10
13th	0.77	0.27	0.59	0.21	0.53	0.19	0.51	0.18	0.49	0.17
17th	2.89	0.77	2.48	0.67	2.36	0.63	2.32	0.62	2.29	0.61
19th	1.17	0.28	0.95	0.23	0.88	0.21	0.86	0.20	0.83	0.20
23rd	2.65	0.52	2.28	0.45	2.17	0.43	2.13	0.42	2.11	0.41
25th	1.4	0.25	1.16	0.21	1.09	0.20	1.06	0.19	1.04	0.19
29th	2.52	0.4	2.17	0.34	2.07	0.33	2.03	0.32	2.01	0.31
THD (%)	4.93	36	4.24	49.70	3.94	31.54	3.84	22.35	3.79	18.74

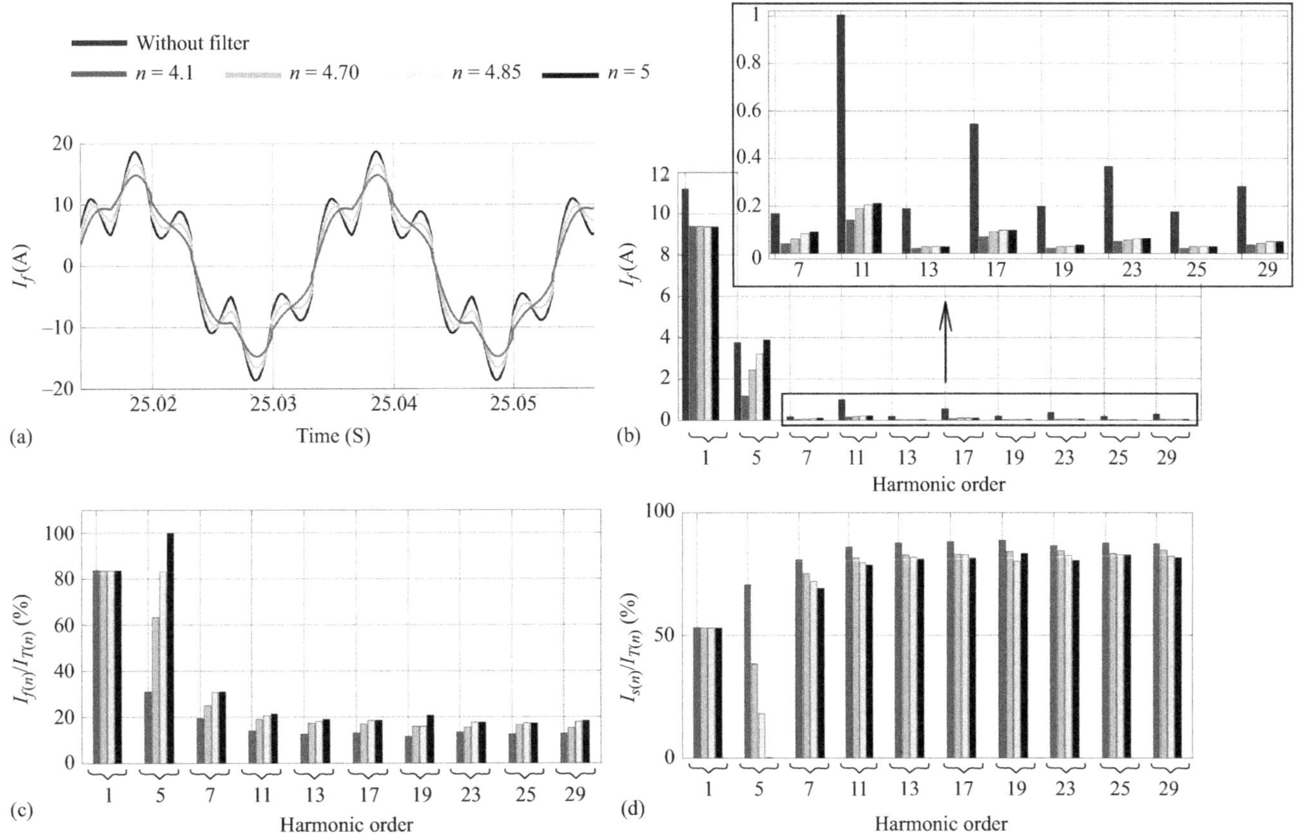

Figure 3.10 (a) Filter current and (b) its spectrum; (c) and (d) filter effectiveness

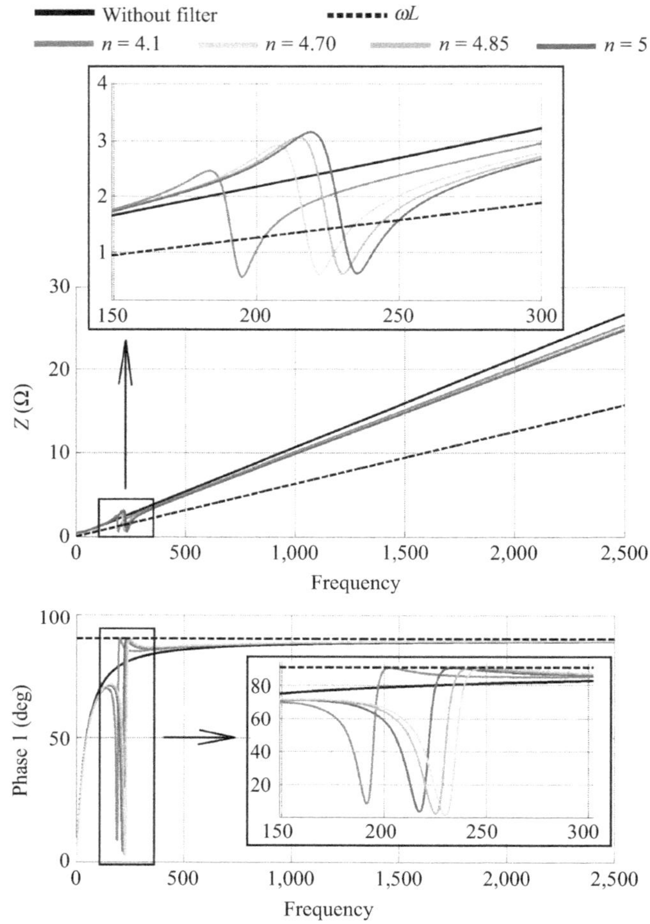

Figure 3.11 Characteristics of power system impedance measured at the thyristor bridge input, ωL – inductor reactance

is mostly related to the filter reactor resistance. The q-factor of air core reactor is between 50 and 150 (MV and HV) and for the iron core reactor, it is between 20 and 50 (LV).

With increase of single filter resistance, the filter impedance of filtered harmonic (fifth) has increased (Figure 3.12) and the parallel and series resonances are damped (Figure 3.13(a)).

Figure 3.13(b) presents the power system impedance characteristics for different grid inductance (different short circuit power at PCC). With the increase of grid inductance (decreasing of the short circuit power), the system impedance increases at the parallel resonance, decreases at the series resonance and decreases the frequency of parallel resonance.

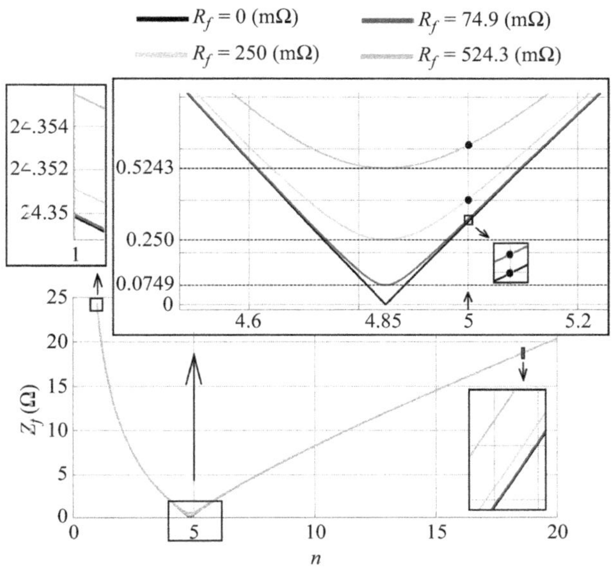

Figure 3.12 Impedance frequency characteristics for different filter resistance

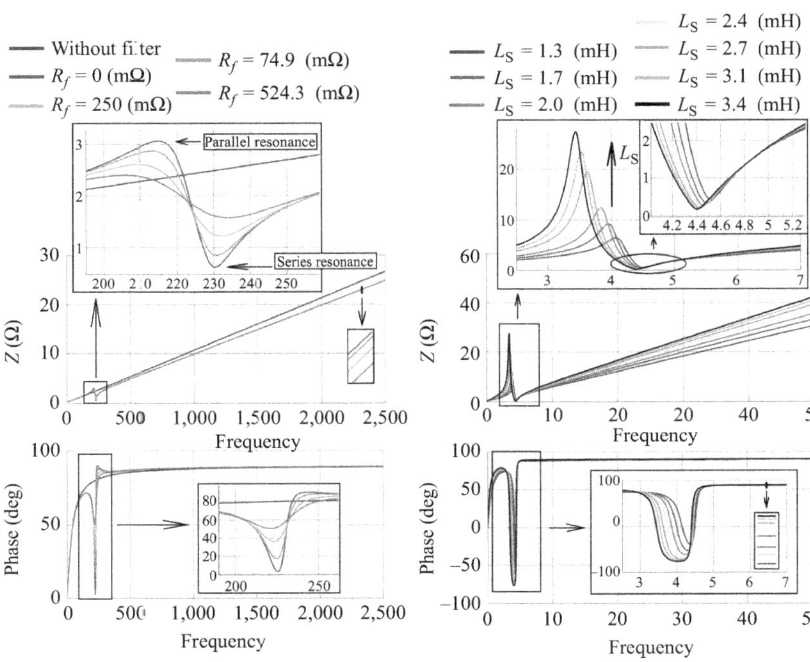

Figure 3.13 Power system impedance characteristics measured at the thyristor bridge input terminals for (a) different values of filter resistance, and (b) different values of grid inductance

The group of single-tuned filters is needed to prevent more than one harmonic from entering the electrical grid [4,9]. Table 3.2 presents six exemplary methods used to calculate the parameters of the filters group (e.g. two single filters of fifth and seventh harmonics are considered). Each method presents different total reactive power compensation division (between the parallel-connected single-branch filters). The distribution of total reactive power (in the filters group according to each method) has an influence on the frequency characteristics of filters group and filtration efficiency as well [1,7,10].

In [4], it is demonstrated that adopting the comparison criteria as the amplitude of grid voltage and current harmonics and power losses of filters for each of the considered methods in Table 3.2, different filtration efficiency is obtained and in general sense it is difficult to indicate which of the presented methods is the most effective. The efficiency of each method depends on the supply network, the load parameters and on the analysis of filter cost.

3.4.2 Double-tuned filter

The double-tuned filter presented in Figure 3.14(a) is constituted of series part (Z_{f1}) and parallel part (Z_{f2}). As its name indicates, it can be tuned to two different frequencies. The functionality of double-tuned filter is almost the same as the one of the group of two single-tuned filters [20,24]. Many different topologies of double-tuned filter exist [14], and Figure 3.14(a) presents the most common one.

Figure 3.14(b) presents the impedance frequency characteristics of the parallel (Z_{f2}) and series (Z_{f1}) parts and whole filter (Z_f). In the considered example, the filter is tuned to the frequencies of 242.5 and 342.5 Hz and is capacitive for the fundamental harmonic and the harmonics smaller than the order 4.85 (n_1). It is inductive for the harmonics between 4.85 and 6 (n_b). For the harmonics between 6 and 6.85 (n_2), it is again capacitive. For the harmonic higher than the 6.85, it is inductive. The example of filter parameters is evaluated in Table 3.3 (the resistance of filter reactors – R_{Lf1} and R_{Lf2} – are considered).

The fifth- and seventh-order harmonics (PCC) have been considerably decreased after the filter connection (Figure 3.15(b) and (d)). The double-tuned filter has improved the grid current waveform (Figure 3.15(c)) and reduced the depth of commutation notches (Figure 3.15(a)). In Figure 3.15(b) and (d) and Table 3.4, it can be observed the reduction of harmonics higher than the fifth and the seventh after the filter connection.

At the power system impedance frequency characteristic of Figure 3.16, the parallel resonances between the filter and the grid can be seen.

3.4.3 Broad-band filters

3.4.3.1 First-order filter

The capacitors bank as shown in Figure 3.17(a) designed for reactive power compensation can be included in this category even though the harmonics filtration is not the goal of such installations. The character of the system equivalent to impedance shows that they possess better filtration properties for higher harmonics and

Table 3.2 Exemplary computation methods for the group of parallel connected single-tuned filters

Method A – equal reactive power for each filter	$h = 2 \quad Q_1 = Q_2 = Q_F/2$
Method B – the reactive power of filters inversely proportional to the harmonic order	$Q_F = Q_1 + Q_2 \quad Q_1/Q_2 = n_2/n_1 \rightarrow 3Q_1 = 5Q_2$ $Q_2 = 3/8Q_F$
Method C – the reactive power of filters inversely proportional to the square of harmonic order	$Q_1/Q_2 = n_2^2/n_1^2 \rightarrow 9Q_1 = 25Q_2 \quad Q_1 = 25/34Q_F$
Method D – the reactive power of filters calculated on the base of the shaping of frequency characteristic of the filter impedance [14,16]	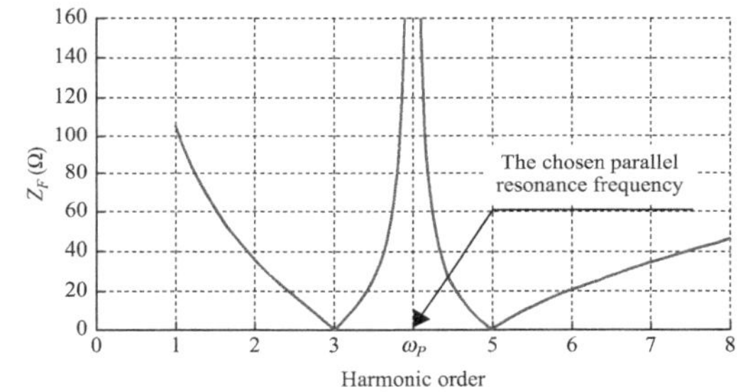
Method E – the reactive power of filters calculated on the base of the assumption that the reactors of filters are identical	$L_1 = L_2 = U^2/\omega Q_F\left[1/\left(1-3^2\right)+1/\left(1-5^2\right)\right]$ $Q_1 = Q_F/\left(1-3^2\right)\left[1/\left(1-3^2\right)+1/\left(1-5^2\right)\right]$
Method F (optimization) – different form of objective function in optimization task can be considered [2], for example the filter's reactive power is computed on the base of the minimum value of the objective function y	$\varphi_1 = U_{a(5)\text{before}}/U_{b(5)\text{after}} = Z_{S(5)}/Z_{x(5)}$ $\varphi_2 = U_{a(7)\text{before}}/U_{b(7)\text{after}} = Z_{S(7)}/Z_{x(7)}$ $y = 1 - (1-\varphi_1)(1-\varphi_2)$

h – the number of filters; n – order of filtered harmonic; Z_F – total impedance of filters group; ω_P – the chosen system parallel resonance frequency (Method D); φ_1, φ_2 – optimization coefficients; $U_{a(5)}$, $U_{a(7)}$, $U_{b(5)}$, $U_{b(7)}$ – voltages of fifth and seventh harmonics after and before the connection of filters group; $Z_{S(5)}$, $Z_{S(7)}$ – the grid impedances for fifth- and seventh-order harmonics; and $Z_{x(5)}$, $Z_{x(7)}$ – the equivalent impedances of the group of filters and electrical grid for fifth and seventh harmonic.

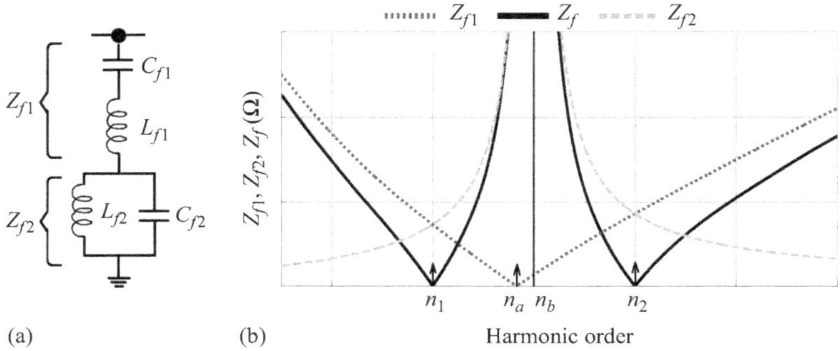

Figure 3.14 Double-tuned filter (a) and (b) impedance frequency characteristics. n_a – resonance frequency order (the series part Z_{f1}); n_b – resonance frequency order (the parallel part Z_{f2}); n_1, n_2 – series resonance frequencies of whole filter (Z_f)

Table 3.3 Example of double-tuned filter parameters

$Q_f = -2{,}172.5$ (Var); $U_f = 230$ (V)									
n	**C_{f1} (μF)**	**L_{f1} (mH)**	**$R_{Lf(1)}$ (Ω)**	**C_{f2} (μF)**	**L_{f2} (μH)**	**$R_{Lf(2)}$ (Ω)**	**$Z_{f(1)}$ (Ω)**	**$Z_{f(5)}$ (Ω)**	**$Z_{f(7)}$ (Ω)**
4.85 6.85	126.00	2.6	0.0087	1,000	275.81	0.91210	24.3498	0.4856	0.4796

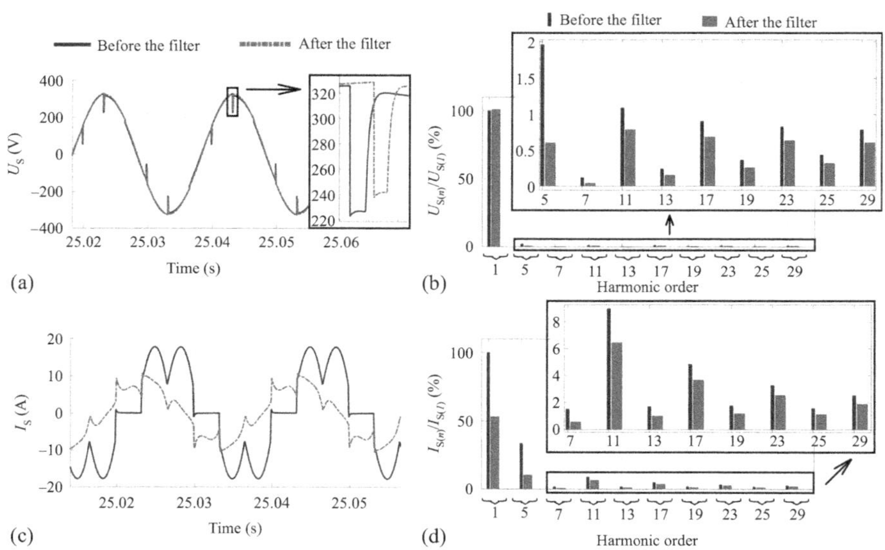

Figure 3.15 (a) Grid voltage and (b) its spectrum, (c) grid current and (d) its spectrum

Table 3.4 Grid voltage and current parameters

n	U_{smax} (V)		I_{smax} (A)	
	Without filter	**With filter**	**Without filter**	**With filter**
First	319.97	322.92	15.86	8.43
Fifth	6.28	1.90	5.34	1.62
Seventh	0.38	0.14	0.24	0.09
11th	3.48	2.50	1.42	1.02
13th	0.76	0.47	0.27	0.16
17th	2.88	2.17	0.77	0.58
19th	1.16	0.80	0.28	0.19
23rd	2.64	2.01	0.52	0.4
25th	1.39	1.00	0.25	0.18
29th	2.51	1.92	0.4	0.3
THD (%)	4.92	3.69	36	25.97

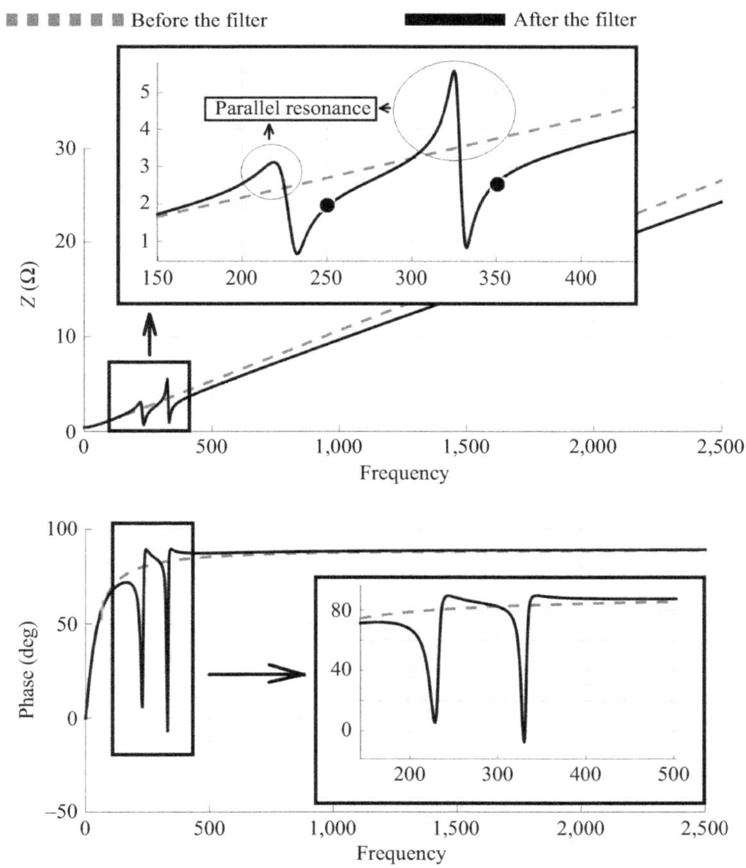

Figure 3.16 Power system impedance frequency characteristics measured at the thyristor bridge input terminals

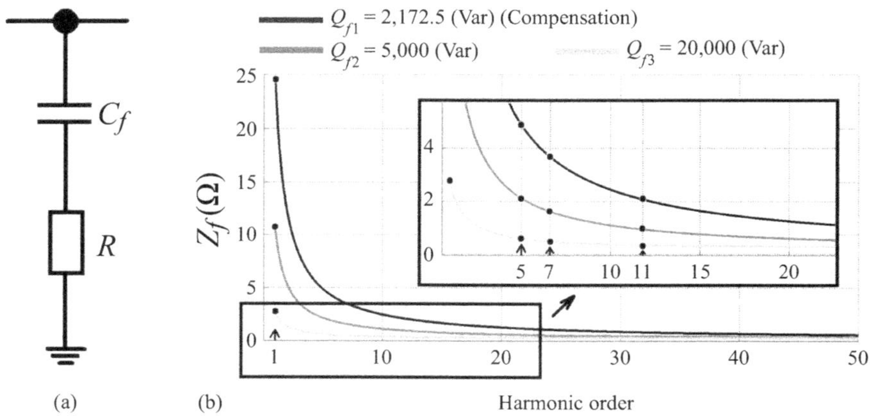

Figure 3.17 (a) First-order filter, (b) frequency characteristics of capacitor reactance for different value of reactive power

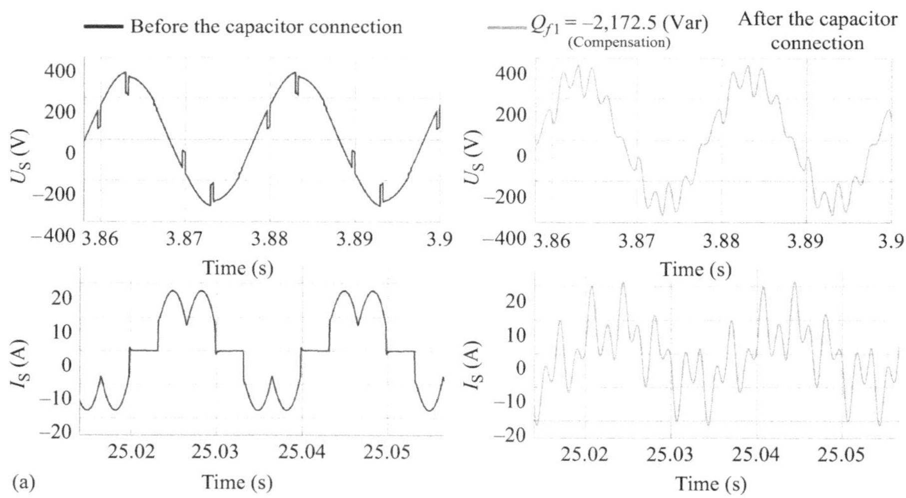

Figure 3.18 Voltage and current waveforms at PCC (a) and their spectrums before and after the capacitor connection (b), capacitor current spectrum (c)

higher capacitor power (Figure 3.17(b)) as well as lower equivalent resistance circuit representing the power losses. In the case of nowadays capacitors, the power losses in conversion to kVar are very small therefore their resistance can be neglected. The waveforms of PCC current and voltage before and after capacitor connection are presented in Figure 3.18(a).

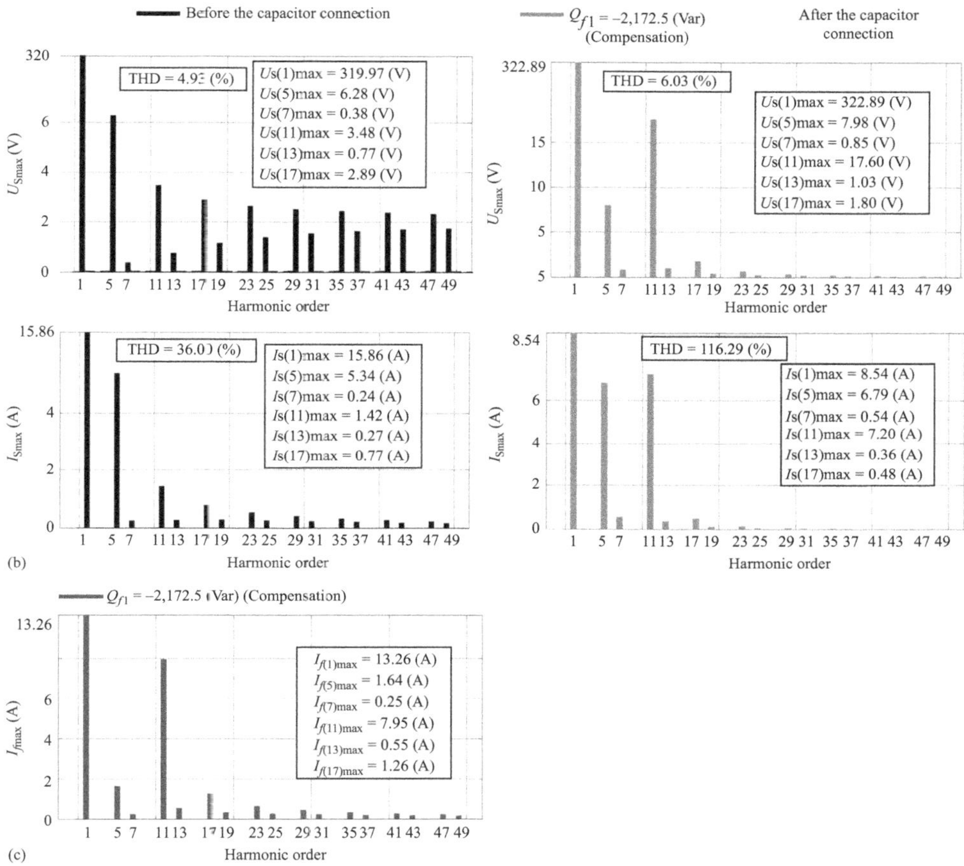

Figure 3.18 (Continued)

In the considered example after the connection of capacitor, the PCC reactive power has been reduced (from 2,140 to 19.33 Var). The compensation was done successfully, but the level of voltage and current distortion remains a problem. The capacitor connection has caused the resonance amplification of PCC voltage and current harmonics especially the 11th (Figure 3.18(b)). After the capacitor connection, the grid current and voltage waveforms have been changed (Figure 3.18(a)) and their THD has increased from 4.93% to 6.03% for the voltage and from 36% to 116.29% for the current (Figure 3.18(b)). This increase is caused by resonance phenomena between the parallel connected filter capacitor and grid inductance. The capacitor has reduced the amplitude of higher order harmonics (from the 17th) of grid current and voltage (Table 3.5, $Q_{f1} = 2{,}172.5$ Var). This situation occurs frequently when the capacitors bank is used (without serial reactors) to improve the

Table 3.5 PCC voltage and current parameters

U_s (V)/I_s (A)				
n	**Without filter**	$Q_{f1}=2{,}172.5$ **Var**	$Q_{f1}=5$ **kVar**	$Q_{f1}=20$ **kVar**
First	319.97/15.86	322.83/8.66	326.31/19.69	338.42/114.56
Fifth	6.28/5.34	7.93/6.75	10.9/9.28	4.48/3.82
Seventh	0.38/0.24	0.73/0.46	0.51/0.32	0.33/0.2
11th	3.48/1.42	11.31/4.63	2.23/0.91	0.53/0.21
13th	0.77/0.27	1.08/0.37	0.25/0.09	0.04/0.01
17th	2.89/0.77	1.79/0.48	0.61/0.16	0.22/0.06
19th	1.17/0.28	0.46/0.11	0.17/0.04	0.05/0.01
23rd	2.65/0.52	0.72/0.14	0.3/0.06	0.14/0.03
25th	1.4/0.25	0.29/0.05	0.13/0.02	0.05/0.01
29th	2.52/0.4	0.4/0.06	0.18/0.03	0.1/0.02
THD (%)	4.93/36	4.35/94.93	3.42/47.41	1.34/3.34

Grey shaded area indicates the frequency (e.g. 11th, $Q_{f1}=5$Kvar) from which the capacitor has started to reduce the higher harmonics (e.g. 13th, 17th etc.) and to which it has amplified the lower harmonics (e.g. 1st, 5th and 7th) in comparison to the case without filter.

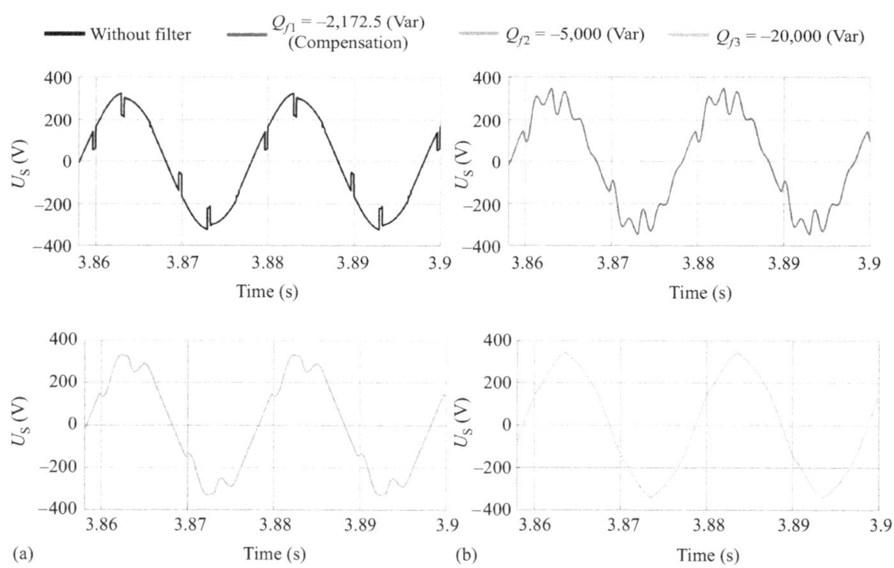

Figure 3.19 Grid voltage (a) and current (b) characteristics for different values of capacitor power

DPF in the network system with distorted voltages and currents. The changing of capacitors bank power changes the frequency of system parallel resonance causing the amplification of different harmonic. The spectrum of capacitor current is show in Figure 3.18(c).

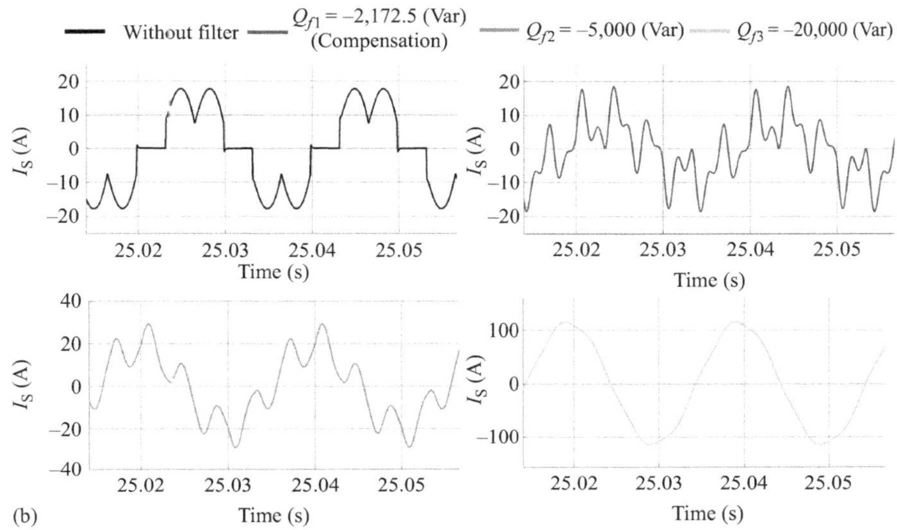

Figure 3.19 (*Continued*)

Figure 3.19 presented the variation of PCC voltage and current waveforms with the increase of capacitor power (see also Figure 3.17(b)). The increase of capacitor reactive power has improved the THD voltage and current (efficient harmonics reduction for 20 kVar) at the PCC, but the overcompensation (basic harmonic amplification) is observed in the power system (Table 3.5).

The impedance of power system (Z) measured from the input terminals of thyristor bridge is presented in Figure 3.20. The parallel resonance has occurred because of the parallel connection between capacitor and grid inductance, and the series resonance has occurred because of the series connection of the reactor (L), the capacitor and grid inductance (parallel connected). With the increase of the capacitor reactive power, the power system impedance at the parallel resonance has decreased. The parallel and series resonances have moved to the lower frequencies (Figure 3.20(a)).

The impedance of power system measured at the thyristor bridge input terminals for different damping resistance is presented in Figure 3.20(b). It can be observed that with the damping resistance increase, the impedance of Z (at the resonance frequency) has decreased from 10.77 to 6 Ω (parallel resonance) and has increased from 1.34 to 4.362 Ω (series resonance). The capacitor resistance has damped the series and parallel resistance.

3.4.3.2 Second-order filter

The single-tuned filter in Figure 3.8(a) has been used to design the second-order filter by connecting the resistance R in parallel with the reactor L_f and resistance R_f (Figure 3.21(a)). The resistance R_f is considered as reactor resistance. The single-branch filter characteristic presents a small value of impedance for the resonance

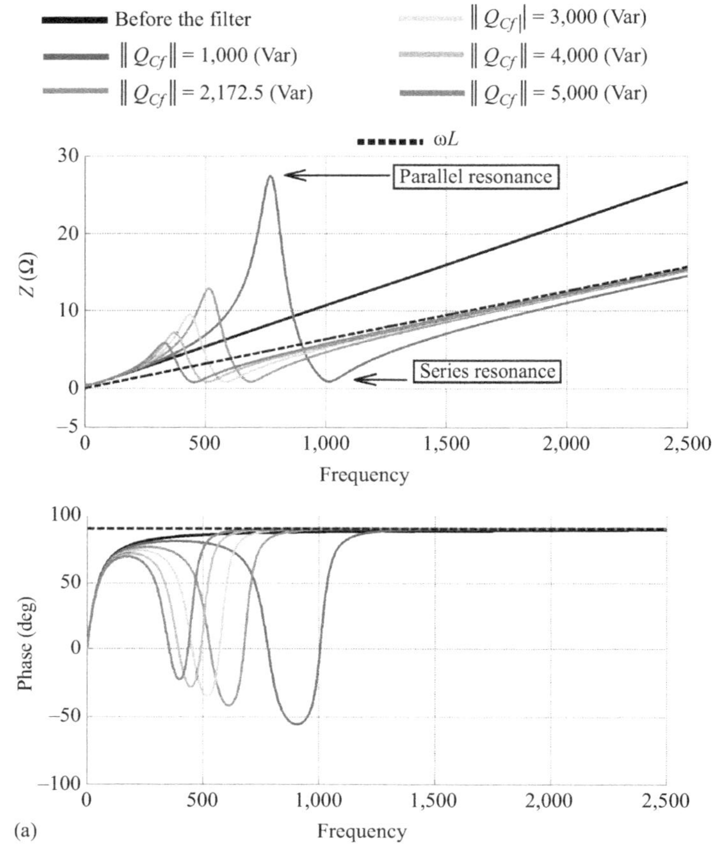

Figure 3.20 Power system impedance characteristic measured at the input terminals of thyristor bridge for different: (a) filter reactive power and (b) filter damping resistance ($Q_{f1} = 2,172.5$ Var)

frequency and frequencies around it (e.g. the fifth harmonic frequency) and high impedance value for the frequencies higher than the resonance frequency. The second-order filter presents not only a small value of impedance for the resonance frequency, but also for the higher frequencies, depending on the damping resistance (Figure 3.21(b)). The lower the resistance R, the better the filtration of higher frequency components (Table 3.6). The filter impedance of for example fifth harmonic has increased with the decrease of damping resistance value (Figure 3.21(b)). The value of impedance for the fundamental harmonic is almost the same.

The waveforms of PCC voltage and current and their spectrum for different damping resistance are presented in Figure 3.22. It can be observed that the depth of the voltage commutation notches is less when the damping resistance decreases (Figure 3.22(a)). The grid current presents better shape after the second-order filter

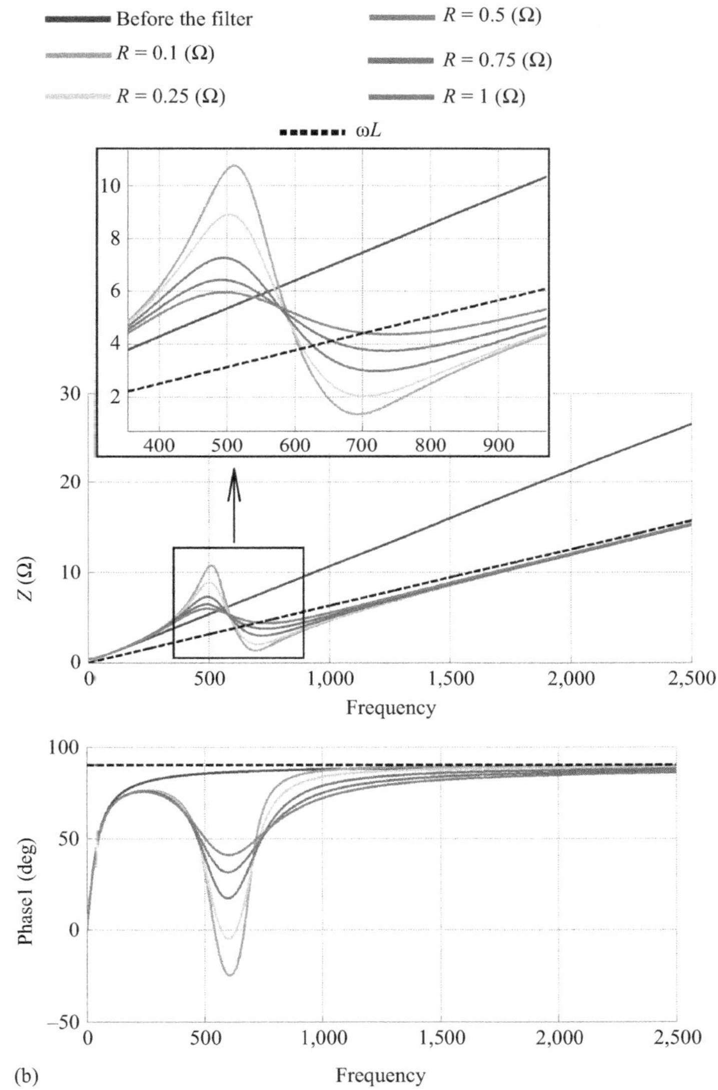

Figure 3.20 (*Continued*)

connection (Figure 3.22(c)). The higher the resistance R, the better the reduction of fifth harmonic components of grid voltage and current (Figure 3.22(b) and (d)). The filter power loss increases with the decrease of damping resistance.

The waveforms of filter current are presented in Figure 3.23(a), (c) and (e). With the decreased of damping resistance R, the filter fifth harmonic has decreased

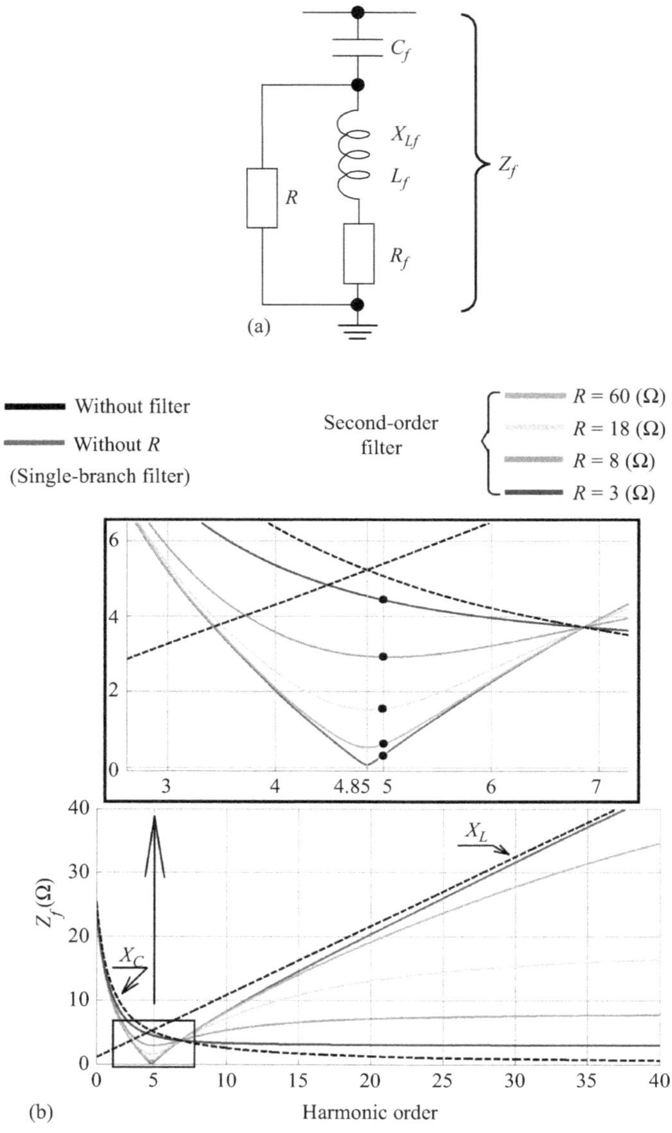

Figure 3.21 Second-order filter (a) and (b) its impedance frequency characteristics for different values of damping resistance R

and the amplitudes of harmonics with frequencies higher than the fifth have increased (Figure 3.23(b)). The filter is then more charged by higher order harmonics than the harmonic to be eliminated. The amplitude of harmonics flowing through the damping resistance has increased, whereas the amplitude of harmonics

Table 3.6 Grid voltage and current parameters

n	U_{smax} **(V)/**I_{smax} **(A)**									
			Single-branch filter		**Second-order filter**					
	Without filter		**Without** R		$R = 60$ **(Ω)**		$R = 18$ **(Ω)**		$R = 8$ **(Ω)**	
	Voltage	**Current**	**Voltage**	**Current**	**Voltage**	**Current**	**Voltage**	**Current**	**Voltage**	**Current**
First	319.97	15.86	322.91	8.47	322.89	8.51	322.87	8.57	322.84	8.62
Fifth	6.28	5.34	1.19	1.01	2.31	1.97	4.59	3.91	6.21	0.29
Seventh	0.38	0.24	0.45	0.28	0.42	0.26	0.37	0.23	0.35	0.22
11th	3.48	1.42	2.73	1.12	2.72	1.11	2.72	1.11	2.79	1.14
13th	0.77	0.27	0.50	0.17	0.50	0.18	0.53	0.18	0.56	0.19
17th	2.89	0.77	2.30	0.62	2.29	0.61	2.27	0.61	2.22	0.59
19th	1.17	0.28	0.84	0.20	0.84	0.20	0.86	0.21	0.85	0.20
23rd	2.65	0.52	2.12	0.42	2.11	0.42	2.07	0.41	1.93	0.38
25th	1.4	0.25	1.05	0.19	1.05	0.19	1.04	0.19	0.98	0.18
29th	2.52	0.4	2.02	0.32	2.01	0.32	1.94	0.31	1.72	0.27
THD (%)	4.93	36	3.82	22.38	3.61	29.68	3.21	49.02	2.92	63.72

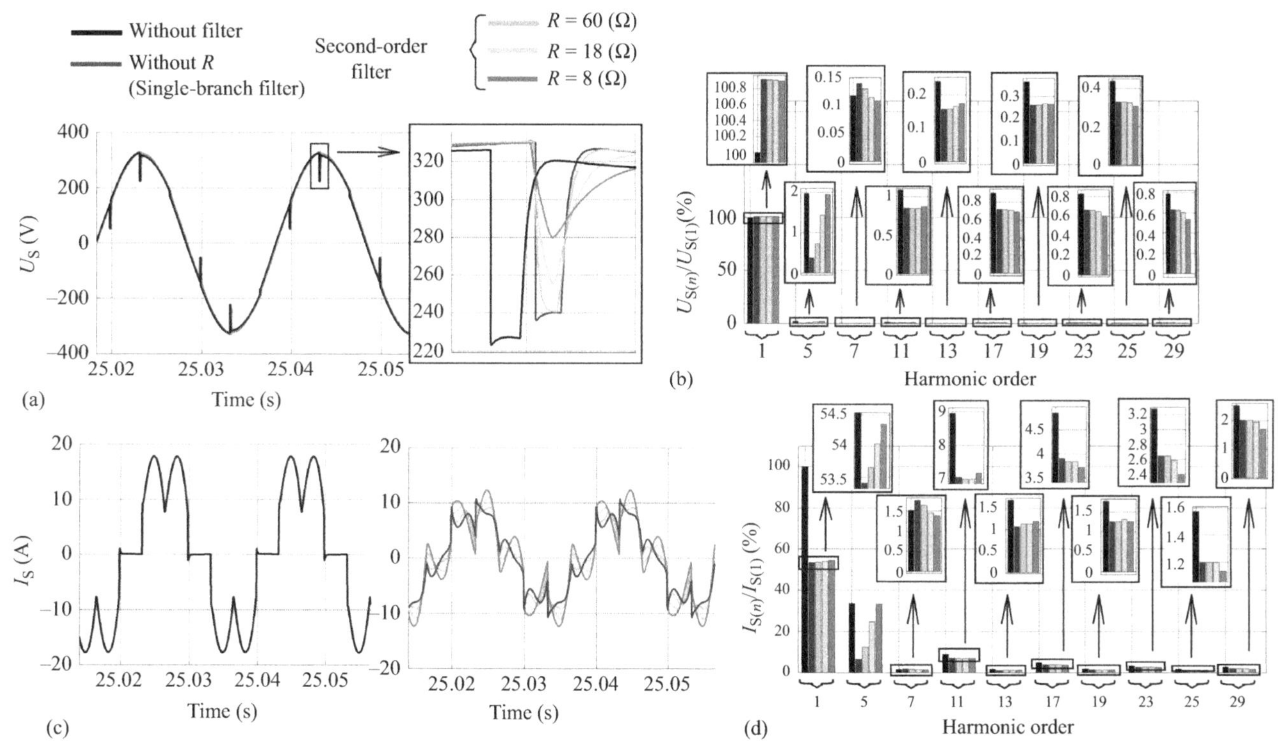

Figure 3.22 Grid voltage (a) and (b) its spectrum (p.u.); grid current (c) and (d) its spectrum (p.u.)

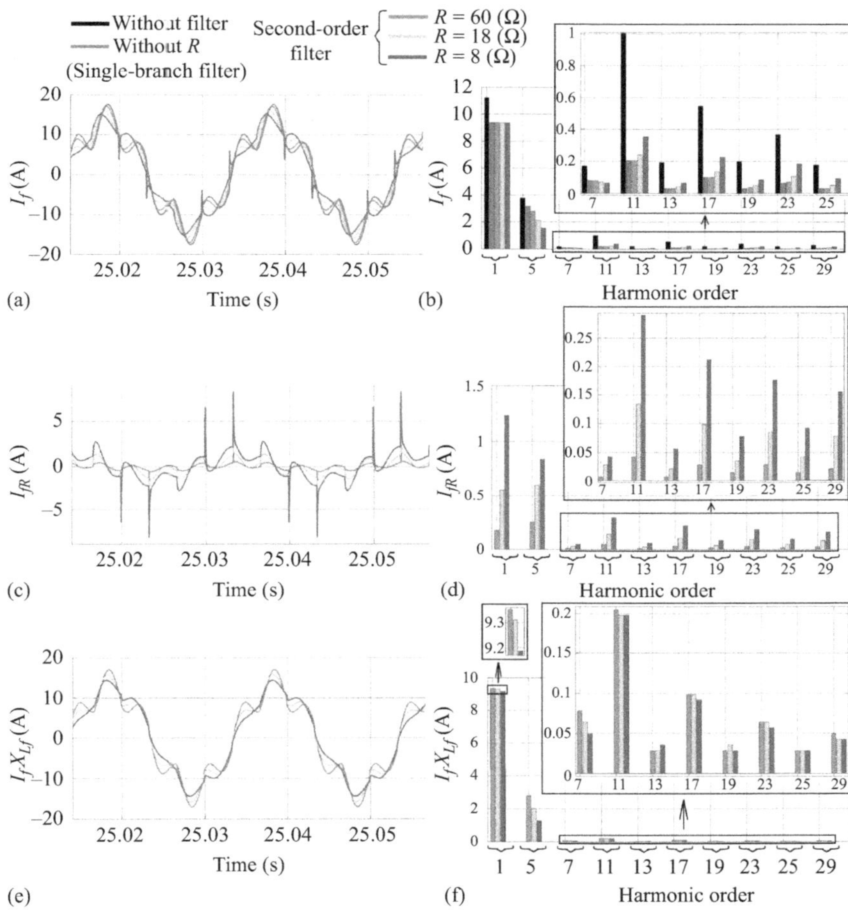

Figure 3.23 Filter current (a) and (b) its spectrum (the harmonics flowing through the filter are compared to the harmonics generated by the load – without filter); (c) resistance current (R) and (d) its spectrum; filter reactor current (e) and (f) its spectrum

flowing through the filter reactor has decreased apart of the 13th (Figure 3.23(d) and (f)).

For the damping resistance equal to 60 Ω (Figure 3.24(a)) the value of filter reactor reactance (X_{Lf}) of each harmonics from 1st to about 55th is smaller than the value of damping resistance. Therefore, the amplitude of harmonics flowing through the damping resistance is smaller than the amplitude of harmonics flowing through the filter reactor. In Figure 3.24(a) ($R = 18$ Ω), the harmonics from 1st to 13th flowing through the filter reactor are higher than the harmonics flowing

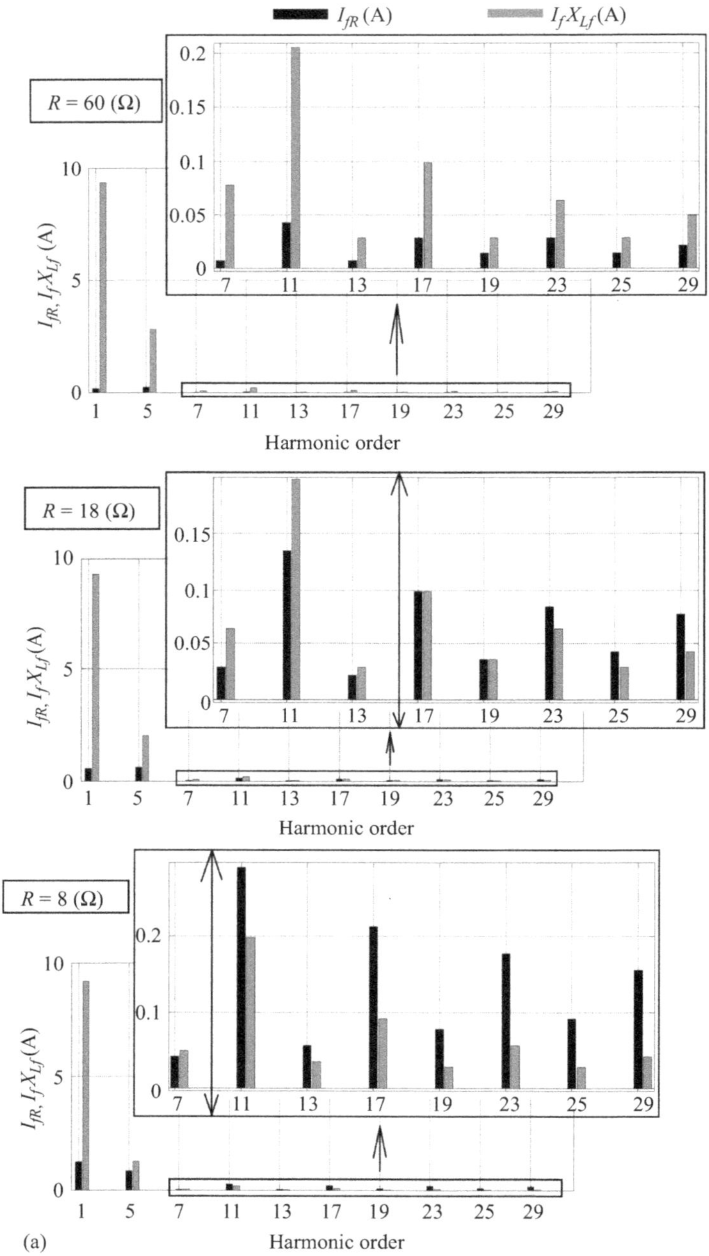

Figure 3.24 (a) Spectrums of filter current flowing through the damping resistance R and reactance X_{Lf}, (b) impedance frequency characteristic measured at the thyristor bridge input terminals for different values of filter damping resistance

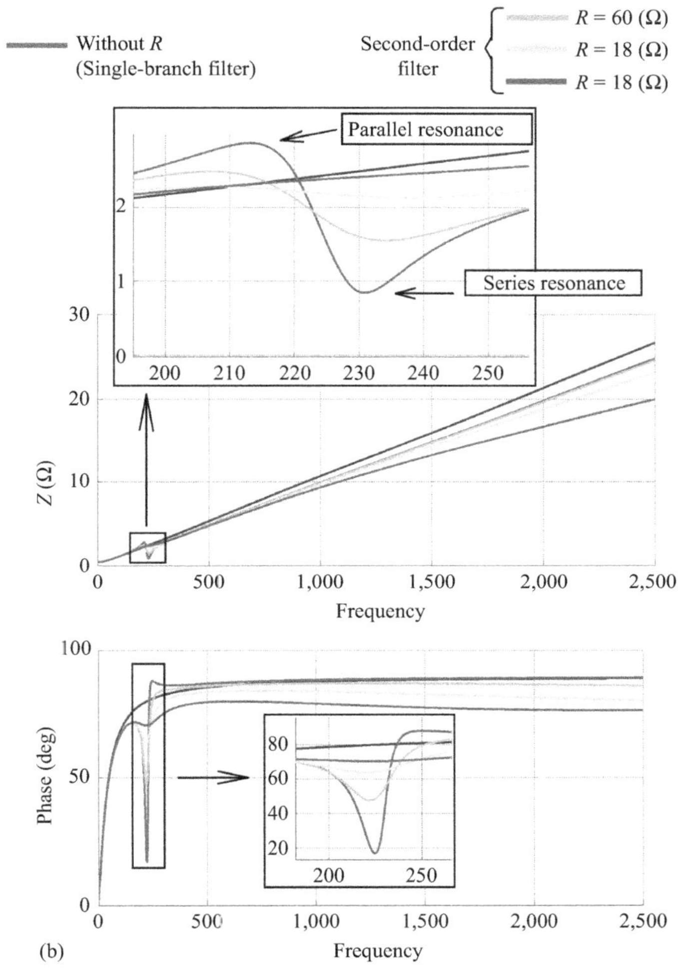

Figure 3.24 (Continued)

through the resistance (because $R > X_{Lf}$) and the harmonics from 23rd to infinity flowing through the reactor are smaller than the harmonics flowing through the resistance (because $R < X_{Lf}$). The damping resistance has generated less power losses (for fundamental harmonic) than the filter reactor ($R = 8\ \Omega$).

The characteristics of power system impedance presented in Figure 3.24(b) show that the parallel and series resonances are damped when the filter damping resistance decreases.

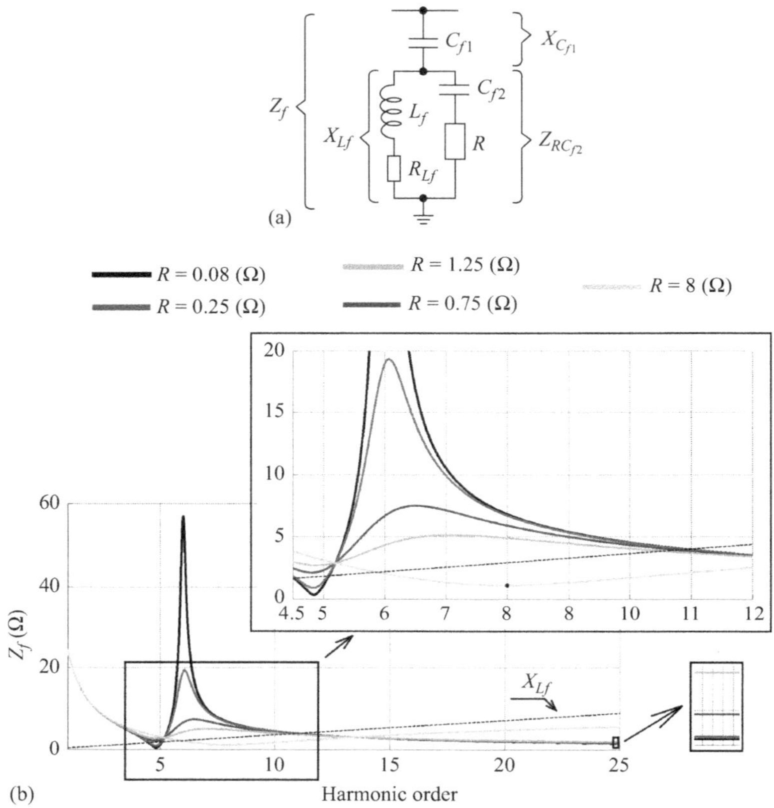

Figure 3.25 Third-order filter (a) and (b) its impedance frequency characteristic for different values of damping resistance R

3.4.3.3 Third-order filter

The filter topology and characteristics are presented in Figure 3.25.

The filter is designed in such a way that it can be decided at which frequency the parallel and series resonance will occur and the resistance (R) is just add to damp the resonances. According to Figure 3.25(b), the parallel resonance has been chosen to occur at the frequency of sixth harmonic and the series at the frequency of the component order 4.85. The filter impedance of fifth harmonic has increased from 1,070.3 to 3,112.0 mΩ with damping resistance increase and the filter fundamental harmonic impedance is almost the same.

The resistance (R) increase has caused the damping of filter series and parallel resonance. The filter impedance at the parallel resonance is reduced and increased at the series resonance (Figure 3.25(b)).

As it can be observed in Figure 3.25(b), with high value of damping resistance (e.g. 8 Ω), the third-order filter is tuned to a single frequency (around the eighth

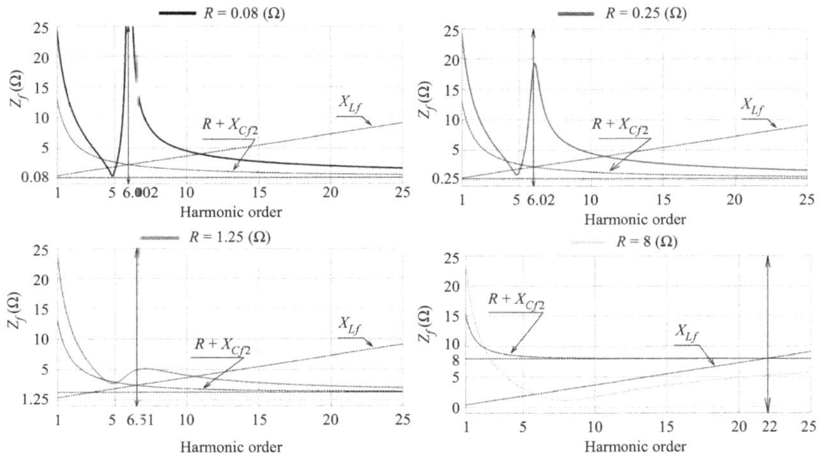

Figure 3.26 Filter impedance frequency characteristics; R = (a) 0.08 Ω, (b) 0.25 Ω, (c) 1.25 Ω and (d) 8 Ω. X_{Cf2} – capacitor reactance, $R + X_{Cf2}$ – impedance (Z_{RCf2}) of the branch containing the damping resistance

harmonic) and its characteristic looks like the one of single-tuned filter. The filter loses the ability to filter the fifth-order harmonic and harmonics in wide range.

The highest value of damping resistance should be chosen in a way to maintain the series and parallel resonances to certain level (not to totally damp the resonances). The smaller is the damping resistance the better is the reduction level of fifth harmonic and harmonics in wide range at the PCC. The seventh harmonic impedance has decreased with the damping resistance increase.

Figure 3.26 represents a comparative study between the filter reactor reactance (X_{Lf}) and capacitor reactance (X_{Cf2}) with damping resistance (R) in term of current harmonic flow and power losses. In Figure 3.26(a), it can be observed that for the harmonics between the first and the sixth, Z_{RCf2} (X_{Cf2}+ R) is higher than X_{Lf}; therefore, the big part of current harmonic will flow through X_{Lf} and Z_{RCf2} will generate less power losses than X_{Lf}. From the sixth harmonic to the 25th, X_{Lf} is higher than Z_{RCf2}, therefore the big part of current harmonic will flow through Z_{RCf2} (more power losses). Figure 3.26(b)–(d) can be interpreted in the same way.

After the filter connection ($R = 80$ mΩ), the amplitude of the 7th, 11th, 13th and 17th PCC voltage and current harmonics has been amplified by the filter (Table 3.7) because they are near the parallel resonance zone according to frequency characteristics of Figure 3.25(b).

With the increase of damping resistance from 0.08 to 1.25 Ω, the grid voltage waveform has improved very small commutation notches (Figure 3.27(a)). But for the large damping resistance (8 Ω), the commutation notches have come back and that means the filter has lost his effectiveness.

Table 3.7 Grid voltage and current parameters

n	U_{smax} (V)/I_{smax} (A)									
			Third-order filter							
	Without filter		$R = 80$ (mΩ)		$R = 250$ (mΩ)		$R = 1{,}250$ (mΩ)		$R = 8{,}000$ (mΩ)	
	Voltage	Current	Voltage	Current	Voltage	Current	Voltage	Current	Voltage	Current
First	319.97	15.86	322.91	8.46	322.91	8.48	322.90	8.52	322.90	8.52
Fifth	6.28	5.34	3.84	3.27	4.10	3.49	6.68	5.69	9.03	7.69
Seventh	0.38	0.24	0.65	0.41	0.62	0.39	0.46	0.29	0.28	0.18
11th	3.48	1.42	8.11	3.32	7.54	3.09	4.88	2.00	1.84	0.75
13th	0.77	0.27	4.29	1.49	3.11	1.08	1.08	0.37	0.41	0.14
17th	2.89	0.77	4.56	1.22	4.27	1.14	2.75	0.73	1.68	0.45
19th	1.17	0.28	0.99	0.24	0.98	0.23	0.84	0.20	0.67	0.16
23rd	2.65	0.52	1.31	0.26	1.30	0.26	1.33	0.26	1.54	0.30
25th	1.4	0.25	0.50	0.09	0.51	0.09	0.57	0.10	0.80	0.15
29th	2.52	0.4	0.66	0.10	0.66	0.10	0.81	0.13	1.42	0.22
THD (%)	4.93	36	3.45	59.97	3.19	58.36	2.82	71.66	3.36	91.15

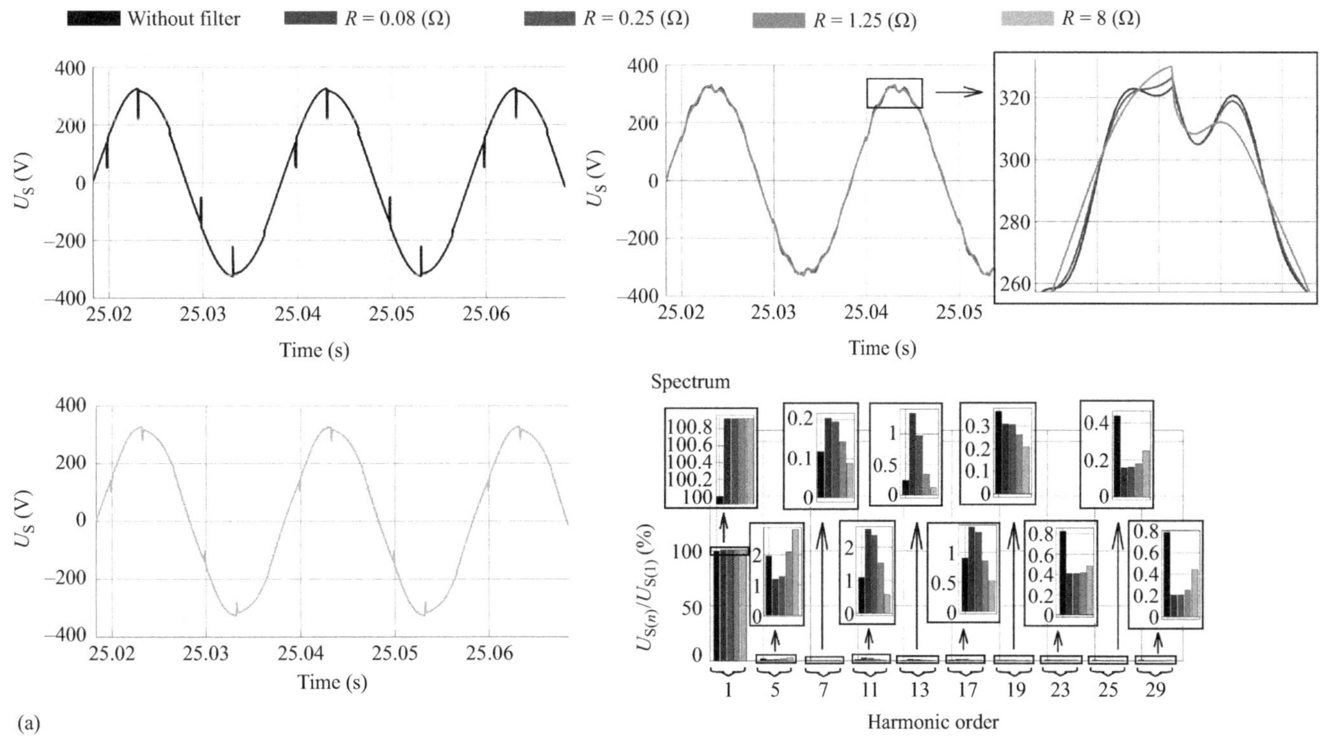

Figure 3.27 PCC voltage (a) and current (b) and their spectrums (p.u.)

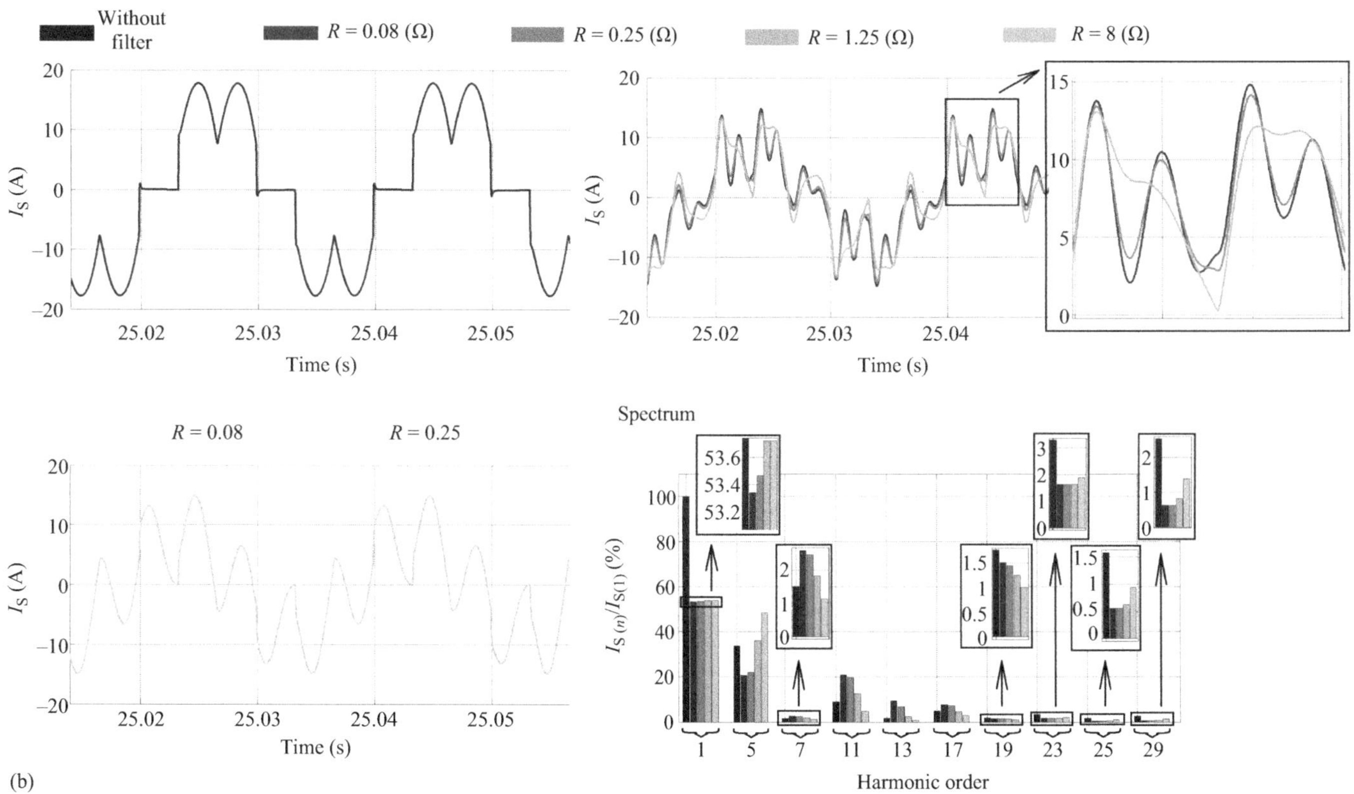

Figure 3.27 (Continued)

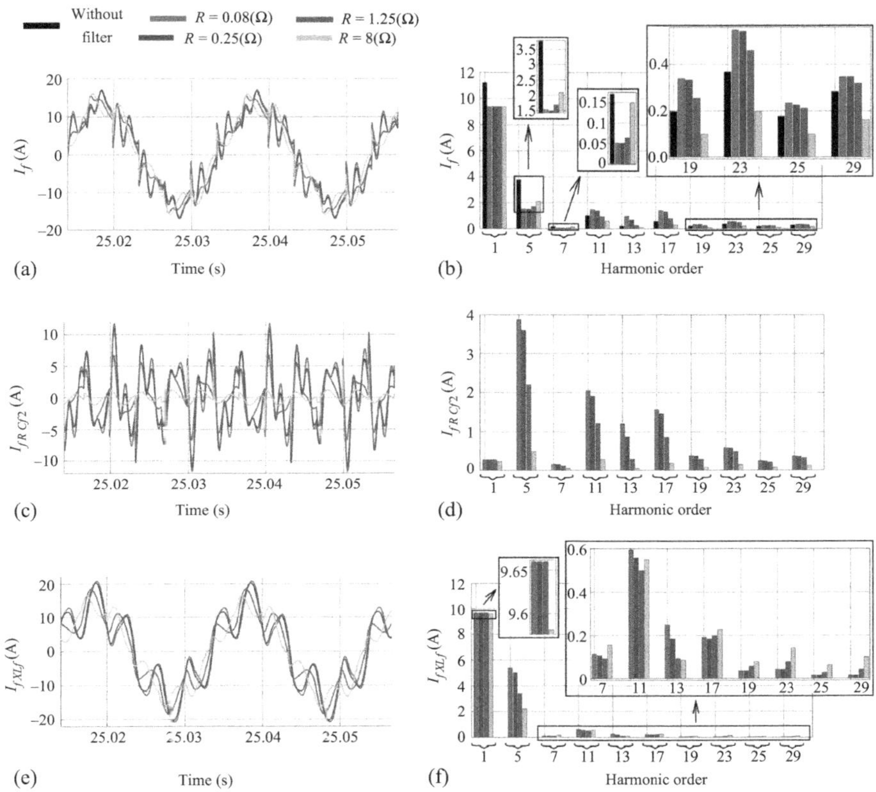

Figure 3.28 *Filter current (a) and (b) its spectrum (the harmonics flowing through the filter are compared to the harmonics generated by the load – without filter); (c) filter damping resistance current (I_{fRCf2}) and (d) its spectrum; (e) filter reactor current (I_{fXLf}) and (f) its spectrum*

From Figure 3.27(a) it can be concluded that the percentage of the PCC voltage (fundamental harmonic) has increased little (not considerably) after the filter connection and is the same during the damping resistance increased.

The percentage of the PCC current (fundamental harmonic) has considerably decreased after the filter connection (because of the compensation) and has not considerably increased during the damping resistance increased (Figure 3.27(b)).

In Figure 3.28 it can be observed the filter current waveforms and their spectrums. The spectrum of the current flowing through the capacitor C_{f1} is presented in Figure 3.28(b), the spectrum of the current flowing through the capacitor C_{f2} in Figure 3.28(d) and the spectrum of current flowing through the reactor L_f in Figure 3.28(f).

The current fundamental harmonic flowing through the filter reactor is higher than the one flowing through the damping resistance as well as the fifth-order

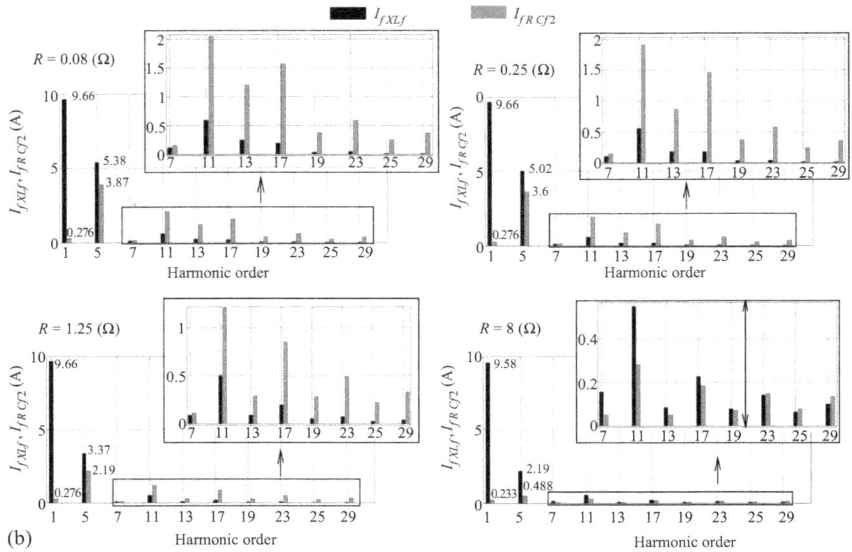

Figure 3.29 Spectrums of current flowing through the filter damping resistance R and reactor reactance X_{Lf} (a) and impedance characteristic measured at the input terminals of thyristor bridge (b) for different filter damping resistance R

harmonic (Figure 3.29(a)). The amplitude of harmonics (from the seventh) flowing through the damping resistance is higher than the amplitude of those flowing through the reactor (for 0.08, 0.25 and 1.25 Ω); therefore, from the power losses point of view, the damping resistance generates more power losses (for the high harmonics).

Between 200 and 250 Hz and between 500 and 1,000 Hz it can be observed parallel and series resonances (Figure 3.29(b)). These resonances are damped when the filter damped resistance increases.

The filter effectiveness presented in Figure 3.30 shows that the filter connection has amplified the amplitude of 11th (for 0.08 and 0.25 Ω), 13th, 17th, 19th 23rd, 25th and 29th (for 0.08, 0.25 and 1.25 Ω) harmonics of current. The third-order filter loses its efficiency for the high damping resistance (i.e. 8 Ω) and with the small damping resistance the harmonic near the parallel resonance can be amplified.

3.4.3.4 C-type filter

The filter which the equivalent circuit and characteristics examples are shown in Figure 3.31 is a special type of third-order filter [5]. Its main advantage is the power losses reduction as well as exploitation costs in comparison to other wideband filter structures [13,21,25]. The filter design must be performed in such a way that less fundamental harmonic current flows through the damping resistance (R) connected

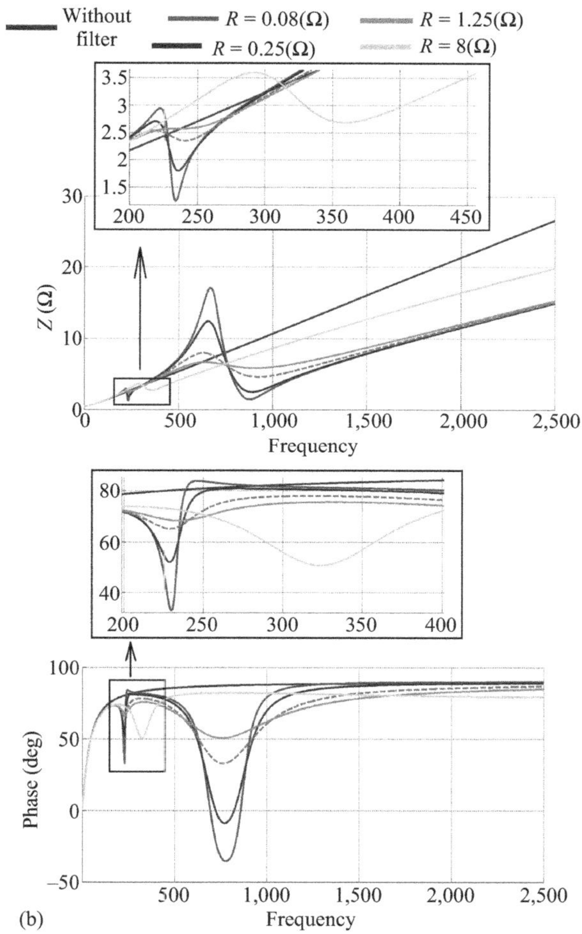

Figure 3.29 (Continued)

in parallel with the branch containing reactor L_f and the capacitor C_{fb}. Less current (or almost no current) of fundamental harmonic will flow through the damping resistance if the impedance Z_{LfCfb} is tuned to resonance frequency of fundamental harmonic (Figure 3.31(b)).

For high damping resistance, the filter impedance frequency characteristic is like the one of single-tuned filter (Figure 3.31(c), 2.5 kΩ) and for the small values, it is similar to the one of first-order filter (Figure 3.31(c), 1.25 Ω). On the one hand, when the damping resistance value is high, the fifth harmonic is better filtered and the higher harmonics amplitude are worse filtered. On the other hand, when the damping resistance is small, the situation is the opposite.

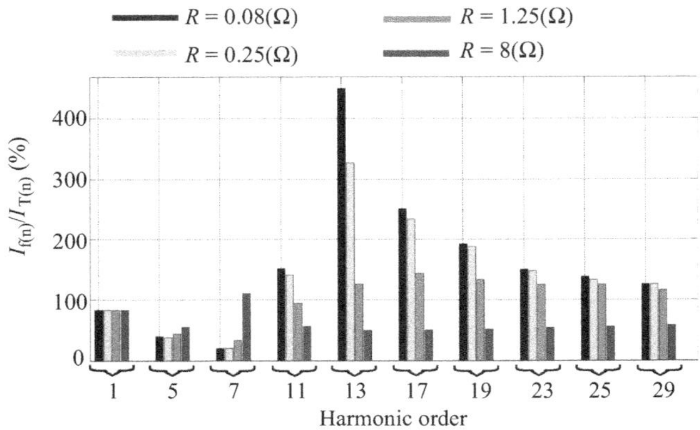

Figure 3.30 Filter effectiveness

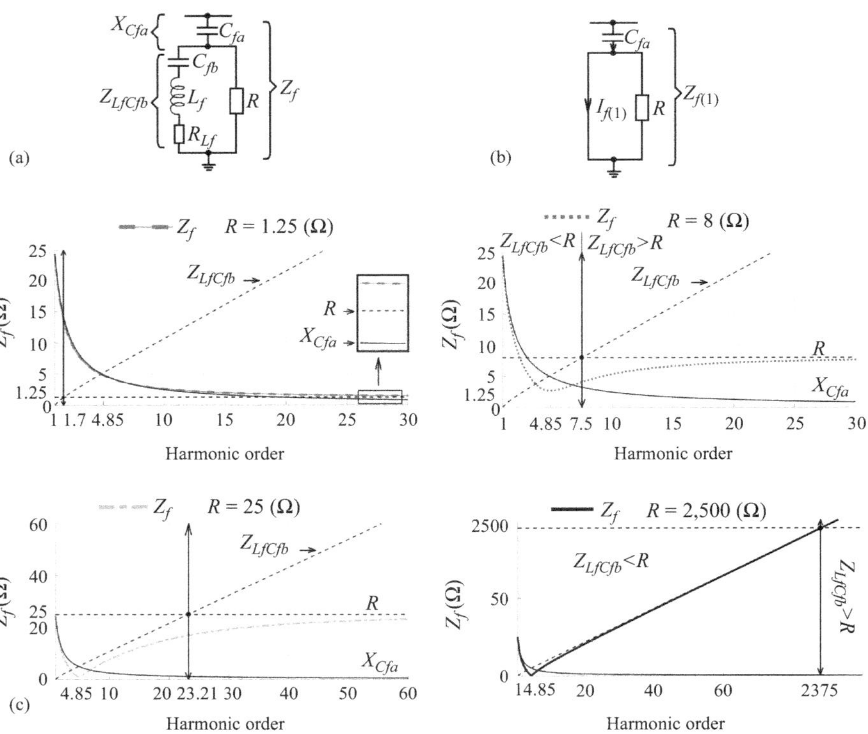

Figure 3.31 C-type filter: (a) equivalent circuit; (b) equivalent circuit for the fundamental harmonic; and (c) impedance frequency characteristics for different damping resistance R

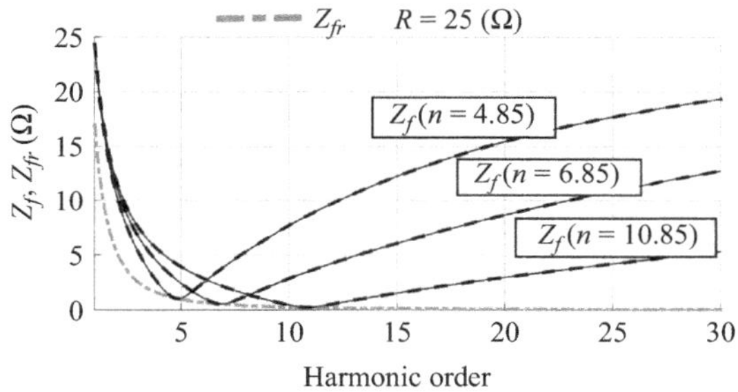

Figure 3.32 Impedance frequency characteristic for different tuned frequencies. Z_{fr} – filter impedance for the resonance frequencies

Figure 3.31(c) sets forth the comparison between the impedance Z_{LfCfb} (impedance of branch containing L_f and C_{fb}) and the damping resistance in point of view of quantity current harmonic which flows. For example, in Figure 3.31(c) (8 Ω), from the first to the seventh harmonic ($Z_{LfCfb} < R$) more current harmonic flows through the L_f C_{fb} branch than the R branch. From the seventh to the 30th harmonic ($Z_{LfCfb} > R$), more current harmonic flows through R branch than the L_f C_{fb} branch. The same interpretation can be formulated for Figure 3.31(c) (25 Ω and 2,500 Ω).

The filter can be tuned to any frequencies higher than the fundamental (ω_r). For the two identical C-type filter tuned respectively to the fifth and the seventh harmonic, the resonance filter impedance for the seventh harmonic is lower than for the fifth harmonic (Figure 3.32).

The PCC voltage waveform in Figure 3.33(b) (without filter) presents the commutation notches. When the damping resistance is small (e.g. 1.25Ω), the commutation notches are reduced considerably (Figure 3.33(b)) as well as the high harmonics (from 17th to 29th). But the fifth-, seventh- and 11th-order harmonics have been amplified (Table 3.8).

For high damping resistance (e.g. 8 Ω, 25 Ω and 2,500 Ω), the filter has difficulties to reduce the commutation notches (Figure 3.33(b)) as well as high harmonics amplitude (from seventh to 29th). But more efficiency is observed for the fifth harmonic amplitude reduction (PCC) and current waveform improvement (Figure 3.33(a)).

The filter is less charged by the higher harmonics (from seventh to 29th) and more charged by the fifth order harmonic (Figure 3.34(b)) when the damping resistance is high (e.g. 2,500 Ω). The fundamental harmonic flowing through the filter flows largely through the L_fC_{fa} branch (Figure 3.34(d) and (f)). The quantity of current harmonic (apart of the fundamental) flowing through the damping resistance is high when its value (e.g. 1.25 Ω) is small, and small when its value is high (Figure 3.34(f)). The L_fC_{fa} branch is more charged by harmonics when the damping resistance increases (Figure 3.34(d)).

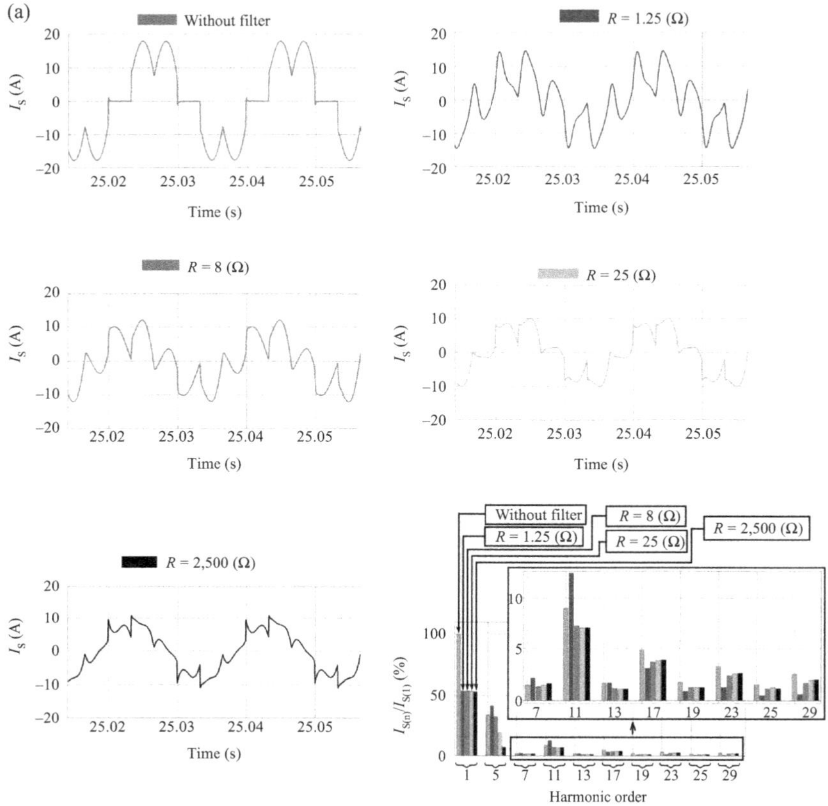

Figure 3.33 PCC current (a) and (b) voltage and their spectrum

In Figure 3.35, it can be seen the variation of current harmonics flow between the parallel filter branches depending on the mutual relation between Z_{LfCfb} and R. Increasing the damping resistance value, increases the harmonic power losses on the L_fC_{fb} branch (Figure 3.35(d)) and decreasing its value, decreases the harmonic power losses on it (Figure 3.35(a)).

Compared to other damped filter, the C-type filter is characterized by smaller power losses for the fundamental harmonic. With high damping resistance, the C-type filter is more capable to reduce the amplitude of harmonic to which it is tuned and then improve the grid current waveform and THD. The grid voltage waveform is then not improving and commutation notches remain almost the same.

With small damping resistance, the C-type filter is more capable to reduce the amplitude of harmonics (in wide range) higher than the one to which it is tuned and then improve the grid voltage waveform and THD (reduction of line notches [5]). The grid current waveform is then not improving because the amplitude of frequency to which it is tuned remains high.

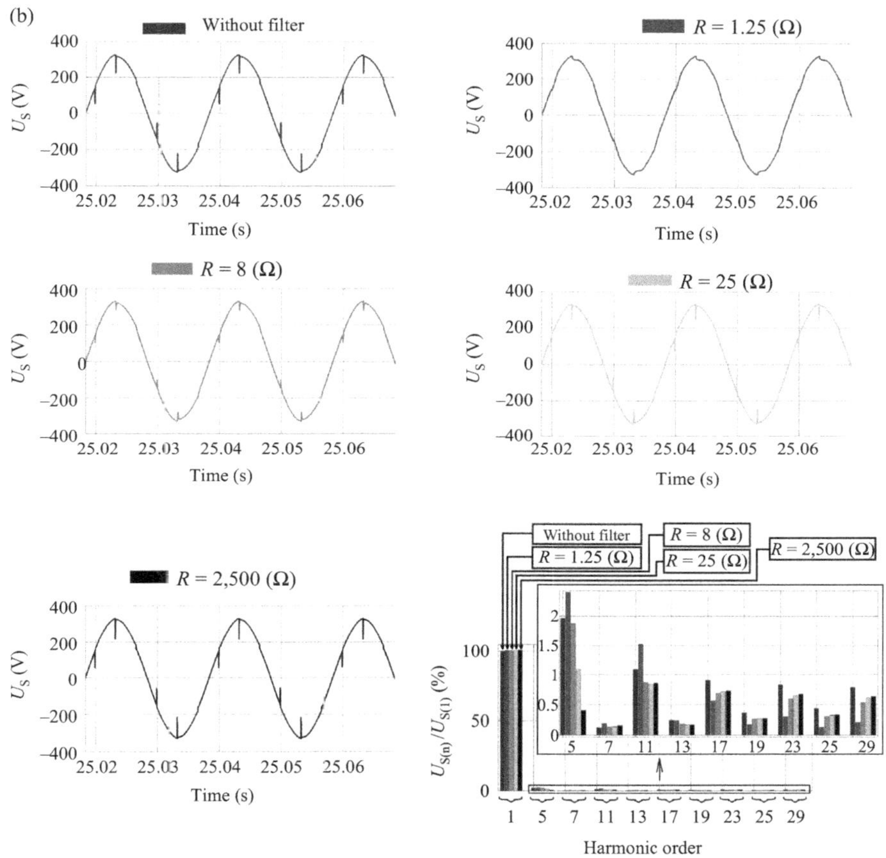

Figure 3.33 (Continued)

3.5 Hybrid passive filter

In high power electronic systems (generally with diodes and thyristor bridge), the passive filter in hybrid configuration is applied as input interface. The HPFs are the result of series and parallel passive filters connection, and their purpose is to reduce the rectifier input current distortions in the range up to 2.5 kHz. The simple configuration of this filter is presented in Figure 3.36. The hybrid passive filter under studies is constituted of SPF and group of single-branch filters. In the considered example, the group of single-branch filter is constituted by two parallel LC filters. The SPF is designed to block the harmonic to be eliminated forcing this latest to flow, throw the filters group which presents lower impedance for this harmonic. The effectiveness of the group of single-branch filters (operating alone

Table 3.8 Grid voltage and current parameters

N	U_s (V)/I_s (A)									
			C-type filter							
	Without filter		**$R = 1.25$ (Ω)**		**$R = 8$ (Ω)**		**$R = 25$ (Ω)**		**$R = 2{,}500$ (Ω)**	
	Voltage	**Current**	**Voltage**	**Current**	**Voltage**	**Current**	**Voltage**	**Current**	**Voltage**	**Current**
First	319.97	15.8	322.76	8.51	322.72	8.50	322.74	8.48	322.78	8.43
Fifth	6.28	5.34	7.69	6.57	5.99	5.12	3.49	3.00	1.30	1.14
Seventh	0.38	0.24	0.61	0.34	0.42	0.22	0.44	0.24	0.50	0.26
11th	3.48	1.42	4.88	1.98	2.77	1.14	2.70	1.11	2.74	1.11
13th	0.77	0.27	0.76	0.27	0.58	0.19	0.53	0.18	0.53	0.18
17th	2.89	0.77	1.80	0.49	2.20	0.59	2.28	0.61	2.33	0.62
19th	1.17	0.28	0.54	0.14	0.84	0.20	0.86	0.20	0.87	0.20
23th	2.65	0.52	0.96	0.20	1.91	0.38	2.07	0.41	2.14	0.42
25th	1.4	0.25	0.41	0.08	0.97	0.18	1.04	0.20	1.07	0.19
29th	2.52	0.4	0.65	0.10	1.71	0.27	1.96	0.31	2.03	0.32
THD (%)	4.93	36	2.96	81.11	2.87	62.70	3.30	39.79	3.84	23.28

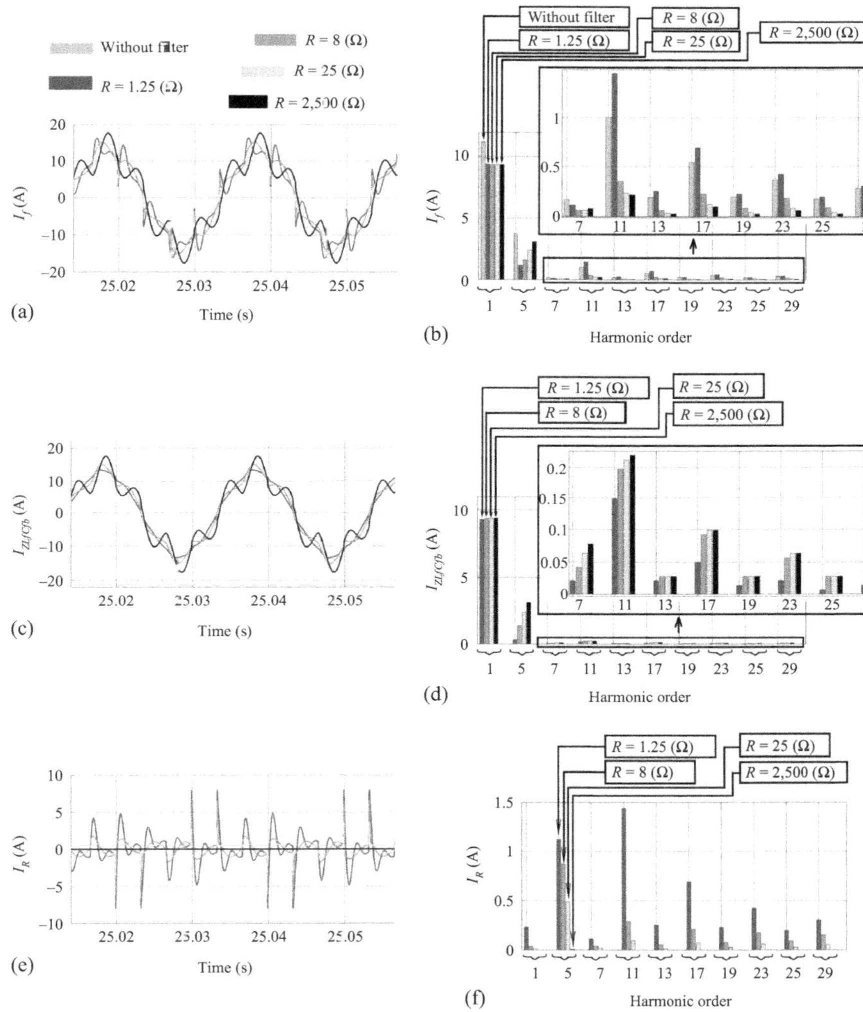

Figure 3.34 Filter current (a) and (b) its spectrum (the harmonics flowing through the filter are compared to the harmonics generated by the load – without filter); (c) filter current of the Z_{LfCfb} branch and (d) its spectrum; (e) damping resistance current (I_{fR}) and (f) its spectrum

in the system) to eliminate the harmonics depends on the grid impedance for these harmonics, but with the SPF this dependence is reduced.

In the considered example, SPF is tuned to the frequency of fifth-order harmonic and the group of parallel connected filters is tuned to the frequencies a little bit lower than the harmonic frequencies to be eliminated (fifth and seventh) (Figure 3.37).

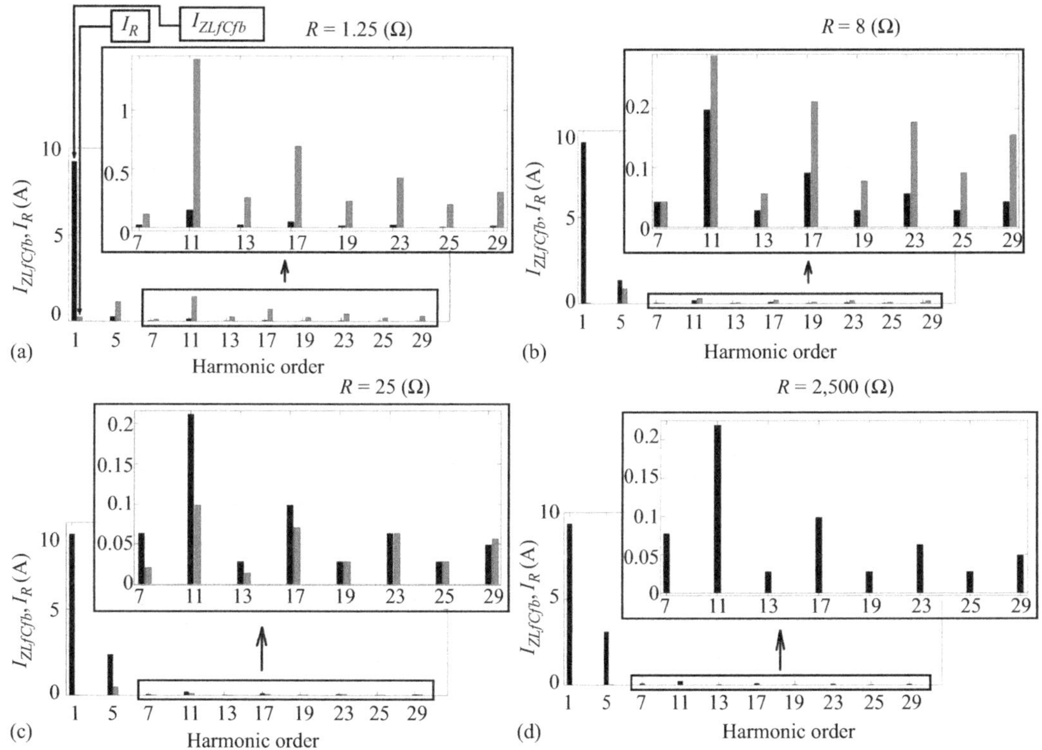

Figure 3.35 Spectrums of current flowing through the resistance R and reactance X_L for R = (a) 1.25 Ω, (b) 8 Ω, (c) 25 Ω and (d) 2,500 Ω

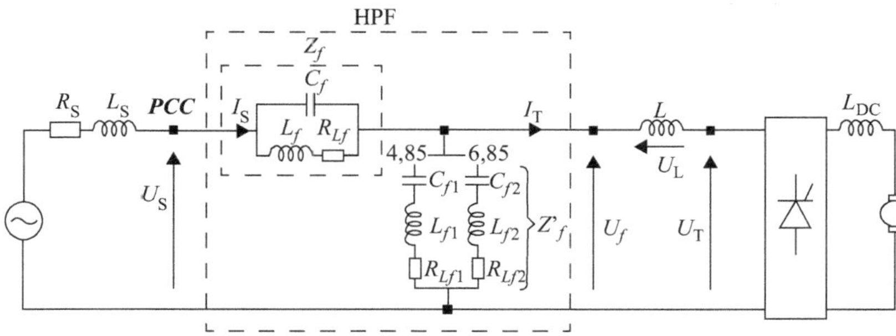

Figure 3.36 System equivalent circuit with the hybrid passive filter

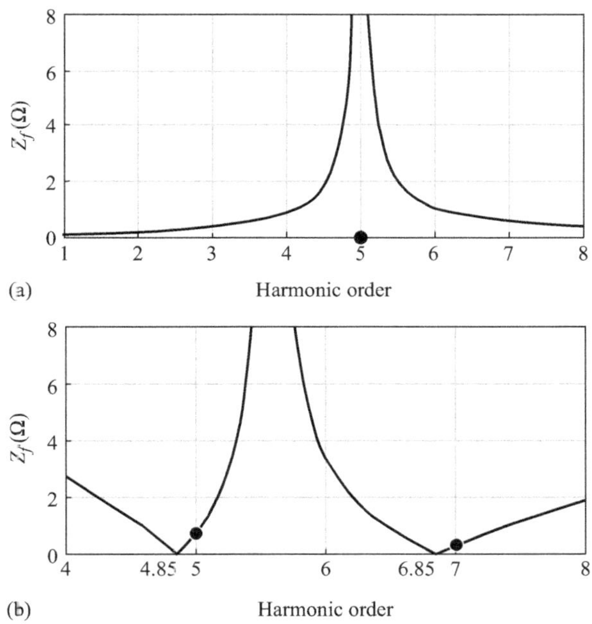

Figure 3.37 Impedance frequency characteristics: (a) SPF, (b) filters group

The comparison (based on the system parameters) between the SPF and the SPF with filters group (HPF) is presented in Figures 3.38–3.40 and Table 3.9.

The parallel connection of the filters group together with the SPF improves the grid current waveform (Figure 3.38(c)); the THD of grid current and voltage at PCC as well as the THD of voltage at the input terminals of thyristor bridge (Figure 3.39(d) and Table 3.9). The harmonics amplification (7th–25th) observed in the case of SPF is not observed in the case of HPF (Figure 3.38(b) and (d) and Figure 3.39(b)) The waveforms of rectifier input voltage for the SPF and for the

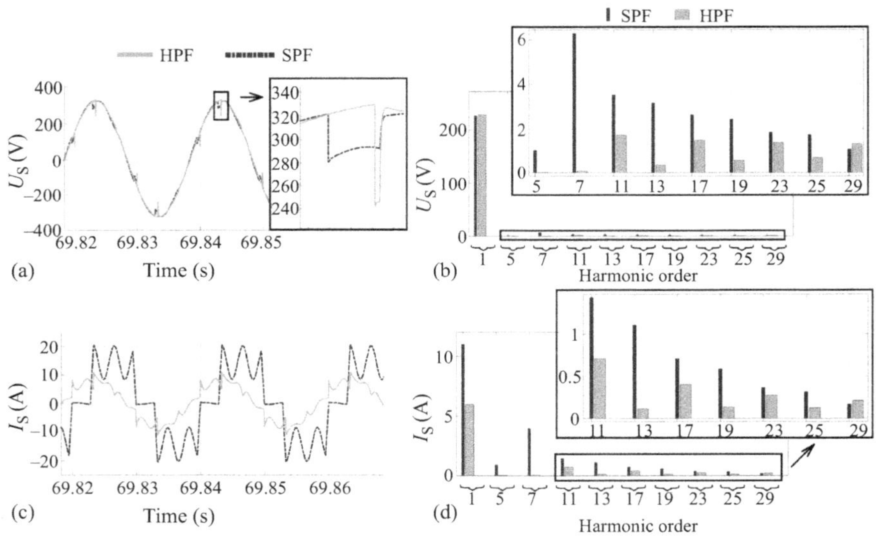

Figure 3.38 *(a) Grid voltage and (b) its spectrum; (c) grid current and (d) its spectrum*

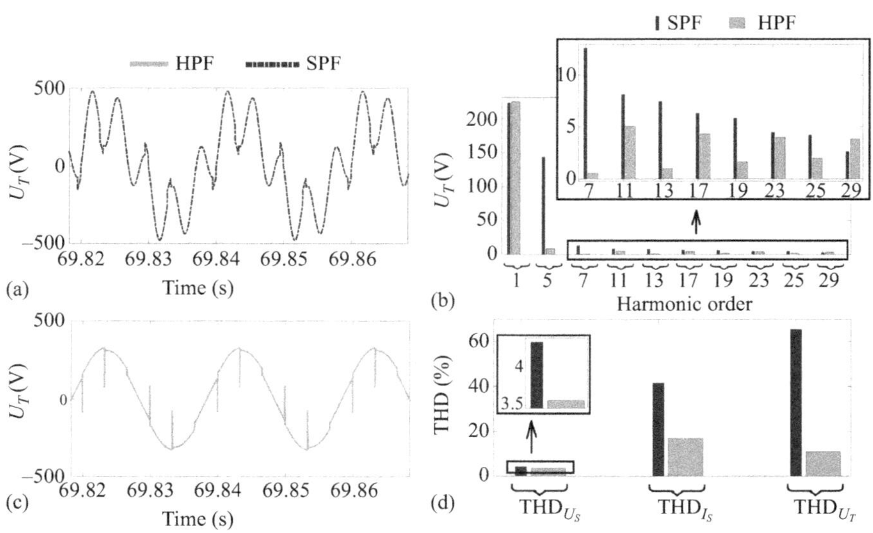

Figure 3.39 *(a) and (c) Voltage at the rectifier input and (b) its spectrum; (d) THD of grid current (I_S) and voltage (U_S) and voltage at the rectifier input (U_T)*

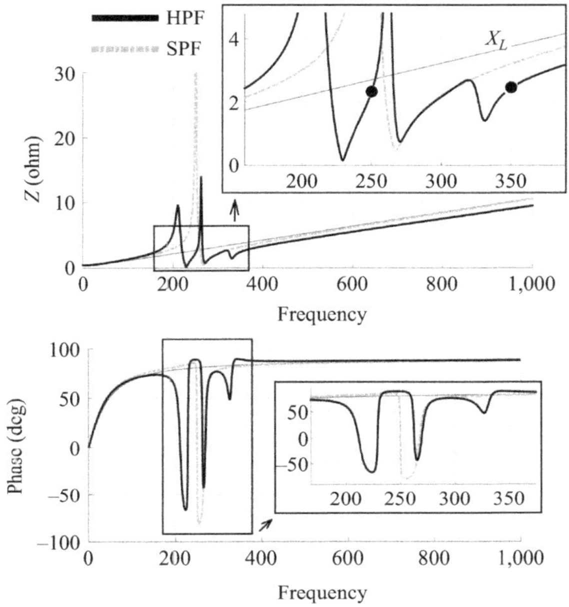

Figure 3.40 Characteristics of system impedance versus frequency

Table 3.9 Grid voltage parameters at PCC

n	U_s (V)		I_s (A)	
	SPF	**HSPF**	**SPF**	**HSPF**
First	319.85	322.91	15.53	8.43
Fifth	1.47	0.03	1.25	0.02
Seventh	8.89	0.11	5.59	0.07
11th	4.97	2.45	2.03	1.00
13th	4.47	0.49	1.56	0.17
17th	3.75	2.13	1.00	0.57
19th	3.45	0.81	0.83	0.19
23rd	2.63	1.97	0.52	0.39
25th	2.47	1.00	0.45	0.18
29th	1.54	1.88	0.24	0.30
THD (%)	4.31	3.56	41.60	17.10

HPF are respectively presented in Figure 3.39(a) and (c). The SPF shows a better reduction of commutation notches than the HPF (Figure 3.38(a)).

The system impedance (Z) and phase versus frequency are shown in Figure 3.40. The SPF presents the highest power system impedance for the parallel resonance.

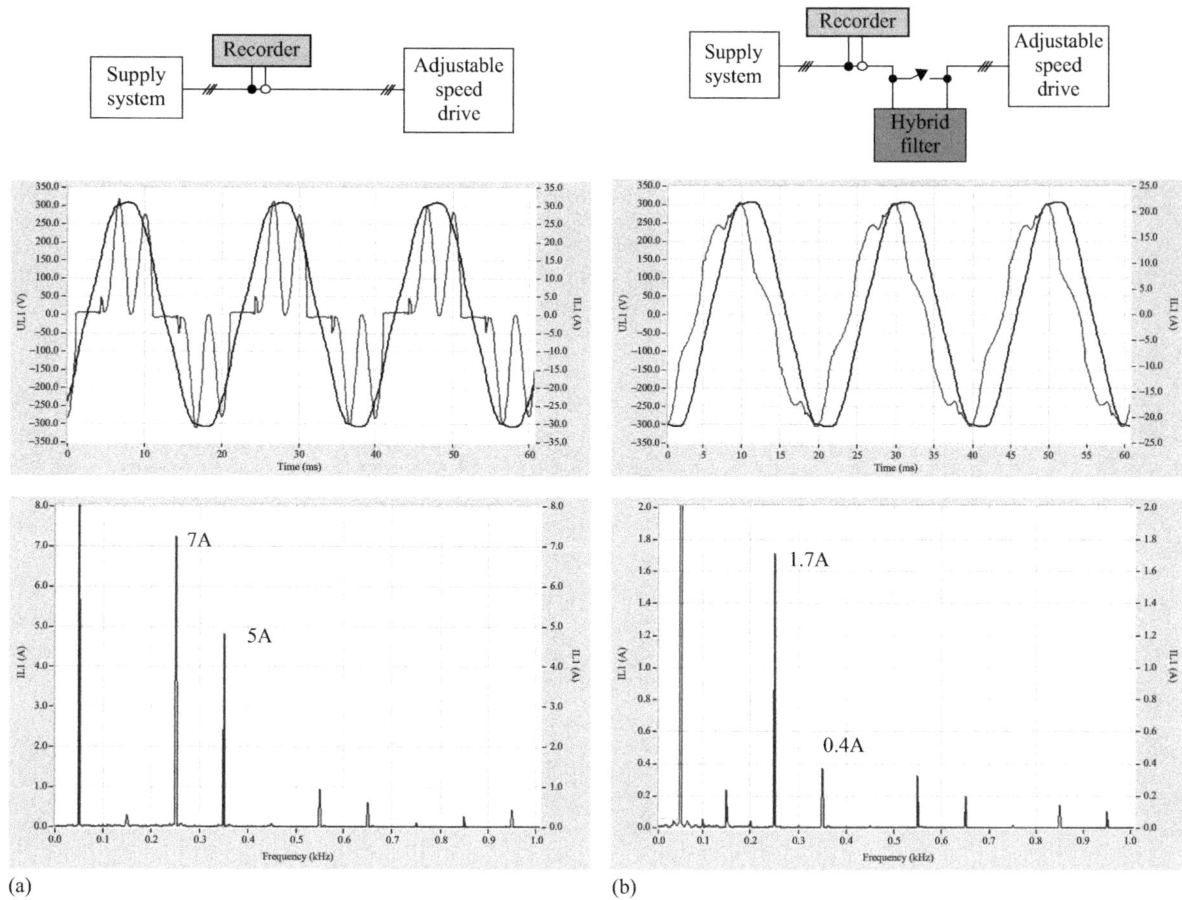

Figure 3.41 Waveforms of voltage and current (spectrum) measured at the power system input without filter (a); (b) with the input HPF

The HPF has better filtered the fifth-order harmonic of grid current and voltage than the SPF and can reduce the problems (harmonics amplification and waveform of grid current and voltage at the rectifier input) caused by the SPF operating alone.

Figure 3.41(a) and (b) presents the waveforms and spectrums of voltage and current measured in the real power system where the thyristor bridge input was equipped with hybrid passive power. It can be observed, a significant reduction of fifth and seventh grid current harmonics (Figure 3.41(b)). The phase shift of voltage and current waveforms after the filter connection (over compensation – fundamental harmonic) is due to lower than the nominal current of the filter load.

3.6 Conclusion

The knowledge of power passive filters has been created for many decades. Although it appears to be very rich, as indicates the experience of the authors, still in the process of design often too little attention is paid to the analysis of changes filtration properties of these systems [11,15]. They vary due to many factors, including for example aging or manufacturing tolerances of the reactors and capacitors during the manufacturing process [20,26]. The aim of this chapter is the presentation of the properties of various topologies of the passive filters using the impedance frequency characteristics [18,19].

It is worth noting that more and more the minimizing of the cost of filtration systems is achieved through the application of a hybrid filters which are a combination of passive and active parts. Presented considerations gain so more contemporary significance.

References

[1] Anthoula M., Thomas Z., Milias-Argitis J.: Optimal sizing and limitations of passive filters in the presence of background harmonic distortion, *Electrical Engineering*, August 2009, 91, 2, pp. 89–100.

[2] Anthoula M., Thomas Z., Milias-Argitis J.: Optimal design of C-type passive filters based on response surface methodology for typical industrial power systems, *Journal of Electric Power Components and Systems*, 2013, 41, 7, pp. 653–668.

[3] Arrillaga J., Bradley D.A., Bodger P.S.: *Power System Harmonic*, John Wiley & Sons, Chichester,1985.

[4] Azebaze M.C.S., Hanzelka Z.: Different approaches for designing the passive power filters, *Przegląd Elektrotechniczny*, November 2015, ISSN 0033-2097, 91, 11, pp. 102–108.

[5] Azebaze M.C.S., Hanzelka Z.: Application of C-type filter to DC adjustable speed drive, *IEEE International Conference – Modern Electric Power Systems*, Wroclaw, Poland, 6–9 July 2015, pp. 1–7.

[6] Czarnecki L.S., Ginn H.L.: The effect of the design method on efficiency of resonant harmonic filters, *IEEE Transactions on Power Delivery*, 2005, 20, 1, pp. 286–291.

[7] Czarnecki L.S., Ginn H.L.: An optimization based method for selection of resonant harmonic filter branch parameters, *IEEE Transactions on Power Delivery*, 2006, 21, 1, pp. 1445–1451.

[8] Fang Z.P., Gui-Jia S., George F.: A series LC filter for harmonic compensation of AC drives, *IEEE Power Electronics Specialists Conference*, August 1999, pp. 213–218.

[9] Hanzelka Z.: Selected problems of passive harmonic filters design, INFOTEC Conference, Jurata, Poland, 23–26 June 2010, pp. 45–60 (in Polish).

[10] He N., Huang L., Wu J., Xu D.: Study on optimal design method for passive power filters set at high voltage bus considering many practical aspects, *IEEE Applied Power Electronics Conference and Exposition*, February 2008, Austin, TX, pp. 396–401.

[11] Hiranandani A.: Calculation of cable ampacities including the effects of harmonics, *IEEE Industry Applications Magazine.* March/April 1998, 4, pp. 42–51.

[12] James K.P.: A transfer function approach to harmonic filter design, *IEEE Industry Application Magazine*, January/February 1997, 3, pp. 68–82.

[13] Klempka R.: A new method for the C-type passive filter design, *Przegląd Elektrotechniczny*, 2012, ISSN 0033-2097, 88, 7, pp. 277–281.

[14] Klempka R.: *Improving the quality of power supply with the use of genetic algorithms (in Polish)*, Krakow, AGH University of Science and Technology Press, 2013 (in Polish).

[15] Klempka R., Hanzelka Z., Varetsky Y.: Bank harmonic filters operation in power supply system – cases studies, Intech. 17 April 2013, power quality issues, ed. Zobaa A. F, *Rijeka Intech*, ISBN 980-953-307-532-2, pp. 201–223.

[16] Klempka R.: Designing a group of single-branch filters taking into account their mutual influence, *Archives for Electrical Engineering*, March 2014, 63, 1, ISSN 1427–4221, pp. 81–91.

[17] Muhammad R., Muhammad I., Danang S.: Single tuned harmonic filter as total harmonic distribution (THD) compensator, CIRED, Lyon, 15–18 June 2015, p. 0056.

[18] Micheal D., Nirod C.S.: On resonance and frequency response characteristics of electrical circuits, *International Journal of Electrical Engineering Education*, October 2013, 50, 4, pp. 368–383.

[19] Sakar S., Abdel Aleem S.H.E., Balci M.E., Zobaa A.F.: A filter design approach to maximize ampacity of cables in nonsinusoidal power systems, *IEEE International Conference*, ISNCC, 15–18 June 2015, Lagow, Poland, pp. 1–4.

[20] Sakar S., Karaoglan D.A., Balci E.M., Aleem A.E.H.S.A., Zobaa F.A.: Optimal design of single-tuned passive filter using response surface methodology, *IEEE International Conference*, ISNCC, 15–18 June 2015, Lagow, Poland, pp. 1–6.

[21] Shady H.E.A.A., Zobaa A.F., Mohamed M.A.A.: Optimal C-type passive filter based on minimization of the voltage harmonic distortion for nonlinear loads, *IEEE Transactions on Industrial Electronic*, January 2012, 59, 1, pp. 281–289.

[22] Singh B., Chandra A., Al-Haddad K.: *Power Quality: Problems and Mitigation Techniques*, John Wiley & Sons Ltd., Chichester, West Sussex, 2015.

[23] Wagner V.E.: Effects of harmonics on equipment, *IEEE Transactions on Power Delivery*, 1993, 8, 2, pp. 672–680.

[24] Yi-hong H.E., Heng S.U.: A new method of designing double-tuned filter, *International Conference on Computer Science and Electronics Engineering*, Hangzhou, 22–23 March 2013.

[25] Yoa X., Jie Z., Shijie M.: Theory for the design of C-type filter, *IEEE International Conference on Harmonics and Quality of Power*, 12–15 September 2004, pp. 11–15.

[26] Zacharia P., Menti A., Zacharis Th.: Genetic algorithm-based optimal design of shunt compensators in the presence of harmonics, *Electric Power Systems Research*, April 2008, 78, pp. 728–735.

Chapter 4
Active harmonic filters

A.M. Sharaf[1]*, Foad H. Gandoman*[2] *and Behnam Khaki*[3]

4.1 Introduction

Over the last few decades, the demand for the use of electric energy on the one hand, and the use of renewable resources as green sources of electricity production on the other, have undergone change power network structure. Besides, using domestic and industrial nonlinear loads in terms of power quality have been affected on power grids. In addition, one of the most important drawbacks of nonlinear loads is injecting harmonic pollution to the electrical systems that has become a challenge for customer and utilities sides.

So far numerous tools and methods have been introduced to identify, assess, and reduce the harmonic pollution in the power systems. Traditionally, passive LC filters have been good solution for filtering harmonic distortion. However, use of passive filters for harmonic mitigation has some disadvantages, such as reactive power overcompensation at fundamental frequency parallel resonances with the network impedance, and finally poor plasticity for dynamic compensation of several frequency harmonic components.

Active power filters (APFs) have gained much more attention due to excellent performance (improve voltage balance, compensate current and voltage harmonics, compensate reactive power, and suppress flicker) to reduction the harmonic distortion and reactive power issues in the electrical power industry.

APFs family in unbalanced distribution network for load balancing, improve power factor, and eliminating current harmonics are used. APFs are classified in different structures:

- Shunt APF
- Series APF
- Hybrid APF.

The chapter will start with the basic principles of industrial loads models and characteristics. Then the discussion leads to APF topologies and design considerations.

[1]Sharaf Energy Systems, Inc., Fredericton, NB, Canada
[2]Technical and Vocational University, Sanandaj Branch, Kurdistan, Iran
[3]Smart Grid Energy Research Center (SGERC), University of California, Los Angeles, CA, USA

Types of APF that include shunt, series, and hybrid application are explained in the next section. Control techniques of APF to improve power quality are considered in the next section. The chapter will close with performance APF and hybrid-APF in two different case studies.

4.2 Industrial load models and characteristics

4.2.1 Dynamic and quasi-static harmonics in modern electrical networks

During last few decades, significant change has occurred in electricity grids, such as

- Increased efficiency and level of control electronic converters (soft starter and adjustable driver) in industrial networks.
- Using green energies in the small-scale modern power plant (distributed generations/DGs), commercial buildings (smart house), and electrical vehicles (EV).

The harmful effects of above-mentioned changes have been increasing harmonic pollution in the modern power systems [1]. Thus, in modern electrical network not only the utility side, but the customer side plays an important role to produce harmonic phenomena. The most common sources of harmonics are given in Table 4.1.

Assessment of harmonic, procedures, and mitigating in modern networks is a complicated issue. On the one hand, there are harmonics injected from utility side and other sources of harmonics of modern distribution networks using conventional methods to assess and reduce harmonic is handicapped [2]. As a result, introducing advanced methods to overcome the mentioned problems has been a big challenge for electrical utility and customer.

Moreover, in the modern electrical grids, harmonic pollution has following undesirable effects:

- Create overheating in electrical equipment such as cables, capacitors, transformers, and others,
- High voltages and circulating currents,

Table 4.1 Overview of harmonic sources

Harmonics sources		
Fast switching of power electronic devices	**Conventional sources**	**Modern equipment (nonlinear loads)**
Controlled rectifiers, uncontrolled rectifiers, inverters, cycloconverter, converters	Electrical rotating, transformers, reactors	Arce frances, static apparent, power controllers, variable speed drives, UPS, PC, TV, florescent lamps, photocopies machine, printer, microwave oven, air conditioner, blenders

- Equipment malfunctions due to excessive voltage distortion,
- Component failure and shortened life span due to increased internal energy losses in connected equipment,
- Wrong tripping of branch breakers,
- Metering errors,
- Fires in wiring and distribution systems due to overheating,
- Resulting in penalties on monthly utility bills due to lower system power factor.

4.2.2 Industrial nonlinear loads types and characteristics

Nowadays, power quality for electric power companies and consumers is of particular importance [3]. Manufacturers in the electrical industry efforts that the generate electricity has suitable power quality. However, in the electrical networks especially modern electrical network, some consumers that have nonlinear behavior have been changed power quality on electrical system. In other words, these types of consumers can destroy sinusoidal shape of current and voltage waveform [4,5].

Nonlinear industrial loads, as sources of harmonics in the electrical networks, play an important role in the elimination of power quality of modern electrical networks. These types of loads due to nonlinear behavior will change current or voltage sinusoidal shape waveform as well as harmonic generation in electrical systems [6]. Table 4.2 shows types of industrial nonlinear loads and contribution of each of them in an unbalanced industrial power quality.

According to Table 4.2, a laptop is the largest and most influential nonlinear loads with 83.8% THD of electric power quality. Laptops are big generator of third-order harmonic current in electrical distribution networks. A printer belongs to the second stage of important generator of harmonic pollution in the power system [8].

In general, nonlinear equipment produces harmonic currents and cause harmonic voltage on the impedance network. The harmonic currents are added together and create devastating effects on network operation as follows:

1. Malfunctioning control equipment and the affiliate system to fundamental frequency of network.
2. High losses in capacitors, transformers, motors, and rotating machines.

Table 4.2 Type of industrial nonlinear loads [7]

Nonlinear load	Power factor	THDi (%)
Fluoresce lamp	0.57	11.1
Freezer	0.44	61.8
Amplifier	0.71	48.1
Television	0.66	72.5
Photostat	0.68	69.9
Laptop	0.52	83.8
PC	0.53	60.1
Printer	0.49	83.6

3. Interference in telephone lines and control circuit that can be controlled remotely.
4. Lack of proper functioning of the measuring devices.

4.3 Active power filter topologies and design considerations

4.3.1 Active power filters use in AC and DC–AC power systems

Today, with the widespread use of electronic equipment and computer processors, as well as economic reasons, the harmonic distortion mitigation is very much taken into consideration [8]. One of the best solutions for filtering harmonic pollution is APF. Study in the field of APF has been started in 1971 by Professor Sasaki [9].

Studies on active filters to improve power quality in AC and DC–AC power systems of the following two aspects can be examined [10]:

1. Methods of detection active filter reference current which are divided into time domain and frequency domain methods.
2. Methods of flow control which can be divided into two main groups—linear and nonlinear methods.

Figure 4.1 illustrates a typical block diagram of an APF configuration, including components and their interconnections. Nonlinear loads play an important role to generate the transmitted harmonic current in the main line that is sensed and supplied to the reference current estimator. The control strategy for the pulse width-modulated (PWM) switching pattern is generated by the reference signal from the current estimator drives the overall system controller. The power circuit in the universal block diagram (Figure 4.1) can be connected in shunt or series as will be considered in APFs classification.

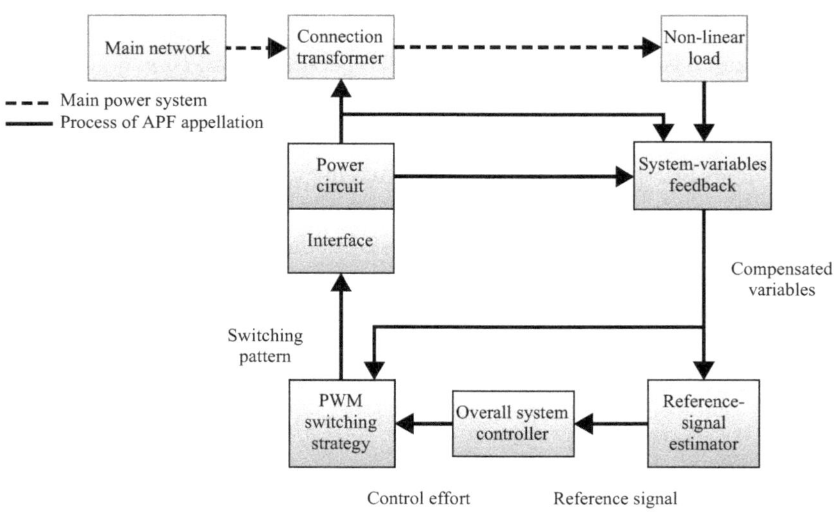

Figure 4.1 Application block diagram for APF

4.3.2 Active power filters—design issues and considerations

APF based on topology of electrical networks are designed and used into four types as follows: (a) two-wire (single-phase) system, (b) three-wire (three-phase) system, (c) three-phase three-wire with zigzag transformer, and (d) three-phase four-wire [11–14]. Figure 4.2 shows that comprehensive application in all parts of an APF will be considered in the design.

First important element of APF is insulated-gate bipolar transistor (IGBT) voltage source inverter (Figure 4.3) that is built three identical IGBT inverter legs. In addition, this part of the APF is designed to be capable of compensating unbalanced and single-phase loads.

The second part is sampling and control reference. The aim of this part is monitoring the current to the nonlinear load and then generating gating signals for the first part to create a current waveform to mitigate harmonics decoration. There are many control methods of APF that have been explained in Section 5 of this chapter. PWM firing pulse generation is the next part that necessary to be considered in the design. It is important to achieve an optimal design of various parameters that influence the mitigation of harmonics, also to discuss the calculation

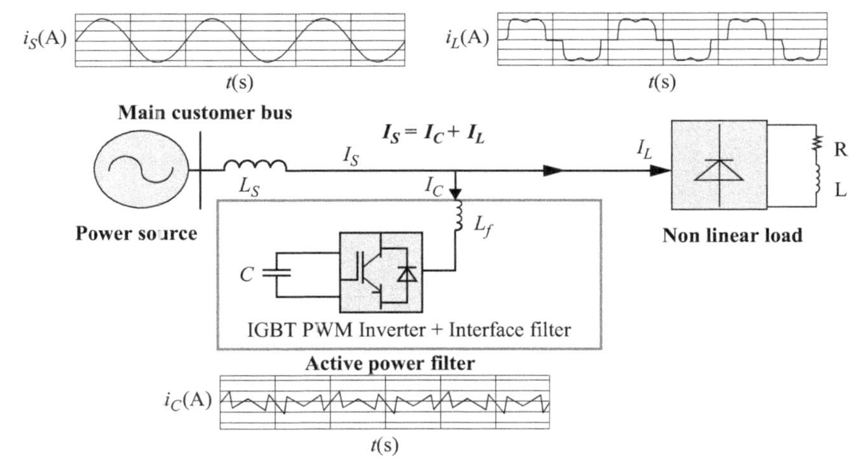

Figure 4.2 Basic compensation principle of APF and waveforms

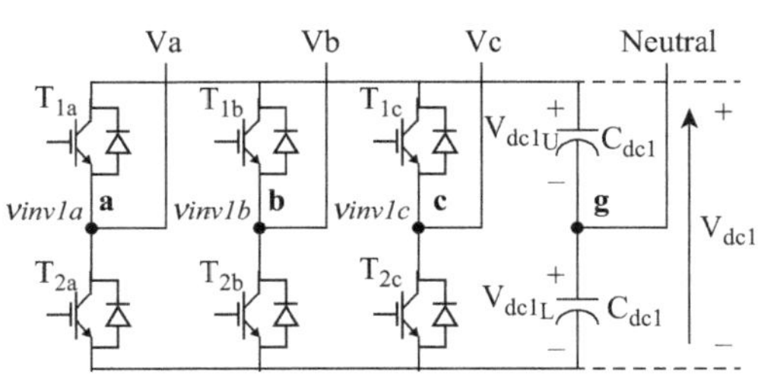

Figure 4.3 IGBT voltage source inverter

circuit current control circuit. Finally, the last element is system interface module that provides isolation between the output of the voltage source inverter and the power system in which the active filter is connected. It should be noted that the isolation consists a high-frequency attenuation filter formed by a small capacitor with two inductors.

4.3.3 Active power filters—industrial applications

During last few decades, in industrial low and medium voltage, APFs have been used to improve the supply quality. Moreover, an industrial site may include different types of linear loads and nonlinear loads. Figure 4.4 is an illustrated application of APF on different types of industrial nonlinear loads [15]. In this table, the

					Active filter rating (rms Amps/kVA of load)	
Type of load	Nonlinear load current waveform	Supply current with active filter	Nonlinear load current THD (%)	Supply current THD (%)	120/208 volt applications	480 volt applications
Single phase power supply	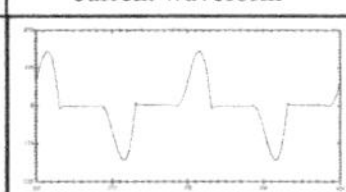	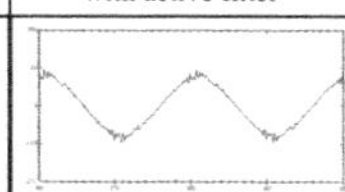	84%	8.2%	1.79	0.77
Semiconverter	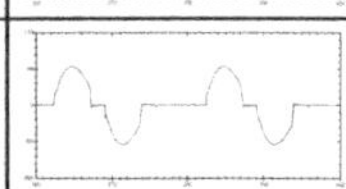	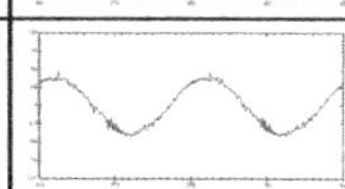	87%	9.6%	1.82	0.79
AC voltage regulator, RL load	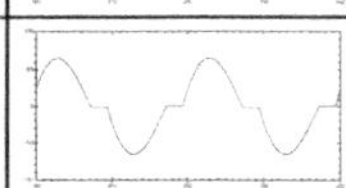	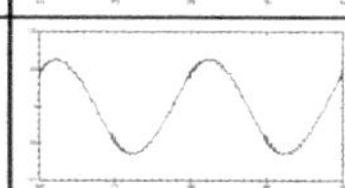	23%	4.3%	0.62	0.27
AC voltage regulator, resistive load	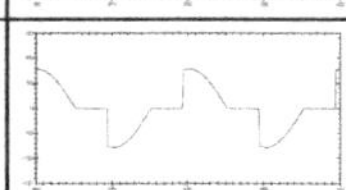	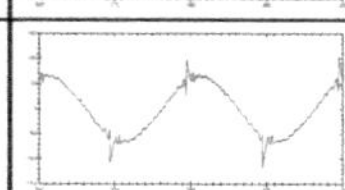	51%	16.7%	1.26	0.54
6 pulse drive, current source inverter	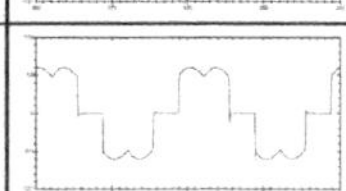	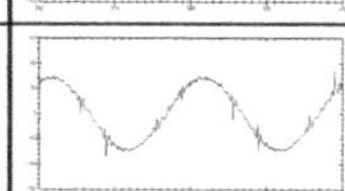	31%	10.3%	0.82	0.36
6 pulse drive, voltage source inverter no series inductance	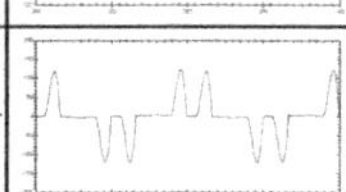	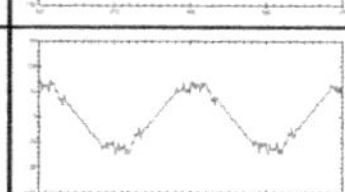	109%	13.2%	2.05	0.88
6 pulse drive, voltage source inverter 3% AC input choke	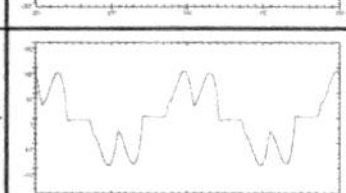	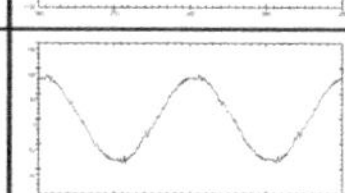	45%	5.0%	1.14	0.49
12 pulse converter, current source inverter	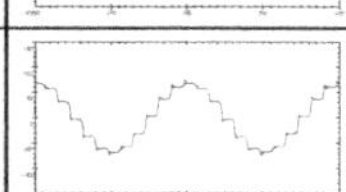	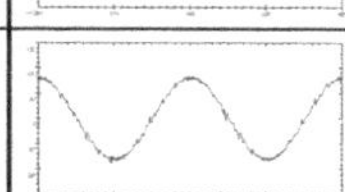	13%	5.6%	0.36	0.15
12 pulse converter, voltage source inverter	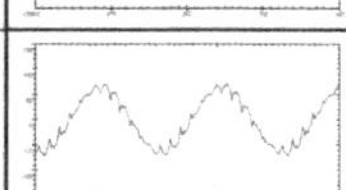	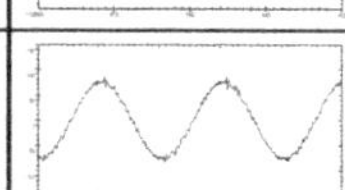	13%	6.3%	0.36	0.15

Figure 4.4 Application of APF on different types of industrial nonlinear loads [15]

first column has introduced industrial loads that have the greatest impact on harmonic pollution. Also, the second column shows the current waveform of the nonlinear loads. The next column illustrates network current waveform after filtering. The THD of nonlinear load and supply are shown in fourth and fifth columns, respectively. Finally, APF ratings for different types of voltage are displayed in last columns.

4.4 Active power filters configurations

4.4.1 Current source active power filters—CSC

This filter is featured by the inductance which is installed on the DC side of the inverter. The inductance is the DC-link energy storage that is connected to the main AC systems through a current source inverter and an interface passive filter.

The size of the inductance depends on the energy variation on the DC-link (ΔE_{DC}) and the variation (ripple) in the DC-link current $\Delta I_{\rm DC}$. If $I_{\rm DC}$ is the DC-link current, when the voltage across the inductance is positive, $I_{\rm DC}$ increases, and when the voltage is negative, $I_{\rm DC}$ decreases. Therefore, energy variation is obtained by [16].

$$\Delta E_{\rm DC} = 0.5 \cdot L_{\rm DC} \cdot \left[(I_{\rm DC} + \Delta I_{\rm DC})^2 - (I_{\rm DC} - \Delta I_{\rm DC})^2 \right] = 2 \cdot L_{\rm DC} \cdot I_{\rm DC} \cdot \Delta I_{\rm DC} \tag{4.1}$$

where $L_{\rm DC}$ is the inductance on the DC side. Now, it can be written as

$$L_{\rm DC} = \frac{\Delta E_{\rm DC}}{2 I_{\rm DC} \Delta I_{\rm DC}} \tag{4.2}$$

where $I_{\rm DC}$ is the DC current, and $\Delta I_{\rm DC}$ is the variation of the DC current. The relation can be also presented by the harmonic power (P_h) which is injected to the AC system by the APF [17].

$$L_{\rm DC} = \frac{P_h}{\omega I_{\rm DC} \Delta I_{\rm DC}} \tag{4.3}$$

In which, ω is the angular velocity of the main current. The current of the energy storage should be at least equal to the peak value of the compensating current. On the contrary, if the current value is too high, then the resistive losses in the inductance degrade the performance of the filter in compensating the harmonics.

As was mentioned, active filters are connected to the grid through passive filters. The most optimal passive filter to interface capacitors of cascade shunt active power filter (CSAPF) with the AC grid is a second-order LC low-pass filter. Despite the fact that this filter contributes to eliminating high-frequency harmonics of the inverter, its parameter values have considerable effect on the APF performance. Care should be taken in the resonance frequency $[f_{res} = 1/\left(2\pi\sqrt{L_f \cdot C_f}\right)]$ of the filter, such that it should be larger than the maximum harmonic frequency compensated by the filter and smaller than the switching frequency of the inverter [17]. It is worth mentioning that filter inductance (L_f) should be larger than the inductance of the main AC grid. However, if its value is too large, then the filter capacitor value (C_f) should be small, which cannot protect the inverter, switched against overvoltage. Accordingly, a compromised should be considered in the interface filter design.

The other important physical part of active filters is the semiconductor switches. These switches in a CSAPF tolerate a unidirectional current, the value of which depends on the DC storage and a bipolar voltage, the peak value of which is dictated by the line voltage. The switches that are installed on the inverter legs should have reverse voltage blocking capability. Thus, using gate turn-off thyristor (GTO) or IGBT and series diodes are two possible options, but both suffer from power loss. The third option is to use reverse-blocking IGBT that is characterized by withstanding bipolar voltages, and it can decrease the loss in the current source inverter.

4.4.2 Voltage source active power filters—VSC

In voltage source active power Filters (VSAPFs), the DC energy storage is realized by a capacitor. The capacitor injects the harmonic current to the main AC grid through an inverter and a passive filter.

Generally, the DC voltage of the capacitor (V_{DC}) must come more than 1.5 times of the peak value of the AC grid voltage. In order to provide better controllability during the rectifier commutation, V_{DC} should satisfy the following inequality [17]:

$$V_{DC} \geq \sqrt{\sqrt{3}\pi V_{AC}\sqrt{\cos(\alpha)^2 + [(1+k)\sin(\alpha) + \delta k]^2}} \tag{4.4}$$

where V_{AC} is the rms value of the main AC voltage, α is the firing angle of the semiconductor switches, and k and δ are as follows:

$$k = \frac{L_f}{L_s + L_l} \tag{4.5}$$

$$\delta = \sqrt{6}\frac{L_s - L_l}{\pi V_{AC}\omega} \tag{4.6}$$

In which, L_f is the inductance of the interface passive filter, L_s is the inductance between the AC grid source and the point at which APF is connected to the grid, and L_l is the inductance between the point at which APF is connected to the grid and the load. After calculation of the minimum possible DC voltage, the capacitor size is obtained by

$$C_{DC} = \frac{P_h}{\omega V_{DC}\Delta V_{DC}} \tag{4.7}$$

In which, C_{DC} is the capacitor value, and ΔV_{DC} is the maximum allowable DC voltage ripple. According to the equation, C_{DC} should be large enough to restrict the DC voltage ripples sufficiently.

VSAPFs interface with the grid through a first-order passive filter. However, a considerably large inductance is needed to eliminate high-frequency harmonics of the inverter, which is costly and results in significant voltage drop across the inductance [18], so its application is limited to several kilowatts. The solution to this issue is to utilize third-order inductance- capacitance- inductance (LCL) passive filter attenuating high-frequency harmonics at the cost of complication of the controller of active filter. The LCL filter elements should be chosen as to position the resonance frequency between ten times of the AC grid frequency and half of the inverter switching frequency [19].

Regarding the inverter, semiconductor switches in VSAPF are subjected to a unipolar voltage imposed by the DC-link capacitor, and a bidirectional current that is determined by the generated active filter current. Accordingly, IGBT switches with antiparallel diodes have been used in VSAPFs.

4.4.3 Shunt-active power filters

Shunt-APFs (SAPFs) are power electronic converter based devices that are connected in parallel with the nonlinear and/or reactive load at the point of common coupling. Figure 4.5 shows the configuration of SAPFs in the network. SAPFs have been widely employed for enhancing power quality, reducing harmonic current, improving power factor, and compensating reactive power [20–22]. Although the concept of shunt-active filters was introduced primarily by Sasaki and Macida in 1971 [23], the SAPF was initially demonstrated in 1976 by Gyugyi and Strycula [24].

The configuration of SAPFs varies on the basis of (1) type of converter source, (2) type of line filter, (3) topology of the converter, and (4) number of phases, and (5) control approach.

1. Type of converter source
   The converter source type can be a voltage source or a current source. However, voltage-source converters are more utilized as they are cheaper, smaller in size, and more efficient compared with current source counterpart. Figure 4.6 depicts voltage-source and current-source converters.
2. Type of line filter
   The line filter also might be altered by different arrangement of L, LC, RLC, or LCL passive circuit according to the application.

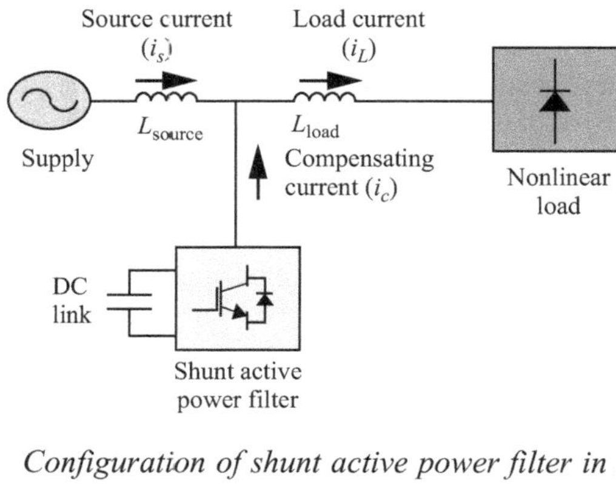

Figure 4.5 Configuration of shunt active power filter in the network

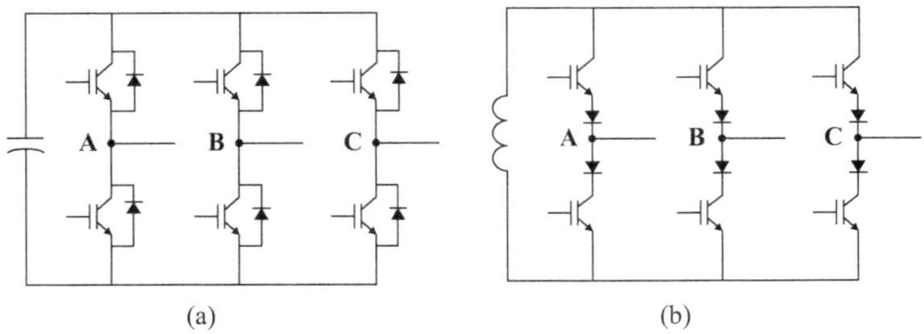

Figure 4.6 Type of converter source (a) voltage-source and (b) current-source

3. Topology of the converter
   Based on the applications of SAPF, such as level of power, type of load, and others, different converter topologies have been introduced in the literature. Three-phase full bridge [25], six-leg conventional topology [26], parallel topologies [27], multilevel configurations [28], hexagram topology [29], interconnected converters [30], and multilevel-cascaded H-bridge [31] are some among all.
   Number of phases
   SAPFs based on the supply and/or the load system can be single-phase (two-wire) or three-phase (three-wire or four-wire) systems.
4. Control approach
   The main unit in SAPFs is the controller. The controller can control the DC-link voltage and compensate the current reference through controlling the switching pulses. Different types of closed-loop controllers are offered in the literature. These approaches can be classified as linear approaches which typically use the PI controller [32] and nonlinear controller such as hysteresis control [33], fuzzy control [32], predictive or dead-beat control [33], and others.

4.4.4 Series-active power filters

Although the power quality concerns related to the harmonic contamination are addressed by employing shunt active power, which are considered as a current source compensator for the harmonic current of nonlinear loads, they have some disadvantages. First, the shunt active filters are relatively expensive. Second, due to direct proportionality of the filter power capacity and the compensated load current, they are not preferable for a large-scale system. In addition, their performance to compensate the voltage-type harmonic source is not as good as current-type harmonic source [34]. Therefore, the series-active filter was introduced by Nabae *et al.*, at the end of the 1980s to solve these challenges [35]. Although series active filters are ideal for eliminating voltage-waveform harmonics and for balancing three-phase voltages, they are still less common in the industry compared with SAPFs.

The series active filter is connected in series with the main power circuit, before the load as shown in Figure 4.7. The series APFs can be coupled into the

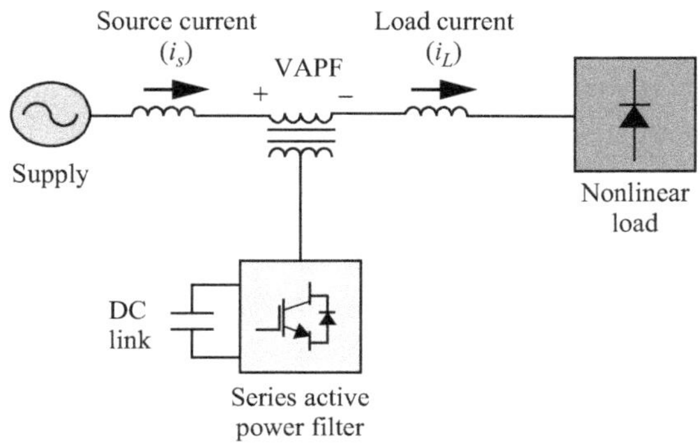

Figure 4.7 Configuration of series active power filter in the network

power circuit either directly or by using a matching transformer. The series active filters provide a zero impedance for the fundamental, while providing a high impedance for the harmonics. That is, they are working as a harmonic isolator rather than a harmonic generator.

Series active filter are normally utilized for voltage-related compensations. Some of the applications include eliminating voltage harmonics, spikes, notches, and sags, regulating and balancing the load terminal voltage or line, reducing negative-sequence voltage and regulate the voltage on three-phase systems, compensating voltage harmonics, and damping harmonic propagation caused by resonance with line impedances and passive shunt compensators.

4.4.5 Hybrid-active power filters

Hybrid active power configuration is proposed to overcome the technical limitations of conventional APFs. Typically, there are three basic topologies for hybrid APFs in form of combination of basic active and passive filters. (1) Series APF and shunt passive filter, (2) SAPF and shunt passive filter, and (3) APF in series with shunt passive filter. Figure 4.8 shows three different hybrid APF topologies.

The hybrid APF as a combination of passive filter and active filter might have different structures based on the active part inverter configuration and passive part type. The active filter inverter can be a voltage source inverter or a current source inverter. The passive filter can be an LC filter, an LCL filter, an RLC filter, a parallel resonance filter, or an injected type filter. Finally, the hybrid APF can be configured as a two-wire, three-wire, or four-wire system.

There are different methods in the literature for generating of reference signal and controlling hybrid APFs that can be categorized in time-domain approaches and frequency-domain techniques. Regarding the time-domain approaches, synchronous reference frame and instantaneous reactive power theory, and sliding mode control are the most common control strategies [36], whereas the frequency-domain techniques are mainly based on fast Fourier transformation (FFT) [37]. In addition, in order to generate the firing signals, several techniques are

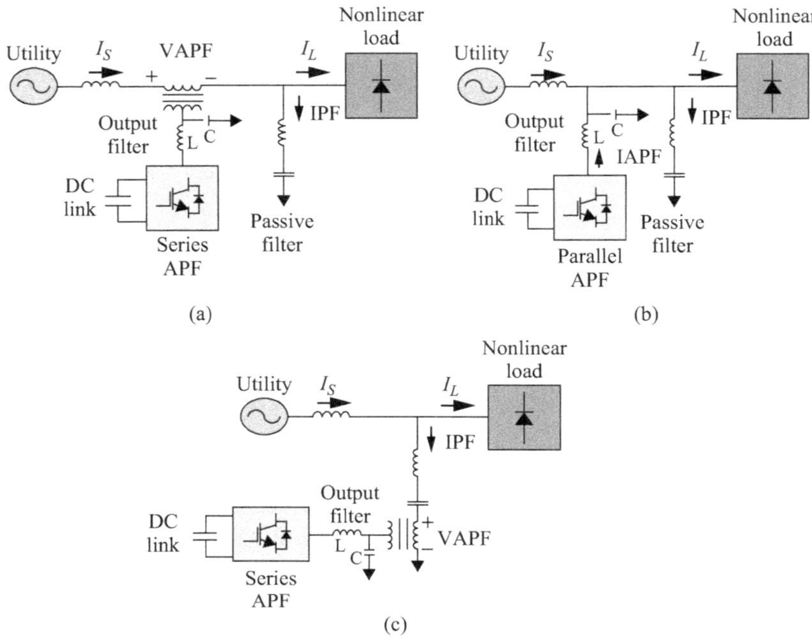

Figure 4.8 Configuration of Hybrid active power filter in the network (a) series active power filter and shunt passive filter, (b) shunt active power filter and shunt passive filter, (c) active power filter in series with shunt passive filter

accommodated in the literature such as PWM, hysteresis controller, and space vector modulation.

Hybrid APFs, benefiting from the both passive and APF advantages, offer better compensation performance as well as lower cost configuration by reducing the voltage and current rating. The idea of developing hybrid APF has been proposed in several publications by different engineers and scientists. The main aim for such a development is to reduce the cost, rating of active filter, switching noise, and electromagnetic interference by using the passive filter-to-filter dominant harmonics. In this regard, a passive high-pass filter with low cost is integrated to the traditional APF in order to filter the higher order harmonics, whereas the active filter terminates the lower order harmonics.

4.4.6 Modern/distributed-active power filter

Recently, due to increase nonlinear loads in modern electrical distribution system, another concept of APF as distributed-active power filter (D-APF) has been introduced to suppress harmonic distortion. Generally, in modern distribution networks to share the workload of harmonic mitigating, several APF units, in shunt, are

Table 4.3 Compare the performance of the types of APFs

Compensation for specific application	Active power filters topology					
	Series-APF	Shunt-APF	Hybrid of active series and passive shunt	Hybrid of active shunt and active series	UPQC	D-APF
Current harmonics	–	Good	Best	Poor	Good	Good
Reactive power	–	Best	Good	Poor	Good	Poor
Load balancing	–	Poor	–	–	–	Poor
Neutral current	–	Good	Poor	–	–	Good
Voltage harmonics	Best	–	–	Poor	Good	Good
Voltage regulation	Best	Poor	Good	Poor	Good	Poor
Voltage balancing	Best	–	Good	Poor	Good	Poor
Voltage flicker	Good	Best	–	Poor	Good	Good
Voltage sag and dips	Best	Poor	Good	Poor	Good	Good

placed at different locations (based on topology of networks) that can perform the harmonic filtering without a direct communication [38]. The important note is that control strategies and parallel operation of D-APF are more emphasized on filtering of harmonics and reactive power compensation in the distribution network.

Moreover, in order to load unbalance compensation encountered as well as capacity enlargement problems by the SAPF as D-APF, the connection topologies for D-APF are proposed as follows [39]:

1. Capacity limitation control mode: In this mode, D-APF units can operate independently and filtering harmonic current according to its own capacity.
2. Central controller mode: In order to compensation of distorted, multiple SAPF units are connected in parallel to different loads which causing harmonic pollution.
3. Common DC-link control mode: DC-link capacitors of the all APF units are common in this configuration. Table 4.3 compares the types of APFs application from different points of view.

4.5 Active power filters—APF control strategies

4.5.1 Overview of APF control techniques

Control techniques play an interesting role in improving the performance of APF. Control strategies of APFs are basically categorized into two types (shown in Figure 4.9), as follows [40–42]:

- *Open-loop system*: In this control technique, the load current and the harmonics values are leaded by open-loop control system. In the other word, APFs sensed the harmonic load current (without any feedback of how much the

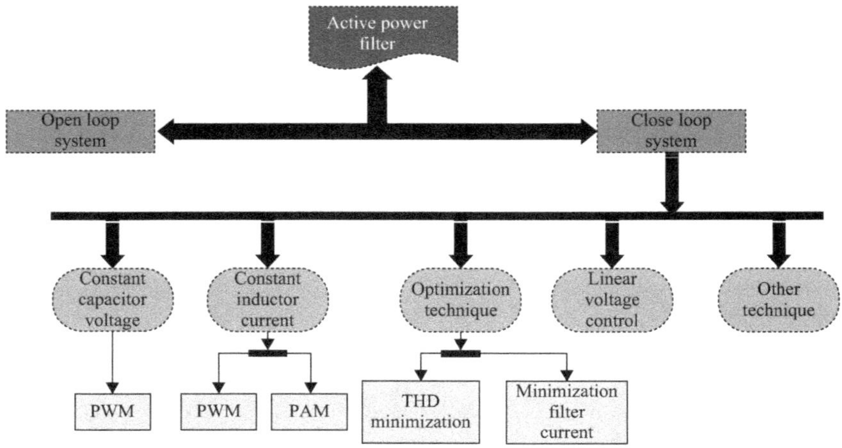

Figure 4.9 Control techniques for APFs

compensation is effective) and inject the amount of current into the power system in order to mitigation the harmonics. Nowadays, these types of control system are not used because of not having any reference to check the performance of APF.

- *Close-loop system*: In order to solve the drawback of open-loop system technique and to sense the under-considered variables, a feedback system is added to control system. Consequently, accuracy of the filter to compensate of reactive power and harmonic filtering in power system are under consideration. This technique consists of constant capacitor voltage and inductor current, optimization techniques, linear voltage control, and other techniques.

4.5.2 Heuristic soft computing-based control methods

As the increased precision control systems are chosen, the best value of the output signal from the computer algorithms/heuristic algorithms (HAs) is used to filter control. Moreover, use of soft computing techniques in control, pattern recognition, relaying, protection, smart grid optimal operation, and sizing is now expanding to smart grid, power electronics, APF/D-APF, and best solution to overcome traditional control system techniques problems [43].

HAs are capable of obtaining a good solution to optimization problems of APFs. Many computer algorithms have been introduced over the last four decades [44–47]. The most common and efficient approaches for solving APFs problems such as genetic algorithm (GA), particle swarm optimization (PSO), ant colony (AC), evolutionary algorithm (EA), harmony search algorithm (HSA), simulated annealing (SA), and bees algorithm (BA) have been used to improve performance of APF/D-APF in modern electrical networks. HAs can be classified into three large categories that are shown in Figure 4.10.

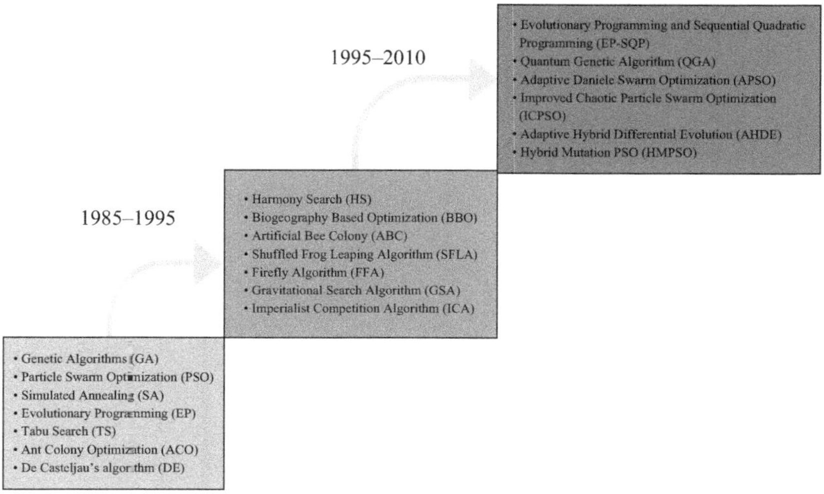

Figure 4.10 Overview of heuristic algorithms

Moreover, the use of APFs in electrical grids confronts with problems such as optimizing control parameters associated control system and optimal setting and sizing of APFs. The uses of HAs such as GA, PSO, ES, SA, and BA are the most common and efficient algorithm techniques for solving above-mentioned problems in power systems. In addition, these methodologies have a better searching ability in finding near global optimal solution compared to mathematical methods and to the standard evolutionary algorithms [48]. As results, HAs are the best solution to the optimal control, and the correct placement of the APFs problems as well as using HAs to locate and control APFs controller devices can be used to help filtering harmonics in modern networks.

4.5.3 Industrial load harmonic mitigation using APF control techniques

The basic scheme of control APF in the industrial load harmonic mitigation is shown in Figure 4.11.

Generally, the aim of control part of the APF is to generate appropriate gating signals based on the estimated compensation reference signals for the switching transistors that play an effective role to filtering harmonic pollution in the power system [49]. As a result, choosing the right controller has a positive effect on APFs performance.

Different control systems to improve the performance of APFs have been introduced, such as linear control, digital deadbeat control hysteresis control, and others. In this chapter, the important APFs control technique has been briefly introduced [50].

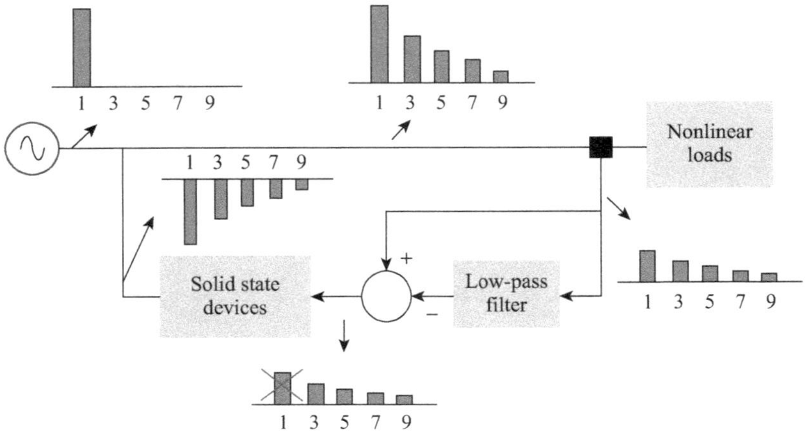

Figure 4.11 Basic application of APF control system

1. Linear control technique
   In this type of controller of a negative feedback network used to evaluate the performance of the filter. Figure 4.12(a) and (b) illustrates a block diagram and waveform of application of the linear controller, respectively [51].
   Figure 4.12 illustrates that the compensation signal (i_f or v_f) is compared with its assessment reference signal (i_{fref} or v_{fref}) by the compensated error amplifier to produce the control signal. After that, the terminating control signal is contrast with a sawtooth signal by a PWM controller to produce the suitable gating signals for the switching transistor of APFs. In the following, the frequency of the repetitive sawtooth signal creates the switching frequency [51,52]. Nevertheless, this frequency is kept stable in a linear control method.
2. Hysteresis control technique
   The next APF control technique is a hysteresis control method. It imposes a bang–bang type immediate control that forces and APF compensation voltage (v_f) or current (i_f) signal to follow its evaluated reference signal (i_{fref} or v_{fref}) within a definite tolerance band. Figure 4.13(a) and (b) shows control scheme of hysteresis control technique in a block diagram [52,53]. In the scheme of control, a signal deviation (H) is created on v_{fref} or i_{fref} to form the upper and lower limits of a hysteresis band as a first step of the control system method After that, the i_f or v_f measured and compared with i_{fref} or v_{fref}. The eventuating error is subjected to a hysteresis controller to determine the gating signals when exceeds the upper or lower limits set by +H/2 or −H/2 (the estimated reference signals) [54]. In other words, no switching action happens until the error remains within the hysteresis band. In addition, whenever the error hits the hysteresis, the switching occurs after that.

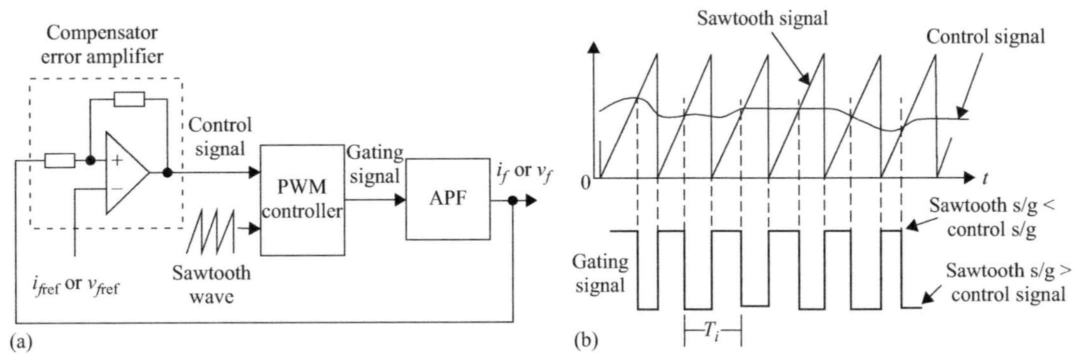

Figure 4.12 (a) Block diagram of linear control technique and (b) waveform of linear control technique

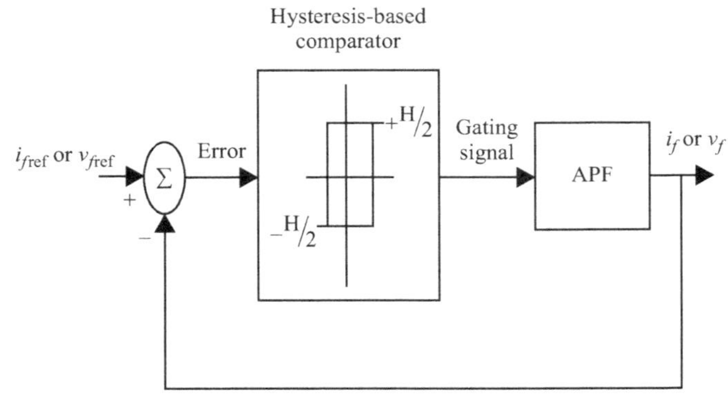

(a)

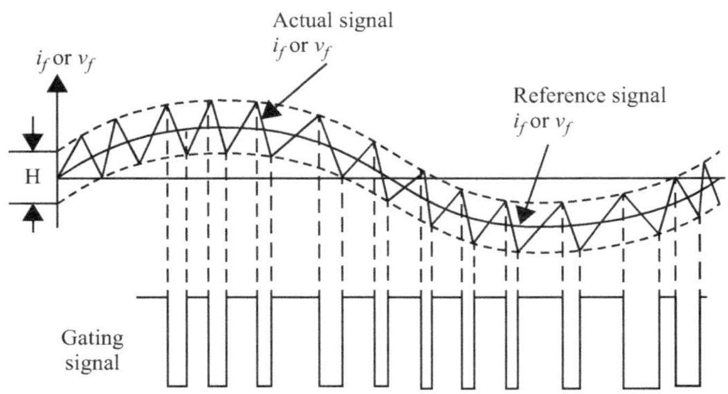

(b)

Figure 4.13 (a) Hysteresis control technique Block diagram and (b) waveform of hysteresis control technique

4.6 Emerging APF—applications and typologies

Addressing fast new switching power electronic devices and intelligent loads controller's strategies with online search and optimization will concern the following new application [55–60], as shown in Figure 4.14.

1. Use of modified topologies and VSC 6 or 12 pulses STATCOM and Distributed-STATCOM devices and controllers for series and series-parallel APFs.
2. Enhanced PWM switching strategies using symmetrical, asymmetrical, and optimal sequence using Mehta–Heretic search as well as optimization technique such as PSO, GA, ANN, and others.
3. Integrated APF and voltage/power factors enhancement function.

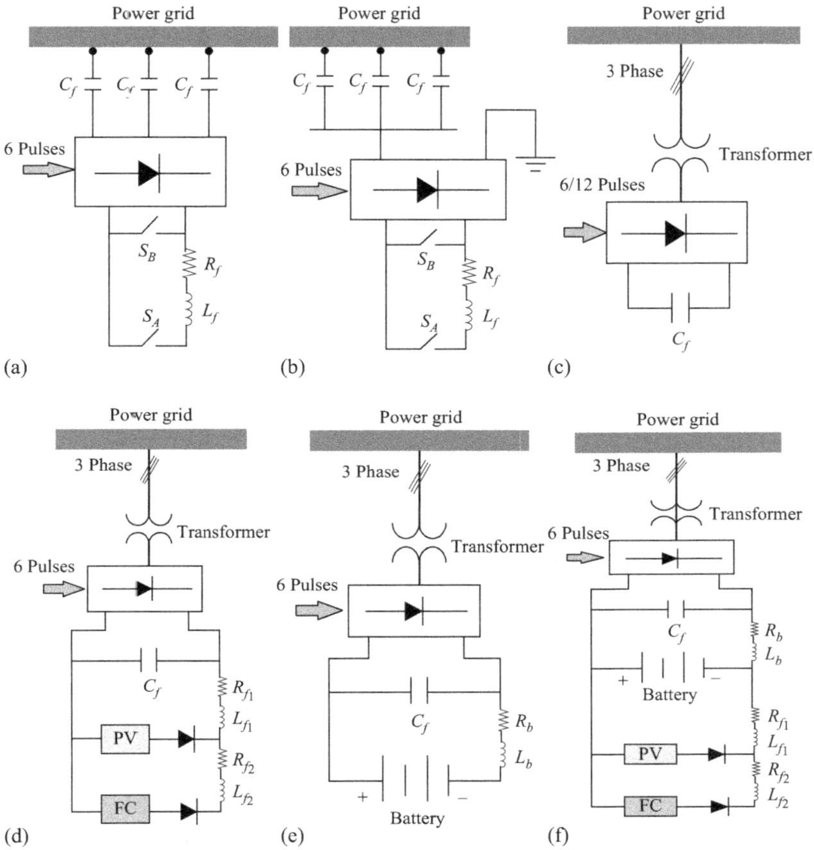

Figure 4.14 Modified APFs topologies

4. Modified dynamic fast acting controller for enhanced power quality and efficient utilization.
5. Use of switch modulated on/off PWM point-on wave switching and modified PWM techniques for decoupled DC–AC interface of renewable energy scheme and improved power quality PQ and energy delivery using molt-level converter for medium-voltage (MV) and low-voltage (LV) smart grid applications.
6. Integrated meth-historic optimization/search techniques in control and swathing of APFs controller in online and off-line appellation.
7. Application of DC side and AC side storage/power filter in emergency DC–AC renewable energy, large industrial drives, V2H, V2G, and battery storage application in DSM-EE efficient energy storage and utilization.

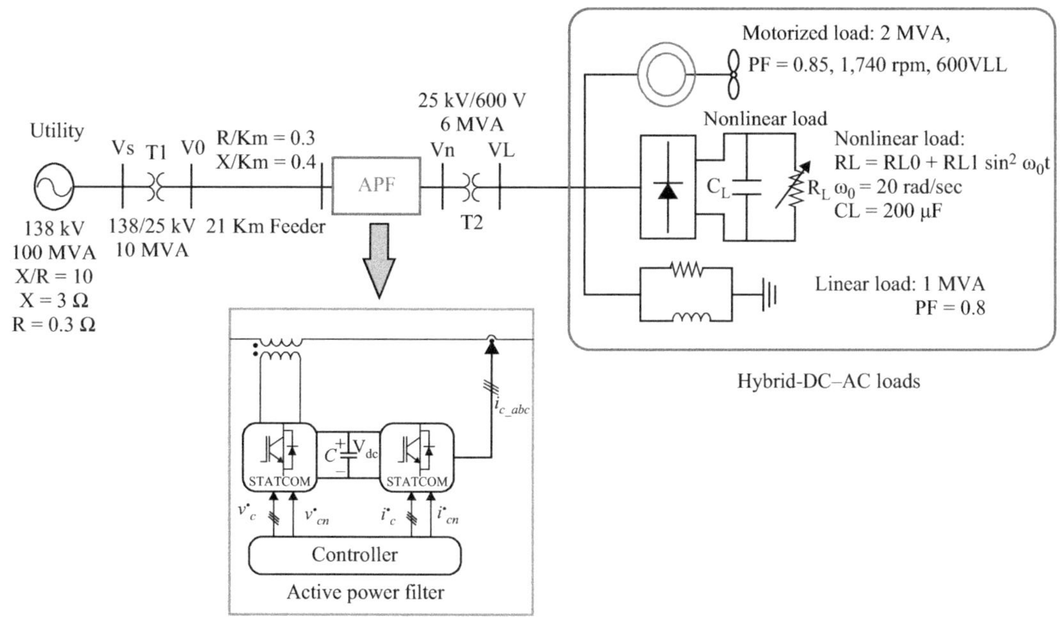

Figure 4.15 Case study I: APF system with hybrid DC–AC loads

4.7 Case studies: design and optimization of an industrial active power filter

4.7.1 Case study I: APF application and control strategies for hybrid AC–DC industrial loads

In order to show the application and the control strategies of APFs in modern electrical networks, the APF family (series-shunt APF) has been under study in MATLAB® software environment. The results of the simulation show that the APF application is very much suitable for filtering the harmonics and compensate reactive power under hybrid AC–DC industrial loads. Three-phase sample study AC system with the APF, which is connected to hybrid DC–AC industrial loads, is given in Figure 4.15.

A block diagram of the APF controller is shown in Figure 4.11. The controller contains a voltage controller, a current reference generator, and finally VSC switching signals generator. In order to calculate active and reactive powers that are used in STATCOM control system, load currents are measured instantly. As shown in Figure 4.16, the DC values in active and reactive powers are removed though the DC component elimination block as well as this part of the control system provides feedback signals and lead it to reference current generator block.

The simulation results of performance of APF to mitigating harmonic decoration in the case study are as follows in Figure 4.17.

Also, the inverter voltage during APF application is given by Figure 4.18.

The FFT analysis of APF in the case study is shown in Figure 4.19. In attention, the simulation results prove that voltage and current harmonics decoration in utility and load buses reduces to 8.34% for utility voltage, 10.71% for load voltage, 10.81% for utility current, and finally 9.92% for load current.

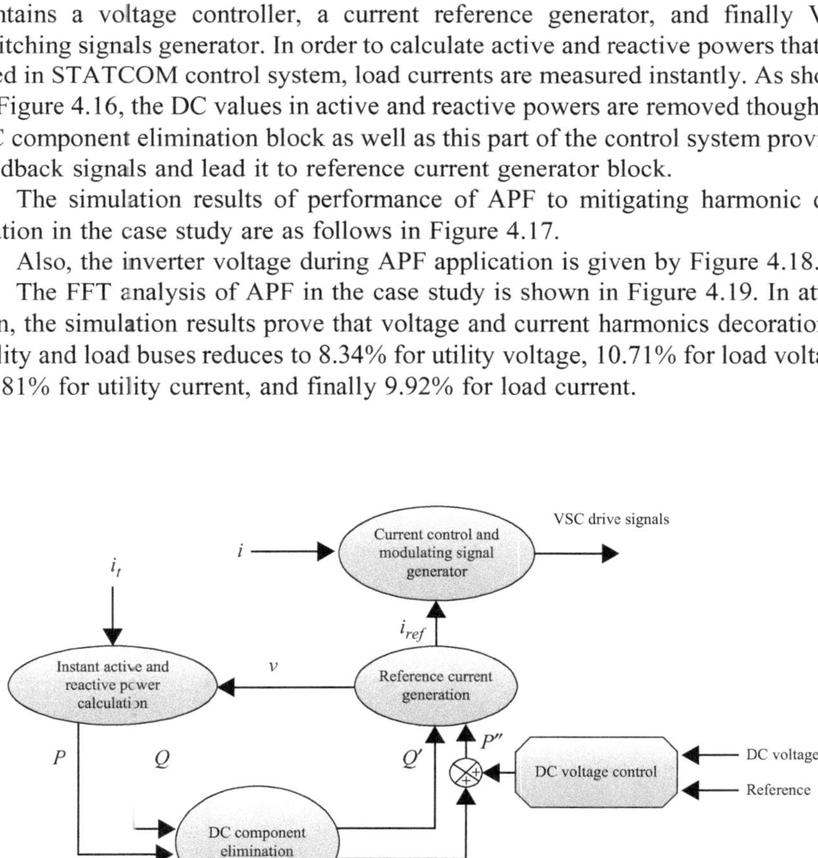

Figure 4.16 The block diagram of an APF for current harmonic filtering in the case study

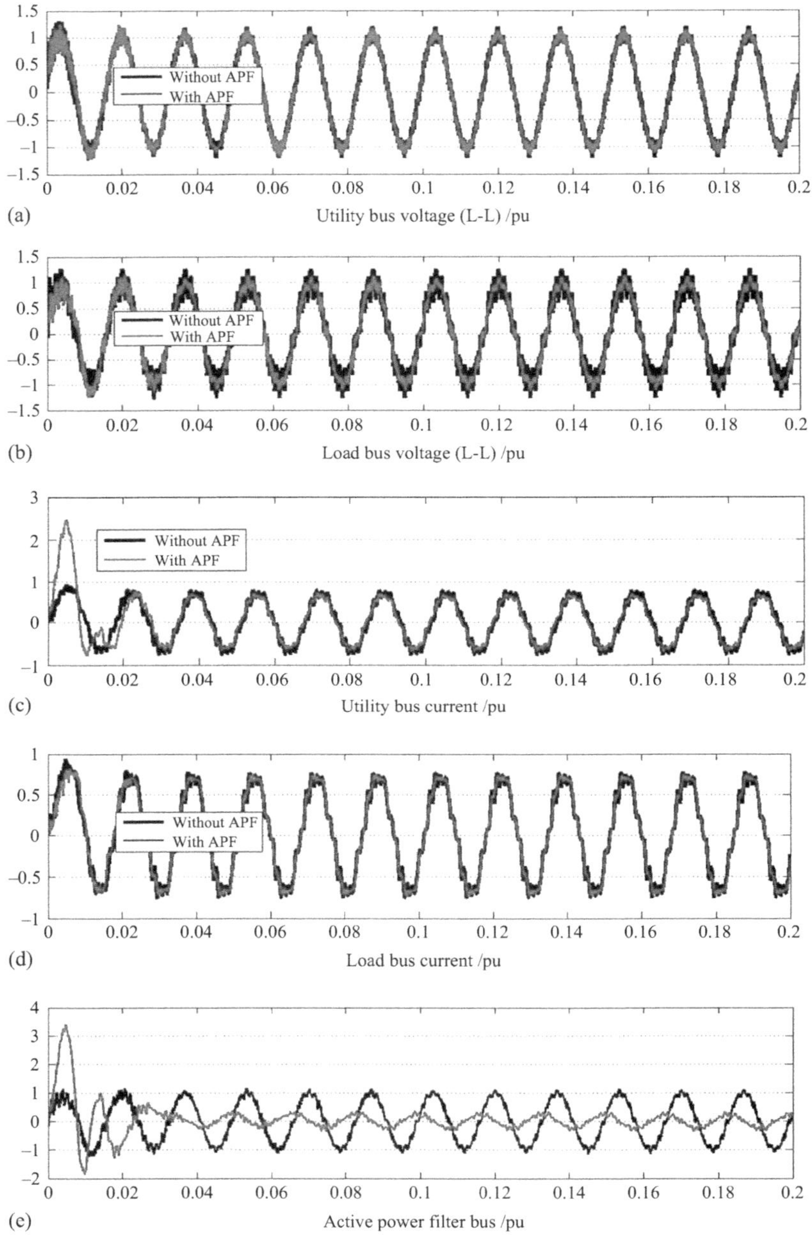

Figure 4.17 *(a) Utility bus voltage wave form with and without APF, (b) load bus voltage wave form with and without APF, (c) utility bus current wave form with and without APF, (d) load bus current wave form with and without APF, and (e) active power filter bus voltage and current waveforms*

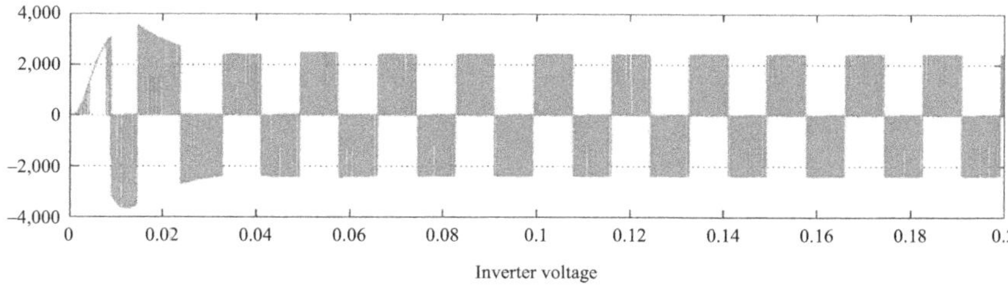

Figure 4.18 Inverter voltage wave form in APF operation position

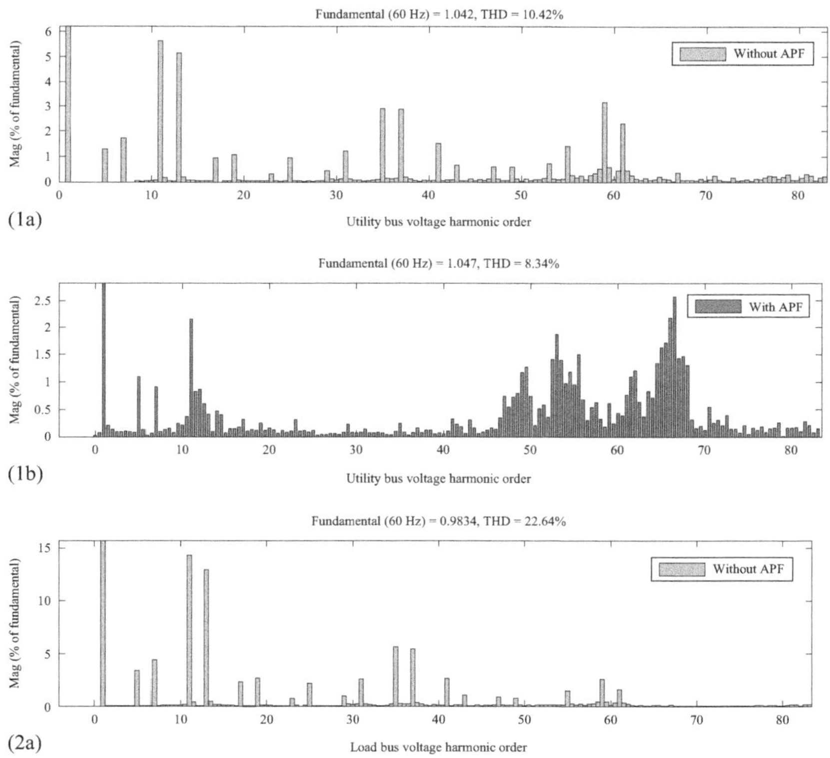

Figure 4.19 (1a and 1b) Utility bus voltage THD with and without APF, (2a and 2b) load bus voltage THD with and without APF, (3b and 3b) utility bus current THD with and without APF, (4b and 4b) load bus current THD with and without APF

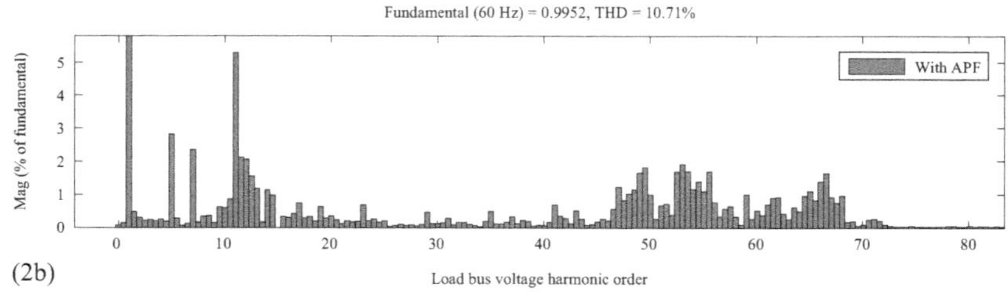

(2b)

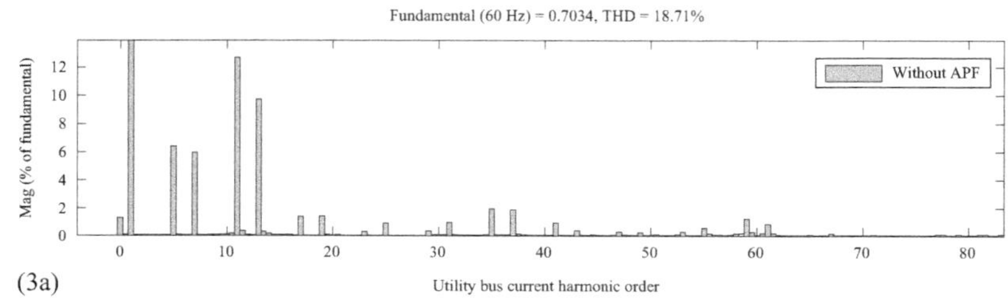

(3a)

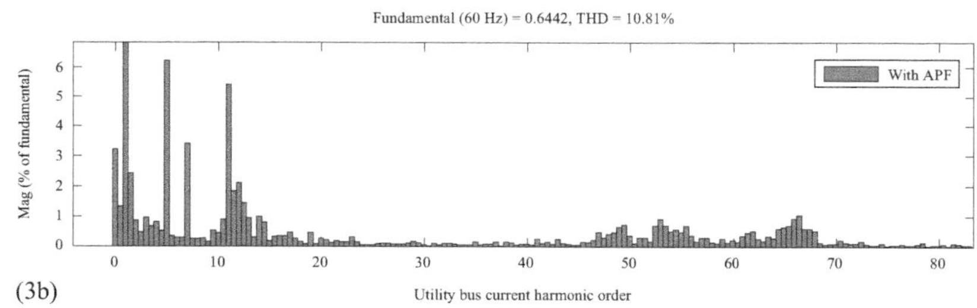

(3b)

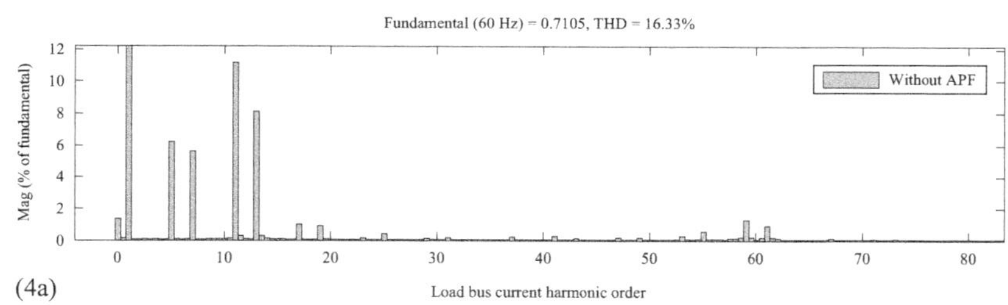

(4a)

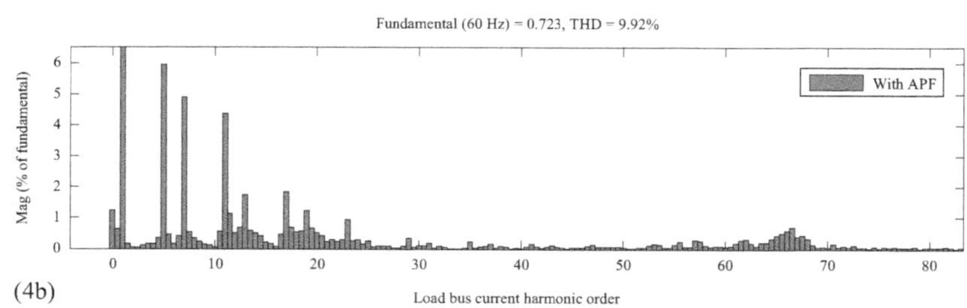

(4b)

Figure 4.19 (Continued)

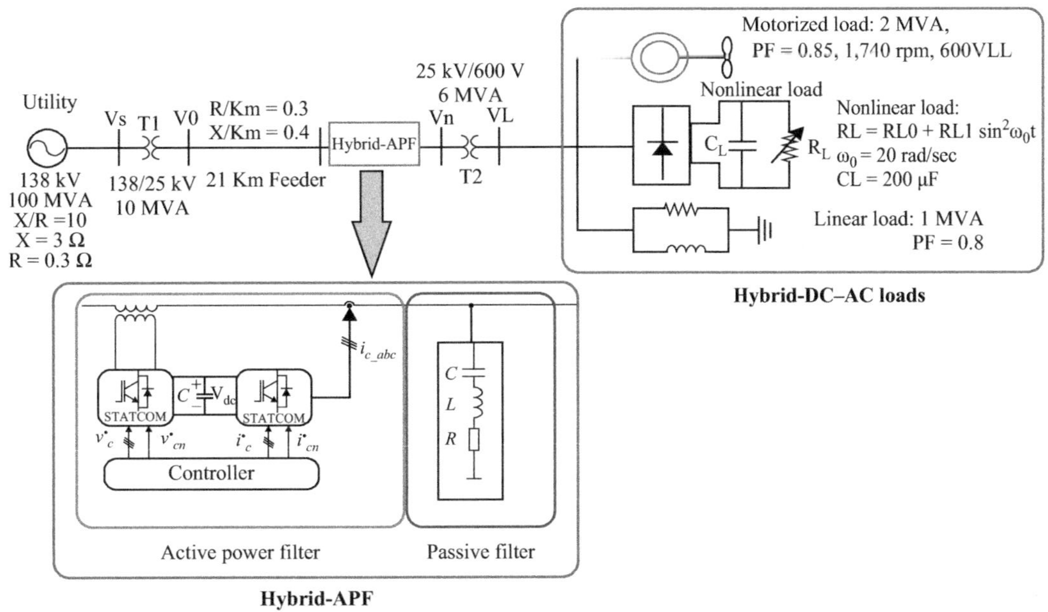

Figure 4.20 Case study II: Hybrid-APF system with hybrid DC–AC loads

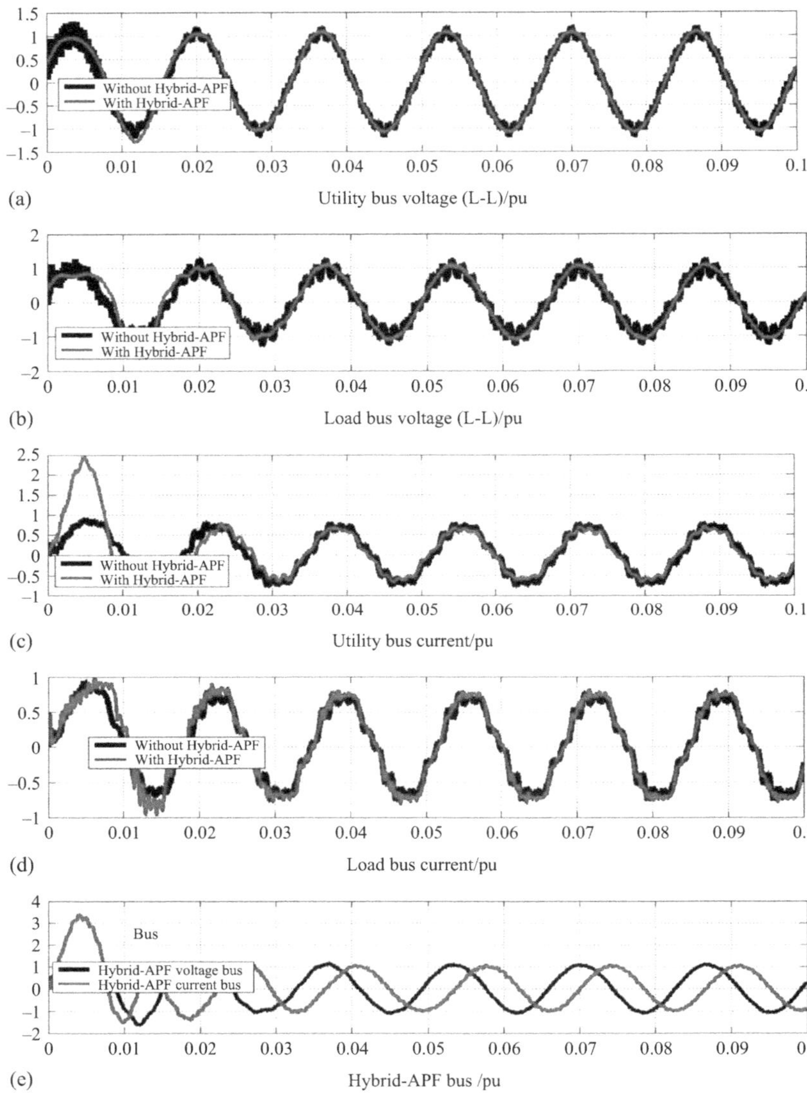

Figure 4.21 (a) Utility bus voltage wave form with and without Hybrid-APF, (b) load bus voltage wave form with and without Hybrid-APF, (c) utility bus current wave form with and without Hybrid-APF, (d) load bus current wave form with and without Hybrid-APF, and (e) active power filter bus voltage and current waveforms

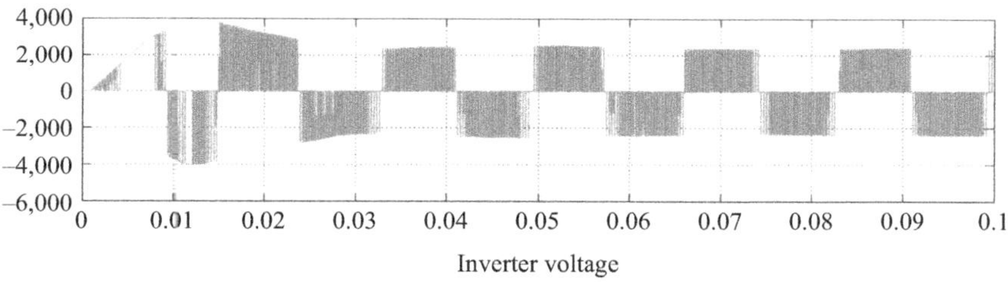

Figure 4.22 Inverter voltage wave form in Hybrid-APF operation position

4.7.2 Case study II: hybrid-APFs for AC–DC system

In this section, in order to show application of combined APFs with passive power filter (hybrid-APFs), a single tuned passive filter has been added to the case study, as shown in Figure 4.20. The results of the simulation show that the hybrid-APF application is very much suitable for filtering the harmonics and compensate reactive power than APF under hybrid AC–DC industrial loads.

The waveform of voltage and current as well as THD in the utility, load, and hybrid-APF bases are given by Figure 4.21.

Moreover, the inverter voltage during hybrid-APF application is given by Figure 4.22.

The FFT analysis of hybrid-APF in the case study is shown in Figure 4.23. In addition, the simulation results prove that voltage and current harmonics decoration in utility and load bases reduces to 1.95% for utility voltage, 3.78% for load voltage, 4.69% for utility current, and finally 2.82% for load current in hybrid-APF position.

Table 4.4 illustrates summary of voltage and current THD in the utility and load buses for all position. The table shown that, THD in the hybrid-APF mode more than the APF mode has been decreased.

4.8 Conclusions

APF can reduce some types of power quality disturbances such as harmonics, reactive power, and unbalanced load currents. The APF uses fast switching solid-state devices and GTO emerging as a viable in smart grid applications, battery charging scheme battery storage system. Smart grid uses distributed-STATCOM (D-STATCOM) for energy efficiency, energy saving, loss reduction, and power factor enhancement. The chapter presents different APF topology with modified control application and case studies for smart grid and renewable energy utilization. New emerging application for APFs in campus intelligent met heuristic optimization and search algorithm for optimal PWM using symmetrical, a symmetrical, inverse sine carrier, and optimum point on wave switching strategy with online search to minimize dominating harmonic and THD at the point of interface with the AC smart grid.

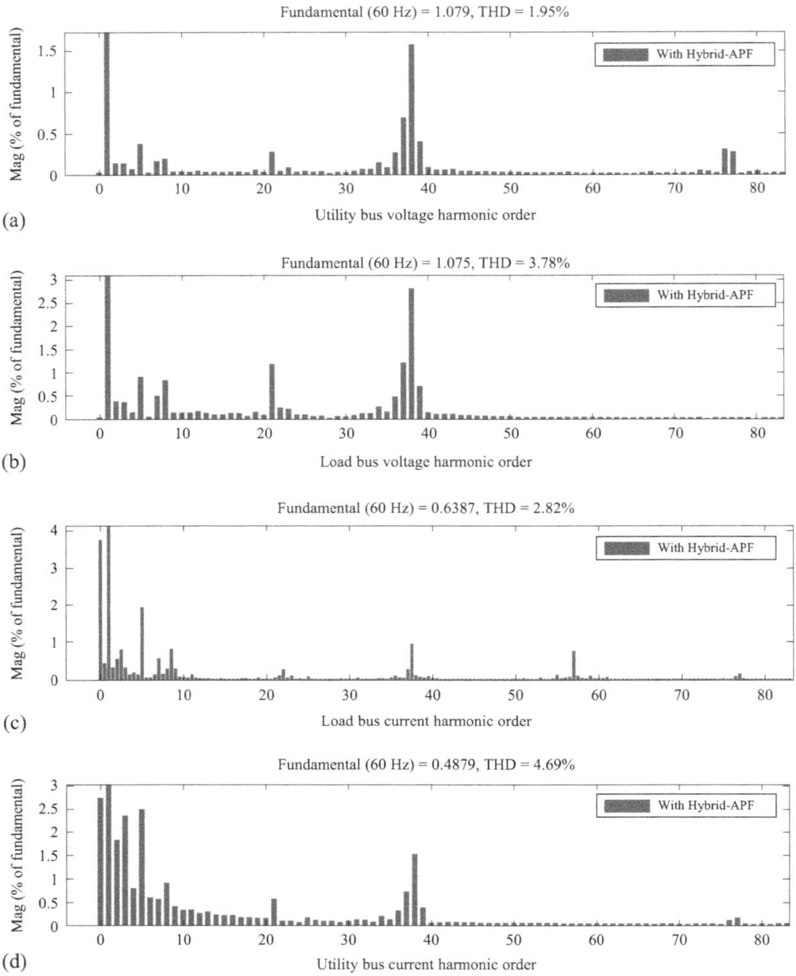

Figure 4.23 (a) Utility bus voltage THD with Hybrid-APF, (b) load bus voltage THD with Hybrid-APF, (c) utility bus current THD with Hybrid-APF, and (d) load bus current THD with Hybrid-APF

Table 4.4 THD results without, with APF, and with hybrid-APF in the case study

Case study	**Total harmonic distortion (THD)/present**			
	Utility voltage (%)	**Utility current (%)**	**Load voltage (%)**	**Load current (%)**
Without APF	10.22	17.60	22.16	15.23
With APF	9.19	9.76	10.65	9.87
With hybrid-APF	1.079	4.69	3.76	2.82

References

[1] Eric J. D., Alexander E. E., David J. P. 'Harmonic pollution metering: theoretical considerations'. IEEE Transactions on Power Delivery. 2000; **15**(1):19–23.

[2] Izzeldin I. A., Rama Rao S. K., Perumal N. 'Harmonics mitigation and power factor correction with a modern three-phase four-leg shunt active power filter'. Proceedings of the IEEE International Conference on Power and Energy; Kuala Lumpur, Malaysia, Nov–Dec 2010. pp. 156–161.

[3] Bhonsle C. D., Kelkar R. B. 'Performance evaluation of composite filter for power quality improvement of electric arc furnace distribution network'. International Journal of Electrical Power & Energy Systems. 2016; **79**(6):53–65.

[4] De Matos E. O., Mota Soares T., Bezerra U. H., Tostes M. E. d. L., Manito A. R. A., Costa B. C. 'Using linear and non-parametric regression models to describe the contribution of non-linear loads on the voltage harmonic distortions in the electrical grid'. IET Generation, Transmission & Distribution. 2016;**10**(8):1825–1832.

[5] Xu W., Liu Y. 'A method for determining customer and utility harmonic contributions at the point of common coupling'. IEEE Transactions on Power Delivery. 2000;**15**(2):804–811.

[6] Chamorro H. R., Ordóñez C. A., Cantor E. L., Ramos A. G. 'Active power filter compensation in presence of several industrial loads based on P-Q theory'. International Journal of Power and Energy Systems. 2012;**32**(2):51–61.

[7] Abu Bakar M. I. 'Assessments for the impact of harmonic current distortion of nonlinear load in power system harmonics'. Proceedings of the Transmission and Distribution Conference and Exposition: Latin America. Bogota, Oct 2008. pp. 1–6.

[8] Barr R. A., Gosbell V. J. 'Power system harmonic voltage limits for the future'. Proceedings of the 2014 16th International Conference on Harmonics and Quality of Power (ICHQP), Bucharest, May 2014. pp. 483–487.

[9] Peng Z. F. 'Harmonic sources and filtering approaches'. IEEE Industry Applications Magazine. 2001;**7**(1):18–25.

[10] Singh B., Al-Haddad K., Chandra A. 'A review of active filters for power quality improvement'. IEEE Transactions on Industrial Electronics. 1999; **46**(5):960–971.

[11] Wagner V. E., Balda J. C., Griffith D. C., Jewell W. T. 'Effects of harmonics on equipment.' IEEE Transactions on Power Delivery. 1993;**8**(2):672–680.

[12] Duffey C K., Stratford R. P. 'Update of harmonic standard IEEE-519: IEEE Recommended Practices and Requirements for Harmonic Control in Electric Power Systems'. Proceedings of the Industrial Applications Society 35th Annual, Dallas, TX, 1988. pp. 249–255.

[13] Liew A. C. 'Excessive neutral currents in three-phase fluorescent lighting circuits'. IEEE Transactions on Industry Applications. 1989;**25**(4):776–782.

[14] Gruzs T. M. 'A survey of neutral currents in three-phase computer power systems'. Proceedings of the Industrial and Commercial Power Systems Technical Conference, Chicago, IL, 1989. pp. 114–122.
[15] McGranaghan M. 'Active filter design and specification for control of harmonics in industrial and commercial facilities'. Electrotek Concepts, Inc., Knoxville TN, USA, 2005.
[16] Emadi A., Nasiri A., Bekiarov S. B. 'Uninterruptible power supplies and active filters'. CRC Press, 2004; pp. 30–40.
[17] Benchaita L., Saadate S., Nia A. S. 'A comparison of voltage source and current source shunt active filter by simulation and experimentation'. IEEE Transactions on Power Systems. 1999;**14**(2):642–647.
[18] Beres R. N., Wang X., Liserre M., Blaabjerg F., Bak C. L. 'A review of passive power filters for three-phase grid-connected voltage-source converters'. IEEE Journal of Emerging and Selected Topics in Power Electronics. 2016;**4**(1):54–69.
[19] Liserre M., Blaabjerg F. Hansen S. 'Design and control of an LCL-filter-based three-phase active rectifier'. IEEE Transactions on Industry Applications. 2005;**41**(5):1281–1291.
[20] Campanhol L. B. G., d. Silva S. A. O., Goedtel A. 'Application of shunt active power filter for harmonic reduction and reactive power compensation in three-phase four-wire systems'. IET Power Electronics. 2014;**7**(11):2825–2836.
[21] Khadem S. K., Basu M., Conlon M. F. 'Harmonic power compensation capacity of shunt active power filter and its relationship with design parameters'. IET Power Electronics. 2014;**7**(2):418–430.
[22] Kanjiya P., Khadkikar V., Zeineldin H. H. 'Optimal control of shunt active power filter to meet IEEE Std. 519 current harmonic constraints under noni deal supply condition'. IEEE Transactions on Industrial Electronics. 2014; **62**(2):724–734.
[23] Sasaki H., Machida T. 'A new method to eliminate AC harmonic currents by magnetic flux compensation-considerations on basic design'. IEEE Transactions on Power Apparatus and Systems. 1971;**90**(5):2009–2019.
[24] Nastran J., Cajhen R., Seliger M., Jereb P. 'Active power filter for nonlinear AC loads'. IEEE Transactions on Power Electronics. 1994;**9**(1):92–96.
[25] Mendalek N., Al-Haddad K., Kanaan Y. H., Hassoun G. 'Sliding mode control of three-phase four-leg shunt active power filter'. Proceedings of the 2008 IEEE Power Electronics Specialists Conference, Rhodes, June 2008. pp. 4362–4367.
[26] Gupta R., Ghosh A., Joshi A. 'Control of 3-level shunt active power filter using harmonic selective controller'. Proceedings of the IEEE Power India Conference, New Delhi, June 2006. pp. 7–14.
[27] Asiminoaei L., Lascu C., Blaabjerg F., Boldea I. 'Performance Improvement of shunt active power filter with dual parallel topology'. IEEE Transactions on Power Electronics. 2007;**22**(1):247–259.

[28] Kouro S., Malinowski M., Gopakumar K., *et al.* 'Recent advances and industrial applications of multilevel converters'. IEEE Transactions on Industrial Electronics. 2010;**57**(8):2553–2580.

[29] Wen J., Zhou L., Smedley K. 'Power quality improvement at medium-voltage grids using Hexagram Active Power Filter'. Proceedings of the Applied Power Electronics Conference and Exposition (APEC), 2010 Twenty-Fifth Annual IEEE, Palm Springs, CA, Feb 2010. pp. 47–57.

[30] Jacobina C. B., Fabricio E. L. L., Menezes A. V. G., Correa M. B. R., Carlos G. A. A. 'Shunt compensator based on three-phase interconnected converters'. Proceedings of the in IEEE Energy Conversion Congress and Exposition, Denver, CO, 2013. pp. 6661–6671.

[31] Ortuzar M. E., Carmi R. E., Dixon J. W., Moran L. 'Voltage-source active power filter based on multilevel converter and ultra-capacitor DC link'. IEEE Transactions on Industrial Electronics. 2006;**53**(2):477–485.

[32] Chourasiya S., Agarwal S. 'A review: control techniques for shunt active power filter for power quality improvement from non-linear loads'. International Electrical Engineering Journal (IEEJ). 2015;**6**(10):2028–2032.

[33] Green T. C., Marks J. H. 'Control techniques for active power filters'. IEE Proceedings-Electric Power Applications. 2005;**152**(2):369–381.

[34] Akagi H. 'New trends in active filters for power conditioning'. IEEE Transactions on Industry Applications. 1996;**32**(6):1312–1322.

[35] Peng F. Z., Akagi H., Nabae A. 'A new approach to harmonic compensation in power systems—a combined system of shunt passive and series active filters'. IEEE Transactions on Industry Applications. 1990;**26**(6):983–990.

[36] Zaveri N., Chudasama A. 'Analysis of different real time reference generation techniques used for harmonic mitigation in three phase shunt active filters'. International Journal of Recent Trends in Engineering. 2009;**2**(7):123–126.

[37] Rahmani S., Al-Haddad K., Kanaan H. Y., Singh B. 'Implementation and simulation of modified PWM with two current control techniques applied to single-phase shunt hybrid power filter'. IEE Proceedings Electric Power Applications. 2006;**153**(3):317–326.

[38] Cheng P. T., Lee T. L. 'Distributed active filter systems (DAFSs): a new approach to power system harmonics'. IEEE Transactions on Industry Applications. 2006;**42**(5):1301–1309.

[39] Kuo S. Y., Lee T. L., Chen C. A., Cheng P., Pan T. C. T. 'Distributed active filters for harmonic resonance suppression in industrial facilities'. Proceedings of the Power Conversion Conference. Nagoya, Apr 2007. pp. 391–397.

[40] Chen D., Xie S. 'Review of the control strategies applied to active power filters'. Proceedings of the 2004 IEEE International Conference on Electric Utility Deregulation, Restructuring and Power Technologies, Apr 2004. pp. 666–670.

[41] Steela K., Rajpurohit B. S. 'A survey on active power filters control strategies'. Proceedings of the 2014 IEEE Sixth India International Conference on Power Electronics (IICPE), Kurukshetra, Dec 2014. pp. 1–6.

[42] Montero M. I. M., Cadaval E. R., Gonzalez F. B. 'Comparison of control strategies for shunt active power filters in three-phase four-wire systems'. IEEE Transactions on Power Electronics. 2007;**22**(1):229–236.

[43] Ouchen S., Betka A., Abdeddaim S., Menadi A. 'Fuzzy-predictive direct power control implementation of a grid connected photovoltaic system, associated with an active power filter'. Energy Conversion and Management. 2016;**122**(2):515–525.

[44] Affiliated M. A. K., Allaoui T., Denai M. 'A hybrid fuzzy sliding-mode control for a three-phase shunt active power filter'. Energy Systems. 2016; **8**(5):1–1064.

[45] Kumar B. S., Reddy K. R., Archana S. 'The application of PSO to hybrid active power filter design for 3 phase 4-wire system with balanced & unbalanced loads'. International Journal of Advances in Engineering & Technology. 2012;**1**(1):32–42.

[46] Badoni M., Singh A., Singh B. 'Adaptive recursive inverse-based control algorithm for shunt active power filter'. IET Power Electronics. 2016; **9**(5):1053–1064.

[47] Hedayati B. K. Rahmati A. 'Genetic algorithm application in controlling performance and power dissipation of active power filters'. Canadian Journal on Electrical & Electronics Engineering. 2010;**1**(1):15–19.

[48] Akagi H. 'Modern active filters and traditional passive filters'. Bulletin of the Polish Academy of Sciences Technical Sciences. 2006;**54**(3):255–240.

[49] Singh B., Al-Haddad K., Chandra A. 'A review of active filters for power quality improvement'. IEEE Transactions on Industrial Electronics. 1999; **46**(6):960–971.

[50] Cichowlas M., Malinowski M., Kazmierkowski M. P., Sobczuk D. L., Rodriguez P., Pou J. 'Active filtering function of three-phase PWM boost rectifier under different line voltage conditions'. IEEE Transactions on Industrial Electronics. 2005;**52**(2):410–419.

[51] Stones J., Collinson A. 'Power quality'. IEE Power Engineering Journal. 2001;**15**(2):58–64.

[52] Zainal S., Cheng T. P., Awang J. 'Harmonics mitigation using active power filter: a technological review'. Elektrika. 2006;**8**(2):17–26.

[53] El-Habrouk M., Darwish M. K. 'Design and implementation of a modified Fourier analysis harmonic current computation technique for power active filters using DSPs'. IEE Electric Power Applications. 2001;**148**(1):21–28.

[54] Malesani L., Mattavelli P., Buso S. 'Dead-beat current control for active filters'. Proceedings of the Industrial Electronics Conference (IECON), Aachen, Germany, Aug 1998. pp. 1859–1864.

[55] Sharaf M. A., Gandoman H. F. 'A flexible FACTS based Scheme for smart grids PV-battery storage systems'. International Journal of Distributed Energy Rescuers. 2014;**10**(4):261–271.

[56] Sharaf M. A., Gandoman H. F. 'A robust FACTS PV-smart grid interface scheme for efficient energy utilization'. International Journal of Power and Energy Conversion. 2015;**6**(4):344–358.

[57] Sharaf M. A., Abdel Aleem E. H. S., Gandoman H. F. 'A robust decoupled DC–AC FACTS based switched filter compensation scheme for PV-Smart grid applications'. Proceedings of the First Future University International Conference on New Energy & Environmental Engineering "ICNEEE", Egypt, Cairo, Apr 2016.

[58] Sharaf M. A., Khaki B. 'A FACTS based switched capacitor compensation scheme for smart grid applications'. Proceedings of the IEEE International Symposium on Innovations in Intelligent Systems and Applications (INISTA), Trabzon, Jun 2012. pp. 1–5.

[59] Sharaf M. A., Khaki B. 'A novel FACTS hybrid modulated filter/capacitor compensator'. Proceedings of the IEEE International Conference on Smart Grid Engineering (SGE), Oshawa, ON, June 2012. pp. 1–6.

[60] Sharaf M. A., Gandoman H. F. 'A switched hybrid filter DVS-green plug for smart grid nonlinear loads'. Proceedings of the IEEE International Conference on Smart Energy Grid Engineering (SEGE'15), UOIT, Oshawa, ON, Aug 2015. pp. 17–19.

Chapter 5

Shunt flexible a.c. transmission

Grazia Todeschini[1]

5.1 Introduction

Shunt FACT devices are widely used to improve transmission system operation by regulating active and reactive power flow during both steady-state and transient conditions [12,17]. This family of devices includes high-voltage d.c. (HVDC) transmission, static VAr compensators (SVCs), and static synchronous compensators (SSC or STATCOM).

The use of power switches makes the response of shunt FACT devices very fast but at the same time introduces nonsinusoidal currents and voltage components into the power system. If any of the voltage or current harmonic components is close to one of the characteristic frequencies of the system, a resonance condition will result, potentially leading to high harmonic voltage and current distortion [1]. To eliminate, or at least minimize, the impact of harmonics generated by shunt FACT devices, harmonic studies are performed at the design stage of a new project. The main purpose of the harmonic studies is to design a device that does not introduce excessive harmonic disturbance in the power system, while being at the same time robust and immune to already existing harmonics. A dedicated computer model is built to perform the harmonic studies, and the model includes both the system representation and the shunt FACT device representation.

Once the model is built, the parameters that impact the harmonic studies are identified, including frequency variation, temperature, and tolerance of power components. The next step is to identify the system configurations that may be encountered during normal operation (switching of capacitor banks) and transient conditions ($N-1$ contingencies). Once the different parameters and operating conditions are identified, two different harmonic studies are performed: harmonic performance studies and harmonic rating studies.

This chapter is intended to discuss the most general approach to harmonic studies, which can be applied to shunt FACT devices and to all devices regulated by means of power electronic switches, such as wind power plants. This approach is supported by international standards such as [29,30,36]. Some specific installations

[1]College of Engineering, Swansea University, Swansea, UK

may require special considerations that are not addressed in detail in this chapter; however, specific references are provided for further reading whenever these special conditions are identified.

5.2 Overview of harmonic concerns for shunt FACT devices and chapter content

The analysis carried out in this chapter addresses harmonic current and voltage components with frequency that is a multiple integer of the fundamental component (50 or 60 Hz). The range of frequency of interest is limited between the 2nd and the 50th harmonic (2,500 Hz for 50 Hz systems or 3,000 Hz for 60 Hz systems). Non-integer harmonics and higher-order harmonics require a different type of modeling,[1] and in general they are not as much of concern as the "integer" harmonics; therefore, they are not discussed in this chapter.

Shunt FACT devices generate harmonics because the operation of power electronic switches results in discontinuous current waveforms. As a result of the installation of numerous power electronics and nonlinear devices, a certain amount of harmonic distortion is present on the power system at all times, which is referred to as "background distortion."

In the great majority of transmission systems, the levels of background distortion and the level of harmonics generated by shunt FACT devices are relatively low, and they do not harm system operation or cause concerns. However, when resonance conditions are met, harmonic current or voltage components may be amplified and thus cause high harmonic current flow, and potential system malfunctioning or damage of equipment. The concept of resonance condition is discussed in Section 5.2.1.

To minimize the risk of encountering resonance conditions, harmonic studies are performed in the planning stage of a FACT project with two objectives [30]:

- To verify that the impact of the shunt FACT device in terms of harmonic generation is within the limits specified by the standards, or by the local transmission system operators or owners.
- To determine the rating of the shunt FACT device components (typically, rms currents for transformers and inductors and rms voltage for capacitors).

Figure 5.1 shows the conceptual circuit model used to perform harmonic analysis for shunt FACT devices. The left side of the figure shows the model of the power system, whereas the right side of the figure shows the model of the shunt FACT device. The point of common coupling (PCC) is the bus to which the shunt FACT device is

[1]The term "interharmonics" refers to current and voltage harmonic components at frequencies that are not integer multiple of 60 Hz. Given that the levels of interharmonics in the power system are quite small and shunt FACT devices in normal operation are not a significant source, interharmonics are not considered for the harmonic studies typically performed for shunt FACT devices. Reference [28] discusses interharmonics sources and effects.

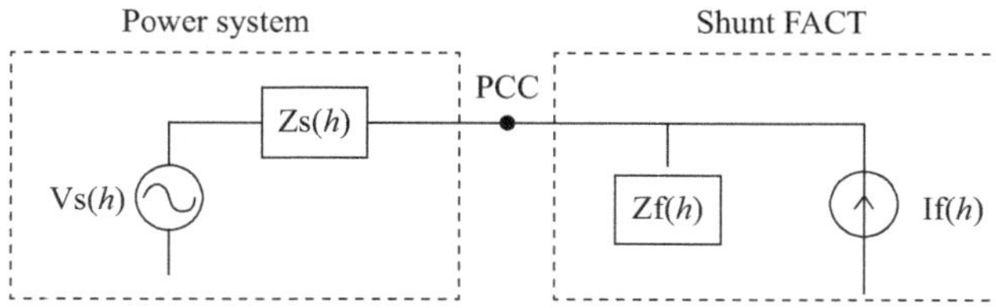

Figure 5.1 Overview of the computer model used to perform harmonic analysis

connected [36]. This bus is sometimes referred to as point of interconnection (POI) in the literature.

The model shown in Figure 5.1[2] is built from these assumptions:

- The background distortion is modeled as a voltage source at multiple frequencies, as discussed in Section 5.3.2.
- The frequency response of the power system is represented by means of impedance sectors, which indicate the variation range of system resistance and system reactance at each frequency. A description of the impedance sectors calculation is provided in Section 5.3.1.2.
- The shunt FACT device is represented by means of harmonic current or harmonic voltage sources, transformer, and elementary branches. Section 5.4 illustrates how to derive the models of some common shunt FACT devices.

The equivalent model is single phase, as in the great majority of systems harmonic conditions apply equally to the three phases.[3]

Sections 5.3 and 5.4 present the methodology and procedure used to obtain the power system model and the shunt FACT device model, respectively. Section 5.5 illustrates the methodology used to perform harmonic performance studies and harmonic rating studies.

5.2.1 Resonance conditions

Two types of resonance conditions can be defined for an electrical circuit, namely series resonance and parallel resonance [1]. These two conditions are illustrated in Figure 5.2.

[2]The system model is sometimes referred to as the "Thevenin model," but this nomenclature is not correct, as discussed in Section 3 of [31].

[3]The analysis presented in this chapter deals only with positive-sequence models and harmonic components. In presence of system unbalance conditions, negative- and zero-sequence harmonic components may be generated. Building a computer model that takes into account the zero-sequence impedance of the system is quite laborious and is generally not justified. In the case of excessive harmonics caused by unbalance conditions external to the shunt FACT device, the best approach is to mitigate the unbalance conditions at the system level. If small levels of unbalance are present, the shunt FACT device can be programed to compensate for negative- and zero-sequence harmonic components [12].

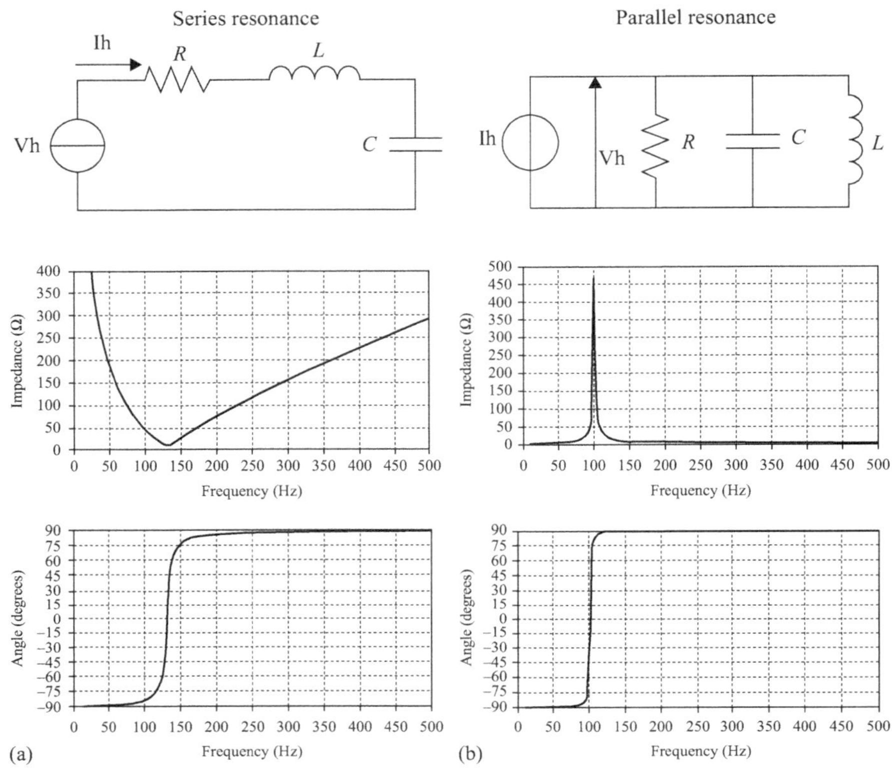

Figure 5.2 Series and parallel resonance circuits and frequency scans showing impedance magnitude and phase angle

Series resonance means that, at a particular frequency, the total impedance of the circuit is equal to the resistance. In other words, the inductive reactance and the capacitive reactance cancel each other. This concept can be illustrated mathematically by referring to Figure 5.2(a). This figure shows a series circuit in which a resistor, an inductor, and a capacitor are connected to a voltage source. The circuit impedance as a function of the frequency f, and the harmonic order h is expressed by the formula:

$$\overline{Z}(h) = R + j\omega L - j\frac{1}{\omega C}\ [\Omega]$$

where $\omega = 2\pi hf$. The condition in which the inductive reactance is equal to the capacitive reactance is

$$\omega L = \frac{1}{\omega C}$$

Figure 5.2(b) shows a parallel circuit in which a resistor, an inductor, and a capacitor are connected to a voltage source. The circuit admittance as a function of the frequency is

$$\overline{Y}(h) = \frac{1}{R} - j\frac{1}{\omega L} + j\omega C \text{ [S]}$$

The condition when the inductive susceptance is equal to the capacitive susceptance is expressed as follows:

$$-\frac{1}{\omega L} = \omega C$$

Based on the derivations above, it can be easily verified that the resonance frequency for both series resonance and parallel resonance conditions is expressed by the same formula:

$$f_h = \frac{1}{2\pi\sqrt{\mathrm{LC}}} \text{ [Hz]}$$

If the circuit contains a harmonic source with frequency f_h, there is a potential for high harmonic current and voltage distortion. The type of harmonic source that is amplified by the resonance condition is different for the two circuits shown in Figure 5.2:

- Series resonance amplifies harmonic voltage sources: at harmonic frequency h, the circuit current is limited by the resistance only, and it is calculated as

$$I_h = \frac{V_h}{Z_h} = \frac{V_h}{R} \text{ [A]}$$

- Parallel resonance amplifies harmonic current sources: in this case, at harmonic frequency h, the circuit voltage is obtained as the product between the current source and the resistance, which is a large number:

$$V_h = \frac{I_h}{Y_h} = \frac{I_h}{G} \text{ [V]}$$

The formulae above indicate that different values of circuit resistance may significantly reduce the effect of a resonance condition. Circuit resistance is an indication of the damping of a power circuit, and therefore, it is very important to model this parameter accurately to be able to assess correctly the levels of resonance conditions that may be encountered in a complex power system (Section 5.3.1).

5.2.2 Frequency scans

Frequency scans are a powerful tool used to visualize the frequency variation of the impedance (or admittance) of a circuit. Two examples of frequency scans are shown together with the simple circuits of Figure 5.2. The first graph shows the variation of the impedance magnitude as a function of the frequency. The second graph shows the variation of the impedance phase angle as a function of the frequency.

For the example shown in Figure 5.2(a), at 0 Hz the impedance of the circuit is approximately 400 Ω, which corresponds to the capacitive reactance. With increasing frequency, the capacitive reactance is compensated by the inductive reactance, and therefore, the total impedance of the systems reduces, until reaching a minimum value at a frequency close to 130 Hz. The minimum value of impedance corresponds to the circuit resistance, which in this case is 10 Ω. For frequencies above 130 Hz, the total impedance of the circuit increases, until stabilizing to an asymptotic value that corresponds to the circuit inductive reactance. The phase angle variation reflects the impedance variation with the frequency: for very low frequencies, the phase angle is close to −90 degree, whereas for higher frequency, it is close to 90 degree. The phase angle is equal to zero at the resonance point, where the circuit impedance is resistive.

For the example shown in Figure 5.2(b), the resonance condition is obtained at approximately 100 Hz. In this case, it is more difficult to detect the variation of the impedance at frequencies that are far from the resonance point. The phase angle plot helps visualizing the variation of the impedance behavior, which is capacitive at low frequencies and inductive at high frequencies.

For simple circuits, frequency scans can be obtained by writing the expression of the circuit impedance and admittance and calculating the values at different frequencies. For circuits consisting of numerous circuit components, this approach is unpractical, and computer models are used to calculate the frequency scans and identify the resonance conditions.

Software tools used for harmonic studies calculate the frequency scan at a specific point of a circuit by applying a harmonic source. In general, the circuit impedance is of interest; therefore, a voltage source is applied. The frequency of the harmonic source is varied (e.g., between 60 and 3,000 Hz), and for each frequency, the current amplitude at the output of the source is measured. The harmonic impedance at each frequency is calculated in complex form as $\overline{Z}(h) = \overline{V}(h)/\overline{I}(h)$, and two plots are obtained: (1) harmonic impedance amplitude versus frequency, and (2) harmonic impedance phase angle versus frequency.

5.3 Power system model

Figure 5.1 indicates that the power system model used for harmonic studies consists in a series branch in which the impedance is a function of the frequency. For each frequency, the system impedance is not simply represented by a series connection of inductance and capacitance, therefore describing a point in the impedance plane, but rather by an "area." This area, referred to as "impedance sectors," represents the variation of system impedance as a result of varying load and generation levels, switching of shunt elements and lines outages [1].

Calculating the harmonic impedance sectors for the power system surrounding the shunt FACT device is probably the most challenging and critical part of building the harmonic model shown in Figure 5.1. Performing this task actually requires building a separate model that is used exclusively for the purpose of deriving the harmonic impedance of the system.[4]

The next sections will describe in detail the process that leads to the calculation of the impedance sectors for a power system. This process consists of the following steps:

- Build power system model
- Select operating conditions
- Calculate frequency scans for each operating conditions to determine system impedance and phase angle as a function of the frequency
- Calculate resistance and reactance for each operating condition and for each frequency by means of impedance points
- Describe the variation of resistance and reactance at each frequency by means of impedance sectors.

5.3.1 Power system components

The extent of the network to be modeled for calculating the impedance scans depends on the characteristics of the power system, and it is difficult to generalize. One approach is to start by modeling a relatively small area of the system, retaining sufficient elements to incorporate all the contingencies to be studied. The analysis is then repeated by adding a portion of the network, and this process is repeated until there is no significant change in the harmonic impedance characteristics [31]. This approach also provides a sort of sanity check to the model, as drastic variations of the impedance scans may indicate a problem with the model in a certain area.

Sometimes, it is difficult to gather accurate information about all system components, but this task should be carried out diligently because small variations of the component parameters (capacitance, inductance, and resistance) and a consequent shift of the resonance points in the frequency scan may have a large impact on the system harmonic behavior.

The following paragraphs provide only general guidelines for the modeling power system components, whereas more detailed information can be found in [8,26].

- Lines and cables

  In terms of circuit components, transmission lines are combination of series resistance and inductance and shunt capacitance. If the information is available, it is preferred to model the transmission line geometry to accurately determine the equivalent inductance and capacitance.

[4]In some cases, it may not be possible to retrieve the detailed system information that is necessary to calculate the frequency scans. For these situations, Reference [30] introduces the use of "generic" impedance sectors based on the system short circuit level and the harmonic order. This approach is highly conservative and should be considered only when no other information on the system is available. An alternate method of determining the system impedance consists in taking harmonic impedance measurements at the PCC [13]. Although being capable of providing valuable information [39], however, harmonic impedance measurements are rarely performed in the practice.

In some cases, the information provided for modeling of transmission lines consists of the lumped inductive reactance and lumped capacitive reactance. In this case, the transmission line can be modeled by means of pi-sections by using the total line inductive and capacitive reactance. In other more extreme cases, the only information available for a transmission line is the total length and the voltage levels. In these cases, typical values of inductance and capacitance per unit length can be used.

The values of charging currents can be used to verify the correct modeling of the capacitance of each line. The value of the charging current should be close (within 1%) to the expected value calculated on the basis of the voltage and the line capacitance.

Cables are mostly capacitive elements, and therefore, they have a significant impact on the frequency scans. Therefore, it is highly recommended to model the geometry of cables to accurately model their equivalent capacitance.

- Transformers
The transformer parameters that mostly impact the frequency scans are reactance and resistance. Modeling stray capacitances is generally not required as they are quite small, and they contribute to the total system impedance only at high frequencies. Saturation data are not included in the model as inrush current is a transient phenomenon, whereas harmonic studies assume steady-state conditions.[5]

Transformer reactance should be modeled by taking into account the impact of tap position, as neglecting this parameter may results in errors up to 30% [31]. This error is minimized if the software used to calculate the frequency scans performs a load flow prior to solving the network.

Transformer losses, and therefore resistance, vary with the frequency according to this formula [9]:

$$P_h = R_{dc}I^2 + P_s[Wh^q + (1 - W)h^p]$$

where h is the harmonic order, P_h is the loss at harmonic frequency, W (0.25–0.38) is the winding stray loss, P_s is the total stray loss, $q = 1.9$–2, $p = 1$ or 1.4. In general, the software tools used to calculate the frequency scans take into account the frequency dependency of the transformer resistance by using typical values of the parameters above.

- Shunt capacitors and shunt reactors
Shunt capacitors have a large impact on frequency scans, as their large capacitance affects the total system impedance at low frequencies. In addition, these components are switched on and off very frequently and therefore cause significant shifts of the resonance points. As a consequence, it is very important to identify all shunt capacitors within a few busses from the PCC and model each individual bank separately.

[5]A saturated transformer generates harmonic currents in the low order (between 2nd and 7th harmonic). This phenomenon is the subject of Insulation Coordination Studies. Under normal operating conditions, transformers are not saturated, and therefore harmonic generation from transformers is not modeled.

Shunt reactors impact the frequency scans by shifting the resonance frequency and by adding damping to the system. Therefore, it is important to model accurately the quality factor (QF) of the reactors, which is frequency dependent:

$$\mathrm{QF}(h) = \frac{\omega L}{R(h)}$$

where L is the bank inductance, and R is the bank resistance.

- Loads
  Loads are an important component of the system model, as they significantly contribute to the damping of the frequency scans and, to a smaller extent, to the shift of the resonance frequencies. Loads are modeled as a combination of a resistive component and an inductive component. Typically, the utilities provide two data to model loads: active power and reactive power (MW and MVAr), or apparent power and power factor (MVA and PF). From these two values, the inductive and resistive components of the load can be calculated. In some cases, loads may be modeled as connected directly to the transmission system. In reality, loads are connected to lower voltage levels; therefore, if the information is available, it is preferred to model the coupling transformer.

  For the purpose of harmonic studies, the equivalent load model can be either a series combination of inductance and capacitance, or a parallel combination [21]. Series representation is suitable at low frequencies but doesn't provide enough damping at high frequencies. A combination of a parallel resistance and inductance is more appropriate for residential and similar loads dominated by resistive devices such as incandescent lights and some power electronic devices. When detailed information on the nature of the load is not available, the best approach consists in dividing the load in a parallel component and a series component, as shown in Figure 5.3.

  After choosing the most appropriate model, it is necessary to study different load levels, such as 100% load, 50% load, and 10% load. Load levels on the transmission system keep changing, and they impact the frequency scans in terms of both the damping and the resonance frequencies, as shown in Figure 5.4 [31].

- Nearby FACTs devices
  If there are other FACT devices electrically close enough to have a significant impact on the network impedance, they must be modeled explicitly, rather than being included as a lumped element within the network, with its associated shunt filters, together with the variation in the number and types of filters which may be connected with varying load [31].

5.3.1.1 Selection of operating conditions and calculation of frequency scans

The power system configuration keeps changing through the day due to numerous conditions such as outages for maintenance or for failure, varying load levels and shunt capacitors in service. As a result, the system harmonic impedance changes in time. To capture the variation of the system harmonic impedance and determine its

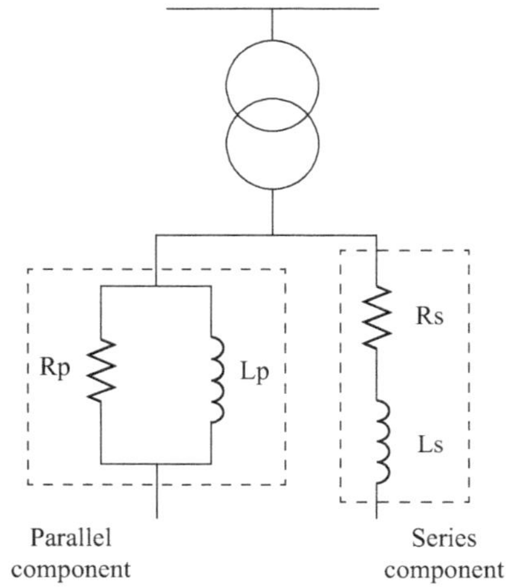

Figure 5.3 Example of load model including a parallel component and a series component [21]

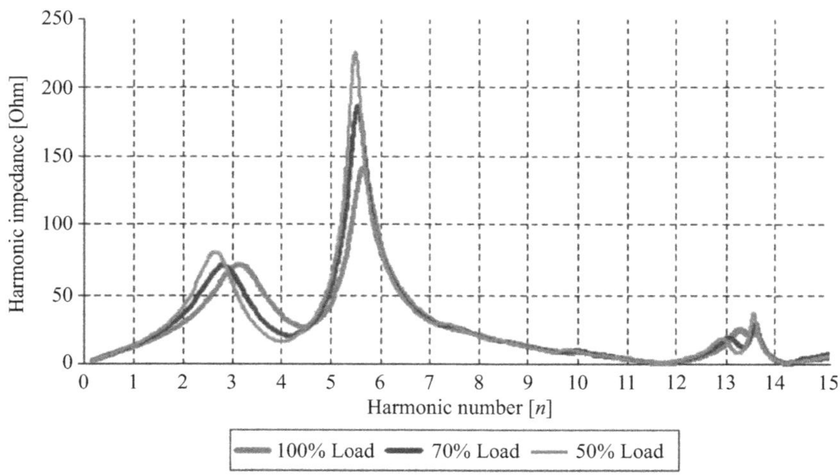

Figure 5.4 Impedance variation as a function of the load level [8]

impact on the harmonic levels, it is important to calculate the impedance scans for a comprehensive set of system conditions.

Examples of system conditions to be considered for the purpose of harmonic studies include the following [31]:

- System load and generation for maximum demand condition
- System load and generation for minimum demand condition
- System load/generation variation for intermediate demand conditions

- Status of reactive compensation devices (switched capacitors and reactors). All possible combinations of shunt reactive compensation at or close to the PCC must be considered because they may result in significantly different resonance conditions
- Transmission outages (contingency and planned). The outages that need to be considered are a function of the way in which the utility operates the network (i.e., $N-1$, $N-1-1$, $N-2$ criteria). For different outages, the components should be rated to withstand additional harmonic loading while not necessarily achieving performance limits.[6] The classification of the contingencies under which harmonic limits may be exceeded should be defined by the utility.

The combination of various operating conditions such as the ones listed above may easily result in hundreds of different system configurations. For each configuration, a frequency scan (Section 5.2.2) is calculated to determine the harmonic impedance variation.

Depending on the tool that is used for the analysis, the calculation of frequency scans can be automated by using a script that automatically switches in and out different elements (such as lines and shunt banks) and changes the load levels. The automation allows studying a large number of system configurations and reduces the time it takes to calculate the frequency scans.

5.3.1.2 Calculation of impedance points

Frequency scans provide the information on the frequency response of the system in terms of impedance and phase angles for each case. For the purpose of harmonic studies, this information included in the frequency scans needs to be organized in terms of resistance and reactance for each frequency [30].

The process of converting the impedance information between these two formats will be explained by means of a very simple example, where it is assumed to have three frequency scans, and the results are considered for four frequencies, labeled as $f1$, $f2$, $f3$, and $f4$.

1. For each frequency of interest, the frequency scans provide the impedance Zk_n and phase angle θk_n, where k is the frequency of interest, and n is the frequency. In the considered example, 12 points are obtained, as shown in Table 5.1.
2. The resistance and reactance of each point is calculated, and the new values are shown in Table 5.2.

$$Rk\_n = Z\cos(\theta k\_n)$$
$$Xk\_n = Z\sin(\theta k\_n)$$

3. The results are grouped by frequency as shown in Table 5.3.
4. For each frequency, the results can be displayed in the form of RX plots. Figure 5.5 shows an example of scattered plot for the frequency $f1$. The resistance is shown on the x-axis, whereas the reactance is shown on the y-axis.

[6]Section 5.5 will explain in more detail the distinction between the operating conditions for harmonic performance studies and the operating conditions for harmonic rating studies.

Table 5.1 Results of frequency scans as impedance and phase angle versus frequency

f (Hz)	Frequency scan 1		Frequency scan 2		Frequency scan 3	
	Z (Ω)	θ (rad)	Z (Ω)	θ (rad)	Z (Ω)	θ (rad)
$f1$	$Z1\_1$	$\theta1\_1$	$Z1\_2$	$\theta1\_2$	$Z1\_3$	$\theta1\_3$
$f2$	$Z2\_1$	$\theta2\_1$	$Z2\_2$	$\theta2\_2$	$Z2\_3$	$\theta2\_3$
$f3$	$Z3\_1$	$\theta3\_1$	$Z3\_2$	$\theta3\_2$	$Z3\_3$	$\theta3\_3$
$f4$	$Z4\_1$	$\theta4\_1$	$Z4\_2$	$\theta4\_2$	$Z4\_3$	$\theta4\_3$

Table 5.2 Results of frequency scans as resistance and reactance versus frequency

f (Hz)	Frequency scan 1		Frequency scan 2		Frequency scan 3	
	R (Ω)	X (Ω)	R (Ω)	X (Ω)	R (Ω)	X (Ω)
$f1$	$R1\_1$	$X1\_1$	$R1\_2$	$X1\_2$	$R1\_3$	$X1\_3$
$f2$	$R2\_1$	$X2\_1$	$R2\_2$	$X2\_2$	$R2\_3$	$X2\_3$
$f3$	$R3\_1$	$X3\_1$	$R3\_2$	$X3\_2$	$R3\_3$	$X3\_3$
$f4$	$R4\_1$	$X4\_1$	$R4\_2$	$X4\_2$	$R4\_3$	$X4\_3$

Table 5.3 Results of frequency scans grouped by frequency

f (Hz)	R (Ω)	X (Ω)
$f1$	$R1\_1$	$X1\_1$
$f1$	$R1\_2$	$X1\_2$
$f1$	$R1\_3$	$X1\_3$
$f2$	$R2\_1$	$X2\_1$
$f2$	$R2\_2$	$X2\_2$
$f2$	$R2\_3$	$X2\_3$
$f3$	$R3\_1$	$X3\_1$
$f3$	$R3\_2$	$X3\_2$
$f3$	$R3\_3$	$X3\_3$
$f4$	$R4\_1$	$X4\_1$
$f4$	$R4\_2$	$X4\_2$
$f4$	$R4\_3$	$X4\_3$

The information provided in the *RX* plot allows visualizing quickly the variation of the system impedance at a specific point for each frequency [30]:

- Points that are close to the origin indicate points which are series resonances. Points which have a large resistance correspond to parallel resonances.
- Points with positive reactance are inductive, while points with negative reactance are capacitive.

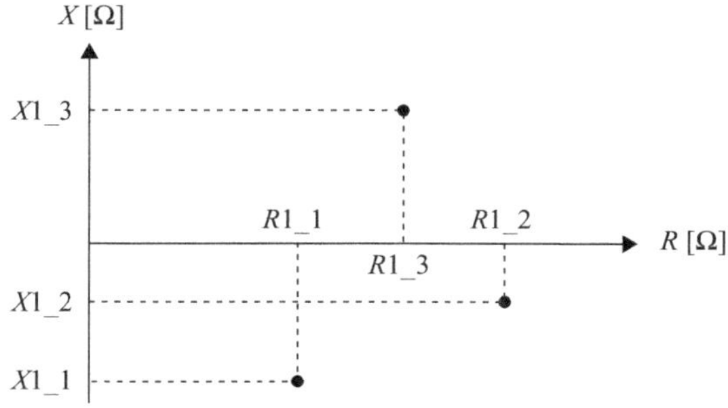

Figure 5.5 Example of RX plot for frequency f1, from the data points of Table 5.3

- At low frequencies, the impedance points mostly lie in the upper portion of the diagrams, indicating that the frequency response is inductive.
- With increasing harmonic orders, more points are expected to be found in the lower portion of the diagrams, due to additional influence of the capacitance.
- Points which are too close to the origin may be indicators that the damping of the system is not modeled correctly.
- If the impedance points cover only a very small area of the diagram, this can be due to a high short-circuit level: the system is very stiff, and therefore, outages and variations of the load levels are not impacting significantly the frequency response.

5.3.1.3 Calculation of impedance sectors

For harmonic studies, it is the norm to study from hundreds to hundreds of thousands of system configurations, meaning that in general, the number of points included in an *RX* plots for each frequency is extremely high. This is the reason why it is unpractical to determine the levels of harmonic distortion for each impedance point. Instead, another solution is typically implemented and endorsed by standards such as [25,30,31] which consist in using a geometric shape to include all points at a specific frequency. These shapes are referred to as "impedance sectors" in this chapter.

Two of the advantages of using impedance sector are as follows:

- They can be easily programmed in the computer tool used for harmonic analysis.
- They include some additional area around the impedance points, thus taking into account for model inaccuracies.

Some examples of impedance sectors typically used for harmonic studies are shown in Figure 5.6. Each impedance sector is defined by different parameters:

1. *Circular sector*: Defined by maximum and minimum impedance (Z_{min} and Z_{max}) and maximum and minimum angle (θ_{min} and θ_{max}).
2. *Circle*: Defined by radius (r), maximum and minimum angle (θ_{min} and θ_{max}) and minimum resistance (R_{min}).

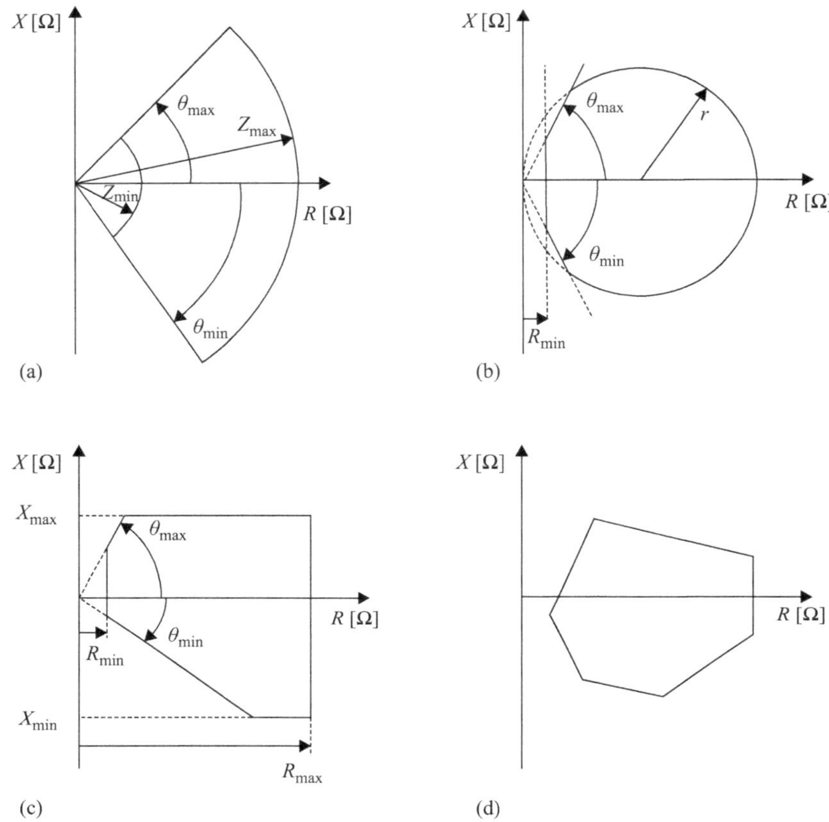

Figure 5.6 Examples of shapes used for impedance data point analysis: (a) circular sector, (b) circle, (c) polygon, and (d) generic shape

3. *Polygon*: Defined by the maximum and minimum angle (θ_{min} and θ_{max}), minimum and maximum resistance (R_{min} and R_{max}), and maximum and minimum reactance (X_{min} and X_{max}).
4. *Generic shape*: If there are not specific requirements, a generic shape that includes all points can be used.

Standards [30,31] include a detailed discussion of the advantages and disadvantages of using different shapes and of the use of this approach in the practice.

Different shapes may be used for harmonic performance studies and for harmonic rating studies. As will be discussed more in detail in Section 5.5, some of the standards on harmonic distortion such as [29,36] allow meeting performance criteria for 95% of the system-operating conditions. In addition, harmonic rating studies may assume a broader range of operating conditions to provide a robust rating of the equipment. Therefore, two sets of harmonic shapes may be used for performance studies and for rating studies, including different sets of operating conditions [31]. This is not always done in the practice because it may be difficult to differentiate between the operating conditions where harmonic limits need to be met from the operating conditions in which the limits can be exceeded. Rather, as a first approach, it is preferred to meet performance under all operating conditions. Only later, if some operating condition results in excessive harmonic distortion and are unlikely to be met in the practice, some considerations may be done to exclude these conditions from the harmonic performance analysis.

5.3.2 Background voltage distortion

The level of harmonic distortion on the transmission system is relatively low, since the harmonic generating loads are installed at lower voltage levels, and transformers area high-pass filter for harmonics. Other sources of harmonics at transmission system levels are given by large industrial loads, wind plants, and FACTs devices [1]. As a result, it is generally difficult to isolate and identify harmonic-producing loads on transmission systems, whereas this task is sometime easier at the distribution level, where the loads are installed.

The background distortion tends to fluctuate in time due to a combination of numerous factors, such as harmonic sources tend to vary due to the changes of load levels and system strength, and the resonance points shift due to outages and switching of shunt elements [6].

Although not common, high levels of background distortion impact the design of a shunt FACT device in two ways. From the point of view of harmonic performance, the network owner or operator may require very low-harmonic emission levels from the shunt FACT device under consideration, to maintain the overall level of harmonic distortion on the system below specified limits (Section 5.5.1.1). This requirement is likely to impact the cost of the shunt FACT device, since reducing harmonic emission levels requires increasing the size of the harmonic filters or installing additional ones. From the point of view of harmonic rating, high level of harmonic distortion in the system is likely to result in higher harmonic currents flowing through the shunt FACT device components, and therefore in an increase of the cost of the equipment to withstand further stresses [31].

Figure 5.1 shows that the model of transmission system background voltage distortion is a simple voltage source at different harmonic frequencies modeled behind the network harmonic impedance. This is the methodology endorsed by most of the standards and used in the industry, although other approaches are possible, such as applying the harmonic source at the PCC [31].

The representation shown in Figure 5.1 implies two simplifications:

1. The background harmonic distortion is modeled as a single source rather than being scattered across the transmission system.
2. The harmonic spectrum is fixed in time rather than varying according to the system conditions.

Concentrating the source of harmonic distortion at one point of the system is indicative of the condition that the shunt FACT device only sees harmonic distortion at the PCC, and therefore, the exact location of the harmonic source is not relevant for the analysis.

The harmonic spectrum typically is representative of the highest level of distortion, and all harmonics are assumed to be in phase. This approach is generally quite conservative and results in a robust rating of the equipment.

5.3.2.1 Determining the harmonic spectrum

The most accurate approach to modeling background harmonic distortion consists in performing harmonic measurements at the PCC, or at the nearest substation, prior to performing harmonic studies, as discussed in Section 5.5.1.3. In the practice, this is rarely done, mostly because the cost of performing such measurements is generally not justified.

A second approach is to gather information on the levels of harmonic distortion across the transmission system. This data may be gathered from recordings at substations where harmonic meters are installed, or from data collected by the transmission system operator. This approach allows identifying issues with particular harmonics which may create problems. For example, the harmonic measurements performed on 28 stations on Hydro-Quebec system in [2] indicate concerns with the fifth and eleventh harmonics. Existing harmonic data can be used to extrapolate level of harmonic distortion at other substations and allow a realistic modeling of the harmonic distortion on the system.

If no information is available, one possible solution is to base the levels of harmonic distortion on the planning limits defined in the standards, such as [28,36], or on the limits defined by the local transmission system operator. This approach is controversial because, while in general the levels of harmonic distortion on the transmission system are well below the limits, in some specific cases, the limits are exceeded, due to the presence of a large disturbing load or due to a resonance condition [2].

5.3.3 Conclusions on system model

The previous sections described the general procedure to derive the transmission system model to be used for harmonic studies.

Describing accurately, the harmonic behavior of transmission system requires an accurate modeling of all system components and investigating a large number of operating conditions. This effort requires the use of dedicated software tools and can be time-consuming but, if performed diligently, can provide accurately insights in the transmission system harmonic behavior and help foreseeing harmonic issues and providing mitigating solutions.

For example, in some cases, frequency scans may be able to indicate harmonic problems caused by switching shunt banks, and therefore, avoiding these operating conditions, if possible, will help reducing the overall levels of harmonic distortion on the system.

5.4 Shunt FACT device model

Building the harmonic model of a shunt FACT device is easier than building the model of the transmission system. The main reason is that a shunt FACT device includes a limited number of components compared to the transmission system, and therefore, a smaller number of configurations are studied. For harmonic studies, detailed information on the control algorithms is not required, since the harmonic currents or voltages generated by the shunt FACT devices are calculated in steady-state operation.[7]

The generic model of a shunt FACT device for harmonic studies shown in Figure 5.1 consists of:

- Elementary circuit components, such as transformers, inductors, capacitors, and resistors
- One or more harmonic sources, depending on the configuration of the power electronic switches

Some shunt FACT devices are equipped with filters or branches which are switched on and off based on the operating point. The switching of capacitive branches have an impact on the resonance point of the system and therefore on the results of the harmonic analysis.

The next sections illustrate how to derive harmonic models of some common shunt FACT devices. The approach explained in the next sections can be used to derive the model of any devices controlled by means of power electronic switches.

5.4.1 Static VAr compensator (SVC)

SVCs are used to regulate the transmission system voltage by absorbing or injecting reactive power, and they consisted of large inductors and capacitor banks switched by means of thyristor valves [17]. A typical SVC includes the elements shown in Figure 5.7:

- Coupling transformer, which reduces the transmission voltage to medium voltage levels (typically below 30 kV).
- Thyristor controlled reactor (TCR): consists of an antiparallel connected pair of thyristor valves in series with a linear air-core reactor. The valves act like a bidirectional switch, and their conduction angle can be adjusted between minimum conduction and maximum conduction. Therefore, the amount or reactive power absorbed by the TCR is continuously adjustable.
- Harmonic filters: they are installed to absorb the harmonics generated by the TCR(s). The number of filters and their tuning depend on the design and the network conditions, but in general, the harmonics which are more of concern

[7]Steady-state operation in this contest includes the study of transient conditions such as overvoltage or frequency variation, which may last up to a few cycles [16]. From the point of view of the control system, this time range can be considered "steady state," since the response time of shunt FACT devices is in the order of some 10 ms [12].

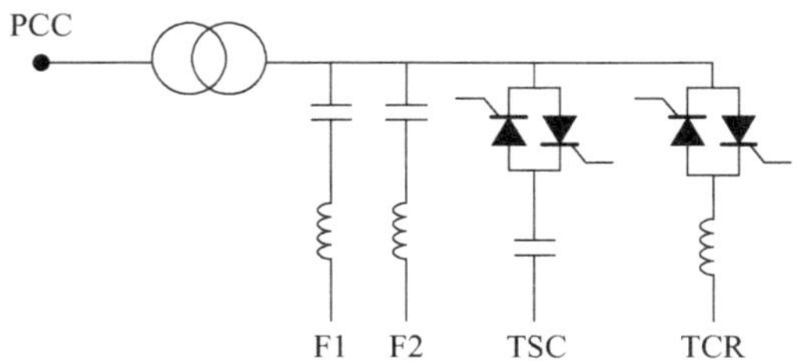

Figure 5.7 Single line diagram of a static VAr compensator consisting of four branches: two filters (F1 and F2), one Thyristor Switched Capacitor (TSC), and one Thyristor Controlled Reactor (TCR)

for SVCs are the third, fifth, and seventh harmonic order. SVCs typically employ single-tune filters, but other filter topologies are also used (Chapter 4). The second purpose of harmonic filters is to inject capacitive reactive power into the system. The filter can be either fixed or switchable.

- Thyristor switched capacitor (TSC): consists of an antiparallel connected pair of thyristor valves in series with a capacitor and a current limiting reactor. The thyristor switch allows the conduction for an integer number of half-cycles and the capacitor is not phase controlled. The current limiting inductor allows reducing switching transients and is typically tuned to create a series resonant frequency between the fourth and the fifth harmonic. The TSC acts as a special type of filter which is controlled by means of thyristor valves.

An SVC may include multiple TSCs, TCRs, or filters or may not include some of these components. In general, TSCs and TCRs are delta connected, while harmonic filters are wye connected.

5.4.1.1 TCR harmonic model

In an SVC, harmonics are generated by the TCRs. As described in the previous section, TSCs are controlled to be fully on or fully off and therefore do not generate harmonics. Since the reactive power output of an SVC needs to be controlled continuously, TCR modulation is used to finely tune the reactive power output when the TSCs are switched in [17].

As a result of the modulation, the amplitude of the harmonic currents generated by the TCR varies in time. This concept is explained by means of an example, based on the simple single-phase circuit shown in Figure 5.8. The circuit consists of a series branch including a voltage source, a resistor, and a thyristor.

The thyristors can be turned on at any point on wave, and they turn off naturally when the current is zero [18]. Figure 5.9 illustrates the branch current "I" for two different conduction angles: 170 degrees and 90 degrees (where the conduction angle is measured from the time the thyristor is switched on to the time it naturally turns off at zero crossing).

The harmonic content of the waveform shown in Figure 5.9 is calculated by applying the Fourier transformation. Since the waveforms are symmetric with

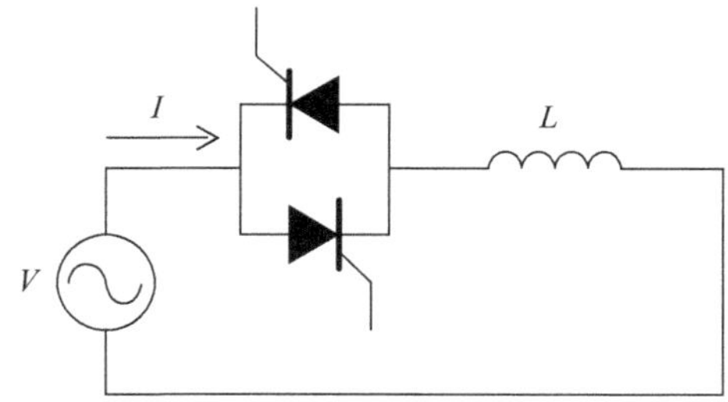

Figure 5.8 Simple series branch for illustration of TCR operation. The branch includes a sinusoidal voltage source, a pair of antiparallel connected thyristors, and an inductor

respect to the y axis, only odd harmonics are expected, and they are calculated as follows:

$$I_h = \frac{V}{\omega L}\frac{4}{\pi}\left[\frac{\sin\alpha\cos(h\alpha) - h\cos\alpha\sin(h\alpha)}{h(h^2-1)}\right]$$

where $h = 2k + 1$, with $k = 1, 2, 3 \ldots$

Figure 5.10 presents the harmonic content for the two current waveforms shown in Figure 5.9, up to 900 Hz. These graphs indicate that the harmonic content of the current waveforms depends on the conduction angle. The harmonic content is lower for the first current waveform, which is closer to a sinusoid.

For a three-phase TCR, under ideal and balanced conditions, the triplen harmonics (3rd, 9th, 15th, ...) [1] circulate within the TCR branch and do not flow on the MV bus.

In the practice, numerous operating conditions impact the levels of harmonic distortion caused by a TCR, including tolerances of components, frequency deviation, unbalance in the firing pulses generation between the three phases. Given that, these factors add to the variation of harmonic generation based on the operating condition, the harmonic spectrum is calculated by simulations rather than in closed form, by using dedicated tools.

In general, the largest harmonic currents injected by a TCR are the third, fifth, and seventh components. The passive filters (Chapter 7) installed on the secondary bus are tuned to absorb the highest harmonics.

5.4.1.2 Consistent and nonconsistent harmonic spectrum

The concept of consistent and nonconsistent harmonic spectrum applies to all shunt FACT devices, and in general to all devices controlled by power-electronics switches [30].

A *consistent* spectrum consists of harmonics generated at a single and feasible operating condition for the device under consideration. A *nonconsistent* spectrum is the combination of the highest harmonic values calculated across the full range of operating conditions. A nonconsistent set, therefore, does not correspond to a real operating condition but should be rather considered a "worst-case scenario."

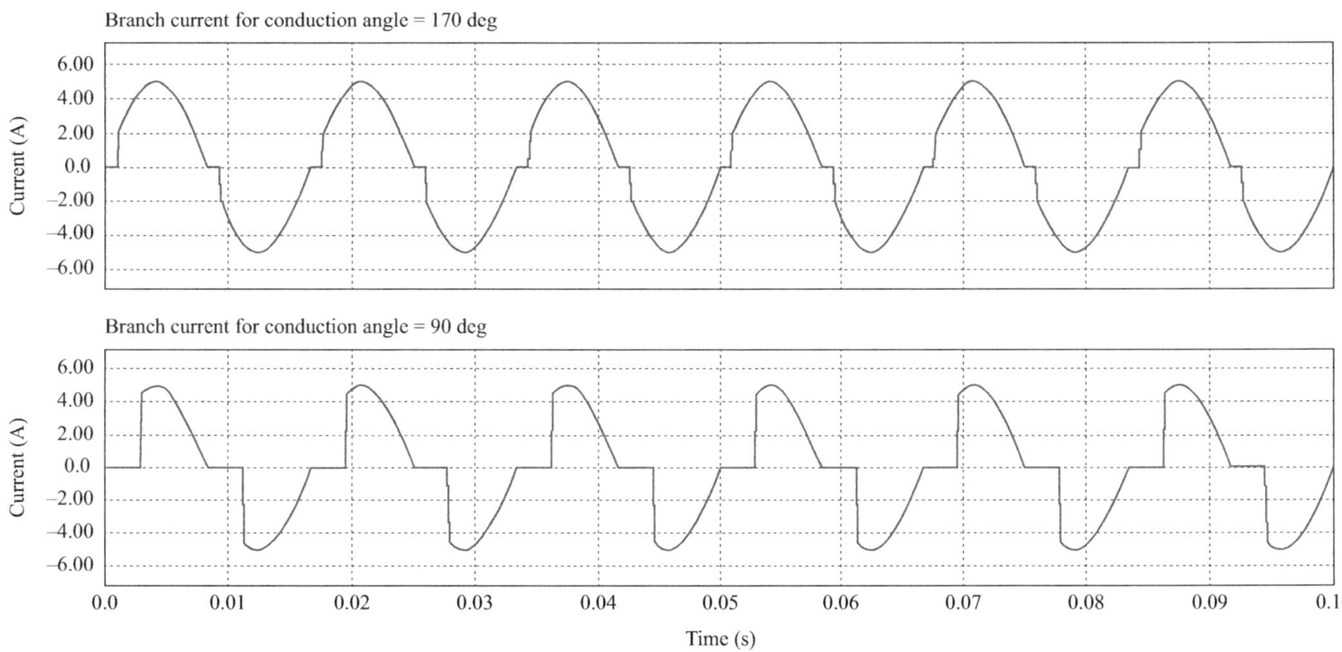

Figure 5.9 Current waveform in the series branch of Figure 5.8 for two thyristor conduction angles

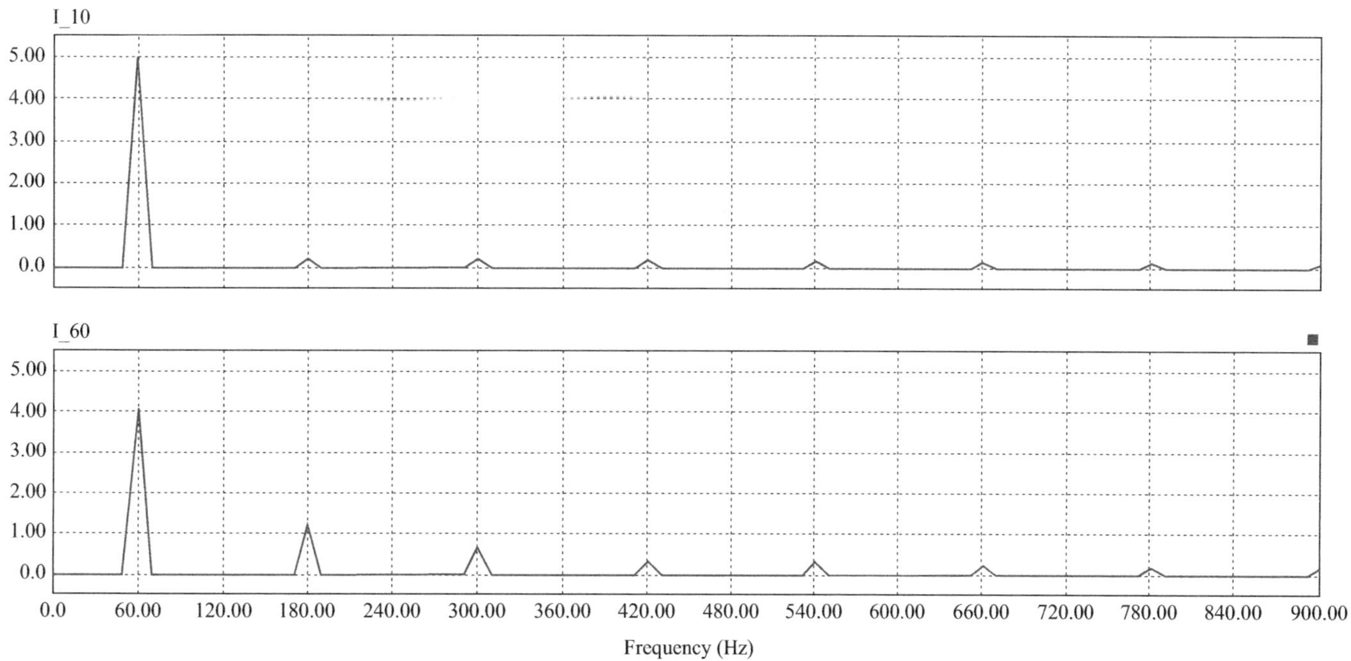

Figure 5.10 Fourier analysis for the current waveforms of Figure 5.9

If the first approach is used, then the harmonic analysis is typically performed for numerous operating conditions, and for each case, the model of the shunt FACT device under consideration includes a different harmonic source. If the second approach is chosen, then the analysis is performed only once, and the harmonic model consists of one single spectrum.

Both approaches are used in the practice, and they impact the results of the harmonic studies, as it will be discussed in Section 5.5.

5.4.1.3 SVC harmonic model

The harmonic model for the SVC shown in Figure 5.7 is illustrated in Figure 5.11:

- The SVC transformer is modeled following the guidelines included in Section 5.3.1.
- Filters and TSCs are modeled by means of basic circuit elements. In many installations, filters are wye connected, while TSC are delta connected. This needs to be taken into account when calculating the equivalent impedance for the single-phase model.
- The TCR is modeled by means of a harmonic current source.

The model shows a switch added in series with the TSC branch. This switch is included because the TSC is switched on only when the capacitive MVAr injection from the SVC is above a certain threshold. When the harmonic study is performed, different operating conditions of the SVC are studied, with and without TSC, and they can be derived from the same model if switches are included. In this case, it is assumed that filters are always kept in service, and therefore, the switch is not required, as shown in Figure 5.11. To model the case where the TCR is out of service, the harmonic current source is removed.

5.4.2 Static synchronous compensator (SSC or STATCOM)

The STATCOM (or SSC) is a reactive power compensation device based on the voltage source converter (VSC) equipped with a d.c. capacitor [12]. An advantage

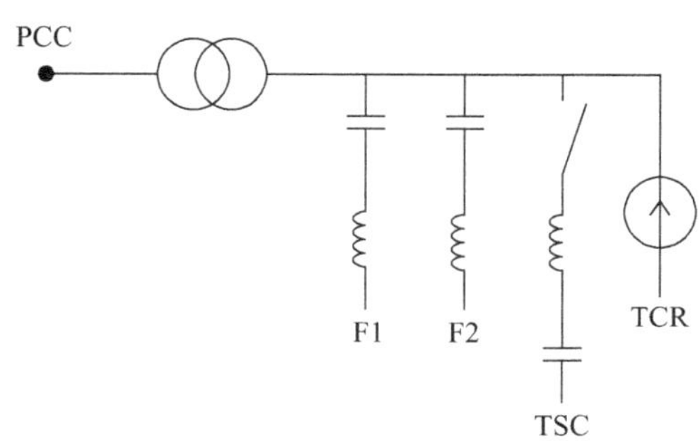

Figure 5.11 Harmonic model of the static VAr compensator shown in Figure 5.7

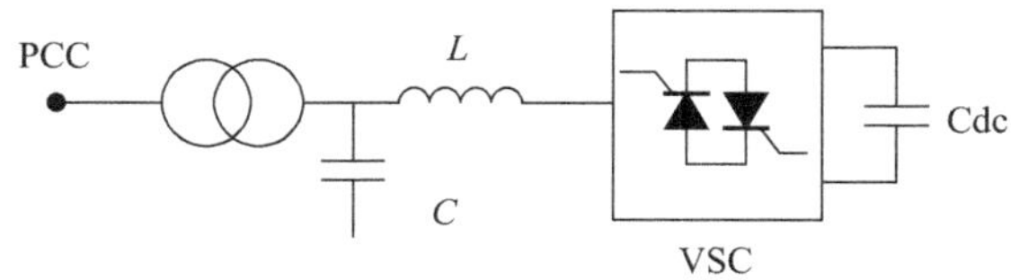

Figure 5.12 Single-line diagram of a STATCOM

of the STATCOM compared to the SVC is that it does not require large inductive and capacitive components to provide reactive power [23].

A typical STATCOM is made by the elements shown in Figure 5.12:

- Coupling transformer, which reduces the transmission voltage to medium voltage levels
- A d.c.–a.c. converter, which is composed by numerous elementary converters. The switches used in the elementary converters are typically IGBTs (insulated gate bipolar transistors) or GTOs (gate turnoff thyristors).
- A d.c. capacitor, which provides continuous d.c. voltage. The d.c. voltage is used to generate the three-phase output voltage at the converter terminals.
- A high-frequency filter in series with the transformer (LC in this specific example).

5.4.2.1 STATCOM as source of harmonics

The STATCOM operates by regulating the voltage at the converter terminals to follow a sinusoidal waveform. In order to do this, the power electronic switches are turned on and off following a predetermined sequence. Due to the high number of elementary converters and the high switching frequency of the GTOs, the three-phase output voltage of a STATCOM resembles quite well a sinusoid [3].

Numerous methodologies are available to control the switching operation of the elementary converters [5,18]. Figure 5.13 shows an example of an output waveform for one branch of a simple STATCOM device made of four full bridges. The configuration under study is shown in the left side of the figure. The middle part of the figure shows the reference voltage, the voltage across each bridge and the output voltage for the case of square modulation. The right part of the figure shows the same waveforms for the case of pulse-width modulation (PWM).

Square-wave modulation consists in switching the power electronic switches only when the reference voltage is above or below a certain threshold. PWM consists in switching the power electronic switches at much higher frequency, in the order of kHz. The switching frequency is determined by the frequency of the triangular waveform shown in Figure 5.13, referred to as "carrier." A high switching frequency reduces the levels of harmonic generation but increases the power losses [22]. Figure 5.13 shows that for both switching methods, the output voltage waveform is symmetrical, and therefore, only odd harmonics are generated.

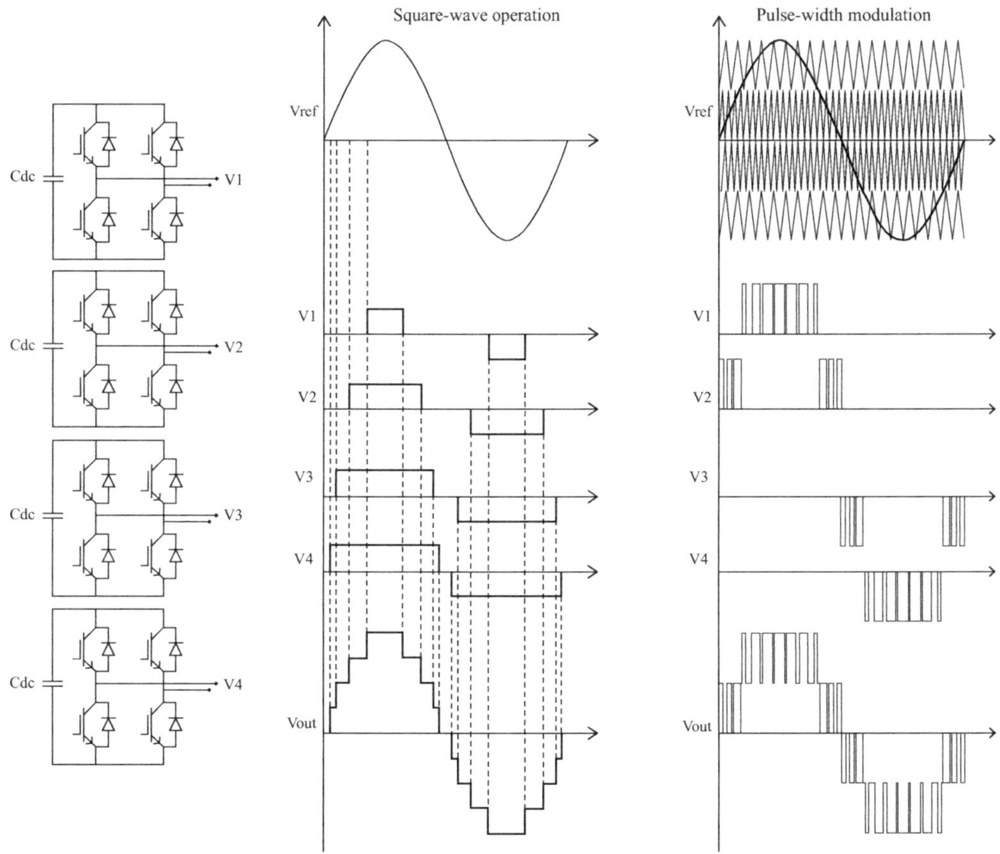

Figure 5.13 Phase of an inverter consisting of four elementary bridges: output waveforms for square-wave operation and pulse-width modulation (PWM)

For square-wave operation, the harmonic components for a STATCOM including $2N + 1$ level are calculated as follows [14]:

$$V_h = \frac{V_{dc}}{h}\frac{4}{\pi}\sum_{k=1}^{N}\cos(h\theta_k)$$

where $h = 2k + 1$, with $k = 1, 2, 3$. The conducting angles $\theta_1, \theta_2, \ldots, \theta_s$ can be chosen to minimize the voltage total harmonic distortion (THD). One common choice is to select the conducting angles so that the lower frequency harmonics (5th, 7th, 11th, and 13th) are eliminated.

For PWM, the output harmonics are generated only for frequency orders multiple of the switching frequency and for sideband harmonics. The harmonic of interest is the following [18]:

$$K_h = np \pm m$$

where K_h is the harmonic order generated, n is an integer number, p is the carrier frequency, and m is the modulation index (in general, $p \geq 21$ and $0 \leq m \leq 1$).

The formulae used to calculate the amplitude of sideband harmonics for PWM are quite complex and many textbooks, such as [18], include tables showing the amplitude of harmonic components as a function of the frequency. Table 5.4 shows an example of this representation, for a case where the carrier frequency is $p = 21$. The table indicates that all harmonics up to the sideband frequencies are negligible. In the practice, the harmonic spectrum is calculated by means of simulations for different operating conditions.

Table 5.4 Harmonic components for STATCOM controlled by means of PWM and carrier frequency = 21

Harmonic order	Amplitude (pu)
1	0.90
2	0
3	0.2
4	0
5	0
16	0
17	0.007
18	0
19	0.251
20	0
21	0.713
22	0
23	0.283
24	0
25	0.02
26	0

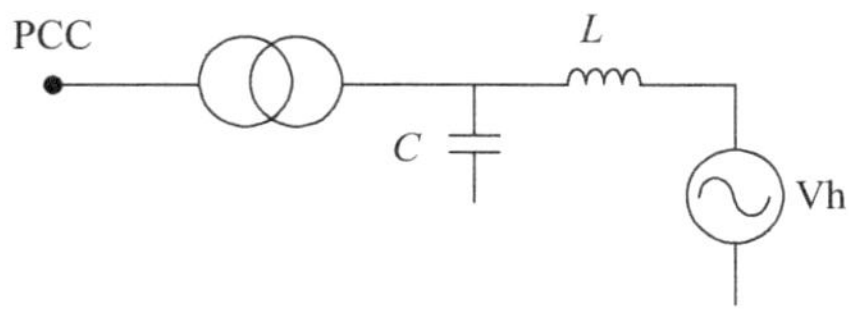

Figure 5.14 Harmonic model of the STATCOM shown in Figure 5.12

Given the high number of modules, from the point of view of harmonic generation, the STATCOM presents less concern for harmonic distortion at the "power frequency"[8] compared to an SVC. As a result, the filters used for the STATCOM are quite small, and tuned to higher harmonic orders [40].

5.4.2.2 STATCOM harmonic model

The harmonic model for the STATCOM shown in Figure 5.12 is presented in Figure 5.14:

- The STATCOM transformer is modeled according to the same guidelines included in Section 5.3.1.
- The high-pass filter is modeled by means of elementary components, according to the guidelines included in Section 5.3.1.
- The VSC is shown as a voltage source at several harmonic frequencies. As for the SVCs, the spectrum is a function of the operating point, and either a "consistent" or a "nonconsistent" set of harmonics may be used.

5.4.3 High-voltage dc (HVDC) transmission

HVDC transmission is used to transfer power between ac networks by means of a dc transmission line. The number of applications of HVDC includes connecting offshore wind plants to inland transmission systems or for interconnecting asynchronous networks [11].

Different configurations of HVDC links are possible, and one example is shown in Figure 5.15:

- On each ac bus, two transformers are connected in parallel, and they have two different phase shifts on the secondary. The purpose of this configuration is to cause voltage cancelation in the harmonics generated by the inverters.
- Harmonic filters are installed at the ac bus to sink the harmonics and to provide reactive power. The harmonic filters for HVDC are equipped with a breaker, since they are switched in and out at different output levels.
- The inverter and rectifier on each side are split into several modules, similarly to the STATCOM.

[8]Some concerns for STATCOM harmonic distortion are related to PLC frequency and telephone interference. As discussed in Section 5.1, these types of studies are outside the scope of this chapter.

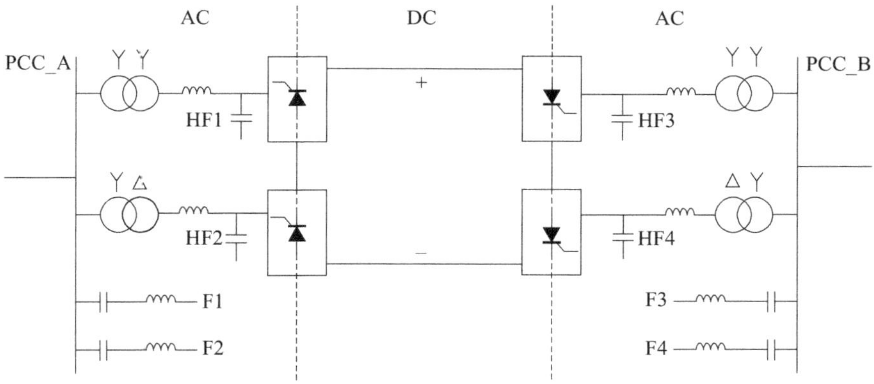

Figure 5.15 Example of HVDC configuration

Two main technologies exist for HVDCs [19,22]:

- Line commutated converter (LCC): the line voltage of the ac system forces the commutation of the switching devices. Controlled switched such as thyristors are typically used for this technology so that the commutation can be controlled to regulate the power exchanged between the ac and the dc system.
- Voltage source converter (VSC): this is the same converter topology described for the STATCOM and is referred to as "self-commutated converter" because it does not rely on the ac system voltage to control the commutation. The VSC operates with a smooth dc voltage provided by a large capacitor. Either GTOs (gate turnoff thyristors) or IGBTs can be used as semiconductor switches: GTO valves allow higher current ratings but lower voltage ratings than IGBT valves.

In the latest years, VSC HVDCs have become the standard topology because they provide a better control of the power exchange between the ac and dc network.

5.4.3.1 HVDC as source of harmonics

For a HVDC based on VSCs, the harmonic generation process is similar to a STATCOM and has been already described in Section 5.4.2.1.

However, since a HVDC involves higher power transfers, higher levels of harmonics are expected. Therefore, as shown in Figure 5.15, both high-frequency filters and low-frequency filters may be installed.

5.4.3.2 HVDC harmonic model

Since an HVDC is connected between two different busses of the power system, which are typically quite distant from each other, typically two different harmonic studies need to be performed at the two different busses. This implies calculating different impedance sectors and deriving two different HVDC models [15].

Figure 5.16 shows the harmonic model for one of the two busses shown in Figure 5.16. Each transformer is modeled independently, with the converter model

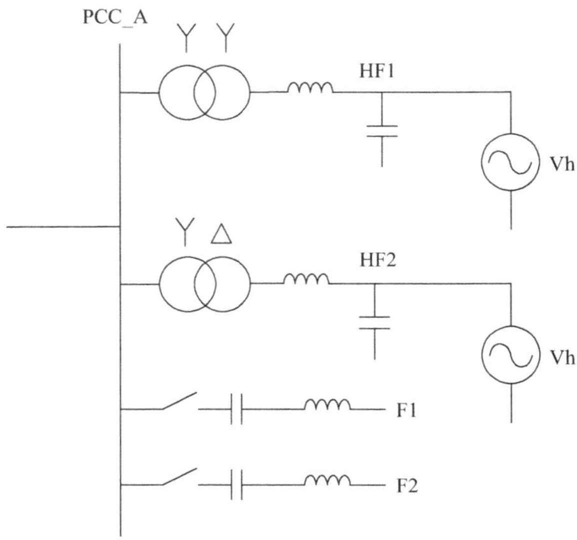

Figure 5.16 HVDC harmonic model for one of the two PCCs shown in Figure 5.15

and the high-pass filter model. The low-pass filters are modeled with a switch connected in series, so that they can be switched in and switched out for different configurations.

5.4.4 Conclusions on shunt FACT device model

Deriving the harmonic impedance model of a shunt FACT device is quite simple since the number of components is limited and only a few components are switched. Determining the harmonic spectrum is more complex, since the amplitude of each harmonic component varies across the operating range of the device. In addition, variables such as component tolerances and frequency variation further impact harmonic generation. Therefore, the harmonic spectrum is generally calculated by means of computer simulations for different operating conditions and for different combination of variables.

The examples provided in this section show an approach to modeling which can be extended to other shunt FACT devices and in general to all devices controlled by means of power electronic switches.

5.5 Harmonic studies

The harmonic model described in the previous sections can be used to perform two different types of harmonic studies [30]:

- *Harmonic performance study* (Section 5.5.1): the purpose of the harmonic performance study is to determine the impact of the shunt FACT device operation on existing equipment. This study is intended to demonstrate that the level of harmonics generated by the device(s) under consideration is below predefined limits.

- *Harmonic rating study* (Section 5.5.2): the purpose of the harmonic rating study is to determine the rating of equipment, based on the highest levels of harmonic distortion that the equipment may experience at any point during its lifetime. The harmonic current flow in transformers, capacitors, and inductors results in an increased equipment rating, which is higher than the rating calculated at fundamental frequency.

These two types of studies are performed independently because they cover different sets of conditions even if they start from the same model. Harmonic-performance studies are limited to study normal operating conditions, while harmonic rating studies typically address a largest set of operating conditions which the equipment may be subjected to during the lifetime of a project.

5.5.1 Harmonic-performance studies

5.5.1.1 Motivation and quality indices

The harmonic performance study is performed to guarantee that each device installed on the transmission system does not impact adversely the operation of other existing equipment. This is done by comparing the levels of harmonic emissions against limits imposed by standards or local system operators or owners.

The following paragraphs provide a short overview of the most common limits imposed to the equipment. For each project, other performance criteria may be applied. A more detailed discussion on the power quality indices is provided in Chapter 1.

- Voltage and current distortion
  Voltage distortion limits are imposed by utilities and transmission system operators to guarantee a safe and efficient operation of the power system. The limits are based on the values included in power quality standards such as [29,36] and national grid codes such as [4,7].

  Voltage distortion limits are presented as a percentage of the fundamental voltage component and are function of the voltage level and frequency order. With increasing voltage levels and frequency, the limits become stricter.

  Table 5.5 shows the harmonic limits included in [36], where the maximum distortion allowed for each harmonic is a function of the transmission system only. Table 5.6 shows the planning levels included in [29], where the term "planning levels" indicates the harmonic levels imposed to the consumers. High voltage (HV) systems operate within 35 and 230 kV, while extra high voltage (EHV) systems include all voltage levels greater than 230 kV. The limits described by [29] are more loose than the limits described by [36] for harmonics which are more common in transmission systems, while they are stricter for harmonics which are less of a concern, thus being more representative of the typical harmonic profiles encountered on transmission systems.

Table 5.5 Voltage distortion limits from [36]

Voltage levels at the PCC	Maximum level for individual harmonics* (%)
1 kV ≤ Vr ≤ 69 kV	3
69 kV ≤ Vr ≤ 161 kV	1.5
Vr ≥ 161 kV	1.0

**Note:* These values apply only to harmonics whose frequencies are integer multiple of the power frequency.

Table 5.6 Harmonic planning levels for HV and EHV systems from [29]

Harmonic order	Planning level (%)
2	1.4
3	2.0
4	0.8
5	2.0
6	0.4
7	2.0
8	0.4
9	1.0
10	0.47
11	1.5
12	0.43
13	1.5

Table 5.7 THD limits from [36]

Voltage levels at the PCC	THD (%)
1 kV ≤ V ≤ 69 kV	5.0
69 kV ≤ V ≤ 161 kV	2.5
V ≥ 161 kV	1.5

A second figure that is used to quantify the levels of harmonic distortion is a power quality index named THD. This quantity combines the harmonics at different frequencies, typically from the 2nd to the 50th harmonic order, and is expressed as follows:

$$\mathrm{THD} = \sqrt{\sum_{h=2}^{50} {V_h}^2 \%}$$

where V_h is the individual voltage distortion. For HV and EHV systems, Technical Committee 77 [29] recommends an indicative planning level of THD equal to 3%. Table 5.7 presents the THD limits included in [36].

For harmonic-performance studies, the limits on individual voltage distortion and THD are typically evaluated at the PCC shown in Figure 5.1. In other cases, harmonic performance limits may be evaluated at other busbars.

The standards and the national grid codes do not require that voltage distortion limits are to be met under all operating conditions. Both [29] and [36] discuss that the voltage distortion limits (or planning levels) must be met under conditions which cover 95% of the time. These are referred to as "normal operating conditions." Therefore, the standards acknowledge that under unusual short-term operating conditions, voltage distortion limits may be exceeded. For example, for a system with a low short-circuit level, tripping a transmission line in proximity to the PCC may lead to a temporarily high voltage distortion, but this condition may last only few seconds. The standards allow for this flexibility since the design of equipment which would result in 100% compliance with the harmonic limits may be prohibitively expensive [30].

In some cases, limits are imposed on the harmonic currents generated by the shunt FACT devices. The harmonic current limits are presented as a function of the individual harmonic components and of the total root mean square, thus using the same approach provided for voltage harmonic components (Chapter 1).

- Telephone interference

  Telephone interference is caused by inductive coupling between power lines and telephone lines. The inductive coupling is caused by harmonic currents generated by the shunt FACT device. The harmonics of concern for telephone interference are in the range between 2 and 7 kHz.

  Two different sets of power quality indices are used for assessing telephone interference [30]:

  - Telephone harmonic form factor (THFF) and Ip (equivalent disturbing current)
  - TIF (telephone interference factor) and IT product

  Both sets of indices require determining voltage and current distortion at the PCC, and they are calculated by applying weighting coefficients to each harmonic component [4].

  Telephone interference is a concern when FACT devices are installed in close proximity to telephone lines. In the last decade, the concerns for telephone interference have been decreasing because of the replacement of metallic telephone conductors with fiber optic, the widespread use of mobile cellular telephone system, and, in general, the evolution of telephone communication technology [30]. However, in some countries where FACT devices may be installed in rural areas, strict limits for telephone interference are still enforced [4,32].

- Other criteria

  Harmonic performance for a shunt FACT device may require evaluating other type of disturbances to the system [30]:

  - Railway signaling interference
  - Television and radio interference
  - Interharmonics
  - Audible noise

The criteria listed above require the use of different models from the ones approached in this chapter as described in [16,24,32], and they are not as universal as voltage distortion. They are mentioned in this section to note that harmonic performance evaluation may not be limited to the analysis discussed in this section, but may require additional modeling and evaluation.

5.5.1.2 Procedure

The evaluation of harmonic performance is generally based on the levels of harmonics generated by the shunt FACT device alone, neglecting the effect of preexisting harmonics on the transmission system. This approach consists in determining the "incremental" value of distortion only and is appropriate where either the preexisting levels are low and/or difficult to determine [25].

When background distortion is known and is significant in comparison to the permitted limits, the performance criteria should be defined in terms of the "aggregate" distortion due to the shunt FACT device and background harmonic distortion. This is because preexisting distortion may change as a result of magnification (or attenuation) by the connection of the shunt FACT device filters.

The standards are not always clear in the definition of performance criteria; therefore, the assumptions made for each harmonic study should be stated explicitly. When the model shown in Figure 5.1 is used, if the incremental value of distortion is calculated, the harmonic source behind network impedance is removed. If the aggregate distortion is calculated, then the complete model is used, where both harmonic distortion from the shunt FACT device and background distortion are applied simultaneously.

Once the approach to harmonic performance evaluation is clarified (either incremental or aggregate), the harmonic performance evaluation is straightforward and requires the execution of steps in sequence, as discussed in the following paragraphs.

1. Determine quantities to monitor
   The quantities which are monitored include the individual voltage components at the PCC and the THD. If current harmonic limits or telephone interference limits are defined, then individual harmonic current components and coefficients such as TIF or IT are also monitored.

2. Determine range of variables
   The second step consists in determining the variables which impact filter detuning and therefore harmonic performance:
   i. Frequency variation: under normal operating condition, the system frequency typically varies within ± 0.5 Hz from the rated value (50 or 60 Hz).
   ii. Temperature variation: typically, the average maximum temperature and the average minimum temperature at a location are considered. If the extreme temperatures are significantly higher than the average, it may be preferred to use these values to determine filter detuning. These values may be retrieved by historical temperature data at the location of

interest. Temperature variations have an impact on the value of the capacitance of the shunt FACT device considered.

iii. Tolerances of components: all components (transformers, filters, and reactors) are manufactured within a defined tolerance, which need to be taken into account.

iv. Capacitor fuse failures: capacitor banks consist of many individual capacitor cans connected in series and in parallel, and they can operate even when some of the cans fail. However, operation with failed cans will change the total bank capacitance; therefore, this added variable needs to be considered for the performance study.

In addition to filter detuning, other factors may be taken into account. For example, if the shunt FACT device consists of branches which can be switched on and off during normal operation (such as filters for HVDC or TSCs for SVCs, Section 5.4), then different configurations are analyzed.

3. Determine number of cases to study
The total number of cases to study depends on the combination of the variables identified in the previous section. In general, the tools used to perform harmonic studies are capable to study different combinations of variables such as frequency and tolerances of components. In some cases, the user has to specify some particular operating conditions, such as filters in or out of service.

   The number of cases to be investigated depends also on the use of a "consistent" or "nonconsistent" harmonic set for the shunt FACT device (Section 5.4.1.2). If multiple "consistent" harmonic sets are used, then it is clear that the number of cases increase further, since performance analysis needs to be repeated for each harmonic set.

4. Run each case and record the variables of interest for each case
For each case and for each harmonic, the tool used for harmonic analysis performs a "sweep" of the harmonic impedance sectors to identify the system impedance which results in the highest level of harmonic distortion. In general, the worst case harmonic distortion is encountered on the perimeter of the harmonic shape, where the damping is lower.

   The quantities of interest are monitored and stored for each case, and the maximum values are identified.

5. Comparing the maximum harmonic levels with the limits
If the values of the monitored quantities are within the limits, then the proposed shunt FACT device configuration is compliant with the specifications, and the model already built can be used to perform the harmonic rating study.

   If the limits are exceeded, it is necessary to alter the device configuration by modifying filter size or tuning order, by adding other filters at different harmonics, or by changing the transformer impedance. After doing these modifications, a new shunt FACT model is built, and the performance analysis is repeated.

   In some cases, multiple configurations which meet harmonic performance criteria may be determined, and the final configuration is chosen based on other criteria, such as harmonic rating, cost, or substation layout. As it will be discussed with more details in Section 5.5.2, in some cases, the harmonic rating study may require modifications of the design. In this case, harmonic performance has to be repeated for the new design.

5.5.1.3 Verification of compliance

Once the shunt FACT device is commissioned, power quality measurements (Chapter 7) are used to verify compliance with the performance criteria.

Ideally, power quality meters should be installed at the substation prior to commissioning, and the level of harmonic distortion should be measured for a long period of time (at least a few weeks) to collect a baseline harmonic distortion. After commissioning, power quality measurements are compared with the preexisting results to assess the impact of the shunt FACT device. This approach allows determining the source of harmonics in particular when high harmonic levels are identified when a shunt FACT device is put in operation.

A second advantage of power quality measurements prior to commissioning is that they provide accurate information on the background distortion, and they can be used to build the system harmonic model (Section 5.3.2).

In the industry practice, power quality measurements are only performed after commissioning. Therefore, it is critical to gather as much information as possible at the early stage of a project to build an accurate harmonic model and guarantee meeting performance criteria. Mitigating measures implemented after commissioning are typically very expensive and may require a considerable long outage before being implemented.

5.5.2 Harmonic rating studies

5.5.2.1 Motivation and standards

Harmonic rating studies[9] determine the maximum stress caused by harmonics and the equipment size. Specific standards have been developed to evaluate the impact of harmonics on the rating of the equipment, and the components used in FACT applications need special design to take into account increased peak voltages and rms currents due to harmonics [20].[10]

If a "consistent" set of harmonic is used when performing harmonic rating and performance studies, the results are going to be more conservative, since the rms current and voltage are going to be lower than using a "nonconsistent" set of harmonics.

Transformers are typically the most expensive component of a shunt FACT device installation, and the results of the harmonic study may have a large impact on the overall cost of a project. The main impact of harmonics for transformers

[9]Harmonic rating studies investigate operating conditions that are in the steady-state domain. Routine switching operation and during fast-front surge condition, the components may experience very high stresses. The analysis of these transient phenomena requires different models and studies as discussed in [10, 16].

[10]The power electronic switches (thyristors, IGBTs, or GTOs) require a separate rating study, which is not addressed in this section.

consists in increased losses. Even if harmonic currents magnitude are only a fraction of the fundamental current, the impact on losses may be significant due to skin effect which causes increased winding resistance at harmonic frequency (Section 5.3.1).

Standards such as [34,37,38] define coefficients for reducing fundamental current amplitude for a transformer supplying nonsinusoidal loads. The same principle is used for transformer supplying shunt FACT devices. The coefficients define the ratio of fundamental current with respect to the rated current, as a function of harmonics:

- Loss factor (IEEE)

$$F_{HL} = \frac{\sum_{h=1}^{N} [I_h/I_1]^2 h^r}{\sum_{h=1}^{N} [I_h/I_1]^2}$$

where h is the harmonic order, $r = 2$ for eddy losses, and $r = 0.8$ for stray losses.

- K-Factor (UL)

$$K = \left[1 + \frac{e}{1+e}\left(\frac{I_1}{I}\right)^2 \sum_{h+2}^{h=N} \left(h^q \left(\frac{I_h}{I_1}\right)^2\right)\right]^{0.5}$$

where e is the ratio between eddy current loss (calculated at 50 Hz) and copper loss, n is the harmonic order, and q is a factor which depends on winding type and frequency and typically varies between 1.5 and 1.7.

The loss factor is mostly used in the United States, while the K-factor is used in Europe. The correlation between the two factors is explained in Annex B of [34], where it is observed that the loss factor is a function of the harmonic current distribution and is independent of the relative magnitude, while the UL K-factor is dependent on both the magnitude and distribution of the harmonic current.

Reactors used for shunt FACT application are typically dry-type air core. A specific standard for reactors used in filters is not available, and general reactor standards such as [33] are used as a reference. Reactors used for shunt FACT devices applications are rated based on the harmonic current spectrum. The harmonic spectrum has an impact on losses (similarly to transformers) and on noise level. The quality factor must be determined as a balance between limiting losses and providing damping for the filters.

Capacitor rating is based on rms current and peak voltage. Both these quantities are calculated by taking into account the harmonic content. Typically, once the harmonic current content is known, the individual harmonic voltage distortion is calculated from the capacitor harmonic reactance. The capacitor MVAr is then calculated based on the rms current and rms voltage, including the harmonic components. The capacitor MVAr is the figure generally used to quantify the capacitor size and cost.

Table 5.8 Voltage levels for capacitor units [27]

Voltage (pu)	Maximum duration
1.00	Continuous
1.10	12 h
1.15	30 min
1.20	5 min
1.30	1 min

Some international standards such as [27,35] indicate that the capacitors should be able to operate under overvoltage conditions with no deterioration. Table 5.8 indicates the limits included in [27]. These coefficients are useful in determining the capacitor rating under transient harmonic conditions determined from the harmonic study.

5.5.2.2 Procedure

The life of a shunt FACT device project varies between 25 and 40 years. Therefore, the rating of the equipment needs to be sufficiently robust such that a redesign or replacement of equipment causing lengthy outages is not required during the service life [31].

For this reason, even if the harmonic rating studies make use of the same model and of the same procedure discussed in Section 5.5.1.2 implemented for harmonic-performance studies, there are some differences in the approach and in the analysis:

- Background harmonic distortion is always included in the model used for harmonic studies, since in general the rating of the equipment is highly impacted by the harmonics injected by the power system.
- A higher number of quantities are monitored for the purpose of evaluating harmonic rating: typically, harmonic currents in each component such as transformers, capacitors, and reactors are monitored.
- Harmonic rating studies are typically performed for different sets of conditions, which are referred to as "steady-state" conditions and "transient" conditions. In general, the variables used for steady-state conditions are the same used for performance studies. For transient conditions are generally more severe and may include larger variations of frequency or temperature.
- When it is known that the power system in close proximity to the shunt FACT device will evolve significantly, future system conditions may be analyzed. These conditions may be expressed for example by a higher background distortion or larger impedance sectors.

The results of the harmonic rating study consist in rms currents and MVAr ratings for all components. In general, manufacturers can provide components which are able to withstand normal level of harmonics, and therefore, the main impact of harmonics is on cost rather than on functionalities.

However, in some cases, harmonic rating studies may provide insights which will result in a change of the design. As an example, a shunt FACT device including a fifth harmonic filter is considered, where the filter size is determined to meet performance criteria at the PCC. If the harmonic rating study indicates that the level of harmonic currents absorbed by the filter is high, then this result may indicate that the filter is too small for the system under consideration.

In this case, it may be preferred to increase the filter size to avoid long-term damage of the equipment, in particular, if system expansions will occur, thus resulting in higher levels of harmonics. A change in the filter size may require modifying the size of other components to maintain the same harmonic performance and power output at fundamental frequency. After the new design is finalized, the harmonic performance study is repeated first, and if harmonic performance criteria are met, a new harmonic rating study is performed.

5.5.2.3 Verification of harmonic ratings

In general, once the shunt FACT device is in the service, the only indication of the harmonic levels flowing within its components is given by the harmonic voltage distortion at the PCC and at the medium voltage bus where the equipment is connected.

A high-voltage harmonic distortion at these locations may not only result in exceeding voltage distortion limits, but may cause concerns for the stress on the equipment. In these cases, it may be necessary to repeat the harmonic studies using the measured harmonic levels and verify with the manufacturer the rating of the equipment.

References

[1] Arrilaga J. and Watson N. R. *Power System Harmonics*, 2nd Edition. Hoboken, NJ: John Wiley & Sons; 2003.

[2] Beaulieu G. *Planning Levels and Harmonic Measurements on Hydro-Quebec's Transmission System*, Proceedings of the Eighth International Conference on Harmonics and Quality of Power, 1998.

[3] Betz R. E., Townsend C., Summers T. J. and Mirzaeva G. *Simultaneous Estimation of Sequence and Harmonic Components for StatCom Applications*, Presented at IEEE 14th Annual International Power Electronics and Motion Control Conference, Ohrid, Macedonia, 2010.

[4] Canadian Standards Association. *Canadian Electrical Code Part III – Electrical Coordination*, Doc. No. CAN/CSA-C22.3 No. 3-98 (R2013); Toronto, ON, Canada, 2013.

[5] Dash S. S., Palanivel P. and Premalatha S. *Performance Analysis of Multilevel Inverters Using Variable Switching Frequency Carrier Based PWM Techniques*, International Conference on Renewable Energies and Power Quality, Santiago de Compostela, Spain, 2012.

[6] Downs S. and McGoldrick N. *Harmonic Measurements on Transmission Systems*, Published in Proceedings of IEE Colloquium on Sources and Effects of Harmonic Distortion in Power Systems, IET, 1997.

[7] Engineering Directorate of the Energy Networks Association. *Engineering Recommendation, Planning Levels for Harmonic Voltage Distortion and the Connection of Non-Linear Equipment to Transmission Systems and Distribution Networks of the United Kingdom*, Doc. No. UK G5/4-1, London, UK, 2005.

[8] Enslin J. H. R., Wakefield R. A., Hu Y. and Eric S. *Harmonic Impedance Study for Southwest Connecticut Phase II Alternatives*, Technical Report, October 18, 2004.

[9] Girgis R. S. *Proposed Standards for Frequency Conversion Factors of Transformer Performance Parameters*, IEEE Transaction on Power Delivery, vol. 18, no. 4, October 2003, pp. 1262–1267.

[10] Greenwood A. *Electrical Transients in Power Systems*, 2nd Edition. New York: Wiley-Interscience, 1991.

[11] Henderson M., Gagnon J., Bertagnolli D., Hosie B., DeShazo G. L. and Silverstein B. *Building a Plan for HVDC*, IEEE Power and Energy Magazine, vol. 5, no. 2, 2007, pp. 52–60.

[12] Hingorani N. G. and Gyugyi L. *Understanding FACTs: Concepts and Technology of Flexible AC Transmission Systems*. New York: IEEE Press; 2000.

[13] Hui J., Freitas W., Viera J. and Yang H. *Utility Harmonic Impedance Measurement Based on Data Selection*, IEEE Transactions on Power Delivery, vol. 27, no. 4, 2012, pp. 2193–2202.

[14] Khomfoi S. and Tolbert L. M. 'Multilevel power converters', in *Power Electronics Handbook*, 2nd Edition. Atlanta, GA: Elsevier, 2007, Chapter 17, pp. 451–482.

[15] Madrigal M. and Acha E. *Harmonic Modelling of Voltage Source Converters for HVDC Stations*, Published in Proceedings of the Seventh International Conference on AC–DC Power Transmission, 2001, pp. 125–131.

[16] Martinez-Velasco J. A. *Power System Transients: Parameter Determination*. Boca Raton, FL: CRC Press; 2009.

[17] Mathur R. M. and Varma R. K. *Thyristor-Based FACTs Controllers for Electrical Transmission Systems*. New York: IEEE Press; 2002.

[18] Mohan N., Undeland T. M. and Robbins W. P. *Power Electronics: Applications, Principles and Design*, 3rd Edition. Hoboken, NJ: John Wiley & Sons; 2003.

[19] Padiyar K. R. *HVDC Power Transmission Systems: Technology and System Interactions* Hoboken, NJ: John Wiley & Sons; 1990.

[20] Reid W. E. and Puskarich K. A. *Harmonic Filter Application Criteria*, Presented at IEEE/PES Winter Power Meeting, New York, 1994.

[21] Riberio P. F. *Modeling of Power System Components*, retrieved on May 15, 2016, http://www.calvin.edu/~pribeiro/ribeiro/presentations/Modeling-LinearLoad-Harmonics.PPT.

[22] Schettler F., Huang H. and Christl N. *HVDC Transmission Systems Using Voltage Source Converters: Design and Applications*, IEEE Power Engineering Society Summer Meeting, Seattle, WA, USA, 2000.

[23] Shahnia D., Rajakaruna S. and Ghosh A. *Static Compensators (STATCOMs) in Power Systems*. New York: Springer; 2015.

[24] Smede H., Johansson C., Winroth O. and Schutt H. *Design of HVDC Converter Stations with Respect to Audible Noise Requirements*, IEEE Transaction on Power Delivery, vol. 10, no. 2, pp. 747–758.

[25] Substation Committee, Power and Energy Society. *IEEE Guide for the Functional Specification of Transmission Static Var Compensators*, Doc. No 1031-2011, New York: IEEE; 2011.

[26] Task Force on Harmonics Modeling and Simulations. *Modeling and Simulation of the Propagation of Harmonics in Electric Power Networks, Part I: Concepts, Models, and Simulation Techniques*, IEEE Transactions on Power Delivery, vol. 11, no. 1, January 1996, pp. 452–465.

[27] Technical Committee 33, *Shunt Capacitors for a.c. Power Systems having Rated Voltage Greater than 1000 V*, Doc. Number IEC 60871-1. Geneva: IEC; 2014.

[28] Technical Committee 77, *Electromagnetic Compatibility (EMC) – Part 2-1: Electromagnetic Environment for Low-frequency Conducted Disturbances and Signaling in Public Power Supply Systems*, Doc. Number 61000-2-1. Geneva: IEC; 2006.

[29] Technical Committee 77, *Electromagnetic Compatibility (EMC) – Part 3-2: Limits – Limits for Harmonic Current Emissions*, Doc. Number 61000-3-2. Geneva: IEC; 2006.

[30] Technical Committee SC B4, Working Group 14.30. *Guide to the Specification and Design Evaluation of AC Filters for HVDC Systems*, Doc. No. 139. Paris: Cigré; 1999.

[31] Technical Committee SC B4, Working Group 47. *Special Aspects of AC Filter Design for HVDC Systems*, Doc. No. 553. Paris: Cigré; 2013.

[32] Todeschini G., Mueller D. R. and Morris-Young G. *Telephone Interference Caused by Harmonics in Distribution Systems: Analysis and Simulations*, Proceedings of IEEE Power & Energy Society General Meeting, July 2013, Vancouver, BC, Canada.

[33] Transformer Committee, Power and Energy Society. *IEEE Standard for Requirements, Terminology, and Test Code for Dry-Type Air-Core Series-Connected Reactors*, Doc. No C57.16-2011. New York: IEEE; 2011.

[34] Transformer Committee, Power and Energy Society. *IEEE Recommended Practice for Establishing Transformer Capability When Supplying Nonsinusoidal Load Currents*, Doc No. C57.110-1998. New York: IEEE; 1998.

[35] Transmission and Distribution Committee, Power and Energy Society. *IEEE Standard for Shunt Power Capacitors*, Doc. Number 18-2012. New York: IEEE; 2012.

[36] Transmission and Distribution Committee, Power and Energy Society. *IEEE Recommended Practice and Requirements for Harmonic Control in Electric Power Systems*, Doc. Number 519-2014. New York: IEEE; 2014.

[37] Underwriters Laboratories, *Standard for Dry-Type General Purpose and Power Transformers*, Doc. No. UL 1561. ANSI, Northbrook, IL; 2011.

[38] Underwriters Laboratories, *Standard for Transformers, Distribution, Dry-Type – Over 600 Volts*, Doc. No. UL 1562. ANSI, Northbrook, IL; 2013.

[39] Xiao Y., Maun J., Mahmoud H.B., *et al. Harmonic Impedance Measurement Using Voltage and Current Increments from Disturbing Loads*, In Proceedings of the Ninth International Conference on Harmonics and Quality of Power, 2000, pp. 220–225.

[40] Zhao G., Liu J. and Han M. *Design of Input Filters Considering the Stability of STATCOM Systems*, Journal of Power Electronics, vol. 11, no. 6, pp. 904–913, November 2011.

Chapter 6

Power-quality improvement using series FACTS

Salah Kamel[1] and Francisco Jurado[2]

6.1 Introduction

6.1.1 Electricity network and power-quality overview

The electrical energy is considered the most popular form of energy, because its transportation is easy with high efficiency and reasonable cost. The electricity plays an important role in our modern industrialized society. The power systems that provide this electricity become more complicated and sophisticated systems. However, the main components of any power system are the generation side, the transmission system, and the distribution side. Each component is essential to deliver the electrical energy from the generation site to the customer site.

Figure 6.1 shows the major parts of the electrical power systems infrastructure [1,2]. In general they include

- Extra high-voltage transmission system
- High-voltage transmission systems
- Generation systems
  - Centralized power plants
  - Large-scale hydro production power plants
  - Large-scale wind farms
  - Nuclear power plants
- Load subsystem
  - Big factories and consumers
  - Distribution substations

6.1.2 Load-flow analysis

The load-flow calculation is considered the most important computational procedures that used in power system analysis. However, the load-flow problem is based on the nonlinear relationships among the injected power at the buses, load

[1]Department of Electrical Engineering, Faculty of Engineering, Aswan University, 81542 Aswan, Egypt
[2]Department of Electrical Engineering, University of Jaén, EPS, Linares, Linares, Jaén, Spain

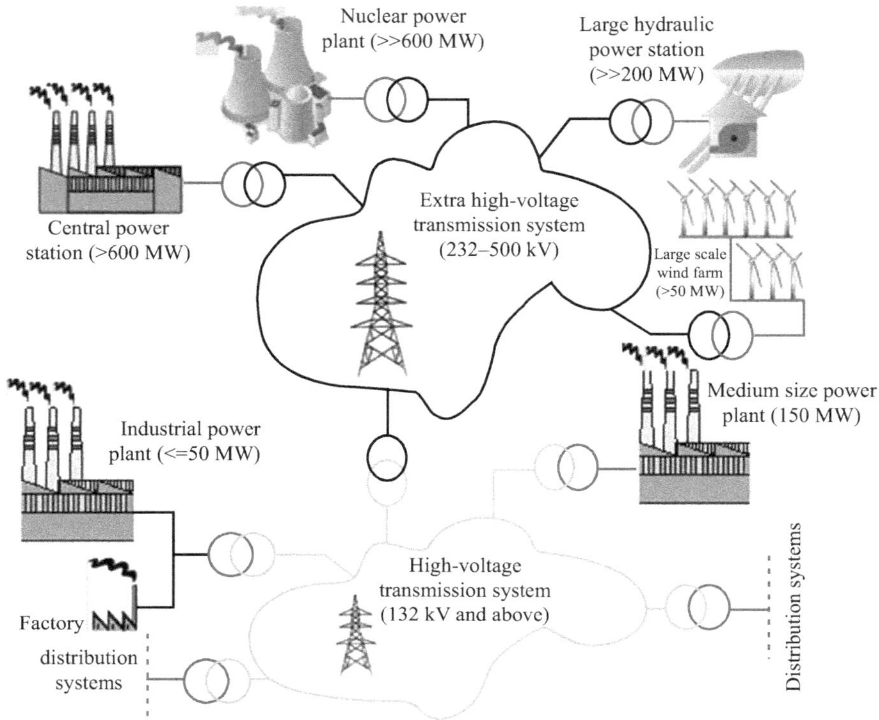

Figure 6.1 Infrastructure of electrical power systems

demands, buses voltage, and system parameters. Load-flow algorithms are important tools for planning future expansion of power systems and determining the best operation of systems. The load-flow study is considered the starting point for other studies such as transient and dynamic stability studies. In general, it can be used to analyze power-quality improvement. Load-flow solution can be performed using recent computer programs designed specifically for that. Load-flow solution can give the voltage magnitude and angle at each bus, active and reactive power of generators and loads, power flows of lines, and the total system losses [3–7].

Nowadays, the complexity in solving large-scale load-flow problem has been solved by using the computer. The application of computers has led to the development of many power system software that contained load flow as one of the analyses besides others such as fault, stability, and contingency analysis. Such examples of available software are EDSA Technical 2000, ETAP Power Station, DigSILENT, and PSS/E [8–11]. By using the computer, it possible to determine all the effects of changing or adding any parameters, devices, or generations in power supply system. The summaries of main requirements for load-flow solution method, which are quoted in [5], are presented in Figure 6.2.

The research on load-flow analysis was started when computer technology is at its infancy. Studies were more invested to develop numerical methods that would

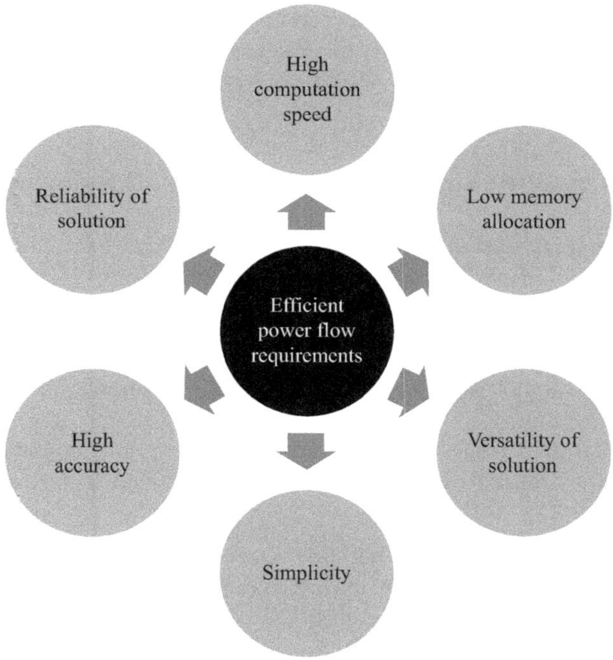

Figure 6.2 The main requirements of efficient power-flow algorithm

make load-flow computation easier and faster under computer constrains in managing memory and speed. Various load-flow methods were developed. The Gauss-Seidel method is considered an earliest load flow method that has low convergence characteristic. Newton–Raphson method that is based on power mismatch approach (NR-PM) was developed to enhance the convergence characteristic of Gauss-Seidel load flow method [3,12].

The Newton–Raphson method is considered the strong load-flow method that is accepted in industry applications. The main disadvantage of this method is the modifications of most elements in the Jacobian matrix during the iterative process. Fast decoupled load-flow approach was developed to reduce the required minimum memory storage and computation time [3,13]. In spite of that, the convergence rate of the fast decoupled method is affected with the Ill conditioned cases like high R/X ratio [14].

The new Newton–Raphson method based on injected current mismatches is developed to solve the mentioned drawback of the conventional Newton–Raphson load flow (NR-PM) [15]. In this approach, the most elements of Jacobian matrix are constant and equal elements of the admittance matrix. Hence, the computation time and memory size are greatly reduced.

The Newton–Raphson method that is based on current injection mismatches is more suitable in representation of load bus (PQ) buses. It has a low convergence characteristic in the case of voltage controlled bus (PV) representation. A revised current injection mismatch load flow (Newton-Raphson Revised Current Mismatch, (NR-RCIM)) has proposed to this problem and correct the representation of PV buses [16,17]. In this technique, the reactive power is taken as an additional state variable [18–23].

6.2 Power-quality improvement using FACTS devices

The flexible ac transmission system (FACTS) devices can be used to increase the controllability and generally improve power quality [24–29]. However, any load-flow program needs to be modified to implement the FACTS devices due to the following (summarized in Figure 6.3) [30–37]:

- The implementation of FACTS devices in the system requires addition reference buses.
- The FACTS impedances should be implemented in the admittance matrix.
- The powers contributed by FACTS should be added in power mismatches vector.

However, various previous research works were done to implement the FACTS in load-flow programs [30–54]. References [38,39] have developed easy FACTS models based on decoupled technique. The main drawbacks of this technique are that the constraints of FACTS parameters cannot be determined, where these parameters are calculated after the convergence of load-flow program. The technique has developed only in the case of control of the voltage magnitude, active and reactive powers simultaneously. Also, if we assume that the FACTS is the only link between two subnetworks, what will be the solution in this technique.

References [25,27] have proposed comprehensive models to solve the limitations of decoupled FACTS models. This approach presented the FACTS control parameters as independent variables. Hence, the size of the Jacobian matrix is increased in order to add the independent state variables of FACTS. Also, good initial conditions for the FACTS state variables are needed to enhance the convergence characteristic.

References [43–49] have proposed a FACTS model that is based on injected loads at their ends. The models can be implemented easily in any load-flow program.

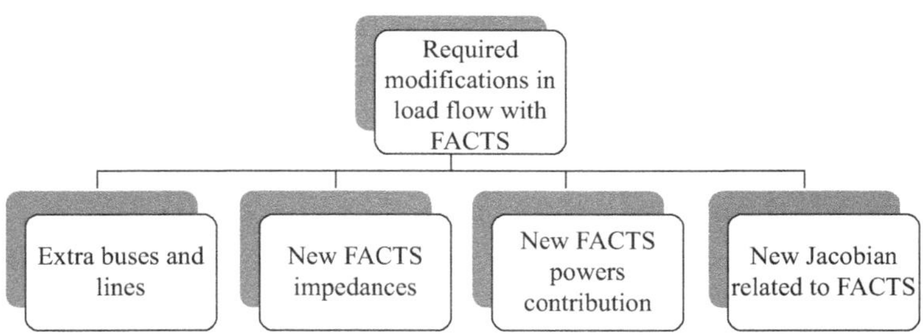

Figure 6.3 Required modifications of load-flow code to implement FACTS in the system

The main drawback of this approach is the required modification of the Jacobian matrix. Haque and Yam [30] presented a unified power flow controller model in load-flow algorithm to avoid the modification in the Jacobian matrix.

Nor *et al.* [31] have used the matrix partitioning technique to model the FACTS devices. The main advantages of the comprehensive approach are collected in this technique; moreover, the changing in the software codes can be done without affecting much the original algorithms. The main drawback of this approach is that a new codes related to the FACTS have to be written.

References [32–35] have developed a new technique, depended on transmitting the FACTS device into an equivalent network to reduce the complexities in load-flow codes. The main disadvantage of this technique is that the size of the Jacobian matrix is increased in order to add the state variables of FACTS.

This chapter presents an efficient modeling for the series FACTS devices in NR-RCIM load-flow algorithm. This is considered a first step to analyze the power quality after adding FACTS devices. The proposed series FACTS (static synchronous series compensator (SSSC) and interline power flow controller (IPFC)) models are based on current injection approach. The expected features of the proposed FACTS models in NR-RCIM load flow method are as follows:

- The developed FACTS models can be implemented into NR-RCIM load-flow method without modification in the original Jacobian matrix. Consequentially, the reusability of computer load-flow program codes is improved.
- The original structure and symmetry of admittance matrix is still kept without change.
- The sparsity technique can be applied in NR-RCIM load-flow algorithm with FACTS.
- The problem of selection of proper initial values of control parameters is solved.
- The controls of active and/or reactive power simultaneously or individually are available.
- The constraints of FACTS parameters can be checked during the iterative process.
- The problem that happened when the series compensator is the only link between two sub-networks is solved.

The NR-RCIM load flow program with the developed FACTS models will be a useful tool for power system planning and operation control.

6.3 Proposed SSSC model

The equivalent circuit of SSSC can be given as shown in Figure 6.4. The SSSC can be represented by a voltage source V_s that can be adjusted to control the power flow of line *i–j*.

The proposed SSSC model depends on transmitting the voltage source to current source as given in Figure 6.5.

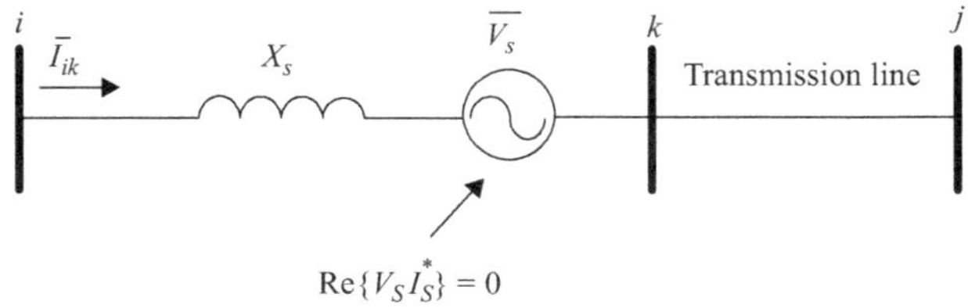

Figure 6.4 Representation of SSSC using voltage source and transformer's reactance

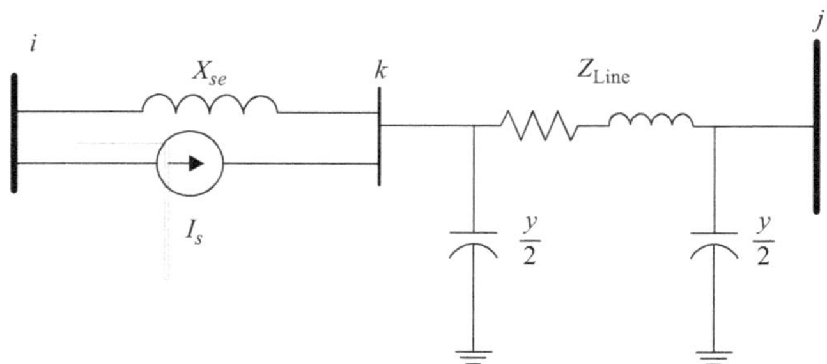

Figure 6.5 Equivalent circuit of SSSC based on current source representation

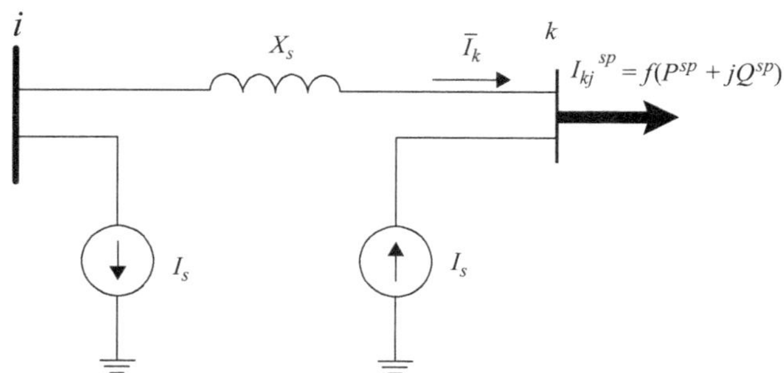

Figure 6.6 Current injection SSSC model (P–Q control)

The current (I_s) can be injected at the terminal of SSSC as shown in Figure 6.6. This can be calculated as a function of the required line flow and bus voltage. However, the active and reactive line flow can be controlled individually or simultaneously as follows.

6.3.1 Case 1: PQ control

In this case, the active and reactive line flow can be controlled at the same time. The developed SSSC model can be presented in Figure 6.6.

The injected current at bus (k) can be calculated as

$$I_s = I_{kj}^{sp} - I_{ik} \tag{6.1}$$

$$I_s = \left(\frac{S_{kj}^{sp}}{V_k}\right)^* - \left(\frac{V_i - V_k}{X_s}\right) \tag{6.2}$$

where

$$S_{kj}^{sp} = P_{kj}^{sp} + jQ_{kj}^{sp} \tag{6.3}$$

$$V_i = V_{ir} + jV_{im} \tag{6.4}$$

$$V_k = V_{kr} + jV_{km} \tag{6.5}$$

However, (6.2) can be given as

$$I_s = \frac{\left(P_{kj}^{sp} V_k^{\text{Re}} + Q_{kj}^{sp} V_k^{\text{Im}}\right) + j\left(P_{kj}^{sp} V_k^{\text{Im}} - Q_{kj}^{sp} V_k^{\text{Re}}\right)}{V_k^2} - \frac{\left(V_i^{\text{Im}} - V_k^{\text{Im}}\right)}{X_s} + j\frac{\left(V_i^{\text{Re}} - V_k^{\text{Re}}\right)}{X_s} \tag{6.6}$$

The real and imaginary components of I_s can be given by (6.7) and (6.8):

$$I_s^{\text{Re}} = \frac{\left(P_{kj}^{sp} V_k^{\text{Re}} + Q_{kj}^{sp} V_k^{\text{Im}}\right)}{V_k^2} - \frac{\left(V_i^{\text{Im}} - V_k^{\text{Im}}\right)}{X_s} \tag{6.7}$$

$$I_s^{\text{Im}} = \frac{\left(P_{kj}^{sp} V_k^{\text{Im}} - Q_{kj}^{sp} V_k^{\text{Re}}\right)}{V_k^2} + \frac{\left(V_i^{\text{Re}} - V_k^{\text{Re}}\right)}{X_s} \tag{6.8}$$

This current can be injected at bus (k) and at bus (i) with the opposite sign.

6.3.2 Case 2: P control

In this case, the active power flow is only controlled, and the reactive power is uncontrolled.

The reactive power flow between bus (k) and bus (j) can be calculated from Figure 6.7 as

$$Q_{kj} = \text{Im}\left(V_k I_{kj}^*\right) \tag{6.9}$$

where the current between bus (k) and bus (j) can be given as

$$I_{kj} = j\left(\frac{B}{2}\right)V_k + Y_{kj}\left(V_k - V_j\right) \tag{6.10}$$

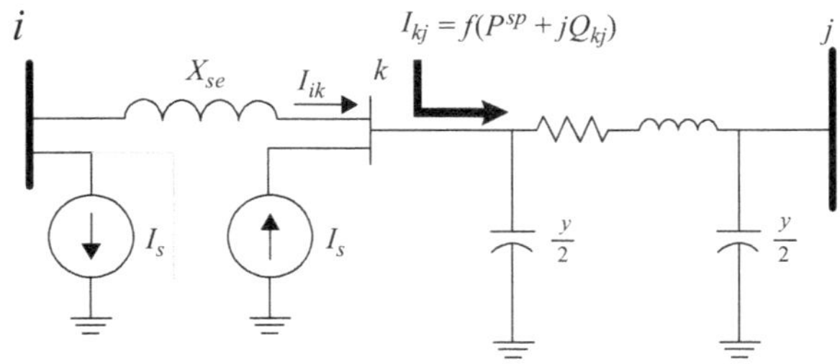

Figure 6.7 Current injection SSSC model (P control)

Finally, Q_{kj} can be calculated as follows:

$$\begin{aligned} Q_{kj} = {} & V_k^{\mathrm{Im}}\left[G_{kj}\left(V_k^{\mathrm{Re}} - V_j^{\mathrm{Re}}\right) - B_{kj}\left(V_k^{\mathrm{Im}} - V_j^{\mathrm{Im}}\right) - V_k^{\mathrm{Im}}\left(\frac{B}{2}\right)\right] \\ & - V_k^{\mathrm{Re}}\left[G_{kj}\left(V_k^{\mathrm{Im}} - V_j^{\mathrm{Im}}\right) + B_{kj}\left(V_k^{\mathrm{Re}} - V_j^{\mathrm{Re}}\right) + V_k^{\mathrm{Re}}\left(\frac{B}{2}\right)\right] \end{aligned} \tag{6.11}$$

Then, the I_s can be calculated as a function of Q_{kj} as

$$\begin{aligned} I_s = {} & \frac{\left(P_{kj}^{sp} V_k^{\mathrm{Re}} + Q_{kj} V_k^{\mathrm{Im}}\right) + j\left(P_{kj}^{sp} V_k^{\mathrm{Im}} - Q_{kj} V_k^{\mathrm{Re}}\right)}{V_j^2} - \frac{\left(V_i^{\mathrm{Im}} - V_k^{\mathrm{Im}}\right)}{X_s} \\ & + j\frac{\left(V_i^{\mathrm{Re}} - V_k^{\mathrm{Re}}\right)}{X_s} \end{aligned} \tag{6.12}$$

The real and imaginary components of I_s can be given as

$$I_s^{Re} = \frac{\left(P_{kj}^{sp} V_k^{\mathrm{Re}} + Q_{kj} V_k^{\mathrm{Im}}\right)}{V_k^2} - \frac{\left(V_i^{\mathrm{Im}} - V_k^{\mathrm{Im}}\right)}{X_s} \tag{6.13}$$

$$I_s^{\mathrm{Im}} = \frac{\left(P_{kj}^{sp} V_k^{\mathrm{Im}} - Q_{kj} V_k^{\mathrm{Re}}\right)}{V_k^2} + \frac{\left(V_i^{\mathrm{Re}} - V_k^{\mathrm{Re}}\right)}{X_s} \tag{6.14}$$

6.3.3 NR-RCIM load-flow method with developed SSSC model

The developed SSSC model can be implemented easily in NR-RCIM load-flow method. Equation (6.15) describes NR-RCIM load-flow algorithm with the developed SSSC model, where the current mismatch vector is only changed at the ends of SSSC.

$$
\begin{bmatrix}
\Delta I_1^{\mathrm{Im}} \\
\Delta I_1^{\mathrm{Re}} \\
\vdots \\
\vdots \\
\Delta I_i^{\mathrm{Im}} - I_s^{\mathrm{Im}} \\
\Delta I_i^{\mathrm{Re}} - I_s^{\mathrm{Re}} \\
\Delta I_k^{\mathrm{Im}} + I_s^{\mathrm{Im}} \\
\Delta I_k^{\mathrm{Re}} + I_s^{\mathrm{Re}} \\
\vdots \\
\vdots \\
\Delta I_f^{\mathrm{Im}} \\
\Delta I_f^{\mathrm{Re}} \\
\Delta V_f^2 \\
\vdots \\
\Delta I_N^{\mathrm{Im}} \\
\Delta I_N^{\mathrm{Re}}
\end{bmatrix}
=
\begin{bmatrix}
B_{11}^{*} & G_{11}^{*} & \cdots & \cdots & B_{1i} & G_{1i} & B_{1k} & G_{1k} & \cdots & \cdots & B_{1f} & G_{1f} & 0 & \cdots & B_{1N} & G_{1N} \\
G_{11}^{**} & B_{11}^{**} & \cdots & \cdots & G_{1i} & -B_{1i} & G_{1k} & -B_{1k} & \cdots & \cdots & G_{1f} & -B_{1f} & 0 & \cdots & G_{1N} & -B_{1N} \\
\vdots & \vdots & \iddots & \iddots & \vdots & \vdots & \vdots & \vdots & \iddots & \iddots & \vdots & \vdots & \vdots & \iddots & \vdots & \vdots \\
\vdots & \vdots & \iddots & \iddots & \vdots & \vdots & \vdots & \vdots & \iddots & \iddots & \vdots & \vdots & \vdots & \iddots & \vdots & \vdots \\
B_{i1} & G_{i1} & \cdots & \cdots & B_{ii}^{*} & G_{ii}^{*} & B_{ik} & G_{ik} & \cdots & \cdots & B_{if} & G_{if} & 0 & \cdots & B_{iN} & G_{iN} \\
G_{i1} & -B_{i1} & \cdots & \cdots & G_{ii}^{**} & B_{ii}^{**} & G_{ik} & -B_{ik} & \cdots & \cdots & G_{if} & -B_{if} & 0 & \cdots & G_{iN} & -B_{iN} \\
B_{k1} & G_{k1} & \cdots & \cdots & B_{ki} & G_{ki} & B_{kk} & G_{kk} & \cdots & \cdots & B_{kf} & G_{kf} & 0 & \cdots & B_{kN} & G_{kN} \\
G_{k1} & -B_{k1} & \cdots & \cdots & G_{ki} & -B_{ki} & G_{kk} & -B_{kk} & \cdots & \cdots & G_{kf} & -B_{kf} & 0 & \cdots & G_{kN} & -B_{kN} \\
\vdots & \vdots & \iddots & \iddots & \vdots & \vdots & \vdots & \vdots & \iddots & \iddots & \vdots & \vdots & \vdots & \iddots & \vdots & \vdots \\
\vdots & \vdots & \iddots & \iddots & \vdots & \vdots & \vdots & \vdots & \iddots & \iddots & \vdots & \vdots & \vdots & \iddots & \vdots & \vdots \\
B_{f1} & G_{f1} & \cdots & \cdots & B_{fi} & G_{fi} & B_{fk} & G_{fk} & \cdots & \cdots & B_{ff}^{*} & G_{ff}^{*} & \dfrac{V_{rf}}{V_f^2} & \cdots & B_{fN} & G_{fN} \\
G_{f1} & -B_{f1} & \cdots & \cdots & G_{fi} & -B_{fi} & G_{fk} & -B_{fk} & \cdots & \cdots & G_{ff}^{**} & B_{ff}^{**} & \dfrac{-V_{mf}}{V_f^2} & \cdots & G_{fN} & -B_{fN} \\
0 & 0 & \cdots & 0 & 0 & 0 & 0 & 0 & 0 & \cdots & 2V_{rf} & 2V_{mf} & 0 & \cdots & 0 & 0 \\
\vdots & \vdots & \iddots & \vdots & \vdots & \vdots & \vdots & \vdots & \iddots & \iddots & \vdots & \vdots & \vdots & \iddots & \vdots & \vdots \\
B_{N1} & G_{N1} & \cdots & \cdots & B_{Ni} & G_{Ni} & B_{Nk} & G_{Nk} & \cdots & \cdots & B_{Nf} & G_{Nf} & 0 & \cdots & B_{NN}^{*} & G_{NN}^{*} \\
G_{N1} & -B_{N1} & \cdots & \cdots & G_{Ni} & -B_{Ni} & G_{Nk} & -B_{Nk} & \cdots & \cdots & G_{Nf} & -B_{Nf} & 0 & \cdots & G_{NN}^{**} & B_{NN}^{**}
\end{bmatrix}
\begin{bmatrix}
\Delta V_1^{\mathrm{Re}} \\
\Delta V_1^{\mathrm{Im}} \\
\vdots \\
\vdots \\
\Delta V_i^{\mathrm{Re}} \\
\Delta V_i^{\mathrm{Im}} \\
\Delta V_k^{\mathrm{Re}} \\
\Delta V_k^{\mathrm{Im}} \\
\vdots \\
\vdots \\
\Delta V_f^{\mathrm{Re}} \\
\Delta V_f^{\mathrm{Im}} \\
\Delta Q_f \\
\vdots \\
\Delta V_N^{\mathrm{Re}} \\
\Delta V_N^{\mathrm{Im}}
\end{bmatrix}
\tag{6.15}
$$

When the system buses are connected with only constant power loads, the diagonal elements in above equation are given in [15].

The steps of modeling the proposed SSSC model in NR-RCIM load-flow algorithm can be summarized as follows:

- ***Step I:*** *Bus and line data modification*

New PQ bus with zero active and reactive loads and line for every SSSC device has to be added. The impedance of the added line equals SSSC parameters.

- ***Step II:*** *Current mismatches modification*

The real and imaginary components of the current mismatches at PQ buses can be can be calculated as

$$\Delta I_k^{\mathrm{Re}} = \frac{V_k^{\mathrm{Re}}\Delta P_k + V_k^{\mathrm{Im}}\Delta Q_k}{V_k^2} \tag{6.16}$$

$$\Delta I_k^{\mathrm{Im}} = \frac{V_k^{\mathrm{Im}}\Delta P_k - V_k^{\mathrm{Re}}\Delta Q_k}{V_k^2} \tag{6.17}$$

In the case of PV buses, three mismatches equations are needed as follows:

$$\Delta I_f^{\mathrm{Re}} = \frac{V_f^{\mathrm{Re}}\Delta P_f}{V_f^2} \tag{6.18}$$

$$\Delta I_f^{\mathrm{Im}} = \frac{V_f^{\mathrm{Im}}\Delta P_f}{V_f^2} \tag{6.19}$$

$$\Delta V_f^2 = V_f^{sp} - V_f^{cal} \tag{6.20}$$

The addition injected currents at the terminals of SSSC have to be added as Sending bus (i):

$$\Delta I_i^{\mathrm{Re}} = \mathrm{Re}(I_i^{sp}) - I_i^{\mathrm{Re}} - I_s^{\mathrm{Re}} \tag{6.21}$$

$$\Delta I_i^{\mathrm{Im}} = \mathrm{Im}(I_i^{sp}) - I_i^{\mathrm{Im}} - I_s^{\mathrm{Im}} \tag{6.22}$$

Auxiliary bus (k):

$$\Delta I_k^{\mathrm{Re}} = \mathrm{Re}\left(I_k^{sp}\right) - I_k^{\mathrm{Re}} + I_s^{\mathrm{Re}} \tag{6.23}$$

$$\Delta I_k^{\mathrm{Im}} = \mathrm{Im}\left(I_k^{sp}\right) - I_k^{\mathrm{Im}} + I_s^{\mathrm{Im}} \tag{6.24}$$

- ***Step III****: Load flow solution*

Solve (6.15) using the SuperLU library to determine all sparse matrices without any changes in the original Jacobian matrix.

- ***Step IV:*** *Computing the SSSC parameters*

The parameters of SSSC *(V_s, δ_s)* are calculated as follows:

$$V_s = jI_sX_s \tag{6.25}$$

$$V_s^{\mathrm{Re}} = -X_{se1}I_s^{\mathrm{Im}} \tag{6.26}$$

$$V_s^{\mathrm{Im}} = X_sI_s^{\mathrm{Re}} \tag{6.27}$$

$$|V|_s = \sqrt{V_s^{\mathrm{Re}} + V_s^{\mathrm{Im}}} \tag{6.28}$$

$$\delta_s = \tan^{-1}\left(\frac{V_s^{\mathrm{Im}}}{V_s^{\mathrm{Re}}}\right) \tag{6.29}$$

6.4 Proposed IPFC model

The IPFC is a combination of two or more SSSCs coupled via a common d.c. link. The simplest IPFC that consists of two back-to-back d.c.–a.c. converters is used in the developed model. However, this proposed model can be applied for an IPFC with any number of series converters.

The equivalent circuit of IPFC is shown in Figure 6.8. Real power can be exchanged between the two series converters via the common d.c. link, and the summation of this exchanged real power should be zero.

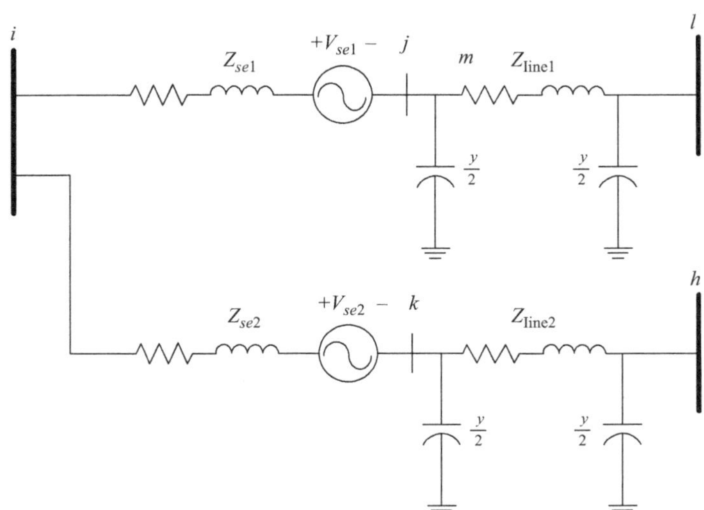

Figure 6.8 Voltage source representation of IPFC with two converters

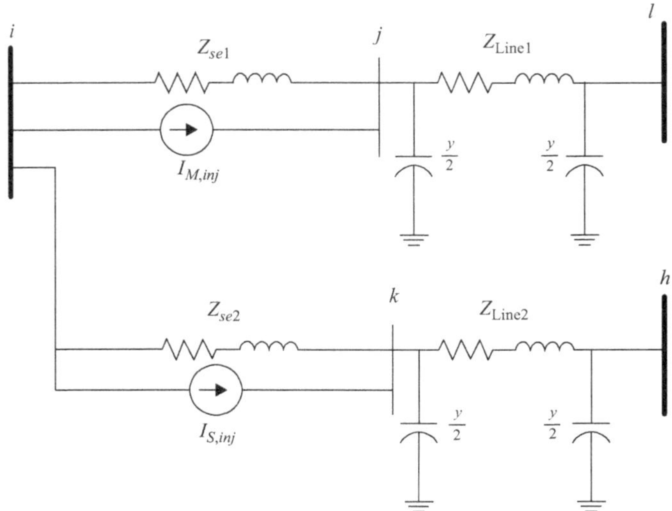

Figure 6.9 Current source representation of IPFC

The IPFC can be used to control the active and reactive power flows of the master line (M) and active or reactive power flow of the slave line (S). The connected IPFC lines can be taken as master or slave according to the actual demand. In the case of slave line, the active power is controlled, whereas the reactive power is uncontrolled.

The developed IPFC model is based on transmitting its voltage sources to current sources parallel with the series impedances as given in Figure 6.9. The current sources can be transmitted to injected currents at the terminals of IPFC as shown in Figure 6.10. These currents are based on the specified line flow and buses voltage. The injected currents can be calculated as follows.

6.4.1 Master line

Figure 6.10 shows the developed IPFC model. The injected current at bus j can be calculated as follows:

$$I_{M,inj} = I_{sp}^{M} - I_{ij} \tag{6.30}$$

$$I_{M,inj} = \left(\frac{S_{sp}^{M}}{V_j}\right)^{*} - \left(\frac{V_i - V_j}{Z_{se1}}\right) \tag{6.31}$$

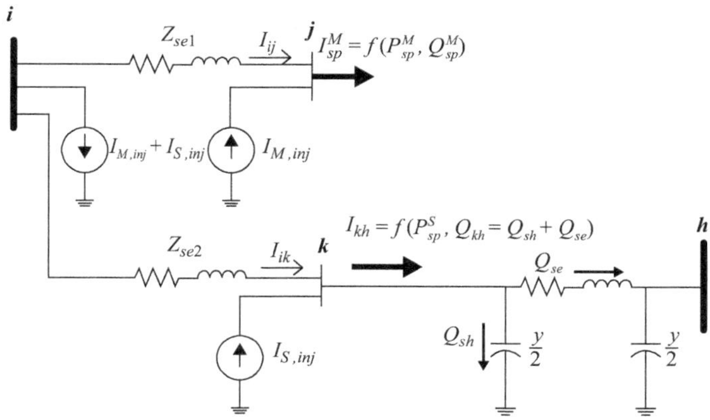

Figure 6.10 Developed current injected model of IPFC

where $S_{sp}^M = P_{sp}^M + jQ_{sp}^M$, $V_i = V_{ir} + jV_{im}$, $V_j = V_{jr} + jV_{jm}$, $Z_{se1} = R_{se1} + jX_{se1}$, the injected current can be rewritten as

$$I_{M,inj} = \frac{\left(P_{sp}^M V_{rj} + Q_{sp}^M V_{mj}\right) + j\left(P_{sp}^M V_{mj} - Q_{sp}^M V_{rj}\right)}{V_j^2} - \frac{\left[R_{se1}\left(V_{ri} - V_{rj}\right) + X_{se1}\left(V_{mi} - V_{mj}\right)\right]}{Z_{se1}^2} - j\frac{\left[R_{se1}\left(V_{mi} - V_{mj}\right) - X_{se1}\left(V_{ri} - V_{rj}\right)\right]}{Z_{se1}^2} \tag{6.32}$$

The real and imaginary parts of $I_{M,inj}$ can be calculated as

$$I_{M,inj}^{Re} = \frac{\left(P_{sp}^M V_{rj} + Q_{sp}^M V_{mj}\right)}{V_j^2} - \frac{\left[R_{se1}\left(V_{ri} - V_{rj}\right) + X_{se1}\left(V_{mi} - V_{mj}\right)\right]}{Z_{se1}^2} \tag{6.33}$$

$$I_{M,inj}^{\mathrm{Im}} = \frac{\left(P_{sp}^M V_{mj} - Q_{sp}^M V_{rj}\right)}{V_j^2} - \frac{\left[R_{se1}\left(V_{mi} - V_{mj}\right) - X_{se1}\left(V_{ri} - V_{rj}\right)\right]}{Z_{se1}^2} \tag{6.34}$$

6.4.2 Slave line

The injected current at bus *(k)* can be calculated as

$$I_{S,inj} = \left(\frac{S_{sp}^S}{V_k}\right)^* - \left(\frac{V_i - V_k}{Z_{se2}}\right) \tag{6.35}$$

where $S_{sp}^S = P_{sp}^S + JQ_{kh}$, $V_k = V_{kr} + jV_{km}$, $V_i = V_{ir} + jV_{im}$, $Z_{se2} = R_{se2} + jX_{se2}$
The reactive power flow (Q_{kh}) can be calculated as follows:

$$Q_{kh} = \mathrm{Im}(V_k I_{kh}^*) \tag{6.36}$$

where

$$I_{kh} = J\left(\frac{B}{2}\right)V_k + Y_{kh}(V_k - V_h) \tag{6.37}$$

Then,

$$\begin{aligned} Q_{kh} = {} & V_{mk}\left[G_{kh}(V_{rk} - V_{rh}) - B_{kh}(V_{mk} - V_{mh}) - V_{mk}\left(\frac{B}{2}\right)\right] \\ & - V_{rk}\left[G_{kh}(V_{mk} - V_{mh}) + B_{kh}(V_{rk} - V_{rh}) + V_{rk}\left(\frac{B}{2}\right)\right] \end{aligned} \tag{6.38}$$

The injected current at bus *(k)* can be calculated as follows:

$$\begin{aligned} I_{S,inj} = {} & \frac{\left(P_{sp}^S V_{rk} + Q_{kh}^S V_{mk}\right) + j\left(P_{sp}^S V_{mk} - Q_{kh} V_{rk}\right)}{V_j^2} \\ & - \frac{[R_{se2}(V_{ri} - V_{rk}) + X_{se2}(V_{mi} - V_{mk})]}{Z_{se2}^2} \\ & - j\frac{[R_{se2}(V_{mi} - V_{mk}) - X_{se2}(V_{ri} - V_{rk})]}{Z_{se2}^2} \end{aligned} \tag{6.39}$$

Then, the real and imaginary parts of $I_{S,inj}$ can be calculated as

$$I_{S,inj}^{Re} = \frac{\left(P_{sp}^S V_{rk} + Q_{kh} V_{mk}\right)}{V_k^2} - \frac{[R_{se2}(V_{ri} - V_{rk}) + X_{se2}(V_{mi} - V_{mk})]}{Z_{se2}^2} \tag{6.40}$$

$$I_{S,inj}^{\mathrm{Im}} = \frac{\left(P_{sp}^S V_{mk} - Q_{kh} V_{rk}\right)}{V_k^2} - \frac{[R_{se2}(V_{mi} - V_{mk}) - X_{se2}(V_{ri} - V_{rk})]}{Z_{se2}^2} \tag{6.41}$$

This current is injected at bus (k) and at bus (i) with the opposite sign as shown in (6.42).

$$
\begin{bmatrix}
\Delta I_{m1} \\
\Delta I_{r1} \\
\vdots \\
\Delta I_{mi} - I_{M,inj}^{\mathrm{Im}} - I_{S,inj}^{\mathrm{Im}} \\
\Delta I_{ri} - I_{M,inj}^{\mathrm{Re}} - I_{S,inj}^{\mathrm{Re}} \\
\Delta I_{mj} + I_{M,inj}^{\mathrm{Im}} \\
\Delta I_{rj} + I_{M,inj}^{\mathrm{Re}} \\
\Delta I_{mk} + I_{S,inj}^{\mathrm{Im}} \\
\Delta I_{rk} + I_{S,inj}^{\mathrm{Re}} \\
\vdots \\
\Delta I_{mf} \\
\Delta I_{rf} \\
\Delta V_f^2 \\
\vdots \\
\Delta I_{mN} \\
\Delta I_{rN}
\end{bmatrix}
=
\begin{bmatrix}
B_{11}^{*} & G_{11}^{*} & \cdots & B_{1i} & G_{1i} & B_{1j} & G_{1j} & B_{1k} & G_{1k} & \cdots & B_{1f} & G_{1f} & 0 & \cdots & B_{1N} & G_{1N} \\
G_{11}^{**} & B_{11}^{**} & \cdots & G_{1i} & -B_{1i} & G_{1j} & -B_{1j} & G_{1k} & -B_{1k} & \cdots & G_{1f} & -B_{1f} & 0 & \cdots & G_{1N} & -B_{1N} \\
\vdots & \vdots & \iddots & \vdots & \vdots & \vdots & \vdots & \vdots & \vdots & \iddots & \vdots & \vdots & \vdots & \iddots & \vdots & \vdots \\
B_{i1} & G_{i1} & \cdots & B_{ii}^{*} & G_{ii}^{*} & B_{ij} & G_{ij} & B_{ik} & G_{ik} & \cdots & B_{if} & G_{if} & 0 & \cdots & B_{iN} & G_{iN} \\
G_{i1} & -B_{i1} & \cdots & G_{ii}^{**} & B_{ii}^{**} & G_{ij} & -B_{ij} & G_{ik} & -B_{ik} & \cdots & G_{if} & -B_{if} & 0 & \cdots & G_{iN} & -B_{iN} \\
B_{j1} & G_{j1} & \cdots & B_{ji} & G_{ji} & B_{jj}^{*} & G_{jj}^{*} & B_{jk} & G_{jk} & \cdots & B_{jf} & G_{jf} & 0 & \cdots & B_{jN} & G_{jN} \\
G_{j1} & -B_{j1} & \cdots & G_{ji} & -B_{ji} & G_{jj}^{**} & B_{jj}^{**} & G_{jk} & -B_{jk} & \cdots & G_{jf} & -B_{jf} & 0 & \cdots & G_{jN} & -B_{jN} \\
B_{k1} & G_{k1} & \cdots & B_{ki} & G_{ki} & B_{kj} & G_{kj} & B_{kk}^{*} & G_{kk}^{*} & \cdots & B_{kf} & G_{kf} & 0 & \cdots & B_{kN} & G_{kN} \\
G_{k1} & -B_{k1} & \cdots & G_{ki} & -B_{ki} & G_{kj} & -B_{kj} & G_{kk}^{**} & B_{kk}^{**} & \cdots & G_{kf} & -B_{kf} & 0 & \cdots & G_{kN} & -B_{kN} \\
\vdots & \vdots & \cdots & \vdots & \vdots & \vdots & \vdots & \vdots & \vdots & \iddots & \vdots & \vdots & \vdots & \iddots & \vdots & \vdots \\
B_{f1} & G_{f1} & \cdots & B_{fi} & G_{fi} & B_{fj} & G_{fj} & B_{fk} & G_{fk} & \cdots & B_{ff}^{*} & G_{ff}^{*} & \dfrac{V_{rf}}{V_f^2} & \cdots & B_{fN} & G_{fN} \\
G_{f1} & -B_{f1} & \cdots & G_{fi} & -B_{fi} & G_{fj} & -B_{fj} & G_{fk} & -B_{fk} & \cdots & G_{ff}^{**} & B_{ff}^{**} & \dfrac{-V_{mf}}{V_f^2} & \cdots & G_{fN} & -B_{fN} \\
0 & 0 & \cdots & 0 & 0 & 0 & 0 & 0 & 0 & \cdots & 2V_{rf} & 2V_{mf} & 0 & \cdots & 0 & 0 \\
\vdots & \vdots & \iddots & \vdots & \vdots & \vdots & \vdots & \vdots & \vdots & \iddots & \vdots & \vdots & \vdots & \iddots & \vdots & \vdots \\
B_{N1} & G_{N1} & \cdots & B_{Ni} & G_{Ni} & B_{Nj} & G_{Nj} & B_{Nk} & G_{Nk} & \cdots & B_{Nf} & G_{Nf} & 0 & \cdots & B_{NN}^{*} & G_{NN}^{*} \\
G_{N1} & -B_{N1} & \cdots & G_{Ni} & -B_{Ni} & G_{Nj} & -B_{Nj} & G_{Nk} & -B_{Nk} & \cdots & G_{Nf} & -B_{Nf} & 0 & \cdots & G_{NN}^{**} & B_{NN}^{**}
\end{bmatrix}
\begin{bmatrix}
\Delta V_{r1} \\
\Delta V_{m1} \\
\vdots \\
\Delta V_{ri} \\
\Delta V_{mi} \\
\Delta V_{rj} \\
\Delta V_{mj} \\
\Delta V_{rk} \\
\Delta V_{mk} \\
\vdots \\
\Delta V_{rf} \\
\Delta V_{mf} \\
\Delta Q_f \\
\vdots \\
\Delta V_{rN} \\
\Delta V_{mN}
\end{bmatrix}
\tag{6.42}
$$

Table 6.1 The calculation of the series voltage parameters of IPFC controller

Master converter	Slave converter
$V_{se}^{M} = I_{M,inj} Z_{se1}$	$V_{se}^{S} = I_{S,inj} Z_{se2}$
$V_{\mathrm{Re}(se)}^{M} = \left(R_{se1} I_{M,inj}^{\mathrm{Re}} - X_{se1} I_{M,inj}^{\mathrm{Im}} \right)$	$V_{\mathrm{Re}(se)}^{S} = \left(R_{se2} I_{S,inj}^{\mathrm{Re}} - X_{se2} I_{S,inj}^{\mathrm{Im}} \right)$
$V_{\mathrm{Im}(se)}^{M} = \left(R_{se1} I_{M,inj}^{\mathrm{Im}} + X_{se1} I_{M,inj}^{\mathrm{Re}} \right)$	$V_{\mathrm{Im}(se)}^{S} = \left(R_{se2} I_{S,inj}^{\mathrm{Im}} + X_{se2} I_{S,inj}^{\mathrm{Re}} \right)$
$\left\| V_{se}^{M} \right\| = \sqrt{V_{\mathrm{Re}(se)}^{M} + V_{\mathrm{Im}(se)}^{M}}$	$\left\| V_{se}^{S} \right\| = \sqrt{V_{\mathrm{Re}(se)}^{S} + V_{\mathrm{Im}(se)}^{S}}$
$\Phi_{se}^{M} = \tan^{-1}\left[V_{\mathrm{Im}(se)}^{M} / V_{\mathrm{Re}(se)}^{M} \right]$	$\Phi_{se}^{S} = \tan^{-1}\left[V_{\mathrm{Im}(se)}^{S} / V_{\mathrm{Re}(se)}^{S} \right]$

The required equations to calculate the IPFC parameters are summarized in Table 6.1.

6.4.3 Incorporating of developed IPFC model in NR-RCIM load flow

The IPFC current injection model can easily be incorporated in NR-RCIM program. If an IPFC is located in the lines *i–l* (*h*) in a power system, the admittance matrix should be modified by adding the reactance equivalent (Xser1 and Xser2) of IPFC. The current mismatches should be calculated including the new values of shunt injected currents at the ends of IPFC model (buses *i*, *j*, and *k*) as described in (6.42). The Jacobian matrix is calculated without changing the basic form. The voltage parameters of IPFC can be calculated during the iterative process as presented in Table 6.1.

Figure 6.11 shows the required steps for implementation of the developed IPFC model into NR-RCIM load flow algorithm.

6.5 Validation of developed series FACTS models

6.5.1 Proposed SSSC model in NR-RCIM

The proposed SSSC model is implemented into NR-RCIM algorithm. The standard WSCC 9-bus, IEEE 30-bus, and IEEE 118-bus systems are used to test the proposed SSSC modeling in NR-RCIM. The series reactance (X_s) of static synchronous series compensator is equal to 0.1 p.u. The maximum mismatch is equal to 10^{-5}.

- **WSCC 9-bus test system**

The SSSC is implemented between buses 8 and 9 to control the power flow to be $40 + j2$ MVA. Bus 10 is added as a reference as shown in Figure 6.12.

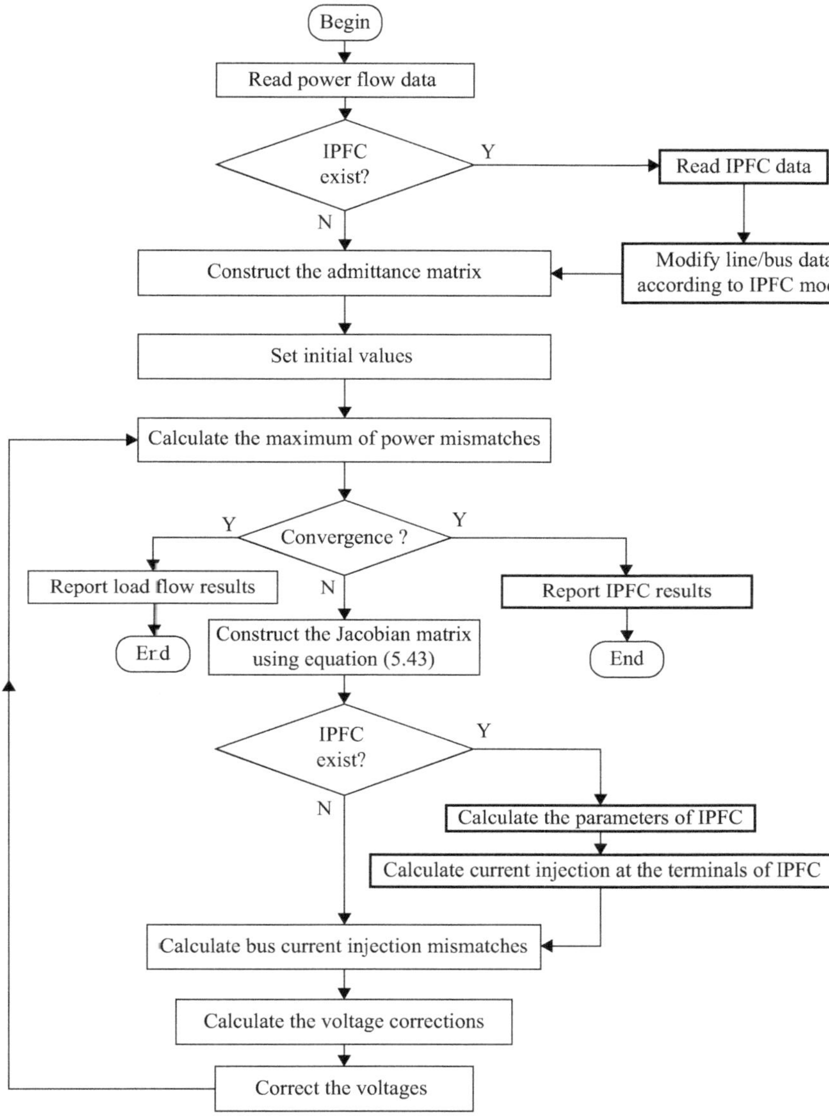

Figure 6.11 Flow chart of NR-RCIM load-flow method with developed IPFC model

Table 6.2 gives the voltage magnitude and phase angle for all buses of test system with and without SSSC. The result of power flow for the system with SSSC is given in Table 6.3. It can be observed that the specified line flow has been achieved.

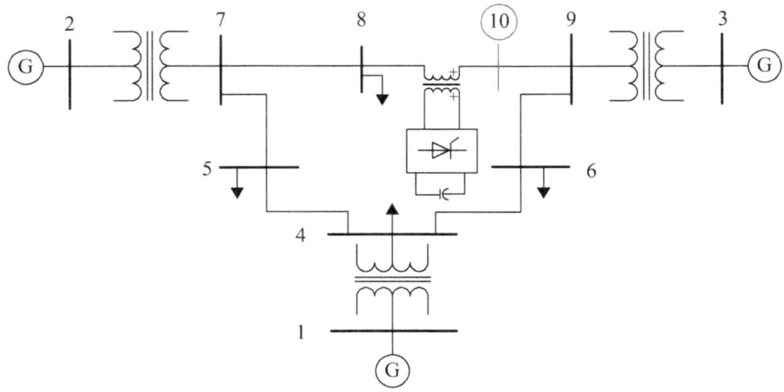

Figure 6.12 Single-line diagram of WSCC 9-bus test system with SSSC control

Table 6.2 Voltage load flow solution of WSCC 9-bus test

Bus no.		**1**	**2**	**3**	**4**	**5**	**6**	**7**	**8**	**9**	**10**
W/O	$\|V\|$ (p.u.)	1.040	1.025	1.025	1.026	0.996	1.012	1.027	1.020	1.031	–
SSSC	φ (deg.)	0.000	9.527	4.387	−2.22	−3.89	−3.789	3.976	1.068	1.686	–
With	$\|V\|$ (p.u.)	1.040	1.025	1.025	1.021	0.991	1.006	1.011	0.984	1.044	1.055
SSSC	φ (deg.)	0.000	1.016	13.645	−2.21	−6.88	−0.513	−4.62	−10.17	10.978	11.950

Table 6.3 Line flows of WSCC 9-bus system including SSSC

Sending bus	**Receiving bus**	**Active power (MW)**	**Reactive power (Mvar)**
5	4	−99.100	−27.960
4	5	100.14	18.990
4	6	−29.000	13.890
6	4	29.220	−28.940
5	7	−25.90	−22.040
7	5	26.130	−7.450
6	9	−119.22	−1.060
9	6	124.810	−12.220
7	8	136.870	21.660
8	7	−135.24	−22.690
8	10	−390.77	5.8000
10	9	40.000	2.0000
9	10	−39.810	−24.220
10	8	390.77	152.10
1	4	71.1400	36.280
4	1	−71.1400	−32.880
2	7	163.000	30.570
7	2	−163.00	−14.210
3	9	85.0000	−31.840
9	3	−85.000	36.440

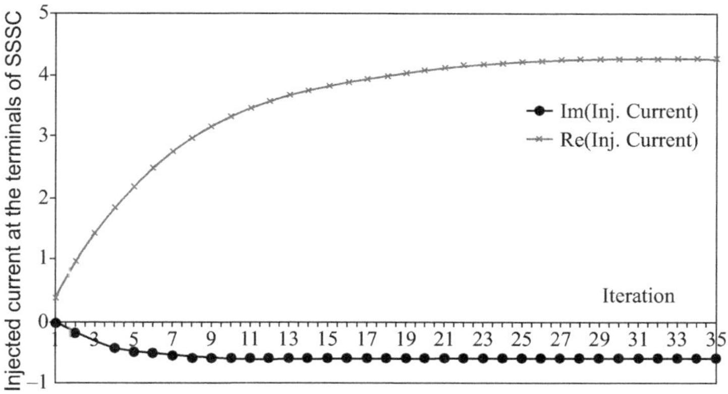

Figure 6.13 Variation of the injected current against number of iterations

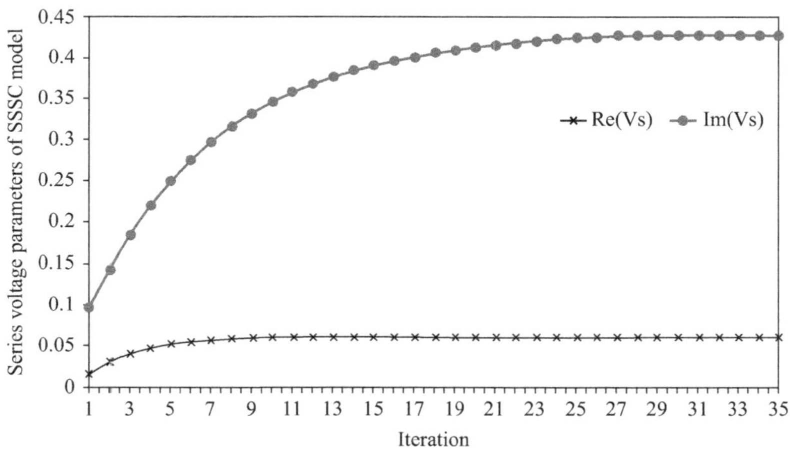

Figure 6.14 Variation of the voltage parameters of SSSC against number of iterations

The convergence characteristic of the developed model is validated by drawing the variation of real and imaginary components of the injected current against number of iterations as shown in Figure 6.13. It is clearly from this figure that the injected current reaches to steady state after a few more iterations compared to the uncontrolled load flow case.

The variation of the real and imaginary components of voltage parameters (V_s^{Re}, V_s^{Im}) against number of iterations is tested and given in Figure 6.14. It can be observed that the voltage parameters change smoothly to the final values without oscillation.

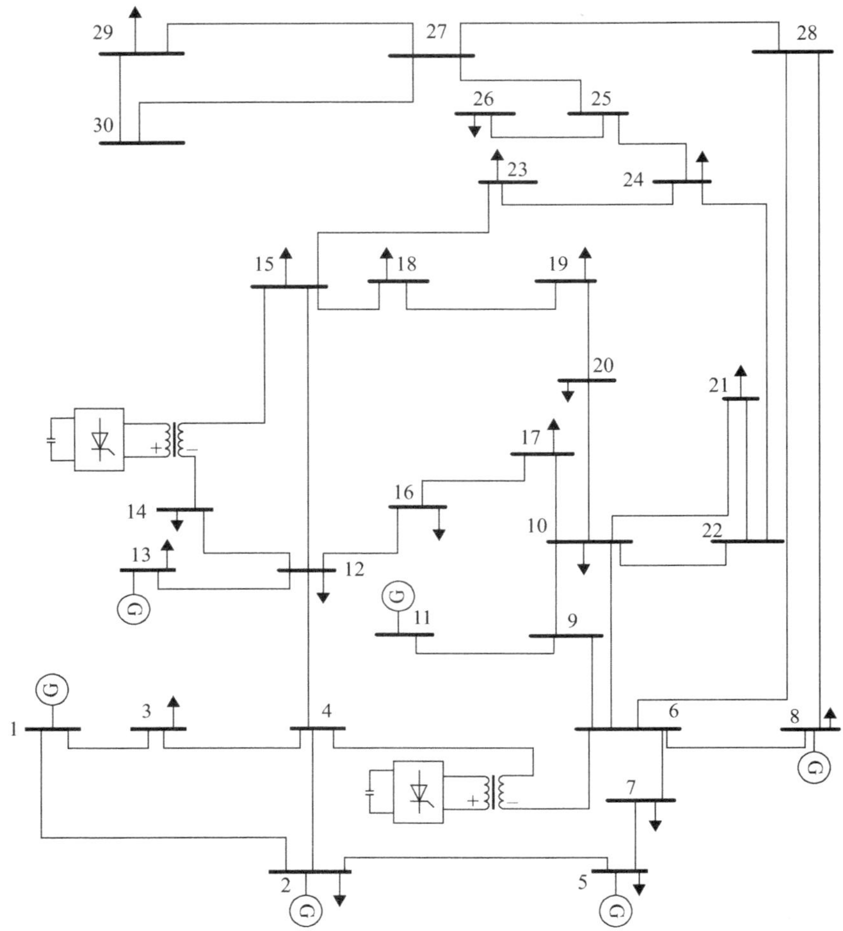

Figure 6.15 IEEE 30-bus test system with two SSSCs

- **IEEE 30-bus test system**

Two SSSC devices are incorporated in the IEEE 30-bus test system as shown in Figure 6.15. The first one is implemented between buses 6 and 4 to control the active and reactive power flows to be $50+j1$ MVA. The second one is implemented between 14 and 15 to set the power flow at $30+j2$ MVA.

The voltages load-flow solution of this system with and without the two SSSC devices is presented in Table 6.4. The results of power flow in the lines of the system with SSSC controllers are presented in Table 6.5. It can be observed that the required power flows are achieved.

Table 6.4 Voltage buses solution of IEEE 30-bus test system

Bus no.	W/O SSSC		With SSSC		Type of bus
	$\|V\|$ (p.u.)	φ (deg.)	$\|V\|$ (p.u.)	φ (deg.)	
1	1.0600	0.0000	1.0600	0.0000	Slack
2	1.0450	−5.3667	1.0450	−6.7111	PV
3	1.0274	−7.6025	1.0416	−4.4603	PQ
4	1.0200	−9.3637	1.0366	−5.4536	PQ
5	1.0100	−14.1147	1.0100	−18.6828	PV
6	1.0142	−11.0886	1.0089	−18.9380	PQ
7	1.0047	−12.8565	1.0010	−19.3673	PQ
8	1.0100	−11.7683	1.0100	−19.6786	PV
9	1.0367	−14.3482	1.0314	−20.1264	PQ
10	1.0259	−16.0518	1.0168	−20.7500	PQ
11	1.0820	−14.3482	1.0820	−20.1264	PV
12	1.0343	−15.3494	1.0389	−16.3211	PQ
13	1.0710	−15.3494	1.0710	−16.3211	PV
14	1.0194	−16.2649	0.9933	−21.0848	PQ
15	1.0151	−16.3505	1.0265	−17.0868	PQ
16	1.0230	−15.9284	1.0185	−18.4549	PQ
17	1.0199	−16.2350	1.0128	−20.2634	PQ
18	1.0065	−16.9697	1.0098	−19.0765	PQ
19	1.0046	−17.1377	1.0036	−20.0787	PQ
20	1.0091	−16.9261	1.0058	−20.3075	PQ
21	1.0127	−16.5109	1.0043	−21.0883	PQ
22	1.0131	−16.4956	1.0049	−21.0316	PQ
23	1.0045	−16.7224	1.0088	−18.8321	PQ
24	0.9990	−16.8575	0.9952	−20.8667	PQ
25	0.9917	−16.3422	0.9829	−21.7773	PQ
26	0.9735	−16.7843	0.9646	−22.2274	PQ
27	0.9959	−15.7521	0.9848	−22.0800	PQ
28	1.0102	−11.6976	1.0063	−19.3905	PQ
29	0.9755	−17.0518	0.9641	−23.4099	PQ
30	0.9636	−17.9865	0.9521	−24.3670	PQ
31	–	–	1.0427	−4.3658	PQ
32	–	–	1.0898	−14.2440	PQ

Figure 6.16 shows the variation of the injected currents against the iterations. Figure 6.17 shows the variation of voltage parameters with number of iterations. The convergence is done after a few more iteration.

The capability of the developed SSSC model at various specified active and reactive power flows is investigated. In this test, an SSSC device is implemented in IEEE 30-bus test system between bus 14 and bus 15 to control the power flow with random values. In the first case, the active power is only controlled to be 50 MW. Table 6.6 gives the voltages of auxiliary, sending and receiving buses, and SSSC parameters.

Table 6.5 Power-flow solution of IEEE 30-bus test system with two SSSCs

Send bus	Receive bus	*P* (MW)	*Q* (Mvar)
1	2	214.85	–33.85
1	3	51.89	–2.64
2	4	–10.82	6.73
3	4	48.40	–3.29
2	5	111.26	1.26
2	6	124.65	–8.14
31	4	50.00	1.00
5	7	11.70	2.22
6	7	11.20	5.20
6	8	28.33	–11.07
6	9	10.38	–10.83
6	10	5.83	–1.34
9	11	–0.00	–25.09
9	10	10.38	13.80
4	12	79.32	6.63
12	13	–0.00	–23.83
12	14	34.97	3.07
12	15	12.69	3.47
12	16	20.46	1.32
32	15	30.00	2.00
16	17	16.59	–1.25
15	18	16.49	0.04
18	19	13.01	–1.43
19	20	3.41	–5.05
10	20	–1.16	5.84
10	17	–7.41	7.66
10	21	13.11	10.90
10	22	5.87	5.20
21	22	–4.48	–0.51
15	23	16.21	1.21
22	24	1.34	4.60
23	24	12.76	–0.90
24	25	5.17	0.78
25	26	3.55	2.37
25	27	1.57	–1.68
28	27	11.74	5.74
27	29	6.20	1.69
27	30	7.11	1.69
29	30	3.71	0.61
8	28	–1.78	0.27
6	28	13.56	–0.07
6	31	–264.71	–0.31
14	32	–128.95	–88.11

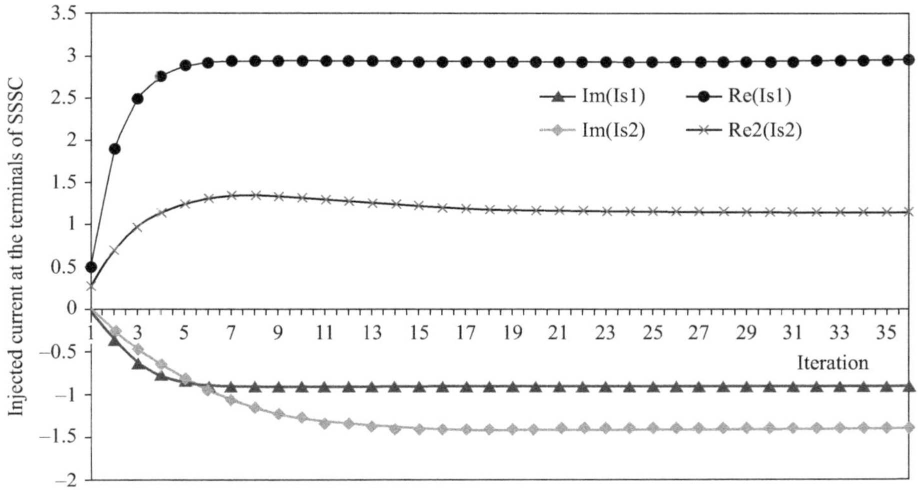

Figure 6.16 Convergence characteristics of Is for IEEE30-bus test system with two SSSCs

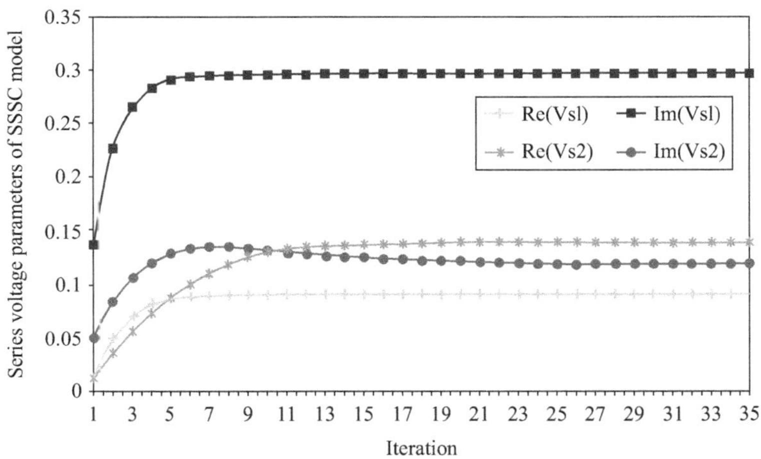

Figure 6.17 Convergence characteristics of Vs for IEEE30-bus test system with two SSSCs

The capability of the proposed SSSC modeling into NR-RCIM load-flow algorithm at various series reactance of SSSC is investigated. In this case, the active power between bus 14 and bus 15 is fixed at 20 MW and the reactive at 1.0 Mvar. The result of this test is presented in Table 6.7.

Table 6.6 The effectiveness of the proposed SSSC model at different required power flows

Required power		Auxiliary bus		Send bus		Receive bus		Parameters	
P^{sp} (MW)	Q^{sp} (*Mvar*)	$\|V\|$ (p.u.)	Φ (deg.)	$\|V\|$ (p.u.)	φ (deg.)	$\|V\|$ (p.u.)	Φ (deg.)	$\|V_s\|$ (p.u.)	δ_s (deg.)
10.0	0.0	1.0390	−14.9	1.011	−17.64	1.018	−15.98	0.065	48.22
50.0	–	1.1052	−8.328	1.009	−24.05	1.0238	−14.21	0.343	59.90
20.0	1.0	1.066	−13.61	0.999	−19.13	1.023	−15.59	0.136	44.26
20.0	−1.0	1.060	−13.32	1.004	−19.25	1.021	−15.55	0.137	50.63
40.0	3.0	1.113	−11.19	0.981	−22.07	1.031	−14.85	0.269	44.28
30.0	2.0	1.090	−12.38	0.989	−20.60	1.027	−15.22	0.203	43.88
50.0	4.0	1.135	−10.04	0.975	−23.51	1.034	−14.49	0.331	44.96

Table 6.7 Performance of the developed SSSC model at various series reactance

*X*s (p.u.)	Auxiliary bus		Sending bus		Receiving bus		SSSC parameters	
	$\|V\|$ (p.u.)	φ (deg.)	$\|V\|$ (p.u.)	φ (deg.)	$\|V\|$ (p.u.)	Φ (deg.)	$\|V_s\|$ (p.u.)	δ_s (deg.)
0.20	1.066	−13.610	0.999	−19.127	1.023	−15.594	0.152	47.719
0.15	1.066	−13.610	0.999	−19.127	1.023	−15.594	0.144	46.088
0.10	1.066	−13.610	0.999	−19.127	1.023	−15.594	0.136	44.257
0.05	1.066	−13.610	0.999	−19.127	1.023	−15.594	0.128	42.190
0.01	1.066	−13.610	0.999	−19.127	1.023	−15.594	0.121	40.339

- **IEEE 118-bus test system**

The IEEE 118-bus test system is selected as a large-scale system. In this test, the capability of the developed model to operate at single and multiple SSSCs is investigated. The control parameter of SSSC and voltages at sending, receiving, and auxiliary for each test case is presented in Table 6.8.

The location of SSSCs in the test system is selected randomly. In all test cases, the specified power flow is $30 + j2$ MVA. The direction of power flow between bus 94 and bus 93 is reflected in case B. In case C, two SSSCs are installed in lines 94–93 and 96–95, respectively, to control the active and reactive power flows at specified values. Three SSSCs are installed in the system in lines 94–93, 96–95, and 101–102. For the last case D, the convergence characteristic of the injected currents at the terminal of SSSCs is presented in Figure 6.18. Also, Figure 6.19 shows the change of voltage parameters of the three SSSCs in the test system.

Table 6.8 Effectiveness of the proposed SSSS model with multiple controllers

Test conditions			Auxiliary bus		Send bus		Receive bus		Parameters	
Case	Send bus	Receive bus	$\|V\|$ (p.u.)	φ (deg.)	$\|V\|$ (p.u.)	Φ (deg.)	$\|V\|$ (p.u.)	φ (deg.)	$\|V_s\|$ (p.u.)	δ_s (deg.)
A	93	94	0.9920	31.41	0.9690	36.29	0.9830	30.15	0.0590	−30.54
B	94	93	1.0010	42.11	0.9770	29.43	0.9920	40.88	0.2500	−59.43
C	94	93	1.0010	43.62	0.9750	32.03	0.9920	42.39	0.2310	−58.20
	96	95	0.9680	32.91	0.9800	22.67	0.961	31.93	0.2050	−58.79
D	94	93	1.001	46.120	0.975	31.897	0.992	44.891	0.276	−55.994
	96	95	0.967	32.781	0.979	22.742	0.961	31.797	0.202	−58.864
	101	102	1.005	46.643	0.991	31.970	0.994	44.764	0.285	−53.093

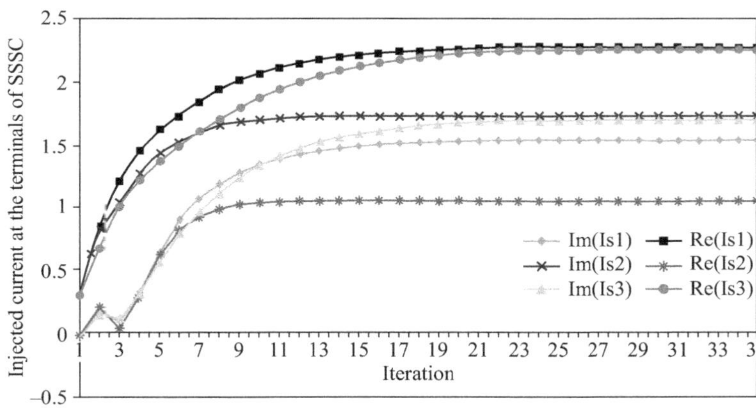

Figure 6.18 Convergence characteristics of Is for IEEE 118-bus test system with three SSSCs

6.5.2 Developed IPFC model in NR-RCIM

The effectiveness and performance of the developed IPFC model is investigated based on different IEEE test system as;

- **IEEE 5-bus test system**

IEEE 5-bus test system is incorporated with IPFC between lines 4–3(5) as shown in Figure 6.20. The purpose of IPFC is to maintain the active and reactive powers leaving IPFC toward buses 3(5) to be 30 MW, 2 Mvar, and 25 MW, respectively. The result of voltage magnitudes and phase angles of system and IPFC sources are presented in Table 6.9. The line flow solution for the system is presented in Figure 6.20.

The power flow results (as a shown in Figure 6.20) have shown the ability of NR-RCIM load flow with the developed IPFC model in the test system.

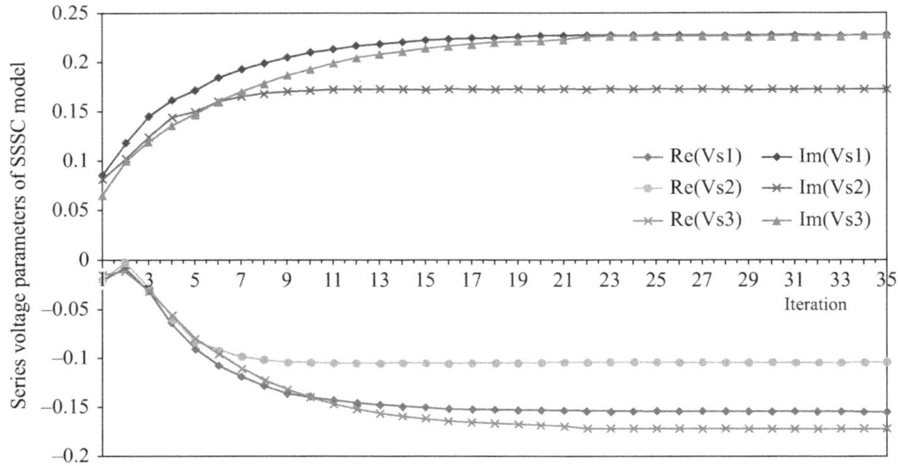

Figure 6.19 Convergence characteristics of Vs for IEEE 118-bus test system with three SSSCs

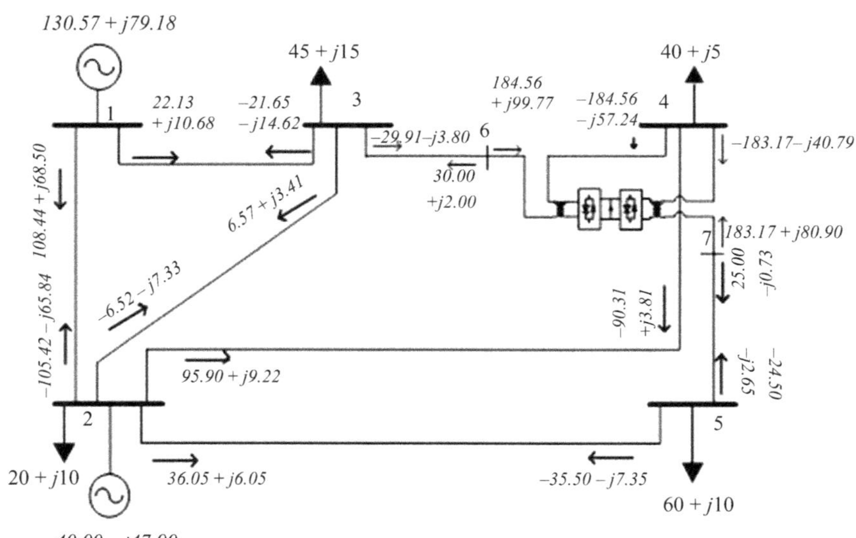

Figure 6.20 Load-flow solution of 5-bus test system with IPFC model

Table 6.9 The voltages of 5-bus test system with IPFC model

Bus no.	|V| (p.u.)	φ (deg.)
1	1.060	0.000
2	1.000	−2.740
3	1.014	−2.258
4	0.937	−12.936
5	0.977	−5.100
Auxiliary bus (6)	1.017	−1.774
Auxiliary bus (7)	1.000	−1.663
Master series source	0.233	62.892
Slave series source	0.223	67.318

- **IEEE 57-bus test system**

The IPFC is implemented in IEEE 57-bus test system in lines 15–13(14) to control the active and reactive powers leaving IPFC toward buses 13(14) to be 35 MW, 5 Mvar, and 25 MW, respectively, whereas the reactance of series transformers is set at different values. Table 6.10 shows the ability of IPFC model to operate at different values of series transformer reactance.

The series voltage parameters of IPFC and shunt-injected current with various specified power flow are given in Table 6.11. The results show the ability of the developed IPFC model to operate at different specified power flows.

Figure 6.21 shows the variation of real and imaginary components of injected currents of master and slave lines against the iterations.

Figure 6.22 shows the changes of the real and imaginary components of series voltage parameters of IPFC against the number of iterations. It can be observed that the series voltages change smoothly to the final values without oscillation.

- **IEEE 118-bus test system**

The IEEE 118 bus is taken as a large-scale test system. The IPFC is located on the lines 94–93(95) to adjust the active and reactive powers leaving IPFC toward buses (93) and (95) at various values as given in Table 6.12. The ability of developed IPFC model to operate at various specified power flows is investigated using this test.

The capability of the developed IPFC model in NR-RCIM load-flow algorithm to solve multiple IPFCs is investigated. The location of IPFCs in this system is chosen randomly. The IPFC is located at lines *i*–*l* (*h*) to control the active and reactive powers leaving IPFC toward buses *l* and *h* to be 60 MW, 3 Mvar, and 30 MW, respectively, whereas the series transformer reactance is taken equal to 0.1 p.u. The results of this test are given in Table 6.13.

Table 6.10 The performance of IEEE 57-bus with IPFC model at various IPFC series reactance

Xe (p.u.)	**Injection currents at IPFCs terminals**				**IPFC parameters**			
					Series source (1)		**Series source (2)**	
	$I_{M,inj}^{Re}$ (p.u.)	$I_{M,inj}^{Im}$ (p.u.)	$I_{S,inj}^{Re}$ (p.u.)	$I_{S,inj}^{Im}$ (p.u.)	$\lvert V_{se}^{M}\rvert$ (p.u.)	Φ_{se}^{M} (deg.)	$\lvert V_{se}^{S}\rvert$ (p.u.)	Φ_{se}^{S} (deg.)
0.20	−0.084	−0.078	−0.212	0.193	0.023	−46.862	0.057	47.726
0.15	−0.223	−0.066	−0.371	0.258	0.035	−73.687	0.068	55.284
0.10	−0.503	−0.040	−0.689	0.386	0.050	−85.508	0.079	60.765
0.05	−1.343	0.037	−1.643	0.773	0.067	88.465	0.091	64.855
0.01	−8.057	0.653	−9.275	3.862	0.081	85.413	0.100	67.427

Table 6.11 IEEE 57-bus system with IPFC model at various specified power flow

Specified power			**Injection currents at IPFCs terminals**				**IPFC parameters**			
							Series source (1)		**Series source (2)**	
P_{sp}^{M} (MW)	Q_{sp}^{M} (Mvar)	P_{sp}^{S} (MW)	$I_{M,inj}^{Re}$ (p.u.)	$I_{M,inj}^{Im}$ (p.u.)	$I_{S,inj}^{Re}$ (p.u.)	$I_{S,inj}^{Im}$ (p.u.)	$\lvert V_{se}^{M}\rvert$ (p.u.)	Φ_{se}^{M} (deg.)	$\lvert V_{se}^{S}\rvert$ (p.u.)	Φ_{se}^{S} (deg.)
50	6	60	0.361	−0.254	0.542	0.129	0.044	54.840	0.056	−76.638
−60	−4	−34	−3.878	1.791	−3.643	1.353	0.427	65.243	0.389	69.660
−40	8.4	−90	−4.211	1.429	−5.468	1.605	0.445	71.298	0.570	73.683
70	−10.5	−110	−1.043	0.319	−4.801	1.206	0.109	73.030	0.495	75.942
0	0	25	−1.499	0.449	−1.056	0.560	0.156	73.357	0.120	62.080

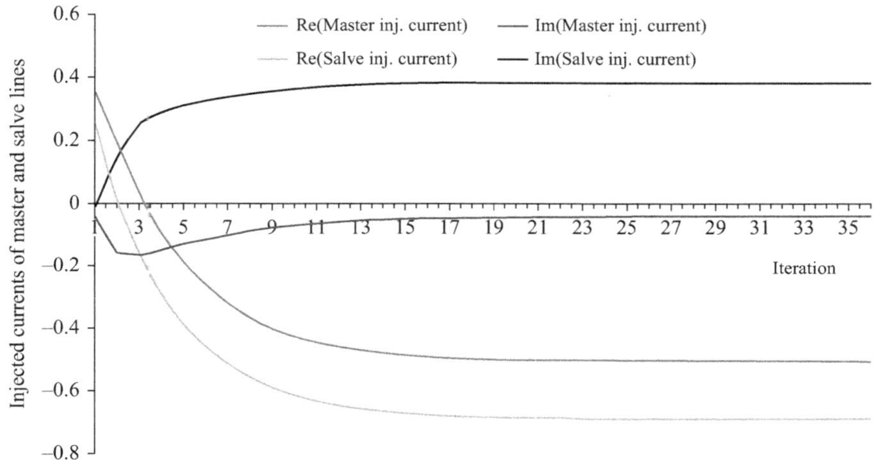

Figure 6.21 Convergence characteristics of injected currents at the ends of IPFC

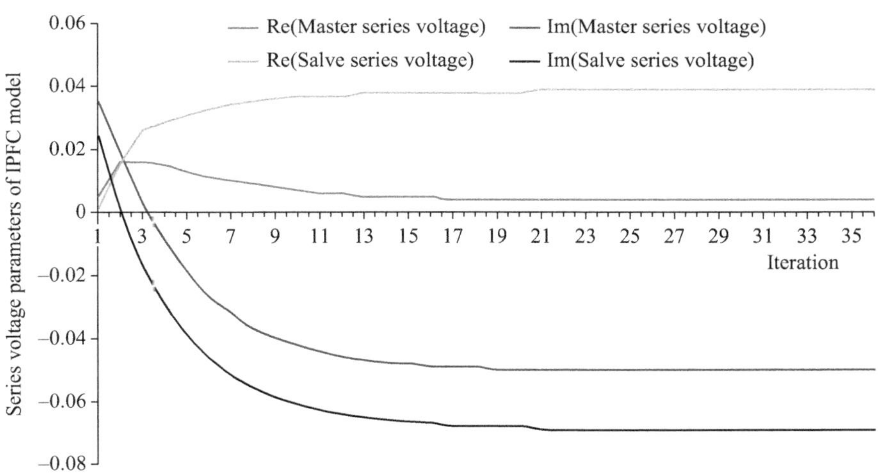

Figure 6.22 Convergence characteristics of series voltage parameters of IPFC

Table 6.12 IEEE 118-bus test system with IPFC at various specified power flows

Specified power			Injection currents at IPFCs terminals				IPFC parameters			
							Series source (1)		Series source (2)	
P_{sp}^M (MW)	Q_{sp}^M (Mvar)	P_{sp}^S (MW)	$I_{M,inj}^{Re}$ (p.u.)	$I_{M,inj}^{Im}$ (p.u.)	$I_{S,inj}^{Re}$ (p.u.)	$I_{S,inj}^{Im}$ (p.u.)	$\lvert V_{se}^M \rvert$ (p.u.)	Φ_{se}^M (deg.)	$\lvert V_{se}^S \rvert$ (p.u.)	Φ_{se}^S (deg.)
130	66	170	5.545	1.758	2.961	−1.206	0.582	−72.445	0.320	67.874
150	72	−57	5.496	2.514	−3.249	−0.136	0.604	−65.451	0.325	−87.643
82	−30	45	2.728	3.235	−0.470	−0.558	0.423	−40.159	0.073	−40.114
−100	37	29	−1.870	−1.403	−1.180	−0.711	0.234	−53.149	0.138	−58.953
−77	−52	−39	−2.537	1.367	−3.181	−0.199	0.288	61.723	0.319	−86.469

Table 6.13 IEEE 118-bus test system with single and multiple IPFCs

Case	IPFC lines	$I_{M,inj}$ (p.u.)	$I_{S,inj}$ (p.u.)	Auxiliary voltage bus		IPFC parameters	
				Bus (j)	Bus (k)	V_{se}^M (p.u.)	V_{se}^S (p.u.)
A	96/82(94)	$1.252 + j0.189$	$2.648 + j0.643$	$0.987\angle 24.961^\circ$	$1.007\angle 34.886^\circ$	$0.127\angle -81.481^\circ$	$0.273\angle -76.399^\circ$
B	96/82(94)	$0.468 - j0.756$	$2.931 + j0.831$	$1.027\angle 20.949^\circ$	$1.005\angle 38.883^\circ$	$0.089\angle 31.774^\circ$	$0.307\angle -75.372^\circ$
	83/82(84)	$-0.896 - j2.584$	$3.028 - j0.573$	$1.023\angle 20.436^\circ$	$1.064\angle 45.960^\circ$	$0.274\angle -19.134^\circ$	$0.081\angle 44.812^\circ$
C	96/82(94)	$2.489 - j0.019$	$8.342 + j4.428$	$1.027\angle 20.127^\circ$	$1.061\angle 61.750^\circ$	$0.249\angle 89.617^\circ$	$0.199\angle -59.604^\circ$
	83/82(84)	$-2.915 - j6.843$	$4.970 - j0.962$	$1.023\angle 19.614^\circ$	$1.161\angle 72.004^\circ$	$0.347\angle -23.089^\circ$	$0.375\angle 81.186^\circ$
	94/93(95)	$1.253 + j3.374$	$-6.809 - j4.485$	$1.013\angle 78.309^\circ$	$0.949\angle 10.257^\circ$	$0.360\angle -20.385^\circ$	$0.264\angle -14.655^\circ$

6.6 Conclusions

The chapter has developed simple modeling for the series FACTS (SSSC and IPFC) into NR-RCIM load flow method that can be used to analyze and improve the power quality. The developed models are based on current injection approach. In this modeling, the series controllers are represented as injected currents at the terminals of their impedances.

These currents are calculated in terms of the specified power flow and buses voltage. The injected currents can easily be implemented in load-flow algorithm without any change in the Jacobian matrix. The parameters of series FACTS can be updated during the iterative process using simple equations based on their basic equivalent circuits.

In these developed series controllers, the real and/or reactive power flow in single- or multiline transmission system can be controlled; the selection of appropriate initial values of FACTS devices is not required; the problem that happens when the series FACTS device is the only link between two areas has been solved; the structure of the admittance and Jacobian matrices has been kept without change. Consequentially, the complexity of load-flow algorithm with series FACTS has been reduced.

References

[1] Luo G-X, Semlyen A. Efficient load flow for large weakly meshed networks. IEEE Transactions on Power Systems. 1990;5(4):1309–16.
[2] Electricity Grid Schematic [Online]. Available: http://en.wikipedia.org/wiki/File:Electricity_Grid_Schematic_English.svg.
[3] Stott B. Review of load-flow calculation methods. Proceedings of the IEEE. 1974;62(7):916–29.
[4] Grainger JJ, Stevenson WD. Power system analysis: McGraw-Hill; 1994.
[5] Arrillaga J, Arnold C. Computer analysis of power systems. Chichester, England: Wiley, 1990.
[6] Murty PSR. Operation and control in power systems, Hyderabad, India: BS Pub; 2011.
[7] Saadat H. Power system analysis. Boston: WCB/McGraw-Hill; 2002.
[8] ETAP Power Station software [Online]. Available: http://www.etap.com/.
[9] EDSA software [Online]. Available: http://www.edsa.com/.
[10] PSS/E software [Online]. Available: http://w3.usa.siemens.com/smartgrid/us/en/Pages/smart-grid.aspx.
[11] DigSILENT software [Online]. Available: http://www.digsilent.de/.
[12] Shahriari A, Mokhlis H, Bakar A. Critical reviews of load flow methods for well, ill and unsolvable condition. Journal of Electrical Engineering. 2012;63(3):144–52.

[13] Mallick S, Rajan D, Thakur S, Acharjee P, Ghoshal S. Development of a new algorithm for power flow analysis. International Journal of Electrical Power & Energy Systems. 2011;33(8):1479–88.
[14] Wu FF. Theoretical study of the convergence of the fast decoupled load flow. IEEE Transactions on Power Apparatus and Systems. 1977;96(1):268–75.
[15] da Costa VM, Martins N, Pereira JLR. Developments in the Newton–Raphson power flow formulation based on current injections. IEEE Transactions on Power Systems. 1999;14(4):1320–6.
[16] Garcia PA, Pereira J, Carneiro S, Vinagre MP, Gomes FV. Improvements in the representation of PV buses on three-phase distribution power flow. IEEE Transactions on Power Delivery. 2004;19(2):894–6.
[17] Ferreira CA. A second order power flow based on current injection equations. International Journal of Electrical Power & Energy Systems. 2005; 27(4):254–63.
[18] Garcia PA, Pereira JLR, Carneiro S, da Costa VM, Martins N. Three-phase power flow calculations using the current injection method. IEEE Transactions on Power Systems. 2000;15(2):508–14.
[19] Da Costa V, Pereira J, Martins N. An augmented Newton–Raphson power flow formulation based on current injections. International Journal of Electrical Power & Energy Systems. 2001;23(4):305–12.
[20] Penido DRR, de Araujo LR, Carneiro S, Pereira JLR, Garcia PAN. Three-phase power flow based on four-conductor current injection method for unbalanced distribution networks. IEEE Transactions on Power Systems. 2008;23(2):494–503.
[21] de Araujo LR, Penido DRR, Júnior SC, Pereira JLR, Garcia PAN. Comparisons between the three-phase current injection method and the forward/backward sweep method. International Journal of Electrical Power & Energy Systems. 2010;32(7):825–33.
[22] Penido D, Araujo L, Carneiro S, Pereira J, Garcia P, editors. Power factor correction on distribution networks including distributed generation. Power Engineering Society General Meeting, 2007 IEEE; 2007: IEEE.
[23] Penido DRR, de Araujo LR, Júnior SC, Pereira JLR. A new tool for multiphase electrical systems analysis based on current injection method. International Journal of Electrical Power & Energy Systems. 2013;44(1):410–20.
[24] Hingorani NG, Gyugyi L. Understanding FACTS: concepts and technology of flexible AC transmission systems. New York: Wiley-IEEE; 2000.
[25] Zhang X-P, Rehtanz C, Pal B. Flexible AC transmission systems: modelling and control. Berlin: Springer; 2012.
[26] Padiyar K. FACTS controllers in power transmission and distribution. Tunbridge Wells: Anshan; 2009.
[27] Acha E, Fuerte-Esquivel CR, Ambriz-Perez H, Angeles-Camacho C. FACTS: modelling and simulation in power networks. Chichester: John Wiley & Sons; 2004.
[28] Acha E, Agelidis V, Anaya O, Miller TJE. Power electronic control in electrical systems. Oxford: Newnes, 2002.

[29] Grigsby LL. The electric power engineering handbook. Boca Raton, FL: CRC, 2001.
[30] Haque M, Yam C. A simple method of solving the controlled load flow problem of a power system in the presence of UPFC. Electric Power Systems Research. 2003;65(1):55–62.
[31] Nor KM, Mokhlis H, Gani TA. Reusability techniques in load-flow analysis computer program. IEEE Transactions on Power Systems. 2004;19 (4):1754–62.
[32] Bhowmick S, Das B, Kumar N. An indirect UPFC model to enhance reusability of newton power-flow codes. IEEE Transactions on Power Delivery. 2008;23(4):2079–88.
[33] Bhowmick S, Das B, Kumar N. An advanced IPFC model to reuse Newton power flow codes. IEEE Transactions on Power Systems. 2009; 24(2):525–32.
[34] Bhowmick S, Das B, Kumar N. An advanced static synchronous compensator model to reuse Newton and decoupled power flow codes. Electric Power Components and Systems. 2011;39(15):1647–66.
[35] Bhowmick S, Das B, Kumar N. An indirect model of SSSC for reducing complexity of coding in Newton power flow algorithm. Electric Power Systems Research. 2007;77(10):1432–41.
[36] Padhy NP, Moamen MA. Power flow control and solutions with multiple and multi-type FACTS devices. Electric Power Systems Research. 2005; 74(3):341–51.
[37] Mokhlis H, Nor KM, editors. Implementation of UPFC model into fast decoupled load flow. TENCON 2004 2004 IEEE Region 10 Conference; 2004: IEEE.
[38] Nabavi-Niaki A, Iravani MR. Steady-state and dynamic models of unified power flow controller (UPFC) for power system studies. IEEE Transactions on Power Systems. 1996;11(4):1937–43.
[39] Gotham DJ, Heydt G. Power flow control in systems with FACTS devices. Electric Machines and Power Systems. 1998;26(9):951–62.
[40] Zhang X-P. Advanced modeling of the multicontrol functional static synchronous series compensator (SSSC) in Newton power flow. IEEE Transactions on Power Systems. 2003;18(4):1410–16.
[41] Fuerte-Esquivel C, Acha E, Ambriz-Perez H. A comprehensive Newton–Raphson UPFC model for the quadratic power flow solution of practical power networks. IEEE Transactions on Power Systems. 2000;15 (1):102–9.
[42] Fuerte-Esquivel C, Acha E. Unified power flow controller: a critical comparison of Newton–Raphson UPFC algorithms in power flow studies. IEE Proceedings-Generation, Transmission and Distribution. 1997;144(5): 437–44.
[43] Noroozian M, Angquist L, Ghandhari M, Andersson G. Use of UPFC for optimal power flow control. IEEE Transactions on Power Delivery. 1997; 12(4):1629–34.

[44] Zhang Y, Zhang Y. A novel power injection model of embedded SSSC with multi-control modes for power flow analysis inclusive of practical constraints. Electric Power Systems Research. 2006;76(5):374–81.
[45] Zhang Y, Chen C. A novel power injection model of IPFC for power flow analysis inclusive of practical constraints. IEEE Transactions on Power Systems. 2006;21(4):1550–6.
[46] Alomoush MI. Derivation of UPFC DC load flow model with examples of its use in restructured power systems. IEEE Transactions on Power Systems. 2003;18(3):1173–80.
[47] Alomoush MI. Exact Pi-model of UPFC-inserted transmission lines in power flow studies. IEEE Power Engineering Review. 2002;22(12):54–6.
[48] Zhang Y, Zhang Y, Wu B, Zhou J. Power injection model of STATCOM with control and operating limit for power flow and voltage stability analysis. Electric Power Systems Research. 2006;76(12):1003–10.
[49] Benabid R, Boudour M, Abido M. Development of a new power injection model with embedded multi-control functions for static synchronous series compensator. IET Generation, Transmission & Distribution. 2012; 6(7):680–92.
[50] Mhaskar U, Mote A, Kulkarni A. A new formulation for load flow solution of power systems with series FACTS devices. IEEE Transactions on Power Systems. 2003;18(4):1307–15.
[51] Wei X, Chow JH, Fardanesh B, Edris A-A. A common modeling framework of voltage-sourced converters for load flow, sensitivity, and dispatch analysis. IEEE Transactions on Power Systems. 2004;19(2):934–41.
[52] Vasquez-Arnez RL, Zanetta LC. A novel approach for modeling the steady-state VSC-based multiline FACTS controllers and their operational constraints. IEEE Transactions on Power Delivery. 2008;23(1):457–64.
[53] Tiwari R, Niazi K, Gupta V. Variable series impedance model of a unified power flow controller for load flow analysis. Electric Power Components and Systems. 2011;40(1):74–92.
[54] Liu J-Y, Song Y-H, Mehta P. Strategies for handling UPFC constraints in steady-state power flow and voltage control. IEEE Transactions on Power Systems. 2000;15(2):566–71.

Chapter 7
Distributed generation systems

Khaled H. Ahmed[1,2] *and Ahmed A. Aboushady*[3]

7.1 Introduction

Power electronics continues to penetrate into more power applications and plays a main part in energy transformation and consuming. It is anticipated in power industry that power electronics will become more involved in energy generation and distribution [1]. The high efficiency and compactness of power converters are the main momentum behind power electronic technologies. It is therefore considered an enabling technology. The earth, wind, sun and plants are the main resources for all renewable energy technologies which provide environmental friendly, numerous, clean energy. In our modern life, energy has a vital role in future economic growth and well-being. While handling climate change problem and guaranteeing the energy supply security, numerous challenges should be taken into account. As part of addressing such challenges is to study to what extent distributed generation (DG) could supplement, or in the longer term potentially provide an alternative to, a centralised system [2]. There are different incentives and barriers that impact on distributed electricity generation that should be taken into account.

The UK government has four long-term energy policy goals [3]:

- Cut carbon dioxide emissions by 60% in 2050 with real progress by 2020;
- Preserve high reliable energy supplies;
- Raise the sustainable economic growth rate for better productivity; and
- Provide adequate and affordable heat for every home.

Government should promote for the DG as it is one of the promising potential solutions to reduce CO_2 emission. However, the reliability and cost are considered as the main challenges. The interests of electricity consumers should be properly taken into account by the government. Cost implications of any changes are a key consideration, as will preserving the integrity of electricity networks. Governments

[1]School of Engineering, University of Aberdeen, Aberdeen AB24 3UE, UK
[2]Department of Electrical Engineering, Faculty of Engineering, Alexandria University, Alexandria 21544, Egypt
[3]School of Engineering, Robert Gordon University, Aberdeen AB10 7GJ, UK

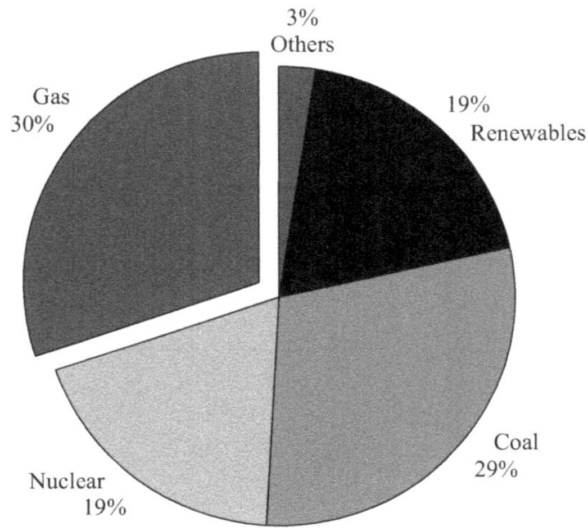

Figure 7.1 Electricity generation in the United Kingdom

are therefore seeking a better understanding of the costs and benefits of various ways in which take-up of DG could be increased. The chart in Figure 7.1 shows the electricity generation mix in 2014.

Basically, the main purpose of the unified national grid is to decrease the generation cost and improve the supply reliability. Building large power stations was the common logic choice until 1990 to gain from the large-scale economies. The location of such power stations was near the fuel source (as coal) or in distant area such as nuclear power station. This leads to that the generation and load-demand systems are far from each other and transmission system becomes inevitable to link the two systems together. The transmission lines are operated with high voltage to reduce the system losses and cross-section area of the cables. The key drivers in future electrical grid development are the new technologies of generation system to address the environmental and economic requirement. Thus, embedding small generation units in distribution local networks is very interesting option, and it may help to push the transmission lines to return to their basic function. This function is to interconnect the generation units and loads in the distribution local networks. However, it is expected that more low-carbon generation mix will be emerged [3]. The domestic generation impacts on the capacity spectrum are already obvious and at the same time the integration of such large offshore wind farms with existing conventional power plant is another big challenge. This is why it is expected that the transmission system will still play a key role as it is working now. On the other hand, distribution networks must be ready for the big change in the coming years of accommodating thousands of generation units. There is no doubt that the DG share is anticipated to expand rapidly in near future. The DG not only provides environmental benefits but also shorter construction time frame can be deemed as a vital factor in accelerating development.

However, there are also obstacles before wider operation of DG is conceivable. One barrier is the connection into the grid especially when considering distribution networks. The issue is new since conventionally there are in practice no generating power plant embedded in the distribution network. For this matter, new rules, regulations and recommendations are required. Furthermore, new technical ideas and solutions are significant to prove that DG is economically acceptable and viable.

7.2 Distributed generation

DG or embedded generation is an electrical power plant which is integrated into low-voltage distribution networks instead of high-voltage transmission networks as shown in Figure 7.2. Generally, they are small generation units from renewable generation sources, thus they includes small hydro, wind, solar power and smaller combined heat and power. The DG development has a vital part to offer in current and future environmental targets [4]. The main goals to be achieved are: increasing electrical network capacity by integrating more productive clients without producing new reinforcement cost to the existing network, to providing better system reliability and improving the supply quality.

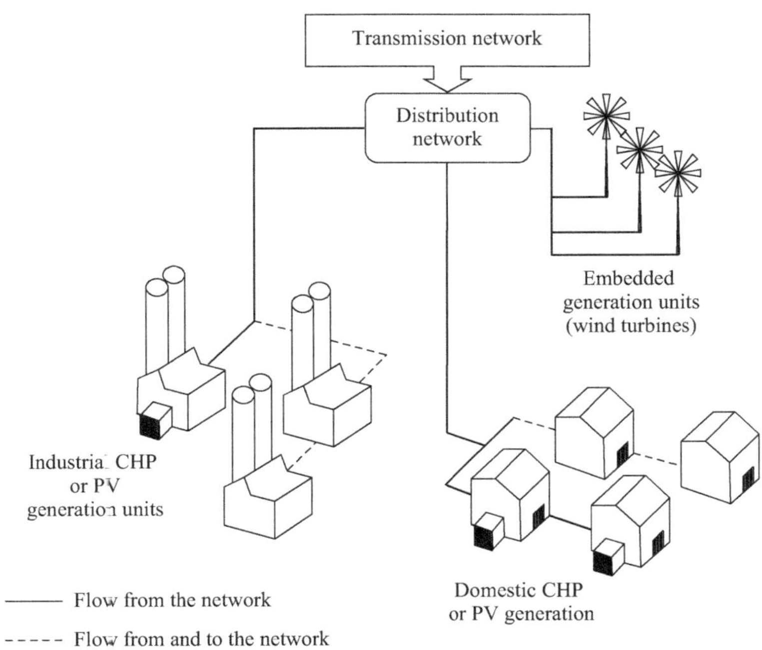

Figure 7.2 Distribution generation structure

7.2.1 Description of the problem

One of the main problems facing DG system deployment is the connection to the network or grid [5]. Unfortunately, the electrical output of many DG resources is incompatible with the fixed frequency electricity supply network; as a result, it is necessary to develop power electronic interfacing methods. Many DG sources are DC. Any method must have features of reliable, fast response, ease of control and hardware interface ability. The best candidate for the application is the PWM converter coupled via a power filter. The system must transfer the energy from the DC link to the three-phase AC system with controlled active and reactive power and without injecting harmonic currents. The requirement for power converters in DG systems and microgrids has presented the importance of developing low distortion, high power quality through power converters. It is worth to be noticed that poor power quality can be delivered by the switching frequency impact and grid voltage distortion. This is why the power filter should be well designed to attenuate the switching frequency components; however, it has negative effects on the controller bandwidth and the presented impedance to network distortion. The system impact features are the following:

- A conventional power system is designed to operate radially without any generation on the distribution network or customer side. The embedding of generation sources such as DGs has significantly impacts on operation, and reverse power flow is a major concern;
- The location of DG can offer system benefits and reduce the need for expensive extension;
- DG can support unintentional islanding; this separation could be operational when a fault occurs upstream on the utility side. In most cases, DG islanding is not desirable in any part of the utility system because it can lead to safety and power quality problems that affect the utility systems and loads;
- Protection systems have traditionally been designed assuming a radial system. After connecting DG, part of the system may no longer be radial, which means coordination may not be achieved [6]. The influence of DG on coordination will depend on size, type and location of the DG;
- DG should permit each unit to synchronise with the grid. A licensing regime for generation and supply must be implemented;
- The relationship with the transmission system must be established [7]; and
- Smaller unlicensed DGs should be rewarded via governmental mechanisms for the energy which they can deliver to the grid.

Functions of power converters for DG systems can be summarised as follows:

- Power conversion from variable DC voltage into fixed AC voltage for standalone applications or AC current output following the grid voltage and frequency for grid-connected applications.
- High-output power quality confirming with low total harmonic distortion (THD), flickering and voltage/frequency deviation.

- Protection of DG generators and electric power systems from abnormal voltage, current, frequency and temperature conditions, with additional functions such as anti-islanding protection and electrical isolation if necessary.
- Controlling the DG systems and achieving certain aims such as maximum power extraction from wind energy conversion system, maximum power point tracking of PV panels, optimum efficiency for fuel cell systems and optimal energy flow control.

7.2.2 Applications of distribution generation

Common shapes of services and benefits are represented from DG [8]. Basic applications can be offered which are shown in this section. The applications of DG include combined heat and power, standby or island mode, peak sharing and grid connected.

7.2.2.1 Islanded mode of operation

In this mode the unit operates as a voltage source with load sharing capability, similar to conventional generator sets. The control system in this mode is based on the voltage and frequency droop as the converter is the main source of power [8]. The output power is conditioned to meet and exceed the requirements of standard regulation where the connected load may make severe demands on frequency, voltage and waveform characteristics. The unit can supply balanced, unbalanced and non-linear loads (three-phase or single-phase) with power factors ranging from 0.7 to 1.0. This mode has some characteristic features as follows:

- No connection with a utility grid.
- The system is the source of power, so determines the voltage and frequency.
- The control system must provide sufficient active and reactive power to match the connected load whilst delivering high power quality (voltage).
- Voltage is the controlled variable so the power quality is determined by the voltage quality.

The converters are designed to share the local loads but also to be controlled from a remote supervisory controller that can set the load contribution of each unit. Figure 7.3 shows the structure of an island mode system.

7.2.2.2 Standby power

In this mode, it is required to supply the power during emergency or power interruption of the main service such as in public health and safety reasons. Also, in some cases, the outage costs are quite high [9]. As storms or accidents are the source of such outages, onsite standby power supplies are built up at sites for example water/waste stations, hospitals, critical industries and electronic based industries. The format of the standby power system is presented in Figure 7.4. The control scheme in this operation mode is similar to island mode.

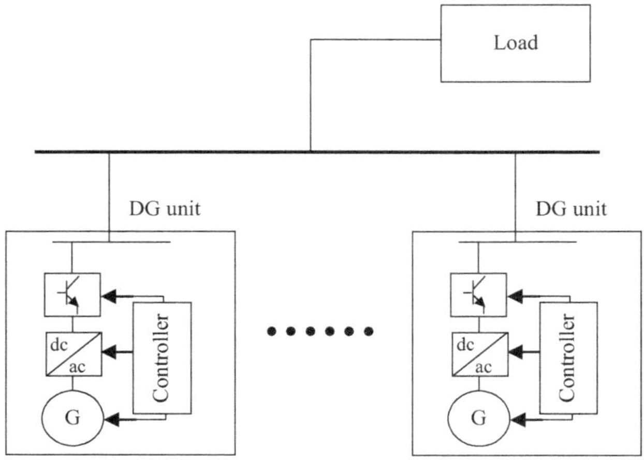

Figure 7.3 Island mode structure

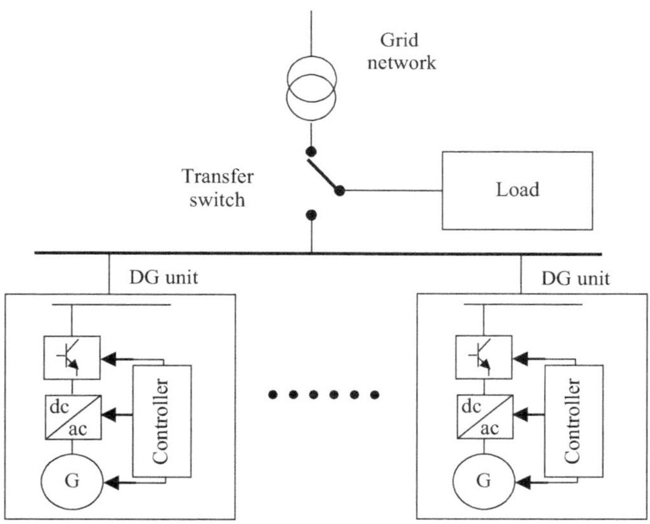

Figure 7.4 Standby power operation

7.2.2.3 Grid-connected mode

The grid network comprises generation units, high-voltage transmission, substations and local low-voltage distribution network. Proper placement selection of DG can offer system advantages and avoid the requirement for high cost extension. The main goal of the grid connection is to establish active and reactive power control in order to interface the system (DG) with the grid [10]. Interfacing the converter to the grid must be through a filter with specific component values to attenuate

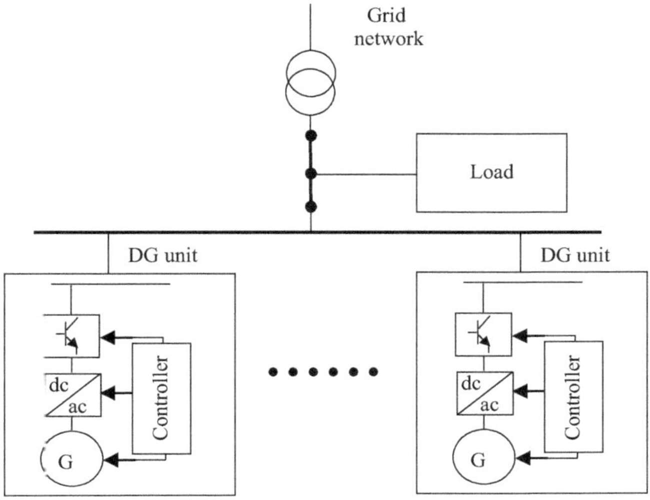

Figure 7.5 Grid-connected mode

harmonics, for power quality. Figure 7.5 shows the structure of the grid-connected mode. The features of the grid-connection mode are as follows:

- The grid determines the frequency and the voltage level in the system.
- The three-phase system voltage can be distorted or unbalanced.
- The control system should offer active and reactive powers with sufficient power quality (current).
- Current is the controlled variable so the power quality is determined by the current quality.

In the shown structure, the power units are paralleled to the network and delivered all or part of their power. In this case, excess generation is delivered to the grid, as well as the local loads. The converters can be designed to take part of the load but also to provide a programmed load demand from a remote supervisor. One of the applications with grid connection is peak sharing. Power costs change every hour as it depends upon demand and generation availability. Such variations are turned into daily and seasonal time of use classes, such as on-peak and off-peak.

7.3 Voltage source converters

In power converters, energy conversion can be classified into four main types such as DC to DC, DC to AC, AC to DC and AC to AC. The aim of a power inverter is to generate an AC output waveform from a DC source. There are main applications of inverters in industry including uninterruptible power supplies (UPS), adjustable speed drives, active power filters, static VAR compensators, high-voltage direct current, voltage compensators and flexible AC transmission systems

(FACTS). The magnitude, phase and frequency should be controlled to produce a sinusoidal output. Depending on the type of AC output waveform, the used power topologies can be voltage source converters, where the independently controlled AC output is a voltage waveform. Alternatively, these topologies can be current source converters, where the independently controlled AC output is a current waveform. In the case of three-phase systems, the pulse-width modulated VSI shown in Figure 7.6(a) is commonly used. Unlike motor drives, most sinusoidal power system supply applications have significant imbalance between the three output phases, and a four wire load connection is typical. Two possible approaches are a centre-tapped DC supply and a four-legged inverter, as shown in Figure 7.6(b) and (c), respectively. If isolation is required, as is frequently the case, a more conventional three-phase inverter can be used with a Δ/Y output transformer, as illustrated in Figure 7.6(d). The transformer leakage inductance can be designed to function as filter inductance. In this topology, the three-wire system on the delta side has two degrees of freedom, and no neutral current path exists, which makes

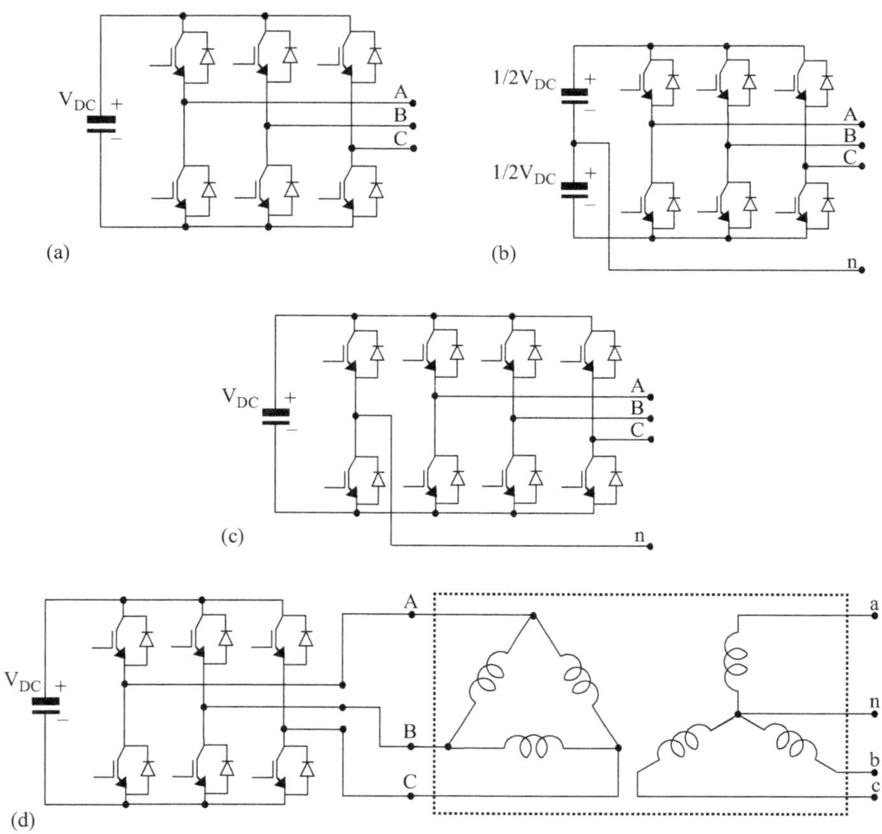

Figure 7.6 Different three-phase inverter topologies

the system relatively easy to control. The drawback is the existence of the costly, heavy and bulky low-frequency transformer.

7.4 Control techniques in DG systems

In this section, a survey of various control techniques associated with converter-based distribution generation system is presented. As alternative energy sources produce DC powers whilst the grid accepts AC power, grid-connected converters are essential and inevitable. Two different modes of operation are in DG systems, namely islanded (standalone) mode and grid-connected mode. However, the controller requirement in each mode of operation is different.

In the grid-connected mode, the system must control and handle disturbances from the network such as unbalanced or distorted grid voltages. The converter module supplies the grid with the required controlled active and reactive powers. Several control methods have been proposed for DG systems integrated into a grid [11–38]. In an islanded mode, the main source of power is the DG plant, so, it shall be capable of delivering its domestic grid with the required frequency and voltage. Frequency and voltage controls in islanded grids, which can be powered by single or several converters, can be achieved by different control techniques, with/without communication links [39–93]. Some control techniques are derived from UPS operation [39–49,63,64]. Control methods based on local measurements offer superior redundancy as no communication is required for operation. Some parallel module approaches require interconnection to achieve balanced load sharing [56–59]. Some methods of ensuring proper load sharing use a master/slave configuration [60–68] and distributed current methods. The output voltages of the parallel units must be synchronise frequency, phase and amplitude to guarantee the quality of output power sharing, otherwise current circulates between the units.

7.4.1 Grid connection

The main grid connection goal is to maintain active and reactive power control from the converter unit DG. The grid determines the frequency and the voltage level in the system. The three-phase grid voltage can be distorted or unbalanced. As current is the controlled variable, power quality is determined by current quality. Converter control techniques are important in terms of the harmonics injected into the grid by the converter. The different control technique classification is shown in Figure 7.7 for grid-connected mode. There are two control approaches based on whether or not a communication link exists between units.

7.4.1.1 Without (no) communication link

Using only local feedback variables is the main advantage of such control techniques which can be collected at the power converter and no communications are required between the power converters. The most common technique is the droop control. In frequency and voltage droop control [11–17], the converter attempts to mimic a synchronous generator. The system frequency and output active power has

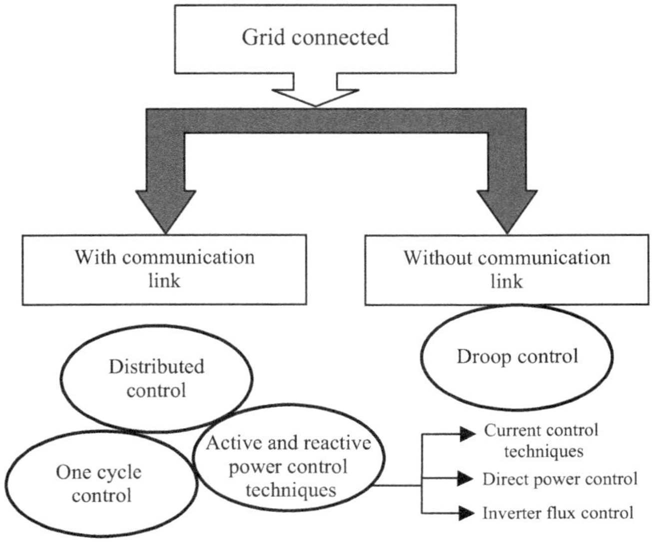

Figure 7.7 Classification of grid-connected mode

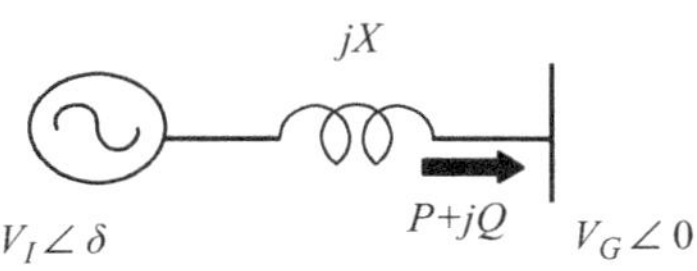

Figure 7.8 Converter connected to infinite bus equivalent circuit

a direct relation between them. Bear in mind that the frequency variation is limited by the generator inertia [12]. The converter control system should be able to include such inherent operating conditions for stable parallel-connected operation. The equivalent converter circuit connected via an inductance to infinite bus is shown in Figure 7.8.

The active and reactive powers injected into the network can be written as

$$P = \frac{3V_I V_G}{X} \sin\delta \tag{7.1}$$

$$Q = \frac{3V_G}{X}(V_I \cos\delta - V_G) \tag{7.2}$$

where X is coupling inductance, δ is the angle between the grid voltage (V_G) and converter output voltage (V_I).

It is obvious from (7.1) and (7.2) that the active power is a function of the angle (δ), whereas the reactive power is a function of the output converter voltage. Thus, the output voltage and frequency droops can be utilised to control the power converter.

$$\omega = \omega_0 - m \cdot P \tag{7.3}$$

$$V = V_0 - n \cdot Q \tag{7.4}$$

where m and n are the droop coefficients for the frequency and amplitude, respectively. The droop-line characteristics in Figure 7.9 can be used to control the converter.

The method gives acceptable effectiveness except if the grid voltage is imbalanced or distorted. Droop control suffers from different obstacles such as there is a trade-off between slow transient response, output regulation, and phase deviation and unallowable frequency. For load sharing, the controller should compensate inter-unit variations and for line impedance effects [13–15]. A control scheme to connect a single-phase converter to stiff AC power is shown in Figure 7.10. It has a classical PWM controller with an inner current loop and an outer voltage loop using PI compensators. The sinusoidal reference for the controller is obtained from a reference block which uses the amplitude signal and frequency signal defined by power droop curves [16]. To solve the power sharing problem and mitigate line impedance effects, artificial intelligent load-dependant droop coefficients have been proposed in the converter frequency and the voltage amplitude [17].

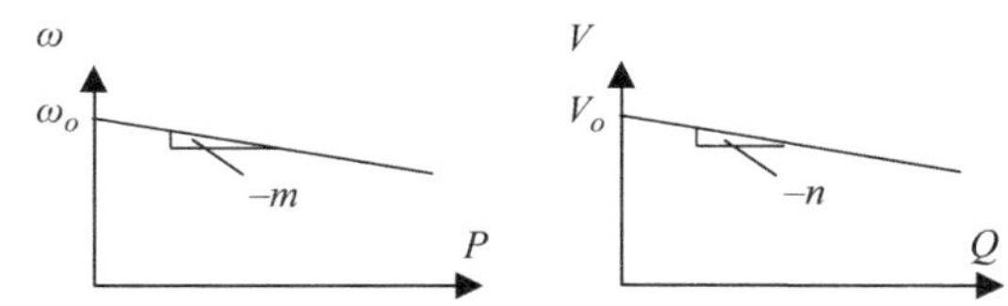

Figure 7.9 Droop curves of frequency and voltage

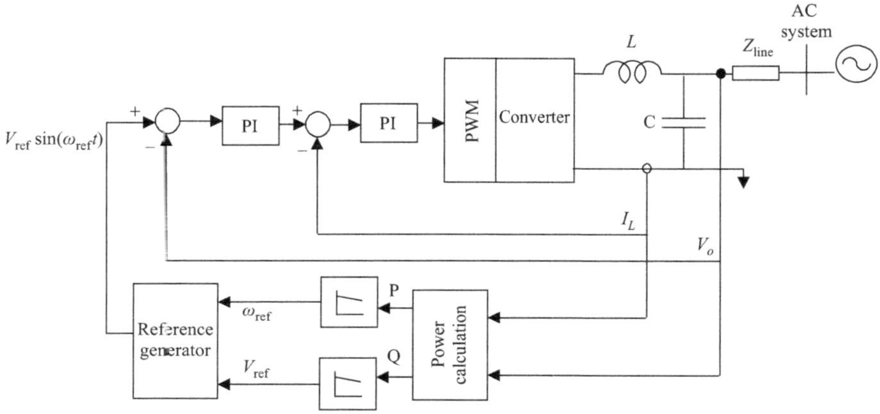

Figure 7.10 Control scheme block diagram of the converter connected to grid

7.4.1.2 With communication link

Some techniques use links among the converters.

However, it is clear that wires/connections are sources of noise and system failure, and they limit the converter unit site. The system performance can be enhanced by using communicating information via power-line communication and internet, but the system must be able to operate without such information to avoid system failure. This implies that converter control should be based on terminal quantities. The features of communication linked based grid connection methods are summarised in Table 7.1.

7.4.2 Islanded mode

An islanded or standalone AC system may be explained as one in which the entire AC power is delivered to the system via converters. In this system there are no synchronous alternators present to provide a reference for the system frequency and voltage. All the system converts need to be operated to offer stable system control for voltage and frequency in the existence of changing loading conditions. Several control technique classification in islanded mode is shown in Figure 7.11. Voltage is the controlled variable, so power quality is determined by voltage quality.

7.4.2.1 Single operation

Three control techniques for converter interfacing in an islanded mode are considered.

1. *Deadbeat control*

   Deadbeat control has been extensively studied [39–49]. At a low sampling frequency, deadbeat control can perform satisfactorily with a fast transient response and low THD. In [39], whoever the load current wave shape, a deadbeat control law was derived. In [40], the converter output voltage control system uses the filter capacitor current and a full-order deadbeat observer compensates the controller lag time loop delay. The output voltage regulation is 0.5%; however, the voltage control is a straightforward controller.

   Disturbance observers can be used [43–48]. A state variable observer and deadbeat control can be combined for real-time digital feedback control of the three-phase PWM converter. System robustness is improved by independently designing the disturbance observer pole and the state variable observer. An alternative method uses a predictive control on both the filter inductor current and output voltage for the LC filter. The system guarantees a rapid transient response and accurate compensation of any disturbance source [48]. A new fast and robust control technique is presented in [49], which gives a voltage THD of 0.62% with a non-linear load. A deadbeat current and voltage controller are used to obtain a fast transient response but may be unstable due to the same pole placement of the double controller. To solve this instability problem, separation of the pole placement is possible [50,51].
2. *Perfect robust servomechanism problem control*

   The advanced control strategy requires a rapid DSP for algorithm implementation. The technique employs perfect robust servomechanism problem (RSP) control theory [52,53], to remove the output voltage unwanted harmonics with serious non-linear loading condition, and to offer rapid recovery from dynamic state.

Table 7.1 Grid connection with communication classification

Method		**Concept**	**Comment**	**Disadvantage**	**Reference**
Distributed control		Sharing active and reactive powers between all modules	Coordination between the units is obtained by exchanging information	Need communication	[18]
One cycle control (OCC)		Combined vector operation and bipolar operation	Offers advantages of straightforward circuit, and solid effectiveness. The circulating current is reduced	Need communication	[19]
Active and reactive power control	Direct power control (DPC)	Based on direct torque control. The torque and flux amplitudes could be replaced by the active and reactive powers	Fast transient response and straightforwardness. The main disadvantage are the variable frequency and spread spectrum	Complex	[20–22]
	Inverter flux control	A flux vector is used as a reference to be tracked using a controller with a tolerant band	To control the inverter flux vector, the technique bounds the error between a reference flux vector and the actual flux vector to be within a specified tolerance zone	Complex	[23,24]
	Current control techniques	Current control to track the required reference	In [3.21], the results are current THD of 10.8% and voltage distortion of 2.8%. Improved results are obtained in [3.22–3.24], with a 1% current THD and 3% voltage distortion	Need communication	[22–38]

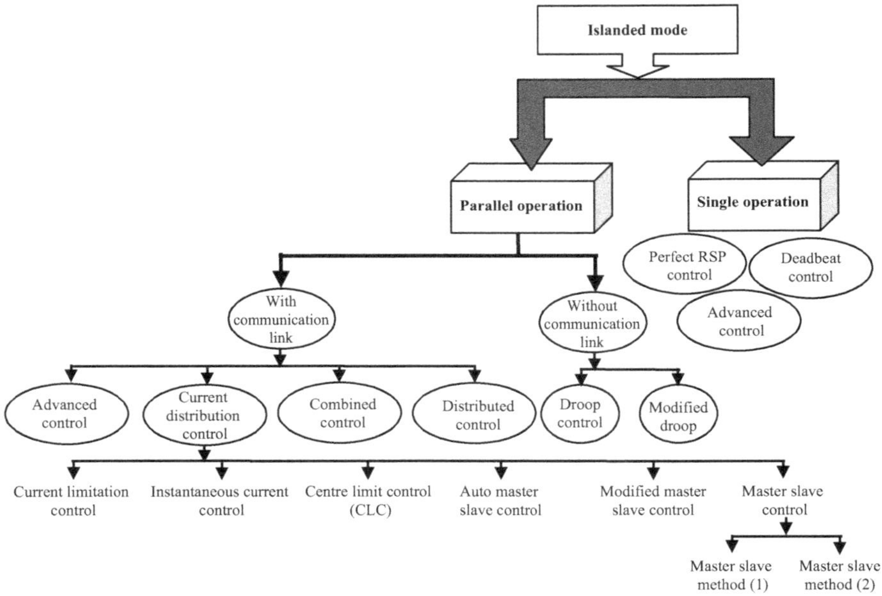

Figure 7.11 Islanded mode operation classification

Perfect RSP theory combines an internal model principle with optimal state feedback to guarantee closed-loop system stability and good transient response [53].

3. *Advanced control techniques*

Nowadays, there is on-going research in new, robust, modern and accurate control current methods, which are more attractive than conventional control due to the classical control method barriers. Sliding mode and discrete optimal control was presented [54]. The control system achieves a fast transient response, good voltage regulation, nearly zero steady-state converter output voltage error and low THD with linear and non-linear loads. In [55], an adaptive controller was proposed and investigated, where the power conversion scheme provides high performance steady-state characteristics, fast system response to load transients and permits bilateral power flow. A DC-link current control scheme is proposed in order to minimise the current source reactance.

7.4.2.2 Parallel operation

The design of the parallel connected converter system should be expandable, reliable and redundant, and the converters must symmetrically participate in the loads.

- **With communication link**

  Information sharing among DG units can be readily implemented due to fast development in the communication field. Many papers therefore consider parallel operation with a communication link. The communications-based methods are summarised in Table 7.2.

Table 7.2 Island mode with communication classification

Method		**Concept**	**Comment**	**Disadvantage**	**Reference**
Distributed control		Synthesising the active and reactive power components of each unit	Current sharing is achieved based with regulation of the output active and reactive powers. The active and reactive powers circulating will increases with increments in phase difference between units	Complex	[56,57]
Combined control		Both average power control and droop control are combined	A remote terminal unit (RTU) for each distribution generation unit, is used for power data communications (real and reactive power)	Need communication	[58,59]
Current distribution control	Master slave	One is came out from parallel connected DC/DC converters and another one from paralleling inverters	The voltage regulation is controlled by the master unit and it produces the current reference for slave units. At steady state the current sharing is good but deviation occurs during transient. There is current overshoot during the transient	Need communication	[60–64]
	Modified master	Instead of using conventional wires for control signals, the wireless communication link can be used	Wireless radio-frequency communication link control is an alternative [3.56], eliminating the need for multiple links/wires among the different units	Expensive	[66,67]

(Continues)

Table 7.2 (Continued)

Method		**Concept**	**Comment**	**Disadvantage**	**Reference**
	Auto master slave control	A good technique, which can be used during faults to choose the master unit	The master unit will be select for the converter with highest output power, while other converters are slaves	Need communication	[65]
	Centre limit control	Derived from DC/DC parallel connected converters.	The central limit produces the reference current for all converter units. The current sharing is obliged at all times. Non-identical power module rating can be connected	The sharing signals are required high communication band	[69,70]
	Instantaneous current control	Based on instantaneous average current sharing	In the switching instant, the output current deviation is forced to zero by the controller. The output reference of the voltage controller instantaneously eliminates unbalanced power	Need communication	[71–76]
	Current limitation control	The limited and unlimited current command difference is delivered to the next unit	The control link number can be minimised, thus improving system reliability. The master voltage controller produces the current reference for its module	Need communication	[77]
Advanced control		Optimal, sliding mode and fuzzy logic controller	Good results	Complex	[78–83]

- **Without communication link**

The power units are running totally independently when communication is not used. Several control techniques have been proposed without control wire interconnections, all based on the droop method [84–93].

1. *Droop control*

In islanded mode of operation in DG, droop techniques are the simplest control to be implemented, allowing the power units to be controlled independently [85–90]. Droop control is using real and reactive powers as control variables. Each module has its own control loops and the only interconnection between modules is the paralleling to the AC power line. Because of the droop characteristics, the system frequency and voltage will drop such that all units operate at a lower frequency and voltage, eliminating any circulating current. The challenge is to minimise the deviation in frequency and voltage from the nominal values. In theory, these deviations can be made arbitrarily small but practical concerns will put a lower bound on how small the droop factors can be. The main problem is poor non-linear load sharing. The control loop must make tight adjustments over the converter output voltage frequency and amplitude, in order to compensate the active and reactive power imbalances [85]. Load sharing is proposed in [86] to droop the output harmonic voltage as a function of the distortion power.

A higher droop leads to better sharing, but this means a proportionate reduction in the voltage loop bandwidth, thus a reduction in waveform quality. A trade-off is made between accuracy of reactive power sharing and output voltage quality. Line impedance affects the accuracy of load sharing. If the system line impedance is known, its effects can be compensated [87]. There is a trade-off between active power sharing accuracy and phase synchronisation [88]. The control technique in [88] enforces linear and non-linear load sharing under different conditions. The controller automatically compensates for converter parameter variations and line impedance imbalances. A small AC voltage control signal is injected into the system. The quantity to be shared, controls the frequency of the injected control signal. The real power control signal is measured and used to adjust the control variable. The power lines are used as the control interconnection among the paralleled inverters. When the system settles, all of the inverters have the same frequency. The frequency of the control signal is drooped according to the sharing power quantity.

To study the influence of line impedance on the power sharing between units, two inverters are connected in parallel. Figure 7.12(a) and (b) shows the output apparent power of the two inverters, with line impedance variation. As shown, the output power is sensitive to the line impedance change.

Figure 7.13 shows the circulation current variation between the two inverters, with varying phase and magnitude difference between the inverter voltages. The circulating current that flows between modules increases the output current of both modules.

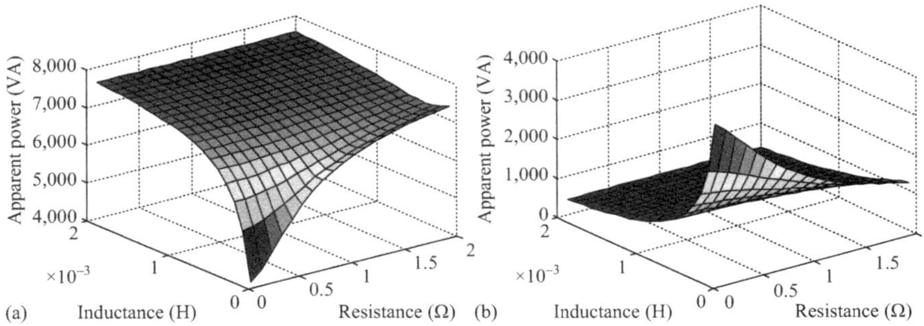

Figure 7.12 The output apparent power: (a) inverter 1 and (b) inverter 2

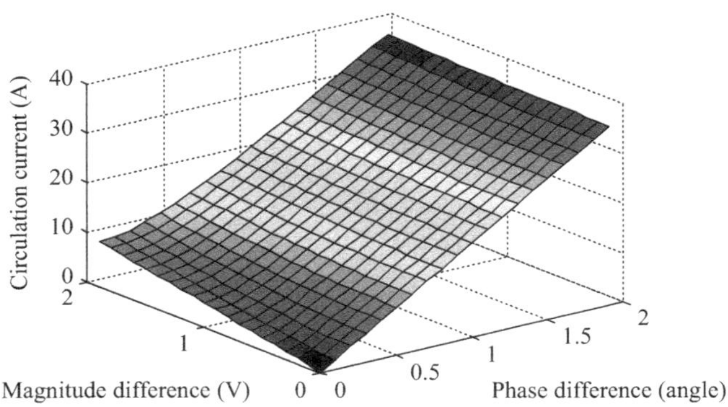

Figure 7.13 The circulation current

2. *Modified droop control*

   A modified droop control was proposed in [91], in order to enrich the parallel connected converters transient performance. Conventional voltage and frequency droop are slow at achieving load sharing and have an oscillating transient response. There is no possibility of modify the transient response without a loss of power-sharing precision or output voltage and frequency accuracy.

   The characteristics of each unit active and reactive powers were achieved by phase difference and voltage amplitude, respectively.

   The authors have proposed to us phase shift (load angle) instead of using the system frequency in the droop control.

   The converter output impedance is emulated by means of a high-pass filter to share linear and non-linear loads [91]. The modified method in [92] achieves a low THD and perfect current sharing. In [93], a droop scheme was proposed to control the sharing of real and reactive powers. A controller is developed, by adding a supplemental transient droop

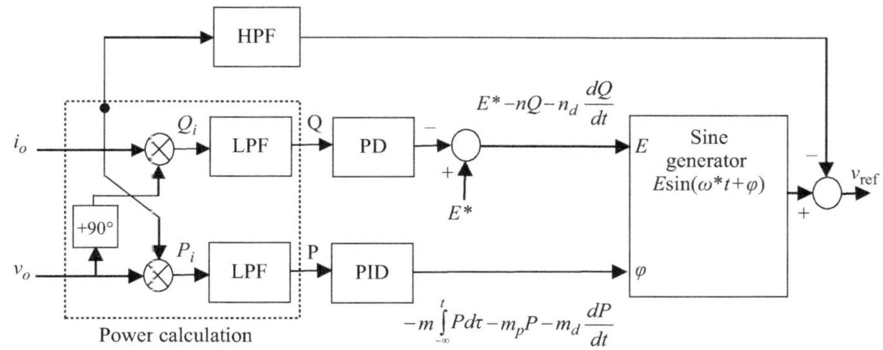

Figure 7.14 The block diagram of the control system [93]

characteristic to the conventional static droop approach, which improves the paralleled system dynamics. In Figure 7.14, n_d is the derivative coefficient of the reactive power Q; m, m_p and m_d are the integral, proportional, and derivative coefficients of the active power P. The coefficients m and n fix the steady state droop function while m_p, m_d and n_p are selected to guarantee stability and good transient response.

7.5 Power quality in DG

The rapid proliferation of DG into the distribution power system has impacted on the distribution power quality. From the point of view of power quality, there are two aspects, current or voltage quality. Power quality can be achieved by the system configuration. The converter-based distribution–generation resource becomes a source for eliminating the harmonics not their source [94]. Control techniques in both grid and island modes of operation aim to improve power quality. There are different standards in the literature, which specify the accepted amount of harmonic distortion in the voltage and current, in grid and islanded modes of operation [95–97].

7.5.1 Grid connected

Harmonic distortion is the main power quality aspect and is assessed using THD. British Standard [95] and IEEE-519 [96] are the standards used in the literature. IEEE-519 is the main reference used in thesis. In grid-connected mode, if the connection is into a stiff network, then the voltage cannot be controlled with a shunt connected DG system. The power quality is then defined by the current quality, via harmonic distortion limits. The ratio between the grid short-circuit current (I_{SC}) and maximum fundamental current (I_L) and voltage level of the system are the main factors to determine the limits for currents. The limits are shown in Table 7.3 for voltage levels below 69 kV.

Table 7.3 Distortion limits of current harmonic for general distribution systems (120 V–69 kV)

Maximum harmonic current distortion in per cent of I_L individual harmonic order (odd harmonics)						
I_{SC}/I_L	<11	$11 \leq h < 17$	$17 \leq h < 23$	$23 \leq h < 35$	$h \geq 35$	THD
<20*	4.0	2.0	1.5	0.6	0.3	5.0
20<50	7.0	3.5	2.5	1.0	0.5	8.0
50<100	10.0	4.5	4.0	1.5	0.7	12.0
100<1000	12.0	5.5	5.0	2.0	1.0	15.0
>1000	15.0	7.0	6.0	2.5	1.4	20

Even harmonics are limited to 25% of the odd harmonic limits above.

* All power generation equipment is limited to these values current distortion, regardless of actual I_{SC}/I_L.

7.5.2 Island mode

In this mode of operation, power quality is concerned with the voltage. The DG unit is the main source of power then it determines the voltage and frequency. Different standard specifications describe the requirements for DG interfacing to the low-voltage network [95].

7.5.2.1 Power frequency

The nominal supply voltage power frequency is 50 Hz. The fundamental frequency mean value measured over 10 s under normal operating conditions must be within a range as follows:

- Systems between interconnected systems and synchronous connection
  50 Hz ± 1% (i.e. 49.5 Hz … 50.5 Hz) during 99.5% of a year; and
  50 Hz + 4% or 50 Hz – 6 % (i.e. 47 Hz … 52 Hz) during 100% of the time.
- Systems between interconnected system and no synchronous connection (e.g. power supplies on islands):
  50 Hz ± 2% (i.e. 49 Hz … 51 Hz) during 95% of a week; and
  50 z ± 15% (i.e. 42.5 Hz … 57.5 Hz) during 100% of the time.

7.5.2.2 Supply voltage variation (Un)

$Un = 230$ V is the standard nominal voltage for low-voltage network and for different three-phase wiring systems as follows:

- $Un = 230$ V is between phase and neutral for four-wire systems.
- $Un = 230$ V is between phases for three-wire systems.

The voltage variation should not exceed ±10% under normal operating conditions.

7.5.2.3 Voltage unbalance

Unbalance is inevitable in low-voltage network. However, there are standard limits for permissible amount of unbalance, which can be spread in the low-voltage

Table 7.4 Values of individual harmonic voltages at the supply terminals

Odd harmonics				**Even harmonics**	
Not multiples of 3		**Multiples of 3**			
Order *h*	**Relative voltage (%)**	**Order** *h*	**Relative voltage (%)**	**Order** *h*	**Relative voltage (%)**
5	6.0	3	5.0	2	2.0
7	5.0	9	1.5	4	1.0
11	3.5	15	0.5	6–24	0.5
13	3.0	21	0.5		
17	2.0				
19	1.5				
23	1.5				
25	1.5				

networks unless the network will be significantly affected. The supply voltage negative-phase sequence component rms is 95% of 10 min should be within 0%–2% of the positive-phase sequence component during normal operating conditions. It can reach to 3% in some areas.

7.5.2.4 Voltage harmonic distortion

Each individual harmonic voltage rms is 95% of 10 min should be less than the value in Table 7.4 during normal operating conditions. Table 7.4 shows individual harmonic limits.

No values are given for harmonics of order higher than 25, as they are usually small, but largely unpredictable due to resonance effects.

7.5.2.5 Voltage fluctuation (flicker)

Under normal operating conditions, in any period of one week the long-term flicker severity caused by voltage fluctuation should be less than 1 for 95% of the time.

7.5.2.6 Supply voltage dips

The main impact of the transient or permeant faults in low-voltage distribution networks is causing the voltage dips. Unfortunately, they are random and they cannot be estimated. The location of the measurement or observation and the type of the supply are controlling the value of the annual system frequency. In addition, it is not regular distribution during the year. The predictable voltage dips number over the year is between few tens and one thousand during normal conditions. The duration of voltage dips is less then 1s and the voltage is kept not less than 40% in the majority of the cases. It is worth to be noticed that during the load switching in the low-voltage distribution networks, a voltage dip with voltage drop between 85% and 90% can occur.

7.6 Harmonics and passive filter design for DG

In DG and microgrids, the requirement for power converters is inevitable. This leads to that the power converters should achieve low distortion, and high-quality power export. Figure 7.15 shows the proposed system including the power filters to achieve the acceptable distortion limits, enhance power quality and reduce resonance effect. There are different standards which can be used (e.g. IEEE 519, EU and IEC). The passive LC power filter influences converter harmonic distortion and increase the harmonics especially if is coupled with non-linear load. Several methods can be used to control harmonics flow such as static VAr harmonic compensation, harmonic current injection, parallel/series active filter systems, magnetic flux compensation and DC ripple injection [98–100]. To reduce voltage or current harmonics in DG systems, passive power filters are the best candidate. From literature, the three main harmonic types which are embedded in any power electronics system are as follows:

- High-order harmonics
- Sampling or switching harmonics
- Low-order harmonics.

This section shows the design of the output LC filter for converter-based DG system. The (IEEE-519) standard is used in the design to Figure 7.15 out the permissible current harmonic amount which can be delivered into the network. The LC filter peak inductor ripple current is determined analytically. The allowable switching component level, which can be injected into the network, is used in the LC filter capacitor design.

Moreover, the studying of different various filter damping methods to reduce affects of resonance are presented and assessed. The derived analytical expressions are demonstrated by MATLAB®/Simulink™ simulations.

7.6.1 Power filter configurations

Figure 7.16 shows the common power filter topologies used for three-phase converters. For the whole frequency range, the voltage attenuation is −20 dB/decade for basic inductor filter shown in Figure 7.16(a). It is worth to be noticed that the converter harmonics can be sufficiently attenuated by increasing the converter switching frequency [99], and −40 dB/decade attenuation can be provided by the

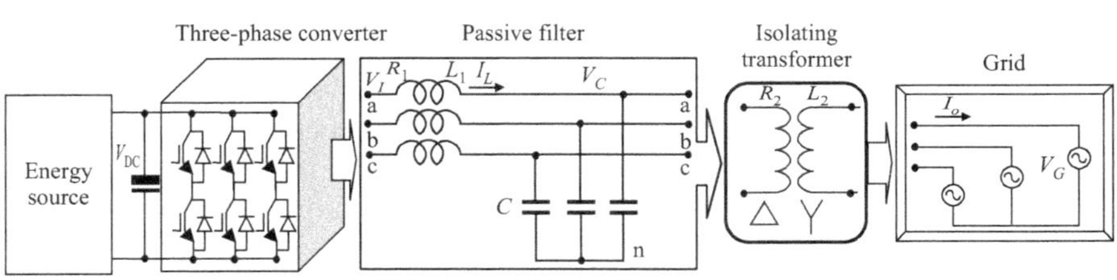

Figure 7.15 The proposed interfacing system

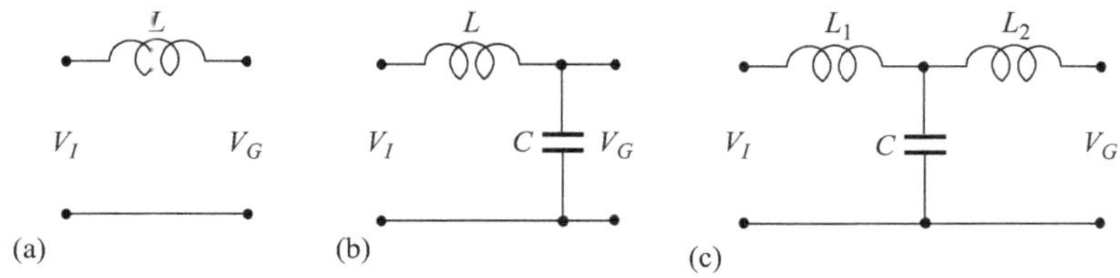

Figure 7.16 Filter circuits

second-order LC-filter in Figure 7.16(b). As mentioned previously that the L-filter provides −20 dB/decade attenuation, a shunt connected capacitor further attenuates the switching frequency components. The selection of such shunt element capacitor is to offer low reactance at high-order frequency components. While it produces high impedance at the control power frequency. The resonant power filter frequency is

$$f_0 = \frac{1}{2\pi}\frac{1}{\sqrt{LC}} \tag{7.5}$$

To confirm the right choice of the LC-filter, the load impedance connected in parallel with the shunt capacitor should be high above the sampling (switching) frequency. The drawback of LC-filter compared to L-filter is higher cost and reactive power consumption. The third filter, Figure 7.16(c), offers better attenuation of converter switching harmonics than the other mentioned filters. The LCL-filter has the following features:

- For the same harmonic attenuation, lower switching frequency can be used.
- Above the resonance frequency, −60 dB/decade attenuation can be achieved.

On the other hand, due to resonance, both steady state and transient input current distortion can be affected by the LCL-filter. The LCL-filter resonant frequency is

$$f_0 = \frac{1}{2\pi}\sqrt{\frac{1}{C}\left(\frac{1}{L_1} + \frac{1}{L_2}\right)} \tag{7.6}$$

7.6.2 Analysis of the three filter topologies

In this section, the DG unit is operated in grid connection. The above three different filter topologies are included in the analysis. The power filters are initially assumed lossless.

7.6.2.1 L-filter

The relation between the output power and the filter inductance (L-filter) is shown Figure 7.17(a), based on

$$P_o = \frac{3V_G}{X_L}\sqrt{{V_I}^2 - {V_G}^2} \tag{7.7}$$

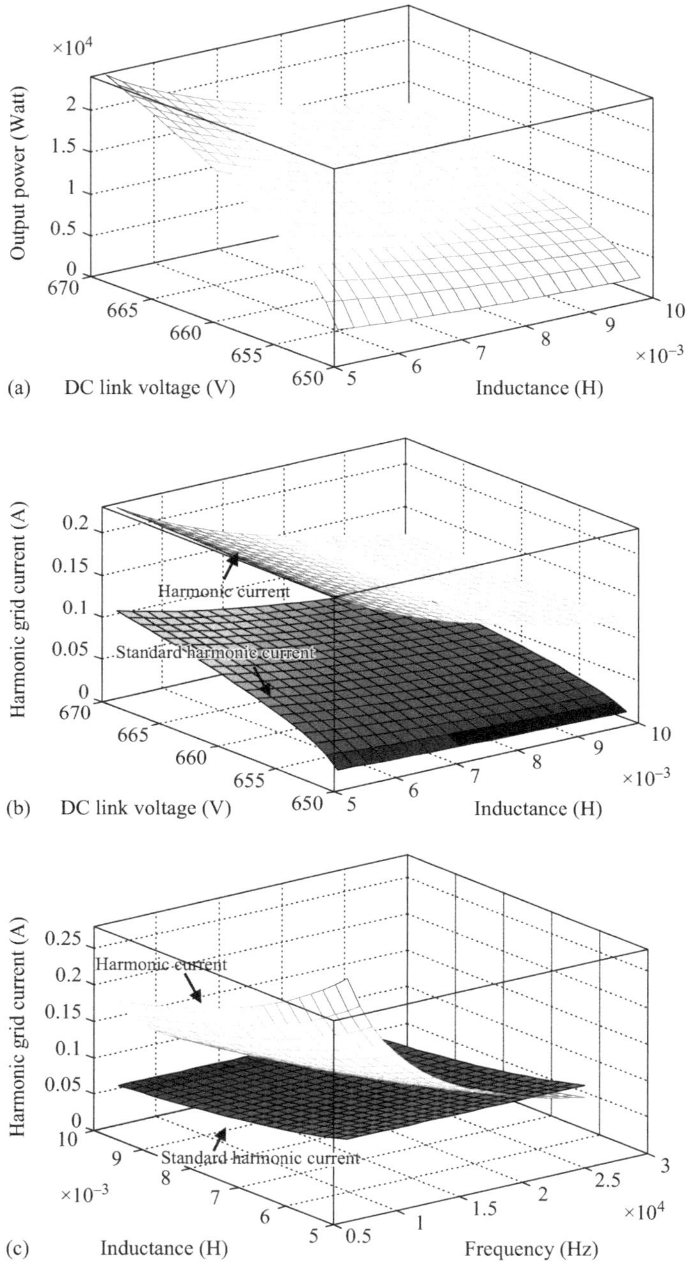

Figure 7.17 L-filter characteristics: (a) output power and filter inductance; (b) harmonic current with variable filter inductance and DC link; and (c) harmonic current with variable filter inductance and switching frequency

It is obvious that the output power is a function of the DC-link voltage (V_{DC}) and filter inductance (L). The injected inductor current into the grid is shown Figure 7.17(b) (the first surface), whereas the injected harmonic current standard limit is presented in the second surface. A 10 kHz switching frequency is used for drawing both surfaces. In the analyses, the grid voltage V_G is assumed an ideal voltage source which includes only fundamental frequency component. It is presented as a short circuit for other frequencies [101]. The harmonic current expression is

$$|I_{o\ har}| = \frac{|V_{I\ har}|}{X_{Lh}} \tag{7.8}$$

From the IEEE standard, the injected harmonic current into the grid must be less than 0.3% I_{rated} (rated current at 10 kW is 14.5 A) if harmonic order is greater than 35 [102]. These leads to inequality (7.9).

$$|I_{o\ har}| < \frac{0.3\% P_{\text{rated}}}{3V_G} \tag{7.9}$$

where P_{rated} is defined from Figure 7.17(a).

Figure 7.17(b) confirms that the L-filter cannot meet the equation (7.9) limitation, hence its failure to effectively diminish the harmonic currents even when the switching frequency is high, 10 kHz. However, if the switching frequency is greatly increased to 20 kHz, they filter will be able to achieve the standard limits as shown in Figure 7.17(c).

7.6.2.2 LC-filter

The limitation of the LC-filter is that the shunt element is theoretically ineffectual when connected to a rigid network due to the grid impedance at the switching frequency is trivial. In this specific case, the attenuation relies only on the filter inductor and the LC-filter mimics the L-filter. The output current and the inductor current ripple (L-filter) are the same.

7.6.2.3 LCL-filter

It is a common case that between the grid (utility) and power filter is an isolating transformer in order to eliminate any DC-current component. This inserts series leakage inductance (L_2), producing an LCL-filter but with constant leakage inductance on the output. The output power and inductor filter inductance (L_1) are shown in Figure 7.18(a), as (7.10). As it can be seen that increasing the DC-link voltage leads to increase the output power.

$$P_o = \frac{3V_G}{X_{L1} + X_{L2} - (X_{L1}X_{L2}/X_C)} \sqrt{V_I^2 - V_G^2\left(1 - \frac{X_{L1}}{X_C}\right)^2} \tag{7.10}$$

The injected output current harmonic for a rigid grid is

$$|I_{o\ har}| = \frac{|V_{I\ har}|}{X_{Lh1} - [X_{Lh2}X_{Ch}/(X_{Lh2} - X_{Ch})]} \frac{X_{Ch}}{X_{Lh2} - X_{Ch}} \tag{7.11}$$

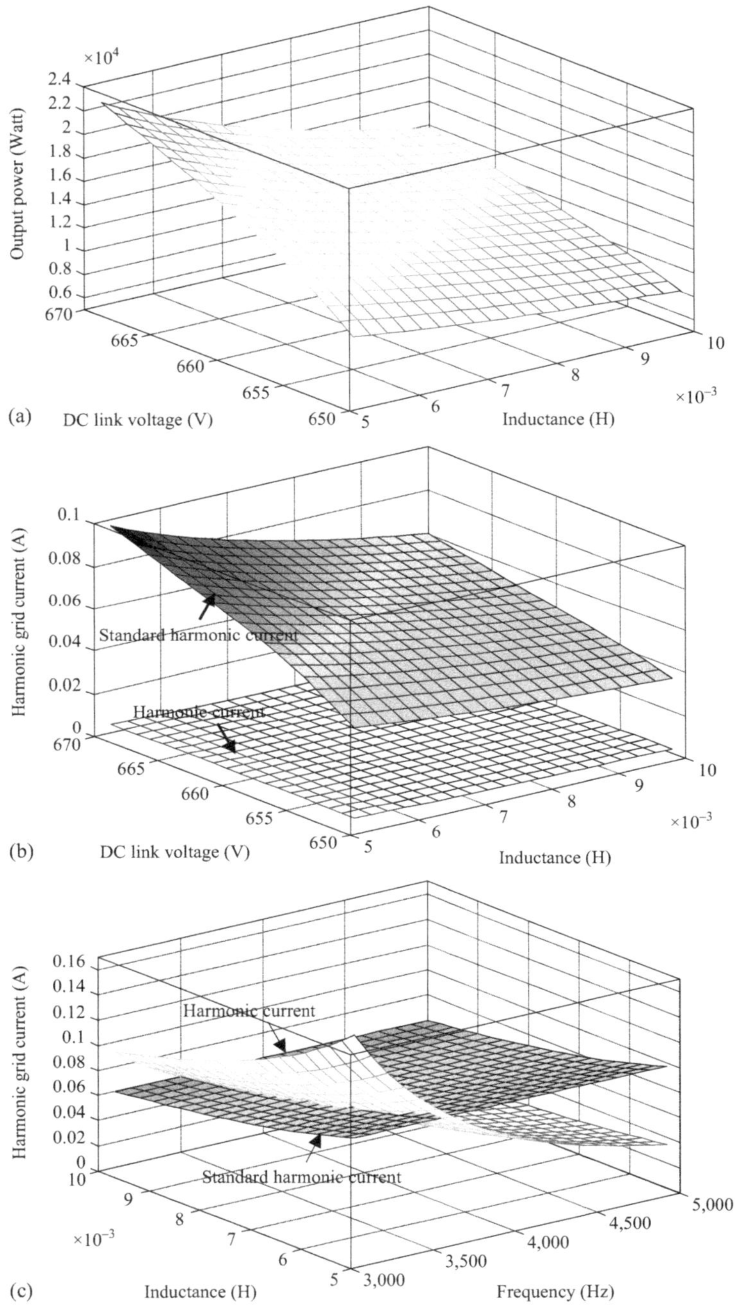

Figure 7.18 LCL-filter characteristics: (a) output power and filter inductance; (b) harmonic current with variable filter inductance and DC link; and (c) harmonic current with variable filter inductance and switching frequency

The output current harmonics and current harmonic standard limits are shown in Figure 7.18(b). The power filter inductance (L_1) and DC-link voltage are used to draw Figure 7.18.

Now, including the inductance of the isolating transformer with the LC-filter, the harmonic limit requirements can be achieved as shown in Figure 7.18(c). The output harmonic currents and the harmonic standard limits are shown in Figure 7.18(c). It is obvious that 3.5 kHz or above switching frequency is sufficient to achieve the harmonic requirement.

Equations (7.12) and (7.13) show the relationship between power filter capacitor (C), inductor (L_1) while the DC-link voltage is constant, isolating interfacing transformer inductance, and output active power. Bear in mind that (L_1) and (C) values have a range, which can be used to achieve the required harmonic limits. The output power will be increased by decreasing the power filter inductance. On the other hand, if the filter capacitance, VAr consumption and harmonic currents will increase. The power converter rating and the control system are affected by the power filter inductor harmonic contents.

$$X_{L1h} = \frac{1000 V_G |V_{I\ har}| X_{Ch} + X_{L2} X_{Ch} P_{\text{rated}}}{(X_{L2h} - X_{Ch}) P_{\text{rated}}} \tag{7.12}$$

where $X_{Ch} = [1/2\pi(f_s - 2f_m)C]$, $X_{L2h} = 2\pi(f_s - 2f_m)L_2$

$$L_1 = \frac{1000 V_G |V_{I\ har}| X_{Ch} + X_{L2} X_{Ch} P_{\text{rated}}}{2\pi(f_s - 2f_m)(X_{L2h} - X_{Ch}) P_{\text{rated}}} \tag{7.13}$$

The link, at different frequencies, between power filter inductance (L_1) and shunt capacitance (C) at different is shown in Figure 7.19(a). The rated filter inductor current, harmonic inductor current (converter current) and harmonic current limit (20%) consistent with acceptable measurement, with capacitance and switching frequency variation are presented in Figure 7.19(b).

The isolating transformer inductance and power filter capacitance can be used to calculate the inductor (converter) as follows:

$$|I_{\text{inv}}| = \sqrt{\left[\frac{P_r}{3V_G}\left(1 - \frac{X_{L2}}{X_C}\right)\right]^2 + \left(\frac{V_G}{X_C}\right)^2} \tag{7.14}$$

The fundamental converter current is independent of switching frequency. At converter switching ($f_s - 2f_m$), the inductor harmonic current is

$$|I_{\text{inv}\ har}| = \frac{V_{I\ har}}{X_{L1h} - [X_{L2h} X_{Ch} / (X_{L2h} - X_{Ch})]} \tag{7.15}$$

The harmonic converter current in percentage of the rated current is shown in Figure 7.20 at different switching frequencies.

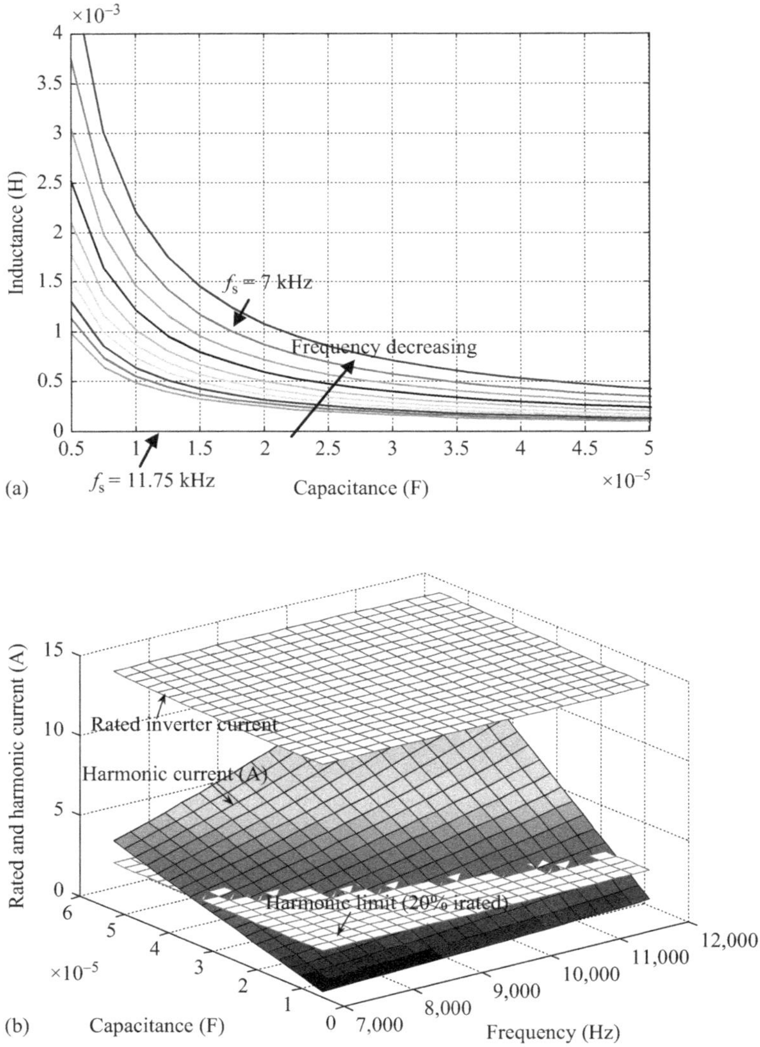

Figure 7.19 Systems characteristics: (a) Inductance L_1 versus capacitance C and frequency, and (b) converter and harmonic currents

7.6.3 Filter design

In [103], the total harmonic inductor current and capacitor voltage of the LC-filter are derived for constant filter component values. The LC values were determined by using minimum LC-filter reactive power. Also, the LC values can be determined by the cost function, the system time constant and THD as in [104]. Another method by connecting in parallel an LC-trap filter in series with the LC-filter was proposed to provide better

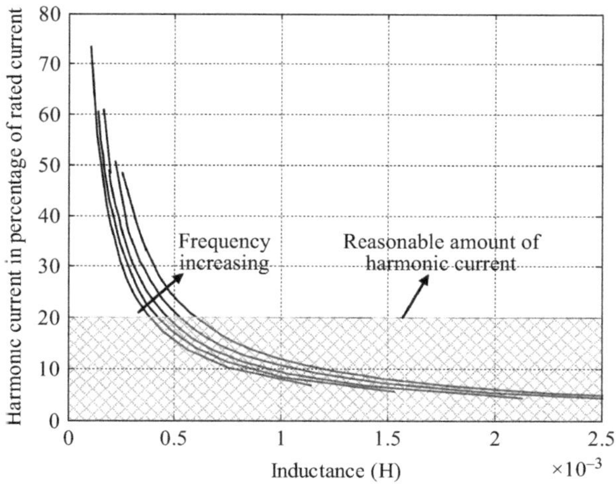

Figure 7.20 Harmonic converter current in percentage of the rated current at different switching frequencies

voltage harmonics filtering [105]. In [106], the transfer function approach was used for filter design. The low-pass LC-filter (with transformer leakage and grid impedance) is capable of mitigating harmonics at low order in the output current.

In high power application, the power converter switching frequency is playing a vital role in determining the converter losses which has a dramatic effect on the whole system efficiency.

In general, as preferable choice, the cost and size of filtering components can be minimised by switching frequency increase. However, the efficiency will be affected with such increase (a design trade off must be made). The converter-phase voltage and its frequency spectrum are illustrated in Figure 7.21. It is obvious from Figure 7.21 that such power converter cannot be integrated into the grid unless attenuating the high-frequency order components. A SPWM power converter running at 2 kHz with a 400 V link is used for the simulation [5].

From Figure 7.15, the converter phase 'a' voltage is V_{Ia}, and the output voltage is V_{Ga}. Thus, the filter inductor voltage is

$$V_L = V_{Ia} - V_{Ga} \tag{7.16}$$

The maximum power filter inductor ripple current is calculated as a function of V_{Ia} and V_{Ga} which are derived from the DC-link voltage as in (7.17) and (7.18) including at maximum output 75% duty cycle.

$$V_{Ia} = \frac{2}{3}V_{DC}, \quad V_{Ga} = \frac{1}{2}V_{DC}, \tag{7.17}$$

$$V_L = \frac{2}{3}V_{DC} - \frac{1}{2}V_{DC} = \frac{1}{6}V_{DC} \tag{7.18}$$

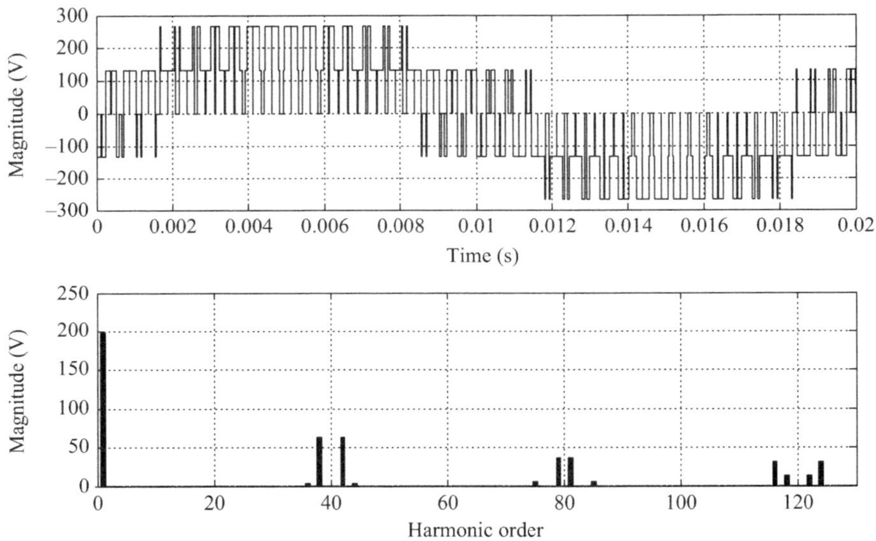

Figure 7.21 Phase voltage of the converter and its frequency spectrum

Using the IEEE harmonic standard [102], the permissible amount of rated current is 15–20%. In the analysis, 20% is presumed.

Equation (7.20) is presented the maximum ripple current. Although the switching frequency and DC-link voltage are constant, the inductance (L_1) can be determined from (7.23).

$$V_L = L\frac{\Delta\hat{I}_L}{\hat{\delta}T_s}, \quad \Delta\hat{I}_L = \hat{\delta}T_s\frac{V_L}{L} \tag{7.19}$$

$$\Delta\hat{I}_L = \hat{\delta}\frac{V_L}{Lf_s} = \frac{1}{6}\frac{\hat{\delta}V_{\rm DC}}{Lf_s} \tag{7.20}$$

$$\hat{\delta} = 1 - \frac{1}{4} = \frac{3}{4} \tag{7.21}$$

$$\Delta\hat{I}_L = \frac{1}{8}\frac{V_{\rm DC}}{Lf_s} \tag{7.22}$$

$$L = \frac{1}{8}\frac{V_{\rm DC}}{xI_Lf_s} \tag{7.23}$$

where V_L is the inductor voltage, f_s is the switching frequency, L is the filter inductor, $V_{\rm DC}$ is the DC-link voltage, $\hat{\delta}$ is the maximum duty cycle, x is the rated inductor current factor.

The high-frequency components of the inductor current have to be diverted from the grid. This is achieved by the shunt impedance (filter capacitor). In grid connection, the main concern in power quality of the injected current as the voltage is fixed. The converter in this case is operated like a current source. While in an islanded mode of operation, the power and frequency are determined by the DG unit. Then, the voltage quality is the primary concern. If V_c is the capacitor voltage, V_I is the converter voltage, and I_{L1}, I_o are the inductor current and converter output current, respectively.

$$\frac{V_c(s)}{V_I(s)} = \frac{Z_c(s)\left[Z_t(s)+Z_g(s)\right]}{Z_c(s)\left[Z_t(s)+Z_g(s)+Z_{L1}(s)\right]+Z_{L1}(s)\left[Z_t(s)+Z_g(s)\right]} \tag{7.24}$$

$$\frac{I_{L1}(s)}{V_I(s)} = \frac{Z_c(s)+Z_t(s)+Z_g(s)}{Z_c(s)\left[Z_t(s)+Z_g(s)+Z_{L1}(s)\right]+Z_{L1}(s)\left[Z_t(s)+Z_g(s)\right]} \tag{7.25}$$

$$\frac{I_o(s)}{V_I(s)} = \frac{Z_c(s)}{Z_c(s)\left[Z_t(s)+Z_g(s)+Z_{L1}(s)\right]+Z_{L1}(s)\left[Z_t(s)+Z_g(s)\right]} \tag{7.26}$$

where Z_c is the filter capacitor impedance, Z_g is the grid impedance, Z_t is the transformer impedance, Z_{L1} is the filter inductor impedance.

$$\frac{I_o(s)}{I_{L1}(s)} = \frac{Z_c(s)}{Z_c(s)+Z_t(s)+Z_g(s)} \tag{7.27}$$

The network can be a rigid utility or a standalone system, so the network impedance range value might be from almost zero (strong network) to infinite (no load, islanded mode). $Z_g = 0$ is the worst-case condition:

$$\frac{I_o(s)}{I_{L1}(s)} = \frac{Z_c(s)}{Z_c(s)+Z_t(s)} \tag{7.28}$$

$$\begin{aligned} |Z_c(f-2f_m)| &= |Z_t(f-2f_m)|\frac{I_o(f-2f_m)}{|I_{L1}(f-2f_m)|-I_o(f-2f_m)} \\ C &= \frac{|I_{L1}(f-2f_m)|-I_o(f-2f_m)}{2\pi(f-2f_m)|Z_t(f-2f_m)|I_o(f-2f_m)} \end{aligned} \tag{7.29}$$

$$\begin{aligned} &|I_{L1}(f-2f_m)| \\ &= \frac{|1+[Z_t(f-2f_m)/Z_c(f-2f_m)]|\cdot|V_I(f-2f_m)|}{|Z_t(f-2f_m)+Z_{L1}(f-2f_m)+[Z_{L1}(f-2f)Z_t(f-2f_m)/Z_c(f-2f_m)]|} \end{aligned} \tag{7.30}$$

At switching frequency, the output current is a portion of the rated output current [102].

$$I_o(f-2f_m) = X\cdot I_{\text{rated}} \tag{7.31}$$

In an island mode, the power converter is the source which delivers the power to the loads. In this case, grid impedance Z_G can be replaced by Z_{Load}.

$$|Z_{\text{Load}}| = \frac{|V_{O\ \text{rated}}|}{|I_{O\ \text{rated}}|} \tag{7.32}$$

$$\frac{V_o(s)}{V_I(s)} = \frac{Z_{\text{Load}}(s)}{Z_t(s)+Z_{\text{Load}}(s)+Z_{L1}(s)+[Z_{L1}(s)(Z_t(s)+Z_{\text{Load}}(s))/Z_c(s)]} \tag{7.33}$$

$$V_o(f\pm 2f_m) = X\cdot V_{\text{Rated}} \tag{7.34}$$

7.6.4 Case study

A 10 kVA three-phase converter is used and connected to a grid via an isolating transformer (represent the output inductor of the power filter) 240/415 Δ/Y with $X = 0.3\ \Omega$. An amount of 400 V is utilised for the DC-link voltage, and 4 kHz is for the power converter switching frequency. Using (7.23), the power filter inductance is 2.5 mH. Thus, the shunt capacitance can be determined by (7.29), $C = 50\ \mu F$. Depending on the location of the DG unit connection with the grid, the grid impedance (L_G and R_G) change. In summary, the final values of the proposed designed filter components are $L_1 = 2.5$ mH, $C = 50\ \mu F$ and $L_t = 1$ mH. $L_G = 0.1$ mH and $R_G = 1\ m\Omega$ are for the grid impedance [5].

7.6.5 Damping filter design

Unfortunately, passive power filters suffer at resonant frequency from low damping, which can lead to serious control instability. In general, the main objective of damping is to restrain resonance with no impact on the attenuation at the switching frequency, nor influencing the fundamental frequency. In literature [107,108], there are different damping techniques which have been proposed to reduce the impact of such problem. They are classified basically to active and passive damping techniques. One of these techniques (active ones) is used to actively damp the resonance effects; however, it adds complexity to the control system. It's worth to be noticed that the power-filter inductor parasitic resistances will offer inherently damping effect. It may not require in some cases to add any damping techniques. Moreover, several passive damping topologies can be embedded in the power circuit; however, each technique has its pros and cons. In all passive damping techniques, the choice of the damping resistor and its location are the main concerns. There is always a trade-off between damping attenuation and power losses. Figure 7.22 presents different resonance frequency damping methods. Resistance can be connected in parallel with the shunt capacitor, as illustrated by Figure 7.22(a). It is a quite simple technique but the losses should be observed in the results. R_d is set to 50 Ω in simulation. Figure 7.22(b) shows another technique called series damping resistance, which is just a series resistor inserted to the shunt path. Increasing R_d will increase the circuit damping as well, but unfortunately two drawbacks can be observed. The rate of attenuation above the resonant frequency can be reduced with larger resistance and the R_d resistance is a source of loss itself. R_d is set to 0.5 Ω in simulation. In Figure 7.22(c), the inductor L_d presents low reactance at power frequencies which the current flow via the parallel resistor can be controlled. The inductive power filter reactance is high at high frequencies, which means the impact of the inductor is low. Figure 7.22(d) illustrates a series R_d–C_d connected in parallel with the shunt capacitor. Both damping capacitance and filter shunt capacitor as one package must be the same capacitance as calculated in (7.29) to offer same reactive power consumption and cut off frequency. R_d is set to 3 Ω in simulation.

In Figure 7.22(e), inductance is added to an R_d–C_d series circuit. At low frequencies, the inductive losses are decreased as the inductor reactance is small. R_d is set to 3 Ω in simulation. The tuned L_d–C_d circuit connected in parallel with the power-filter capacitor, as shown in Figure 7.22(f), offers a current absorber at the resonant

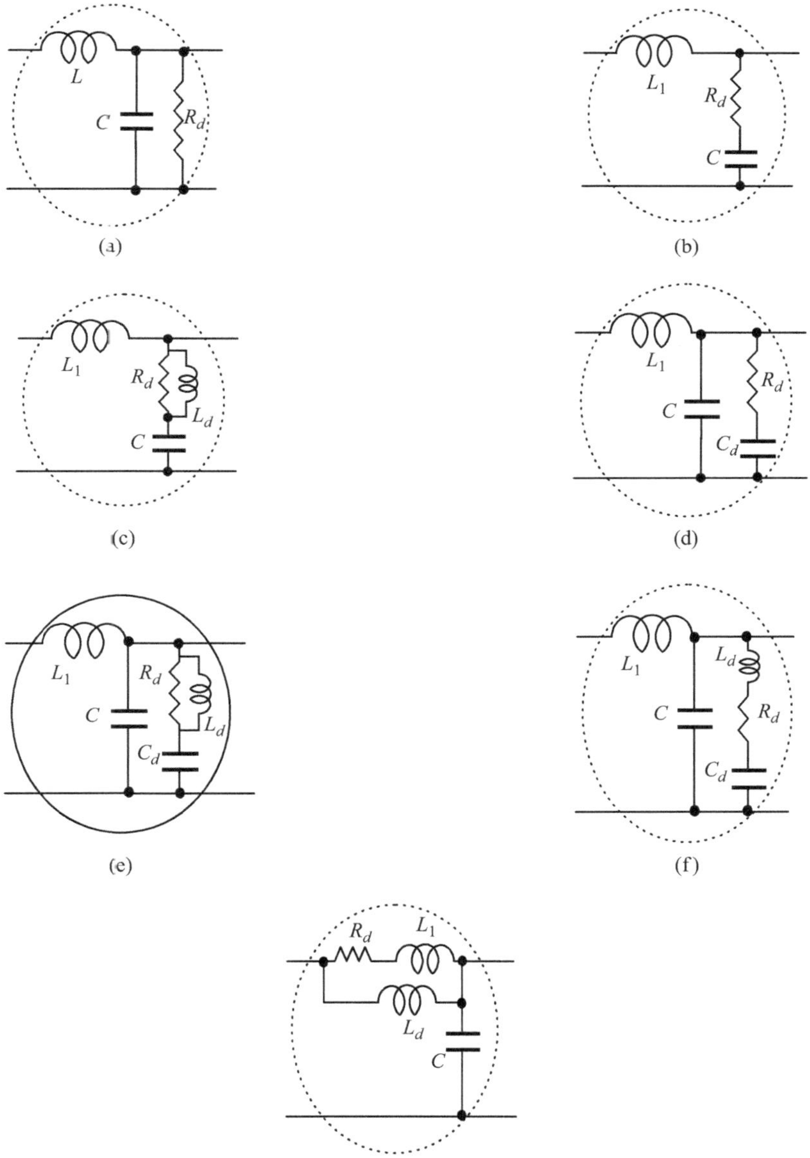

Figure 7.22 Different damping methods: (a) shunt R_d damping, (b) series R with the shunt element, (c) parallel L_d and R_d in series with the parallel filter capacitance, (d) series (C_d and R_d) in parallel with the filter capacitance, (e) (L_d and R_d) in parallel with the series C, all as one combination in parallel with the filter capacitance, (f) series L, R and C in parallel with the shunt item and (g) R_d in series with L_1 which are in parallel with L_d

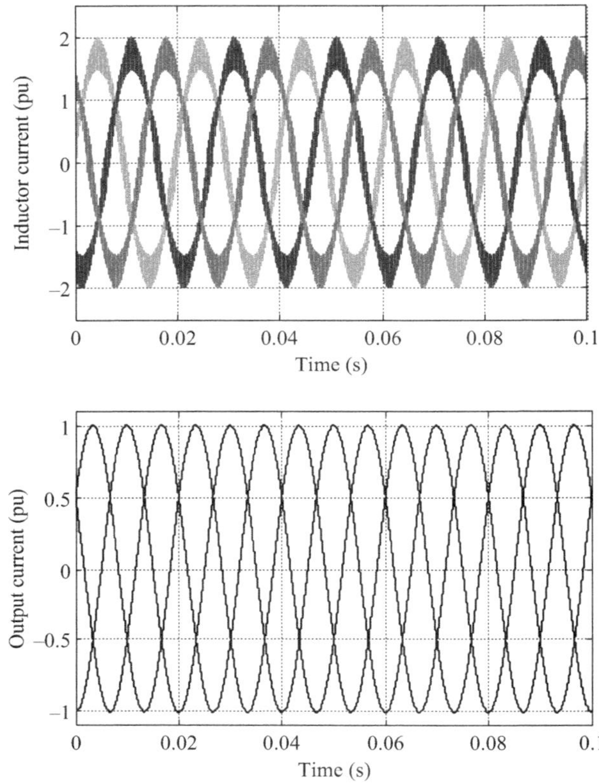

Figure 7.23 Three-phase inductor and output current

frequency. At the power frequency, the system performs like method in Figure 7.22(d), as the inductance impedance is low. The inductor turns effective at the resonant frequency. The damping harmonic filter is chosen to reduce the effect of the 810 Hz resonance on the system. Thus, C_d is chosen as 20 μF and L_d is based on (7.35).

$$L_d = \frac{1}{\omega_o^2 C_d} \tag{7.35}$$

Instead of adding the capacitor in parallel, an inductor can be used. Figure 7.22(g) shows a damping inductor connected in parallel with the power filter inductor. In high-power-density converters, this damping method is preferable. R_d is set to 0.5 Ω in simulation.

7.6.6 Simulation results

The proposed filter design was applied to a three-phase SPWM converter switching at 4 kHz, and connected to a grid network with the specification used in the case study [5]. Figure 7.23 shows the converter three-phase output and inductor currents, whereas Figure 7.24 shows their frequency spectra.

To suppress the impact of the resonance frequency, the above seven damping techniques can be embedded in the power circuit. A comparison between the different damping techniques is presented in Table 7.5, including the power losses at power frequency, with −23 dB attenuation at the resonant frequency. Method 5 offers the minimum power losses compared to the other methods with the same attenuation but it has slight effect on the system gain across the power-frequency range. Figure 7.25 shows the seven damping techniques in one graph to conclude the best method. As it can be seen, method 6 is the best not only for the effective attenuation at the resonance frequency but also it does not affect the system at the power-frequency level.

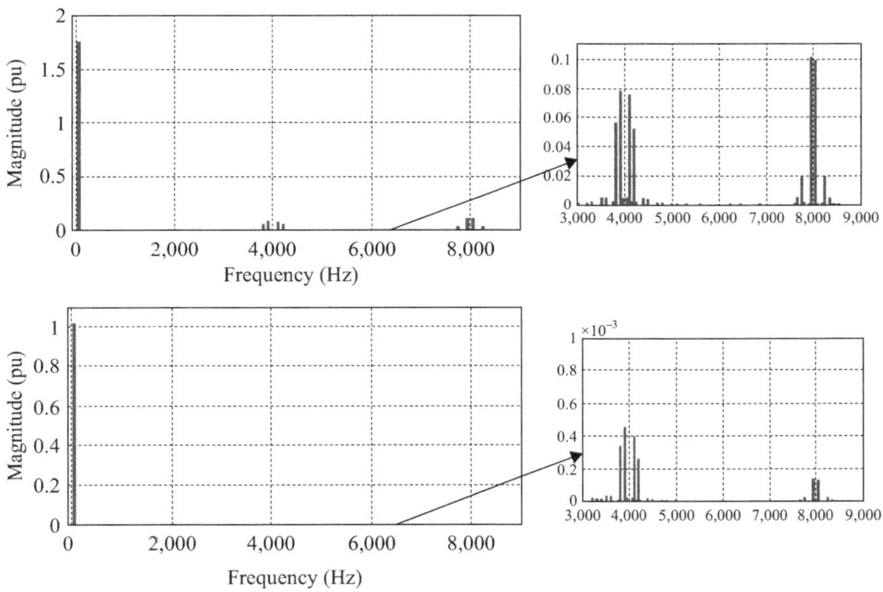

Figure 7.24 The inductor and output currents frequency spectrum

Table 7.5 Power-loss comparison

Damping method	Power loss when transmitted power = 10 kW, 50 Hz
1	1.5 kW
2	6.5 W
3	3.2 W
4	6.9 W
5	1.6 W
6	4 W
7	96 W

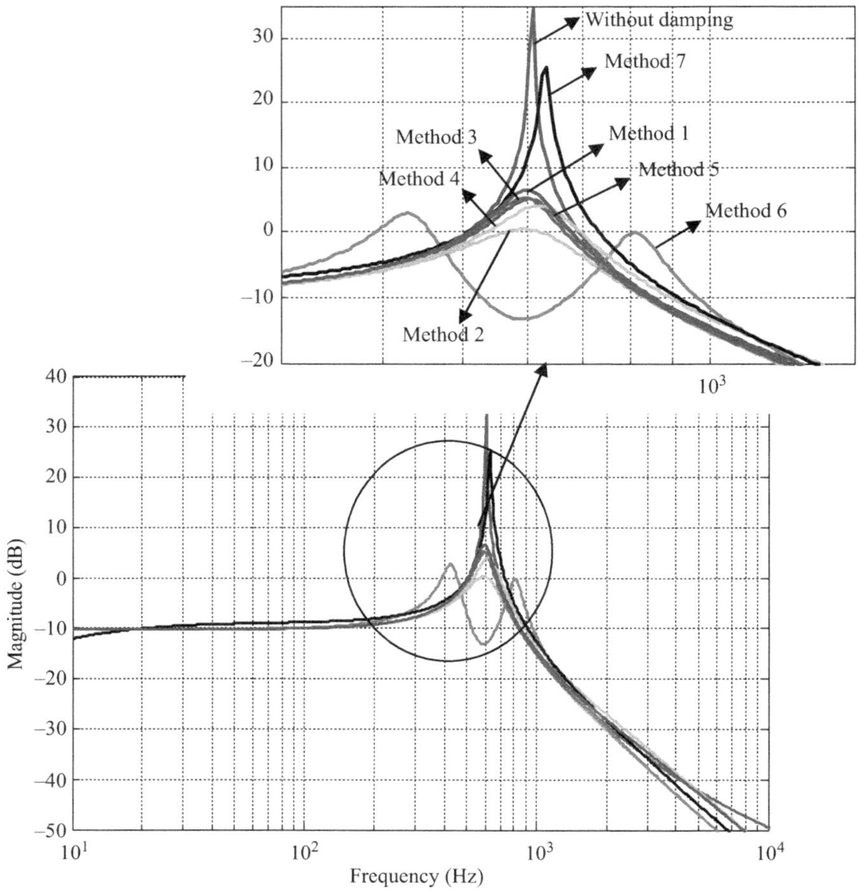

Figure 7.25 Different damping technique characteristics

References

[1] K. H. Ahmed, A. M. Massoud, S. J. Finney, and B. W. Williams, 'A modified stationary reference frame-based predictive current control with zero steady-state error for LCL coupled inverter-based distributed generation systems'. IEEE Transactions on Industrial Electronics, 2011, Vol. 58, No. 4, pp. 1359–1370.

[2] R. H. Lasseter and P. Piagi, 'Microgrid: a conceptual solution'. IEEE PESC in Proc. 2004, pp. 4285–4290.

[3] DTI (Distributed Energy) review report, 'A call for evidence for the review of barriers and incentives to distributed electricity generation including combined heat and power', 2006. http://www.berr.gov.uk/files/file35026.pdf.

[4] S. R. Bull, 'Renewable energy today and tomorrow'. Proceedings of IEEE, Vol. 89, No. 0018-9219, 2001, pp. 1216–1226.
[5] K. H. Ahmed, S. J. Finney, and B. W. Williams, 'Passive filter design for three-phase inverter interfacing in distributed generation'. IEEE Compatibility in Power Electronics Conf., 2007, pp. 1–9.
[6] A. Girgis and S. Brahma, 'Effect of distributed generation on protective device coordination in distribution system'. Large Engineering Systems Conf. Power Engineering LESCOPE, 2001, pp. 115–119.
[7] C. Mitchell, 'The value of distributed generation-policy implications for the UK'. IEE Colloquium on Economics of Embedded Generation (Ref. No. 1998/512), 29 October 1998, pp. 1–5.
[8] 'Distributed Generation: Securing America's Future with Reliable and Flexible Power'. A summary of distributed generation by the Federal Energy Technology Center, October 1999, http://www.distributed-generation.com/Library/FETC.pdf.
[9] 'Assessment of Distributed Generation Technology Applications'. A report prepared by the Resource Dynamics Corporation for the Maine Public Utilities Commission, February 2001, http://www.distributed-generation.com/Library/Maine.pdf.
[10] A. Bertani, C. Bossi, F. Fornari, S. Massucco S. Spelta, and F. Tivegna, 'A microturbine generation system for grid connected and islanding operation'. IEEE Power Systems Conference and Exposition PES, 10–13 October 2004, pp. 360–365.
[11] D. M. Divan, M. C. Chandorkar, and R. Adapa, 'Control of parallel connected inverter in stand-alone ac supply systems'. IEEE-IAS Annul. Meeting in Conf. Rec., 1991, pp. 1003–1009.
[12] P. S. Kundur, 'Power System Stability and Control', New York: McGraw Hill, 1994.
[13] T. Kawabata, N. Sashida, Y. Yamamoto, K. Ogasawara, and Y. Yamasaki, 'Parallel processing inverter system'. Power Electronics, IEEE Transactions on Vol. 6, No. 3, July 1991, pp. 442–450.
[14] T. Kawabata and S. Higashino, 'Parallel operation of voltage source inverters'. IEEE Transactions on Industry Applications, Vol. 24, No. 2, March–April 1988, pp. 281–287.
[15] E. A. A. Coelho, P. C. Cortizo, and P. F. D. Garcia, 'Small signal stability for single phase inverter connected to stiff ac system'. IEEE-IAS Annual Meeting in Conf. Rec., Vol. 4, October 1999, pp. 2180–2187.
[16] M. C. Chandorkar, and D. M. Divan, 'Decentralized operation of distributed UPS systems'. Power Electronics, Drives and Energy Systems for Industrial Growth, 1996, International Conference on Vol. 1, 1996, pp. 565–571.
[17] U. B. Jensen, F. Blaabjerg, and P. N. Enjeti, 'Sharing of nonlinear loads in parallel connected three-phase converters'. IEEE-IAS Annual Meeting in Conf. Rec., Vol. 5, 2000, pp. 2338–2345.

[18] K. J. P. Macken, K. Vanthournout, J. Van den Keybus, G. Deconinck, and R. J. M. Belmans, 'Distributed control of renewable generation units with integrated active filter'. IEEE Transactions on Power Electronics, Vol. 19, No. 5, September 2004, pp. 1353–1360.
[19] Y. Chen and K. Smedley, 'Parallel operation of one-cycle controlled grid connected three-phase inverters'. IEEE 40th Annual Meeting Industry Applications Conf. IAS, Vol. 1, 2–6 October 2005, pp. 591–598.
[20] L. A. Serpa, J. W. Kolar, S. Ponnaluri, and P. M. Barbosa, 'A modified direct power control strategy allowing the connection of three-phase inverter to the grid through LCL filters'. 40th IAS Annual Meeting, Industry Applications Conf., Vol. 1, 6 October 2005, pp. 565–571.
[21] M. Malinowsk and M. P. Kazmierkowski, 'Direct power control of three-phase PWM rectifier using space vector modulation simulation study'. IEEE ISIE International Symposium Industrial Elect., Vol. 4, 2002, pp. 1114–1118.
[22] T. Noguchi, H. Tomiki, S. Kondo, and I. Takahashi, 'Direct power control of PWM converter without power-source voltage sensors'. IEEE Transactions on Industry Applications, Vol. 34, No. 3, May–June 1998, pp. 473–479.
[23] V. G. Török, 'Near-optimum on-line modulation of PWM inverters'. In Proc. IFAC Conf. Control in Power Electronics and Electric Drives, Lausanne, 1983, pp. 247–254.
[24] M. Chandorkar, 'New techniques for inverter flux control'. IEEE Transactions on Industry Applications, Vol. 37, No. 3, May–June 2001, pp. 880–887.
[25] F. A. Magueed and J. Svensson, 'Control of VSC connected to the grid through LCL-filter to achieve balanced currents'. Fortieth IAS Annual Meeting Conference Industry Applications, Vol. 1, 2005, pp. 572–578.
[26] M. Lindgren, 'Control of a voltage-source converter connected to the grid through an LCL filter application to active filtering'. IEEE PESC Proc., 1998, pp. 229–235.
[27] J. Svensson and M. Lindgren, 'Vector current controlled grid connected voltage source converter-influence of nonlinearities on the performance'. IEEE Annual Power Electronics Specialists Conference PESC, Vol. 1, 1998, pp. 531–537.
[28] M. B. Lindgren, 'Feedforward time efficient control of a voltage source converter connected to the grid by low pass filters'. Annual IEEE Power Electronics Specialists Conference PESC, Vol. 2, 1995, pp. 1028–1032.
[29] E. J. Bueno, F. Espinosa, F. J. Rodriguez, J. Urefia, and S. Cobreces, 'Current control of voltage source converters connected to the grid through an LCL-filter'. IEEE 35th Annual Power Electronics Specialists Conference PESC, Vol. 1, 2004, pp. 68–73.
[30] Y. Ito, Y. Kanno and S. Kawauchi, 'Source voltage sensor-less digital control using observer for PWM converter'. IEEE 25th Annual Power Electronics Specialists Conference PESC, Vol. 1, 1994, pp. 160–165.
[31] E. Twining and D. G. Holmes, 'Grid current regulation of a three-phase voltage source inverter with an LCL input filter'. IEEE Transactions on Power Electronics, Vol. 18, No. 3, 2003, pp. 888–895.

[32] N. Uemura and T. Yokoyama, 'Current control method using voltage deadbeat control for single phase utility interactive inverter'. The 25th International on Telecommunications Energy Conference, 19–23 October 2003, pp. 40–45.

[33] B. Bolsens, K. De Brabandere, J. Van den Keybus, J. Driesen, and R. Belmans, 'Model-based generation of low distortion currents in grid-coupled PWM-inverters using an LCL output filter'. IEEE PESC Proc., 2004, pp. 4616–4622.

[34] B. Bolsens, K. De Brabandere, J. Van den Keybus, J. Driesen, and R. Belmans, 'Three-phase observer-based low distortion grid current controller using an LCL output filter'. IEEE 36th Conference on Power Electronics Specialists, June 12 2005, pp. 1705–1711.

[35] A. B. Rey, S. de Pablo, J. M. Ruiz, and F. A. Ruz, 'Two novel current source sliding mode control strategies for a 3-phase grid-connected inverter'. IEEE ISIE Conf., Vol. 2, 2005, pp. 763–768.

[36] M. I. Marei, E. F. El-Saadany, and M. M. A. Salama, 'A novel control algorithm for the DG interface to mitigate power quality problems'. IEEE Transactions on Power Delivery, Vol. 19, 2004, pp. 1384–1392.

[37] M. Liserre, A. Dell'Aquila, and F. Blaabjerg, 'Genetic algorithm-based design of the active damping for an LCL-filter three-phase active rectifier'. IEEE Transactions on Power Electronics, Vol. 19, 2004, pp. 76–86.

[38] S. M. Abu-Sharkh, Z. F. Hussien, and J. K. Sykulski, 'Current control of three-phase PWM inverters for embedded generators'. IEE Eighth International Conference on Power Electronics and Variable Speed Drives, 18–19 September 2000, pp. 524–529.

[39] O. Kukrer, 'Deadbeat control of a three-phase inverter with an output LC filter'. IEEE Transactions on Power Electronics, Vol. 11, No. 1, 1996, pp. 16–23.

[40] T. Kawabata, T. Miyashita, and Y. Yamamoto, 'Dead beat control of three phase PWM inverter'. IEEE Transactions on Power Electronics, Vol. 5, No. 1, 1990, pp. 21–28.

[41] Y. Ito, M. Iwata, and S. Kawauchi, 'Digital control of three-phase PWM inverters for UPSs using dead-beat observer'. Power Conversion Conference, 19–21 April 1993, pp. 79–84.

[42] W. Guo, S. Duan, K. Xuejuan, Y. Kang, and J. Chen, 'A modified deadbeat control for single-phase voltage-source PWM inverters based on asymmetric regular sample'. IEEE 32nd Annual Power Electronics Specialists Conference PESC, Vol. 2, 17–21 June 2001, pp. 962–967.

[43] T. Yokoyama and A. Kawamura, 'Disturbance observer based digitally controlled PWM inverter for UPS application'. Power Conversion Conference, 1993, 19–21 April 1993, pp. 213–218.

[44] S. H. Lee and C. C. Chung, 'Robust control using a state space disturbance observer'. 42nd IEEE Conference Proceedings on Decision and Control, Vol. 2, 9–12 December 2003, pp. 1297–1302.

[45] L. Ben-Brahim, T. Yokoyama, and A. Kawamura, 'Digital control for UPS inverters'. The Fifth International Conference on Power Electronics and Drive Systems PEDS, Vol. 2, 17–20 November 2003, pp. 1252–1257.
[46] T. Yokoyama, and A. Kawamura, 'Disturbance observer based digitally controlled PWM inverter for UPS application'. Power Conversion Conference, 19–21 April 1993, pp. 213–218.
[47] S. El-Barbari and W. Hofmann, 'Control of a 4 leg inverter for standalone photovoltaic systems'. IEEE Fourth International Conference on Power Electronics and Drive Systems, Vol. 1, 22–25 October 2001, pp. 348–354.
[48] P. Mattavelli, 'An improved deadbeat control for UPS using disturbance observers'. IEEE Transactions on Industrial Electronics, Vol. 52, No. 1, February 2005, pp. 206–212.
[49] Y. Ito and S. Kawauchi, 'Microprocessor based robust digital control for UPS with three-phase PWM inverter'. IEEE Transactions on Power Electronics, Vol. 10, No. 2, March 1995, pp. 196–204.
[50] J. S. Cho, Seung-Yo Lee, Hyung-Soo Mok, and Gyu-Ha Choe, 'Analysis and design of modified deadbeat controller for 3-phase uninterruptible power supply'. IEEE International Conference on Power Electronics and Drive Systems PEDS, Vol. 2, 1999, pp. 1003–1009.
[51] J. S. Cho, Seung-Yo Lee, Hyung-Soo Mok, and Gyu-Ha Choe, 'Modified deadbeat digital controller for UPS with 3-phase PWM inverter'. IEEE 34th Annual Industry Applications Conference Meeting IAS, Vol. 4, 3–7 October 1999, pp. 2208–2215.
[52] E. J. Davison and B. Scherzinger, 'Perfect control of the robust servomechanism problem'. IEEE Transactions on Automatic Control, Vol. 32, 1987, pp. 689–702.
[53] M. N. Marwali, and A. Keyhani, 'Control of distributed generation systems-Part I: voltages and currents control'. IEEE Transactions on Power Electronics, Vol. 19, No. 6, November 2004, pp. 1541–1550.
[54] J. W. Jung, M. Dai, and A. Keyhani, 'Optimal control of three-phase PWM inverter for UPS systems'. IEEE 35th Annual Power Electronics Specialists Conf. PESC, Vol. 3, 20–25 June 2004, pp. 2054–2059.
[55] A. J. Baronian and S. B. Dewan, 'An adaptive digital control of current source inverter suitable for parallel processing inverter systems'. IEEE 30th Annual Industry Applications Meeting Conf. IAS, Vol. 3, 8–12 October 1995, pp. 2670–2677.
[56] X. Chen, L. Li, H. Shan, Y. Kang, and J. Chen, 'Parallel operation system of high frequency modularized UPS based on distributed logic control'. IEEE International Conference on Electric Machines and Drives, 2005, pp. 777–782.
[57] X. Chen, Y. Kang, and J. Chen, 'Operation, control technique of parallel connected high power three-phase inverters'. The 4th International Power Electronics and Motion Control Conference IPEMC, Vol. 2, 14–16 August 2004, pp. 956–959.

[58] K. Thyagarajan, A. Davari, and A. Feliachi, 'Load sharing control in distributed generation system'. 37th Southeastern Symposium on System Theory SSST, 20–22 March 2005, pp. 424–428.
[59] M. N. Marwali, J. W. Jung, and A. Keyhani, 'Control of distributed generation systems – Part II: load sharing control'. IEEE Transactions on Power Electronics, Vol. 19, No. 6, November 2004, pp. 1551–1561.
[60] K. Siri and J. Banda, 'Analysis and evaluation of current-sharing control for parallel-connected DC–DC converters taking into account cable resistance'. IEEE Aerospace Applications Conference Proceedings, Part 2, 4–11 February 1995, pp. 29–48.
[61] K. Siri, C. Q. Lee, and T. E. Wu, 'Current distribution control for parallel connected converters: part I'. IEEE Transactions on Aerospace and Electronic Systems, Vol. 28, No. 3, July 1992, pp. 829–840.
[62] K. Siri and C. Q. Lee, 'Current distribution control of converters connected in parallel'. IEEE Industry Applications Society Annual Meeting, 7–12 October 1990, pp. 1274–1280.
[63] T. F. Wu, K. Siri, and C. Q. Lee, 'Reliability improvement in parallel connected converter systems'. International Conference on Industrial Electronics, Control and Instrumentation IECON, 28 October–1 November 1991, pp. 429–434.
[64] J. F. Chen and C. L. Chu, 'Combination voltage-controlled and current-controlled PWM inverters for UPS parallel operation'. IEEE Transactions on Power Electronics, Vol. 10, 1995, pp. 547–558.
[65] Y. Pei, G. Jiang, X. Yang, and Z. Wang, 'Auto-master–slave control technique of parallel inverters in distributed AC power systems and UPS'. IEEE 35th Annual Power Electronics Specialists Conference PESC, Vol. 3, 20–25 June 2004, pp. 2050–2053.
[66] S. K. Mazumder, K. Acharya, and M. Tahir, 'Wireless control of spatially distributed power electronics'. IEEE 20th Annual Applied Power Electronics Conference and Exposition APEC, Vol. 1, 6–10 March 2005, pp. 75–81.
[67] W. Zhang, M. S. Branicky, and S. M. Phillips, 'Stability of networked control systems'. IEEE Control Systems Magazine, 2001, pp. 84–99.
[68] X. Liu and A. Goldsmith, 'Wireless medium access control in networked control systems'. IEEE Proceedings of the American Control Conference, Vol. 4, 2004, pp. 3605–3610.
[69] K. Siri, C. Q. Lee, and T. F. Wu, 'Current distribution control for parallel connected converters: Part II'. IEEE Transactions on Aerospace and Electronics Systems, Vol. 28,No. 3, July 1992, pp. 841–851.
[70] T. F. Wu, K. Siri, and J. Banda, 'The central-limit control and impact of cable resistance in current distribution for parallel-connected DC–DC converters'.IEEE 25th Annual Power Electronics Specialists Conference PESC, 20–25 June 1994, pp. 694–702.
[71] J. Banda and K. Siri, 'Improved central-limit control for parallel-operation of DC–DC power converters'. IEEE 26th Annual Power Electronics Specialists Conference PESC, Vol. 2, 18–22 June 1995, pp. 1104–1110.

[72] C. Q. Lee, K. Siri, and T. F. Wu, 'Dynamic current distribution controls of a parallel connected converter'. IEEE 22nd Annual system Power Electronics Specialists Conference PESC, 24–27 June 1991, pp. 875–881.

[73] X. Sun, Y. S. Lee, and D. Xu 'Modeling, analysis, and implementation of parallel multi-inverter systems with instantaneous average current sharing scheme'. IEEE Transactions on Power Electronics, Vol. 18, No. 3, May 2003, pp. 844–856.

[74] Y. Xing, L. Huang, and Y. Yan, 'Redundant parallel control for current regulated inverters with instantaneous current sharing'. IEEE 34th Annual Power Electronics Specialist Conference PESC, Vol. 3, 15–19 June 2003, pp. 1438–1442.

[75] C. S. Lee, S. Kim, C. B. Kim, *et al.*, 'Parallel UPS with a instantaneous current sharing control'. IEEE 24th Annual Conference of the Industrial Electronics Society IECON, Vol. 1, 31 August–4 September 1998, pp. 568–573.

[76] Y. Xing, L. P. Huang, and Y. G. Yan, 'A decoupling control method for inverters in parallel operation'. International Conference on Power System Technology PowerCon, Vol. 2, 13–17 October 2002, pp. 1025–1028.

[77] S. J. Chiang, C. H. Lin, and C. Y. Yen, 'Current limitation control for multi-module parallel operation of UPS inverters'. IEE Proceedings-Electric Power Applications, Vol. 151, No. 6, 7 November 2004, pp. 752–757.

[78] X. Sun, L.-K. Wong, Y.-S. Lee, and D. Xu, 'Design and analysis of an optimal controller for parallel multi-inverter systems'. IEEE Transactions on Circuits and Systems II: Express Briefs, Vol. 53, No. 1, January 2006, pp. 56–61.

[79] E. A. A. Coelho, B. R. Menezes, P. C. Cortizo, and P. F. D. Garcia, 'Sliding mode controller for parallel connected inverters'. IEEE International Power Electronics Congress CIEP, 12–15 October 1998, pp. 96–102.

[80] R. Ramos, D. Biel, F. Guinjoan, and E. Fossas, 'Distributed control strategy for parallel-connected inverters sliding mode control approach and FPGA-based implementation'. IECON 02, Vol. 1, 5–8 November 2002, pp. 111–116.

[81] R. Ramos, D. Biel, F. Guinjoan, and E. Fossas, 'Design considerations in sliding-mode controlled parallel-connected inverters'. IEEE International Symposium on Circuits and Systems ISCAS, Vol. 4, 26–29 May 2002, pp. 357–360.

[82] D. Biel, R. Ramos, F. Guinjoan, J. Negroni, and C. Meza, 'Sliding-mode control design of parallel-connected switching converters for modular transformerless DC–AC step-up conversion'. IEEE International Symposium on Circuits and Systems ISCAS, 23–26 May 2005, pp. 3067–3070.

[83] A. P. Martins and A. S. Carvalho, 'High dimensional fuzzy controller for an UPS parallel system'. IEEE 24th Annual Conference of the Industrial Electronics Society IECON, Vol. 1, 31 August–4 September 1998, pp. 12–16.

[84] C. C. Hua, K. A. Liao, and J. R. Lin, 'Parallel operation of inverters for distributed photovoltaic power supply system'. IEEE 33rd Annual Power

Electronics Specialists Conference PESC, Vol. 4, 23–27 June 2002, pp. 1979–1983.

[85] U. Borup, F. Blaabjerg, and P. N. Enjeti, 'Sharing of nonlinear load in parallel-connected three-phase converters'. IEEE Transactions on Industry Applications, Vol. 37, No. 6, November–December 2001, pp. 1817–1823.

[86] A. Tuladhar, H. Jin, T. Unger, and K. Mauch, 'Parallel operation of single phase inverters with no control interconnections'. IEEE Allied Power Electronic Conference in Proceedings, Vol. 1, 1997, pp. 94–100.

[87] Y. B. Byun, T. G. Koo, K. Y. Joe, E. S. Kim, J. I. Seo, and D. H. Kim, 'Parallel operation of three-phase UPS inverters by wireless load sharing control'. Applied Power Electronics Conference Exposition, Vol. 1, 2001, pp. 526–532.

[88] A. Tuladhar, Hua Jin, T. Unger, and K. Mauch, 'Control of parallel inverters in distributed ac power systems with consideration of line impedance effect'. IEEE Transactions Industry Application, Vol. 36, 2000, pp. 131–138.

[89] K. De Brabandere, B. Bolsens, J. Van den Keybus, *et al.*, 'A voltage and frequency droop control method for parallel inverters'. IEEE 35th Annual Power Electronics Specialists Conference PESC, Vol. 4, 2004, pp. 2501–2507.

[90] L. Xinchun, F. Feng, D. Shanxu, K. Yong, and C. Jian, 'The droop characteristic decoupling control of parallel connected UPS with no control interconnection'. IEEE International Electric Machines and Drives Conference IEMDC, Vol. 3, 1–4 June 2003, pp. 1777–1780.

[91] J. M. Guerrero, L. G. de Vicuna, J. Matas, M. Castilla, and J. Miret, 'A wireless controller to enhance dynamic performance of parallel inverters in distributed generation systems'. IEEE Transactions on Power Electronics, Vol. 19, No. 5, September 2004, pp. 1205–1213.

[92] S. J. Chiang and J. M. Chang, 'Parallel control of the UPS inverters with frequency-dependent droop scheme'. IEEE 32nd Annual Power Electronics Specialists Conference PESC, Vol. 2, 17–21 June 2001, pp. 957–961.

[93] J. M. Guerrero, L. G. de Vicuna, J. Matas, M. Castilla, and J. Miret, 'Output impedance design of parallel-connected UPS inverters with wireless load-sharing control'. IEEE Transactions on Industrial Electronics Vol. 52, No. 4, August 2005, pp. 1126–1135.

[94] I. El-Samahy, and E. El-Saadany, 'The effect of DG on power quality in a deregulated environment'. IEEE Power Engineering Society General Meeting, 12–16 June 2005, pp. 2969–2976.

[95] British Standards 'Voltage characteristics of electricity supplied by public distribution networks'. BS EN 50160:2007, October 2007.

[96] IEEE Standards 519-1992, Recommended Practices and Requirements for Harmonic Control in Electric Power Systems, 1992.

[97] Recommendations for the Connection of Embedded Generating Plant to the Regional Electricity Companies Distribution Systems, Std. G59/1, Electricity Assoc., 1991, Electricity Assoc. Std.

[98] B. Singh, K. Al-Haddad, and A. Chandra, 'A review of active filters for power quality improvement'. IEEE Transactions on Industry Electronics, Vol. 46, No. 5, 1999, pp. 960–971.
[99] M. El-Habrouk, M. K. Darwish, and P. Mehta, 'Active power filters: a review', Electric Power Applications, IEE Proceedings, Vol. 147, No. 5, 2000, pp. 403–413.
[100] H. Akagi, 'Active harmonic filters'. Proceedings of the IEEE Vol. 93, Issue 12, December 2005, pp. 2128–2141.
[101] D. G. Holmes and T. A. Lipo, 'Pulse Width Modulation for Power Converters: Principles and Practice'. IEEE Press Series on Power Engineering – Wiley-IEEE Press, Edition 1 October 2003.
[102] IEEE Standards 519-1992, Recommended Practices and Requirements for Harmonic Control in Electric Power Systems, 1992.
[103] P. A. Dahono, A. Purwadi, and Qamaruzzaman, 'An LC filter design method for single-phase PWM inverters'. Power Electronics and Drive Systems, Proceedings, Vol. 2, 1995, pp. 571–576.
[104] J. Kim, J. Choi, and H. Hong, 'Output LC filter design of voltage source inverter considering the performance of controller'. Power System Technology, Proceedings, Vol. 3, 2000, pp. 1659–1664.
[105] Y. Sozer, D. A. Torrey, and S. Reva, 'New inverter output filter topology for PWM motor drives'. IEEE Transactions on Power Electronics, Vol. 15, No. 6, November 2000, pp. 1007–1017.
[106] J. K. Phipps, 'A transfer function approach to harmonic filter design'. IEEE Industry Applications Magazine, March–April 1997, Vol. 3, pp. 68–82.
[107] P. A. Dahono, Y. R. Bahar, Y. Sato, and T. Kataoka, 'Damping of transient oscillations on the output LC filter of PWM inverters by using a virtual resistor '. Power Electronics and Drive Systems, Proceedings, Vol. 1, 2001, pp. 403–407.
[108] P. A. Dahono and E. Taryana, 'A new control method for single-phase PWM inverters to realize zero steady-state error and fast response'. Power Electronics and Drive Systems (PEDS), Vol. 2, 2003, pp. 888–892.

Chapter 8

Backward–forward sweep-based islanding scenario generation algorithm for defensive splitting of radial distribution systems

F. Jabari[1] *and B. Mohammadi-Ivatloo*[1]

8.1 Introduction

Nowadays, increasing rate of distributed generators (DGs) penetration within the distribution systems makes it possible to split the radial or meshed distribution networks into several isolated islands prior to an upstream supply grid outage [1–3]. In the literature, several remarkable efforts have been carried out on defensive islanding of distribution systems. El-Zonkoly *et al.* [4] applied the comprehensive learning particle swarm optimization algorithm for splitting of a radial distribution grid in the absence of the upstream grid in order to minimize active and reactive power generation cost and cost of the un-served power considering the operational constraints. The comprehensive learning particle swarm optimizer updates the particle velocity and enables the diversity of the swarm to be preserved to discourage premature convergence [5]. Quevedo *et al.* [6] proposed a two-stage stochastic linear programming algorithm to split a radial distribution network by minimizing total real power losses with respect to nodal voltage limits and load-generation active power balance constrains in the presence of wind production and energy storage systems. In [7], an automatic adaptive identification splitting configuration scheme is introduced to identify the optimum separating scenario in case of loss of upstream network. In this study, a centralized controller forecasts the loads and evaluates the probability of successful islanding operation of all possible islands. The impact of storage devices and load-shedding strategies on microgrid islanded operation is comprehensively discussed in [8]. Brown *et al.* [9] proposed a novel-genetic-algorithm-based optimization strategy entitled "feeder addition problem" to find the best locations to add the interties between feeders in a radial distribution system to improve the reliability of the islanded smart grid considering the power balance criterion. The influence of the cooperative control strategy of μ-sources and the energy storage systems on islanded operation of distribution microgrids is assessed through experimental studies in [10]. In [11], an adaptive

[1]Faculty of Electrical and Computer Engineering, University of Tabriz, Tabriz, Iran

multi-objective harmony search algorithm is applied to reconfigure the distribution microgrids with the aim of reducing the active power losses and increasing load-ability index in the islanding mode. An optimal islanding approach is implemented in [12] for feasibility checking using a serial of tree knapsack problems (TKPs) and islands aggregation process with respect to load priority and controllability, real power balance, voltage stability and transmission capacity criteria. Chen *et al.* [13] investigated an islanding-security-region (ISR)-based control architecture algorithm for smart distribution networks. The ISR make it possible to evaluate the islanding operation mode offline and before the local grid loss. Control measures will then be applied with the aim of system pulling back into the ISR to ensure a successful real-time splitting operation. A review on islanding phenomenon in a distribution network integrated with small-scale hydro power plant has been provided in [14]. An off-line feasibility study of distribution system islanding is presented in [15] by defining which generators and loads remain connected after controlled partitioning to satisfy the load-generation active power balance. In [16], a Monte Carlo and point-estimation-based stochastic contingency analysis is introduced to evaluate the steady-state voltage stability of islanded power system using QV-modal analysis considering the load uncertainty.

A mixed-integer second-order conic programming model is suggested in [17] to apply the reconfiguration process of current distribution systems to minimize energy losses and maximize the customer-oriented reliability indices such as system average interruption frequency index, system average interruption duration index and the energy not supplied. In [18], a bio-inspired metaheuristic artificial-immune-systems based optimization algorithm is presented to solve the reconfiguration problem by minimizing total real power losses and determining the energy losses interval with respect to load and wind production uncertainties. A three-layer artificial neural network (ANN) is introduced in [19] to reconfigure distribution grids to minimize total real power losses and reduce the size of ANN search space using a proposed modified dynamic fuzzy c-means clustering algorithm. Multi-objective hybrid big bang-big crunch has been investigated in [20] for simultaneous reconfiguration and power allocation of dispersed generators to minimize total grid losses, operation cost and greenhouse gas emissions and maximize voltage security margin. In this study, triangular fuzzy number technique is used to model the load uncertainty. Decimal-coded quantum particle swarm optimization is proposed in [21] to reconfigure distribution feeders by reducing the particle length, separating scenarios and increasing the search efficiency considering some impact factors of output reactive power of solar photovoltaic panels. In [22], series of TKPs have been solved by modified shuffled frog leap algorithm to split a radial distribution network into several stable islands considering load priority and controllability, branch capacity and bus voltage constraints.

This chapter proposes a novel two-stage scenario generation strategy for controlled or defensive islanding of smart distribution systems. In the first step, the primary line and bus data matrices have been determined with respect to deterministic and probabilistic nodes and branches for each island. In the second stage, line and bus information matrices can be updated for all islands in each generated scenario using node elimination matrices (NEMs). Backward–forward power flow is then solved in each island according to modified line and bus information matrices. In this study, binary imperialistic competitive algorithm (BICA) is used to find the best splitting points with the aim of minimizing total active power losses and nodal voltage deviations considering voltage limits and feeder capacity constraints. The proposed strategy is evaluated through an extensive test conducted on 12.66 kV, 33-bus radial distribution system with three DG units to validate the speed and the effectiveness of the developed algorithm. The remainder of this chapter is organized as follows: Section 8.2 presents the problem formulation. Simulation results and discussions are provided in Section 8.3. Finally, Section 8.4 concludes the chapter.

8.2 Problem formulation

8.2.1 Proposed backward–forward-sweep-based islanding scenario generation algorithm

As the electrical connections between the distribution network and the upstream grid gets interrupted, the intentional defensive partitioning strategy can be applied to preserve the continuity of electrical demands. In the first stage of the proposed algorithm, the types of all branches and buses should be defined to reduce the islanding search space from the entire network to smaller interface area. As shown in Figure 8.1, it is assumed that the protection system of the upstream transformer trips and splitting is temporarily applied on the stand-alone-isolated distribution system with the following desired coherent DGs:

- DG1 and DG2
- DG3
- DG4 and DG5

DG1 and DG2 are considered as the roots of island A. In the same manner, DG3, DG4 and DG5 are the source nodes of islands B and C, respectively. All branches which directly connect two desired coherent DGs cannot be considered in the interface area. In other words, they should be connected to each other and eliminated from primary huge search space. Others will be considered as boundary branches and belong to the interface area. All buses which are connected to the boundary branches are called boundary nodes and belong to the interface area.

The primary line and bus data matrices of islands A, B and C will be determined with respect to the interface area. As shown in Figure 8.2, islands A, B and C have 20, 15 and 19 buses, respectively.

Equations (8.1)–(8.3) can be considered for the kth boundary branch connecting ith and jth nodes of area A.

$$Z(i, j) = [1 - x(k)]M + Z_0(i, j) \tag{8.1}$$

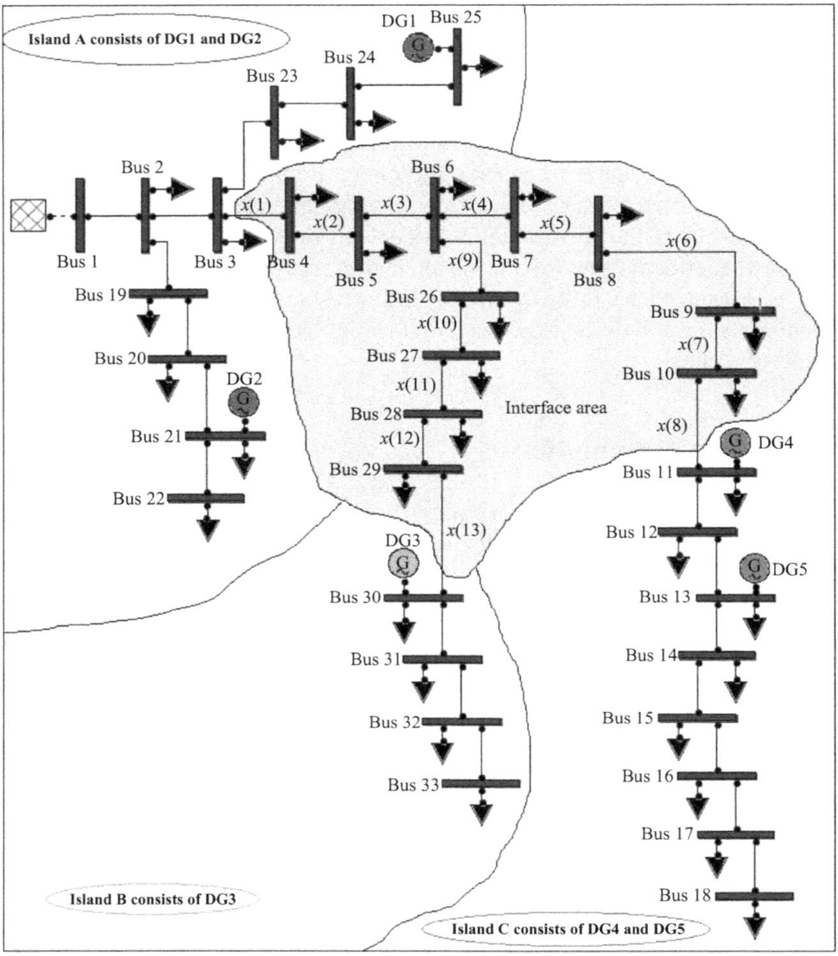

Figure 8.1 The interface area of a 33-bus radial distribution system

$$x(k) = \begin{cases} 0 & \text{if the } k\text{th boundary branch has been opened} \\ 1 & \text{if the } k\text{th boundary branch has been closed} \end{cases} \tag{8.2}$$

$$Z_0(i, j) = R(i, j) + jX(i, j) \tag{8.3}$$

$Z(i, j)$ is the impedance of the branch i to j in primary line data matrix of island A. M is a sufficiently large constant. $Z_0(i, j)$, $R(i, j)$ and $X(i, j)$ are the impedance, resistance and reactance of $i - j$ branch in per unit, respectively. According to (8.2), $x(k)$ shows the state of the kth boundary branch. In a generated scenario, the primary line data matrix can be updated as follows. If $x(k) = 0$, the kth boundary

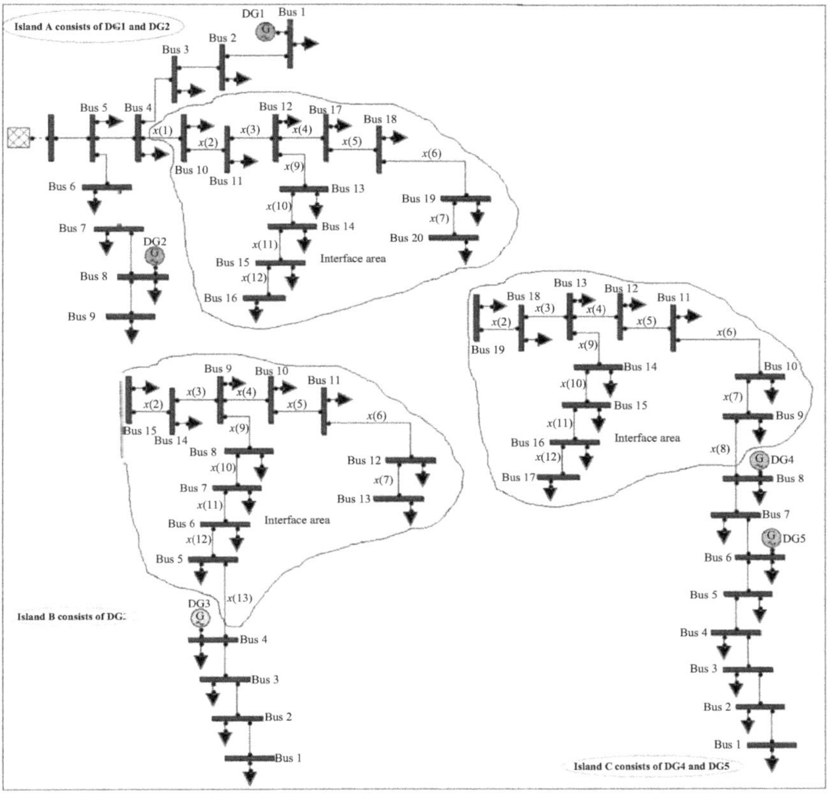

Figure 8.2 Primary islands considering the interface area in the line and bus data matrices

branch is opened and $Z(i, j) \cong M$. If $x(k) = 1$, the kth boundary branch is closed and $Z(i, j) = Z_0(i, j)$.

It is necessary to modify the primary bus information matrices in each possible solution. Therefore, if the ith node of the island A (which is a probabilistic node) is removed from area A to island B, the ith bus of island A will be eliminated from its primary bus information matrix through a NEM. In the other words, the primary bus data matrix will be modified via its NEM. If the ith node of island A does not belong to the interface area, the ith row of $\mathrm{NEM_A}$ will be equal to 1 and $\mathrm{NEM_A}(i) = 1$. Otherwise, the ith row of $\mathrm{NEM_A}$ will be equal to the multiplication of the binary states of all interface probabilistic branches which connect the ith boundary node of island A to its deterministic network. For example, the 5th and 16th nodes of island A are deterministic and probabilistic, respectively. Therefore, the 5th and 16th rows of $\mathrm{NEM_A}$ will be equal to 1 and $x(1) \times x(2) \times x(3) \times x(9) \times x(10) \times x(11) \times x(12)$, respectively. The bus information matrix of island A can be modified as follows. If $\text{NEM}_\text{A}(i) = 0$, the ith bus of island A will be removed. If $\text{NEM}_\text{A}(i) \neq 0$, the ith bus of island A will belong to it.

8.2.2 Objective function and constraints

The objective is to minimize the following function:

$$\text{Objective} = \frac{\sum_{k=1}^{N_{\text{island}}} \sum_{i=1}^{N_B^k - 1} R_{i_k} I_{i_k}^2}{\sum_{j=1}^{N_B} R_j I_j^2} + \sum_{k=1}^{N_{\text{island}}} \sum_{n=1}^{N_B^k} |V_{n_k} - 1| \tag{8.4}$$

where R_{i_k} is the resistance of jth feeder in the kth island, I_{i_k} is the current of jth feeder in the kth island, R_j is the resistance of jth feeder in pre-islanding distribution system, I_j is the current of jth feeder in pre-islanding distribution system, N_B^k is the number of nodes in the kth island, N_{Island} is the number of islands, N_B is the number of nodes in pre-islanding distribution system, and V_{n_k} is the voltage magnitude of nth bus in the kth island in per unit.

The integrity, voltage and feeder capacity criteria have been considered as optimization constraints to find the best islanding scenario.

- Integrity criterion
  The spitting algorithm should not eliminate any node from interconnected distribution system.
- Voltage limits

  $$V_i^{\min} \leq V_i \leq V_i^{\max} \tag{8.5}$$

  where $V_i^{\min}$ is the minimum voltage magnitude of bus I, and $V_i^{\max}$ is the maximum voltage magnitude of bus i.
- Feeder capacity

  $$\sqrt{P_k^2 + Q_k^2} \leq S_k^{\max}$$

  where P_k is the active power flow in the kth feeder, Q_k is the reactive power flow in the kth feeder, and $S_k^{\max}$ is the kth feeder complex power flow capacity limit.

8.2.3 Binary imperialistic-competitive-algorithm-based optimization process

The imperialist competition algorithm (ICA) starts with initial countries similar to the initial population in particle swarm optimization. The countries are then categorized into imperialists and colonies. The imperialists absorb more colonies to their empire. If the colonies which change according to assimilation policy and imperialist competition become stronger, they may take the place of their imperialist [23,24]. In the BICA, basic improvements are applied as in [25]. In this chapter, the BICA consists of an array of binary variables which have to be optimized. In this chapter, the array size is defined as the number of possible probabilistic branches to partition radial distribution network graph. This array is defined as follow:

$$\text{Country} = [\text{Branch}_1, \text{Branch}_2, \text{Branch}_3, \ldots, \text{Branch}_N] \quad (8.6)$$

where N is the number of branches which are potentially suitable for tripping. Branch_i is 1 or 0 according to its condition.

$$\text{Branch}_i = \begin{cases} 0 & \text{If the } i\text{th branch is tripped} \\ 1 & \text{Otherwise} \end{cases} \quad (8.7)$$

More details and steps of used BICA can be found in [25]. The flowchart of the proposed strategy has been depicted in Figure 8.3.

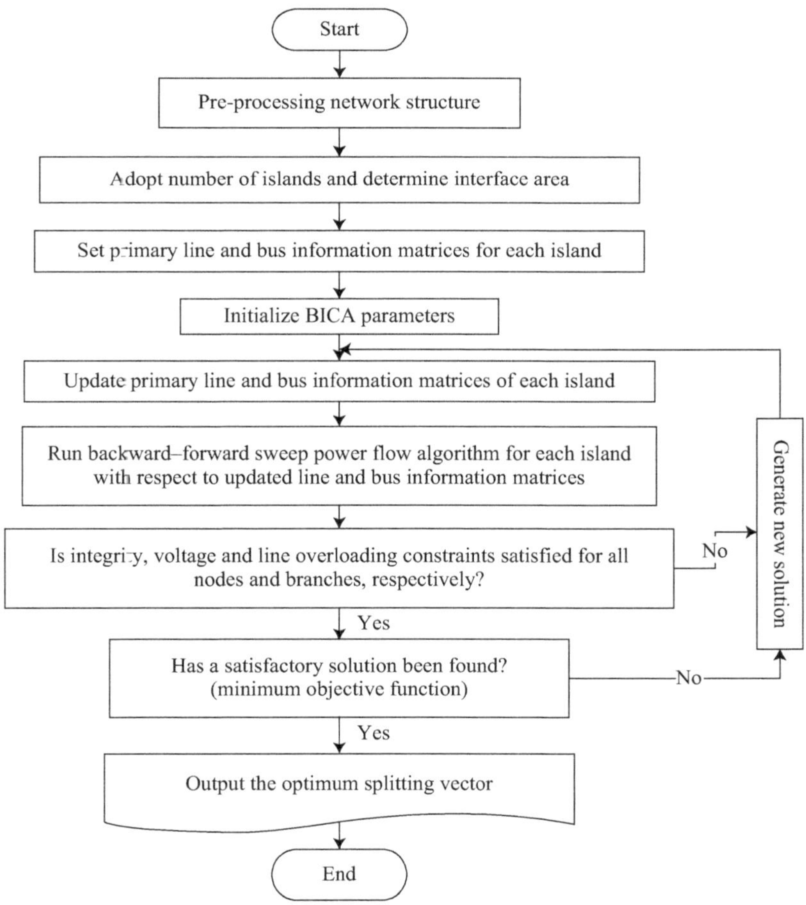

Figure 8.3 The flowchart of the proposed backward–forward-sweep-based islanding algorithm

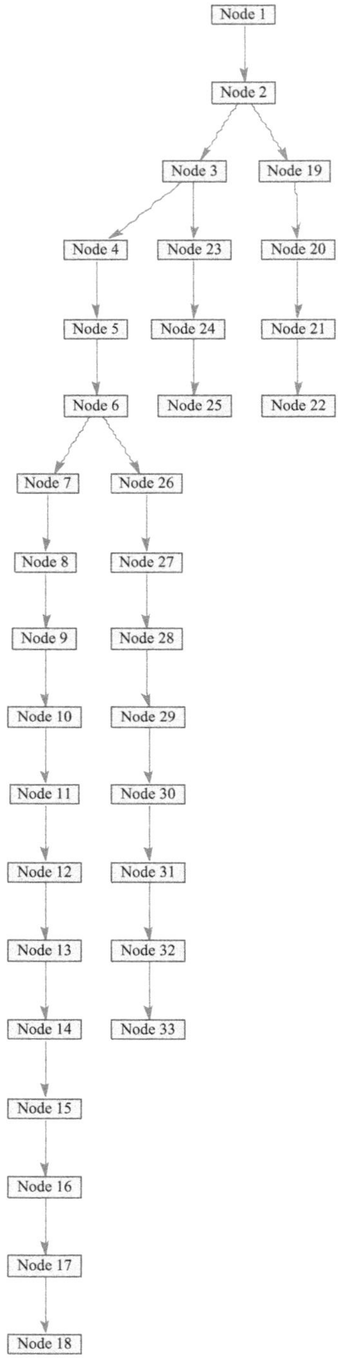

Figure 8.4 The single line diagram of the 12.66 kV and 33-bus radial distribution system [26]

Due to its high computational efficiency, BICA is applied to partition a radial distribution system in this chapter. In the first stage, the number of islands and the interface area are adopted. In the second stage, the primary line and bus information matrices are updated by initial solution vector. The backward–forward sweep is then run for each island to check all mentioned constraints and find the best splitting scenario. If at least one criterion is not satisfied, the new solution will be searched. Otherwise, the objective function will be computed to determine the optimum islanding scenario.

8.3 Simulation studies

To evaluate the speed, robustness and effectiveness of the proposed method, a 12.66 kV and 33-bus radial distribution system [26], are used. The single line diagram of the test system with 32 lines is shown in Figure 8.4.

The system is connected to grid through bus 1. It is considered that three DG units with 0.7100, 1.0220 and 0.7270 MW sizes have optimally been located at buses 25, 30 and 14, respectively [27]. The initial condition derived from backward–forward sweep power flow with and without DG units is shown in Table 8.1 and Figure 8.5.

It is assumed that the breaker of the upstream feeder trips. Hence, the electrical connection between the main network and the downstream microgrid is temporarily lost. In the meantime, the proposed intentional islanding strategy is applied to prevent the partial blackout of the radial distribution system by minimizing total real power losses and nodal voltage deviations and satisfying the operational constraints. The simulation is performed on a Lenovo with 2.10 GHz CPU, 4 GB RAM in MATLAB® software. In the first stage of the proposed method, the generators are grouped into a desired number of coherent units as follows:

- Group A: G1, DG1 and G2
- Group B: G3 and DG2
- Group C: G4, DG3, G5 and G6

According to selected coherent generators, the interface area between three mentioned primary islands will be as illustrated in Figure 8.6.

The primary line and bus information matrices of three mentioned islands with and without DGs can be considered as Tables 8.2–8.7, respectively.

Table 8.1 33-Bus radial test network: initial conditions and results

	Without DGs	**With DGs**
Active losses (kW)	253.9667	105.3662
Reactive losses (kVAR)	169.0311	71.2801
System active load (kW)	3,715	
System reactive load (kVAR)	2,300	

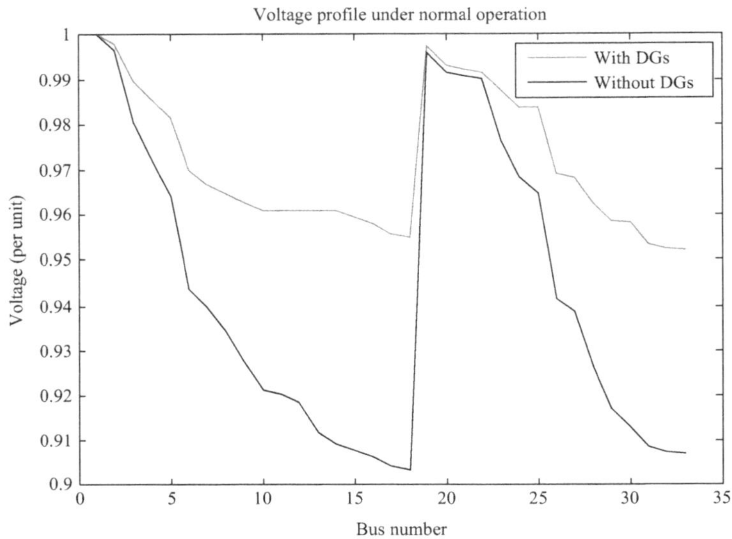

Figure 8.5 Voltage profile under pre-contingency normal operation with and without DG units

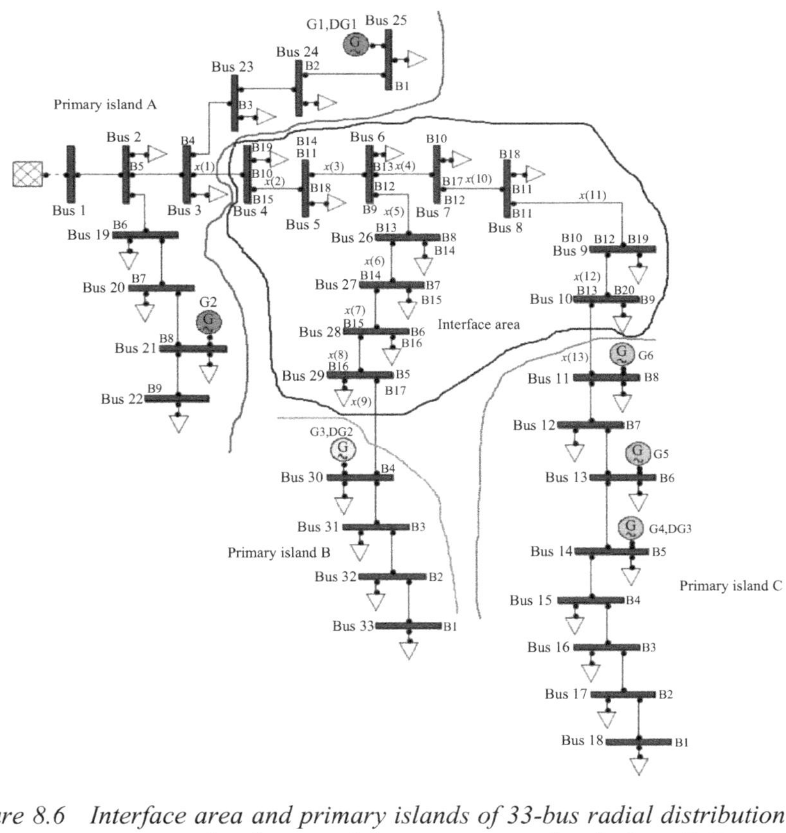

Figure 8.6 Interface area and primary islands of 33-bus radial distribution system in the absence of the upstream grid with three DG units

Table 8.2 Primary bus information matrix of island A without/with DGs

Bus number	Real power (kW)	Reactive power (kVAR)
BD_A = [1	504/−206	240
2	504	240
3	108	60
4	108	48
5	120	72
6	108	48
7	108	48
8	108	48
9	108	48
10	144	96
11	72	36
12	72	24
13	72	30
14	72	30
15	72	24
16	144	84
17	240	120
18	240	120
19	72	24
20	72	24]

Table 8.3 Primary line information matrix of island A

In bus	Out bus	Resistance (Ω)	Reactance (Ω)
LD_A = [1	2	0.8960	0.7011
2	3	0.8980	0.7091
3	4	0.4512	0.3083
4	5	0.4930	0.2511
4	10	$0.3660 + M \times [1 - x(1)]$	$0.1864 + M \times [1 - x(1)]$
5	6	0.1640	0.1565
6	7	1.5042	1.3554
7	8	0.4095	0.4784
8	9	0.7089	0.9373
10	11	$0.3811 + M \times [1 - x(2)]$	$0.1941 + M \times [1 - x(2)]$
11	12	$0.8191 + M \times [1 - x(3)]$	$0.7070 + M \times [1 - x(3)]$
12	13	$0.2030 + M \times [1 - x(5)]$	$0.1034 + M \times [1 - x(5)]$
12	17	$0.1872 + M \times [1 - x(4)]$	$0.6188 + M \times [1 - x(4)]$
13	14	$0.2842 + M \times [1 - x(6)]$	$0.1447 + M \times [1 - x(6)]$
14	15	$1.0590 + M \times [1 - x(7)]$	$0.9377 + M \times [1 - x(7)]$
15	16	$0.8042 + M \times [1 - x(8)]$	$0.7006 + M \times [1 - x(8)]$
17	18	$0.7114 + M \times [1 - x(10)]$	$0.2351 + M \times [1 - x(10)]$
18	19	$1.0300 + M \times [1 - x(11)]$	$0.7400 + M \times [1 - x(11)]$
19	20	$1.0440 + M \times [1 - x(12)]$	$0.7400 + M \times [1 - x(12)]]$

Table 8.4 Primary bus information matrix of island B without/ with DGs

Bus number	Real power (kW)	Reactive power (kVAR)
BD_B = [1	72	48
2	252	120
3	180	84
4	240/−782	720
5	144	84
6	72	24
7	72	30
8	72	24
9	72	120
10	240	120
11	240	24
12	72	24
13	72	24
14	72	36
15	144	96]

Table 8.5 Primary line information matrix of island B

In bus	Out bus	Resistance (Ω)	Reactance (Ω)
LD_B = [1	2	0.3410	0.5302
2	3	0.3105	0.3619
3	4	0.9744	0.9630
4	5	$0.5075 + M \times [1 - x(9)]$	$0.2585 + M \times [1 - x(9)]$
5	6	$0.8042 + M \times [1 - x(8)]$	$0.7006 + M \times [1 - x(8)]$
6	7	$1.0590 + M \times [1 - x(7)]$	$0.9377 + M \times [1 - x(7)]$
7	8	$0.2842 + M \times [1 - x(6)]$	$0.1447 + M \times [1 - x(6)]$
8	9	$0.2030 + M \times [1 - x(5)]$	$0.1034 + M \times [1 - x(5)]$
9	10	$0.1872 + M \times [1 - x(4)]$	$0.6188 + M \times [1 - x(4)]$
9	14	$0.8191 + M \times [1 - x(3)]$	$0.7070 + M \times [1 - x(3)]$
10	11	$0.7114 + M \times [1 - x(10)]$	$0.2351 + M \times [1 - x(10)]$
11	12	$1.0300 + M \times [1 - x(11)]$	$0.7400 + M \times [1 - x(11)]$
12	13	$1.0440 + M \times [1 - x(12)]$	$0.7400 + M \times [1 - x(12)]$
14	15	$0.3811 + M \times [1 - x(2)]$	$0.1941 + M \times [1 - x(2)]]$

In the second stage, the BICA with following parameters is applied to find the optimum splitting points among a huge initial search space with $2^{13} = 8{,}192$ scenarios.

- Number of initial countries: 20
- Number of initial imperialists: 5
- Number of decades: 100
- Revolution rate: 0.5

Table 8.6 Primary bus information matrix of island C without/ with DGs

Bus number	Real power (kW)	Reactive power (kVAR)
BD_C = [1	108	48
2	72	24
3	72	24
4	72	12
5	144/−583	96
6	72	42
7	72	42
8	54	36
9	72	24
10	72	24
11	240	120
12	240	120
13	72	24
14	72	30
15	72	30
16	72	24
17	144	84
18	72	36
19	144	96]

Table 8.7 Primary line information matrix of island C

In bus	Out bus	Resistance (Ω)	Reactance (Ω)
LD_C = [1	2	0.7320	0.5740
2	3	1.2890	1.7210
3	4	0.7463	0.5450
4	5	0.5910	0.5260
5	6	0.5416	0.7129
6	7	1.4680	1.1550
7	8	0.3744	0.1238
8	9	$0.1966 + M \times [1 - x(13)]$	$0.0650 + M \times [1 - x(13)]$
9	10	$1.0440 + M \times [1 - x(12)]$	$0.7400 + M \times [1 - x(12)]$
10	11	$1.0300 + M \times [1 - x(11)]$	$0.7400 + M \times [1 - x(11)]$
11	12	$0.7114 + M \times [1 - x(10)]$	$0.2351 + M \times [1 - x(10)]$
12	13	$0.1872 + M \times [1 - x(4)]$	$0.6188 + M \times [1 - x(4)]$
13	14	$0.2030 + M \times [1 - x(5)]$	$0.1034 + M \times [1 - x(5)]$
13	18	$0.8191 + M \times [1 - x(3)]$	$0.7070 + M \times [1 - x(3)]$
14	15	$0.2842 + M \times [1 - x(6)]$	$0.1447 + M \times [1 - x(6)]$
15	16	$1.0590 + M \times [1 - x(7)]$	$0.9377 + M \times [1 - x(7)]$
16	17	$0.8042 + M \times [1 - x(8)]$	$0.7006 + M \times [1 - x(8)]$
18	19	$0.3811 + M \times [1 - x(2)]$	$0.1941 + M \times [1 - x(2)]]$

- Assimilation coefficient: 2
- Assimilation angle coefficient: 0.5
- Uniting threshold or per cent of search space size, which enables the uniting process of two empires: 0.02

In each generated solution, the primary line and bus information matrices can be modified using the following NEMs:

NEM_A = [1
1
1
1
1
1
1
1
1
$x(1)$
$x(1) \times x(2)$
$x(1) \times x(2) \times x(3)$
$x(1) \times x(2) \times x(3) \times x(5)$
$x(1) \times x(2) \times x(3) \times x(5) \times x(6)$
$x(1) \times x(2) \times x(3) \times x(5) \times x(6) \times x(7)$
$x(1) \times x(2) \times x(3) \times x(5) \times x(6) \times x(7) \times x(8)$
$x(1) \times x(2) \times x(3) \times x(4)$
$x(1) \times x(2) \times x(3) \times x(4) \times x(10)$
$x(1) \times x(2) \times x(3) \times x(4) \times x(10) \times x(11)$
$x(1) \times x(2) \times x(3) \times x(4) \times x(10) \times x(11) \times x(12)$]

NEM_B = [1
1
1
1
$x(9)$
$x(8) \times x(9)$
$x(7) \times x(8) \times x(9)$
$x(6) \times x(7) \times x(8) \times x(9)$
$x(5) \times x(6) \times x(7) \times x(8) \times x(9)$
$x(4) \times x(5) \times x(6) \times x(7) \times x(8) \times x(9)$
$x(4) \times x(5) \times x(6) \times x(7) \times x(8) \times x(9) \times x(10)$
$x(4) \times x(5) \times x(6) \times x(7) \times x(8) \times x(9) \times x(10) \times x(11)$
$x(4) \times x(5) \times x(6) \times x(7) \times x(8) \times x(9) \times x(10) \times x(11) \times x(12)$
$x(3) \times x(5) \times x(6) \times x(7) \times x(8) \times x(9)$
$x(2) \times x(3) \times x(5) \times x(6) \times x(7) \times x(8) \times x(9)$]

$$\text{NEM\_C} = [1 \\ 1 \\ 1 \\ 1 \\ 1 \\ 1 \\ 1 \\ 1 \\ x(13) \\ x(12) \times x(13) \\ x(11) \times x(12) \times x(13) \\ x(10) \times x(11) \times x(12) \times x(13) \\ x(4) \times x(10) \times x(11) \times x(12) \times x(13) \\ x(5) \times x(4) \times x(10) \times x(11) \times x(12) \times x(13) \\ x(6) \times x(5) \times x(4) \times x(10) \times x(11) \times x(12) \times x(13) \\ x(7) \times x(6) \times x(5) \times x(4) \times x(10) \times x(11) \times x(12) \times x(13) \\ x(8) \times x(7) \times x(6) \times x(5) \times x(4) \times x(10) \times x(11) \times x(12) \times x(13) \\ x(3) \times x(4) \times x(10) \times x(11) \times x(12) \times x(13) \\ x(2) \times x(3) \times x(4) \times x(10) \times x(11) \times x(12) \times x(13)$$

As depicted in Figure 8.7, the 33-bus radial distribution grid can optimally be partitioned into three desired stable islands in case of main upstream network loss. The simulation results demonstrate that obtained optimal scenario for separating a radial distribution network will be influenced by the entrance of distributed generations. As well as, the maximum expansion area of the island B is increased in case 2 in comparison with case 1 because the capacity of DG2 is more than DG1 and DG3.

According to Figure 8.8, the voltage profile of islands B and C is improved in the presence of distributed resources.

The minimum real power losses of three best feasible stable islands with and without DGs are reported in Table 8.8. The results show that total real power losses in obtained optimal islanding solutions will be reduced especially in the presence of dispersed generators.

To compare the proposed strategy with other schemes, fast intentional defensive islanding methods such as ordered binary decision diagram (OBDD) [28,29], ant search mechanism (ASM) [30] and spanning tree based Breadth First Search (ST-BFS) [31] algorithms are applied to the test system. The calculation times are compared in Table 8.9, and it is proved that the proposed algorithm is computationally efficient and fast.

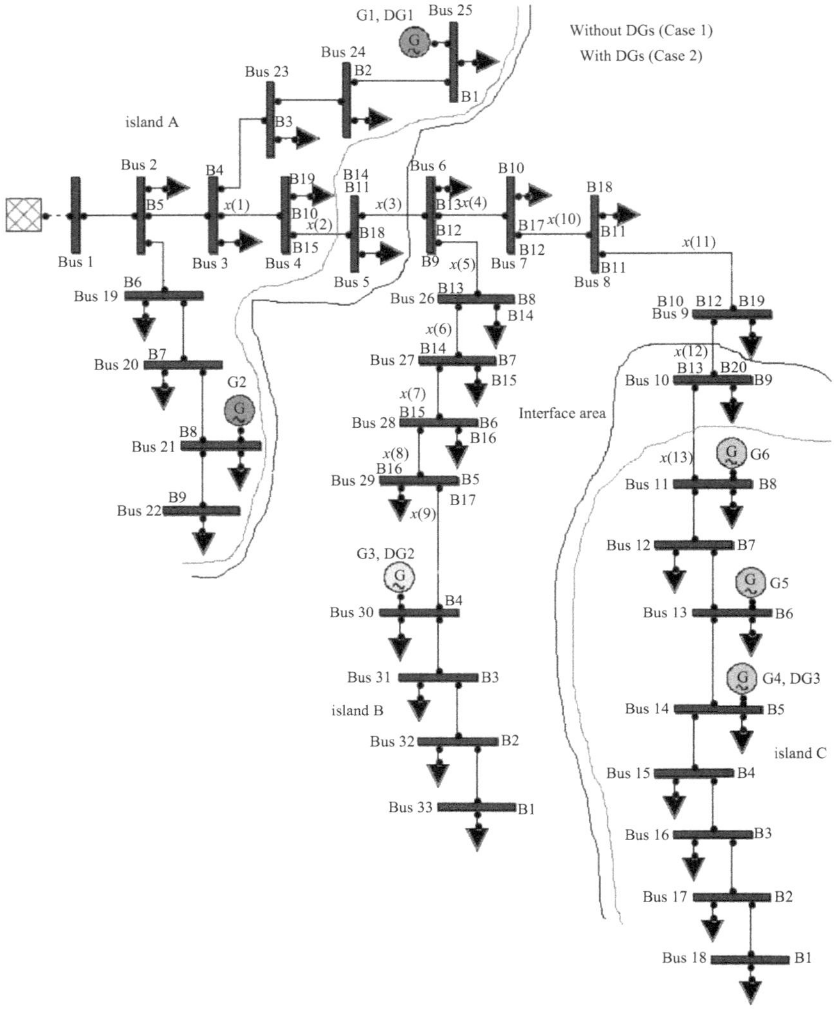

Figure 8.7 Optimally islanded 33-bus radial distribution system following the upstream transformer trip

8.4 Conclusion

In this chapter, a novel backward–forward, sweep-based two-stage strategy was introduced for intentional active islanding of radial distribution systems to avoid any multiple failures happen in the presence of distributed generations. The BICA was applied to minimize total real power losses and nodal voltage deviations considering integrity, voltage limits and feeder capacity criteria. The simulation results demonstrated that the proposed algorithm can successfully split a radial distribution system into several stable islands by tripping obtained optimum

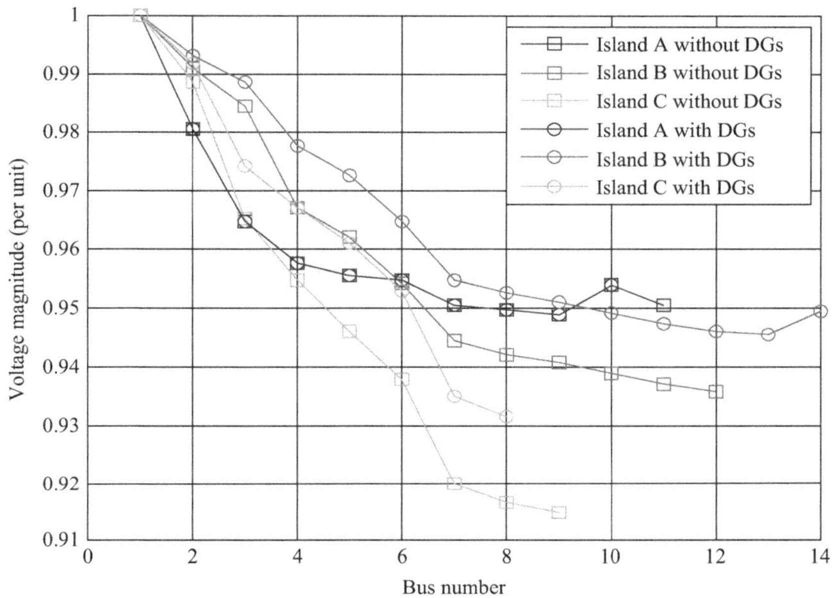

Figure 8.8 The voltage profile of obtained optimal islands with and without DGs

Table 8.8 Minimum active power losses of feasible stable islands in the best scenario with and without DGs

Areas	**Active power losses (kW)**	
	Without DGs	**With DGs**
Island A	99.5674	102.8228
Island B	75.2837	52.2302
Island C	133.5032	101.2438
Total	308.3544	256.2968

Table 8.9 Comparison of the proposed method with OBDD and ASM methods

Calculation time	**Proposed strategy (on a Lenovo with 2.10 GHz CPU, 4 GB RAM)**	**OBDD (on a PC Pentium IV-1.4G CPU and 256M DDRAM)**	**ASM (PC with 2.26 GHz CPU, 2 GB RAM, 3 MB Cache)**	**ST-BFS**
To generate each solution	Less than 0.1 s	Not reported	1 s	Less than 0.4 s
To find the best scenario	3.7715 s	11.7015 s	Not reported	Not reported

separating points. To the best of our knowledge, the contributions of the current chapter can be summarized as follows:

- A novel two-stage islanding scenario generation algorithm has been proposed.
- The backward–forward sweep load flow algorithm has been applied to assess the steady-state stability criteria for each generated solution.
- The impact of DGs on splitting problem has comprehensively been studied. It is proved that the proposed BFS-based optimization process can efficiently reduce total active power losses and improve voltage profiles in the presence of DGs.
- The effectiveness and speed of the presented algorithm has been proved by a comparison with other fast strategies.

Appendix

The line and bus information matrices for 33-bus radial distribution system [26] have been reported in Tables 8.10 and 8.11, respectively.

Table 8.10 Line information for 33-bus radial distribution system [26]

Sending bus	Receiving bus	R (Ω)	X (Ω)
1	2	0.0922	0.0470
2	3	0.4930	0.2511
3	4	0.3660	0.1864
4	5	0.3811	0.1941
5	6	0.8191	0.7070
6	7	0.1872	0.6188
7	8	0.7114	0.2351
8	9	1.0300	0.7400
9	10	1.0440	0.7400
10	11	0.1966	0.0650
11	12	0.3744	0.1238
12	13	1.4680	1.1550
13	14	0.5416	0.7129
14	15	0.5910	0.5260
15	16	0.7463	0.5450
16	17	1.2890	1.7210
17	18	0.7320	0.5740
2	19	0.1640	0.1565
19	20	1.5042	1.3554
20	21	0.4095	0.4784
21	22	0.7089	0.9373
3	23	0.4512	0.3083
23	24	0.8980	0.7091
24	25	0.8960	0.7011
6	26	0.2030	0.1034
26	27	0.2842	0.1447
27	28	1.0590	0.9377
28	29	0.8042	0.7006
29	30	0.5075	0.2585
30	31	0.9744	0.9630
31	32	0.3105	0.3619
32	33	0.3410	0.5302

Table 8.11 Bus information for 33-bus radial distribution system [26]

Bus number	*P* (kW)	*Q* (kW)
1	0	0
2	120	72
3	108	48
4	144	96
5	72	36
6	72	24
7	240	120
8	240	120
9	72	24
10	72	24
11	54	36
12	72	42
13	72	42
14	144	96
15	72	12
16	72	24
17	72	24
18	108	48
19	108	48
20	108	48
21	108	48
22	108	48
23	108	60
24	504	240
25	504	240
26	72	30
27	72	30
28	72	24
29	144	84
30	240	720
31	180	84
32	252	120

References

[1] Brahma, S.M. and A.A. Girgis, *Development of adaptive protection scheme for distribution systems with high penetration of distributed generation.* IEEE Transactions on Power Delivery, 2004. **19**(1): p. 56–63.

[2] Zhang, S., D. Wang, X. Xu, T. Zhao, *Optimal Microgrid Partition Strategy of Distribution Generation Based on Advanced GA.* In *Intelligent System Design and Engineering Application (ISDEA), 2010 International Conference on*. 2010.

[3] Mahat, P., Z. Chen, and B. Bak-Jensen, *Control and operation of distributed generation in distribution systems.* Electric Power Systems Research, 2011. **81**(2): p. 495–502.

[4] El-Zonkoly, A., M. Saad, and R. Khalil, *New algorithm based on CLPSO for controlled islanding of distribution systems*. International Journal of Electrical Power & Energy Systems, 2013. **45**(1): p. 391–403.
[5] Liang, J.J., A.K. Qin, P.N. Suganthan, and S. Baskar, *Comprehensive learning particle swarm optimizer for global optimization of multimodal functions*. IEEE Transactions on Evolutionary Computation, 2006. **10**(3): p. 281–295.
[6] de Quevedo, P.M., J. Allahdadian, J. Contreras, and G. Chicco, *Islanding in distribution systems considering wind power and storage*. Sustainable Energy, Grids and Networks, 2016. **5**: p. 156–166.
[7] Caldon, R., A. Stocco, and R. Turri, *Feasibility of adaptive intentional islanding operation of electric utility systems with distributed generation*. Electric Power Systems Research, 2008. **78**(12): p. 2017–2023.
[8] Lopes, J.A.P., C.L. Moreira, and A.G. Madureira, *Defining control strategies for MicroGrids islanded operation*. IEEE Transactions on Power Systems, 2006. **21**(2): p. 916–924.
[9] Brown, H.E., S. Suryanarayanan, S.A. Natarajan, and S. Rajopadhye, *Improving reliability of islanded distribution systems with distributed renewable energy resources*. IEEE Transactions on Smart Grid, 2012. **3**(4): p. 2028–2038.
[10] Kim, J.Y., J.H. Jeon, and S.K. Kim, *et al.*, *Cooperative control strategy of energy storage system and microsources for stabilizing the microgrid during islanded operation*. IEEE Transactions on Power Electronics, 2010. **25**(12): p. 3037–3048.
[11] Hemmatpour, M.H., M. Mohammadian, and A.A. Gharaveisi, *Optimum islanded microgrid reconfiguration based on maximization of system loadability and minimization of power losses*. International Journal of Electrical Power & Energy Systems, 2016. **78**: p. 343–355.
[12] Jikeng, L., W. Xudong, and W. Peng, *et al.*, *Two-stage method for optimal island partition of distribution system with distributed generations*. IET Generation, Transmission & Distribution, 2012. **6**(3): p. 218–225.
[13] Chen, Y., Z. Xu, and J. Østergaard, *Islanding Control Architecture in future smart grid with both demand and wind turbine control*. Electric Power Systems Research, 2013. **95**: p. 214–224.
[14] Mohamad, H., H. Mokhlis, A.H.A. Bakar, and H.W. Ping, *A review on islanding operation and control for distribution network connected with small hydro power plant*. Renewable and Sustainable Energy Reviews, 2011. **15**(8): p. 3952–3962.
[15] Allahdadian, J., A. Berizzi, and C. Bovo, *Detection of islanding feasibility in subtransmission systems*. International Review of Electrical Engineering (IREE), 2014. **8**(3): p. 1108–1118.
[16] Jabari, F., H. Seyedi, S.N. Ravadanegh, and B.M. Ivatloo, *Stochastic contingency analysis based on voltage stability assessment in islanded power system considering load uncertainty using MCS and k-PEM*. Handbook of Research on Emerging Technologies for Electrical Power Planning, Analysis, and Optimization, IGI, 2016: p. 12.

[17] López, J.C., M. Lavorato, and M.J. Rider, *Optimal reconfiguration of electrical distribution systems considering reliability indices improvement*. International Journal of Electrical Power & Energy Systems, 2016. **78**: p. 837–845.

[18] de Oliveira, L.W., F.d.S. Seta, and E.J. de Oliveira, *Optimal reconfiguration of distribution systems with representation of uncertainties through interval analysis*. International Journal of Electrical Power & Energy Systems, 2016. **83**: p. 382–391.

[19] Fathabadi, H., *Power distribution network reconfiguration for power loss minimization using novel dynamic fuzzy c-means (dFCM) clustering based ANN approach*. International Journal of Electrical Power & Energy Systems, 2016. **78**: p. 96–107.

[20] Esmaeili, M., M. Sedighizadeh, and M. Esmaili, *Multi-objective optimal reconfiguration and DG (Distributed Generation) power allocation in distribution networks using Big Bang-Big Crunch algorithm considering load uncertainty*. Energy, 2016. **103**: p. 86–99.

[21] Guan, W., Y. Tan, H. Zhang, and J. Song, *Distribution system feeder reconfiguration considering different model of DG sources*. International Journal of Electrical Power & Energy Systems, 2015. **68**: p. 210–221.

[22] Oboudi, M.H., R. Hooshmand, and A. Karamad, *Feasible method for making controlled intentional islanding of microgrids based on the modified shuffled frog leap algorithm*. International Journal of Electrical Power & Energy Systems, 2016. **78**: p. 745–754.

[23] Mehdinejad, M., B. Mohammadi-Ivatloo, R. Dadashzadeh-Bonab, and K. Zare, *Solution of optimal reactive power dispatch of power systems using hybrid particle swarm optimization and imperialist competitive algorithms*. International Journal of Electrical Power & Energy Systems, 2016. **83**: p. 104–116.

[24] Mohammadi-Ivatloo, B., A. Rabiee, A. Soroudi, and M. Ehsan, *Imperialist competitive algorithm for solving non-convex dynamic economic power dispatch*. Energy, 2012. **44**(1): p. 228–240.

[25] Mahari, A. and H. Seyedi, *Optimal PMU placement for power system observability using BICA, considering measurement redundancy*. Electric Power Systems Research, 2013. **103**: p. 78–85.

[26] Boor, Z. and S.M. Hosseini, *Optimal placement of DG to improve the reliability of distribution systems considering time varying loads using genetic algorithm*. Majlesi Journal of Electrical Engineering, 2012. **7**(1): p. 16–24.

[27] Nojavan, S., M. Jalali, and K. Zare, *An MINLP approach for optimal DG unit's allocation in radial/mesh distribution systems take into account voltage stability index*. Iranian Journal of Science and Technology Transactions of Electrical Engineering, 2015. **39**(E2): p. 155–165.

[28] Kai, S., Z. Da-Zhong, and L. Qiang, *A simulation study of OBDD-based proper splitting strategies for power systems under consideration of transient stability*. IEEE Transactions on Power Systems, 2005. **20**(1): p. 389–399.

[29] Sun, K., D.-Z. Zheng, and Q. Lu, *Splitting strategies for islanding operation of large-scale power systems using OBDD-based methods*. IEEE Transactions on Power Systems, 2003. **18**(2): p. 912–923.
[30] Aghamohammadi, M.R. and A. Shahmohammadi, *Intentional islanding using a new algorithm based on ant search mechanism*. International Journal of Electrical Power & Energy Systems, 2012. **35**(1): p. 138–147.
[31] Najafi, S., S. Hosseinian, and M. Abedi, *A novel strategy for stability evaluation of islanded power systems*. Iranian Journal of Electrical and Computer Engineering, 2010. **9**(1): p. 43–51.

Chapter 9
Decentralised voltage control in smart grids

Amin Mohammadpour Shotorbani[1],
Behnam Mohammadi-ivatloo[1], *Liwei Wang*[2]
and Saeid Ghassem Zadeh[1]

9.1 Introduction

The conventional electric power grid features on a vertical hierarchical up-down design approach consisting of the generation, transmission and the distribution systems. The new forms of markets with diverse consumers, integration and high penetration of renewable energy resources and the distributed generation call for evolutions of the conventional power grid. The next-generation power grid, namely smart grid, is meant to overcome the existing challenges and to improve the quality of the energy service and generation. Enhanced quality, improved controllability, higher reliability and security, demand-side management (DSM), smart appliances and technologies are some motives for the smart grid [1].

Disassembling and reorganising the conventional grid hierarchy, communication, real-time advanced metering infrastructure, grid optimisation and computational intelligence are the main aspects of the smart grid development to achieve the targeted environmental and the economic benefits.

Smart grid takes the advantages of automation and coordination within the distribution system, energy efficiency, integration of the renewable energy resources, active participation of the consumers in power grid operation, distributed generation and storage, smart appliances and communication and information technologies. The smart grid concept supports and necessitates automated operations in power grid such as automatic reactive power control, fault diagnosis, optimal network switching, loss minimisation and the coordinated control of distributed generation using DSM.

Smart grid exploits information technology, communication, widely dispersed sensors and collection of measurements for advanced power quality monitoring. These features enhance the accuracy and response speed of the gird control and protection systems leading to advanced self-healing grid. Mismatched participation of active power requirements between distributed energy resources (DERs) and disorganised

[1]Faculty of Electrical and Computer Engineering, University of Tabriz, 29 Bahman Blvd., Tabriz, Iran
[2]School of Engineering, University of British Columbia, Okanagan, BC, Canada

operation of the microgrids in islanded mode degrade the quality of grid service. Employment of smart grid technologies can enhance the quality of the supplied power by synchronising the islanded microgrids and improve the reliability by coordinating the operation between the DERs, storage devices, loads and the utility.

9.1.1 Voltage profile as power quality index

Power quality is the capability of the grid to compensate or reject the disturbances. Loss minimisation, voltage profile improvement and extending the lifetime of the expensive grid equipment fall into the constraints of power quality [2] and grid reliability [3]. Power quality concerns the qualification of the service provided by power utility and mainly emphasises on the features of the voltage quality. Power-quality disturbances consist of voltage sags, voltage surge (overvoltage), short-term power interruption, brownout, blackout, voltage transients (spike and impulse), frequency variation, harmonic distortion, noise and voltage unbalance. These disturbances are primarily characterised by service permanency, voltage transients, harmonics and voltage magnitude variations [1–3].

Failure of the power grid to meet the reactive power requirements is the main cause of voltage instability [4]. Flow of the power through the system changes the voltage magnitude and phase angle of the buses. Dynamic generation planning of the DERs [4] and optimisation techniques [5,6] support the management of the reactive power flow to improve the voltage profile and minimise the transmission/distribution loss. Furthermore, stochastic dynamic optimal power flow can handle the problem of real-time updating of the operational constraints [1].

The transmission lines in the distribution grid have lower X/R ratios than those of the transmission grids which cause voltage magnitude variation to be influenced by flow of active power in addition to reactive power [3]. The distribution network operators (DNOs) maintain the voltage level of the distribution nodes within the standard range (i.e. typically within the tolerance of ±5 to $\pm10\%$ of the nominal voltage) for secure customer service. The DNOs control the voltage magnitude in low- and medium-voltage distribution network. The conventional distribution network assumes up-down power flow for the design of control and protection systems. Distribution voltage regulation involves a 'regulation point'. The voltage-regulation equipment regulates the voltage at this point to a pre-set value. The nodes between the substation and the mentioned regulation point experience lower voltage by decreasing the load and vice versa. Still the nodes after the regulation point experience higher voltage by decreasing the load and vice versa.

In the conventional distribution network, the voltage of a node is adjusted by installing under load tap changing (ULTC) transformers (also named as on-load tap changer), dynamic voltage regulators, static VAR compensator or switched capacitor banks [4]. The secondary voltage of the transformer at the distribution substation is automatically regulated by means of a ULTC system to regulate the voltage of a pre-specified node within the standard range. The reference secondary voltage of the ULTC is set by the DNO. The ULTC mechanically varies the turn ratio of the transformer in discrete steps and regulates the secondary side voltage to minimise its difference with the reference value. The tap changer has a slow inner control loop to guarantee the stability of the system.

Moreover, for fulfilment of the loads' reactive power requirements locally and for power factor correction, pre-installed capacitor banks are switched on/off. The number of switched capacitors is proportionally related to the drop in the voltage magnitude. All capacitors of a capacitor bank are switched on when the voltage drops below the standard limit (i.e. below 0.90–0.95 pu) and are switched off if the voltage is at its nominal value (i.e. 1.0 pu). New technologies using power-electronic devices such as thyristor-based ULTC transformer, distribution static synchronous compensator (D-STATCOM) and unified power quality conditioner provide fast response for voltage regulation although imposing higher operational power losses [7].

Figure 9.1 shows the per-phase representation of a distribution feeder including ULTC transformer, DERs, loads and extended feeder branches. The nodes are numbered as $N_0, N_1, \ldots, N_N$ in which the node N_0 is the substation. The load and the DER connected node i is represented by the subscripts L_i and G_i, respectively. The active and reactive powers at each node are aggregated to P and Q, respectively. The impedance of the sections of the feeder between two nodes is shown by $Z = R + jX$, in which the shunt conductance and capacitance of the feeder are ignored. The resistance and reactance of the distribution cable are presented by R and X, respectively.

The power-flow direction may alter through the intermittency of the power injection by DERs. Therefore, the presence of the DERs in the low/medium voltage distribution grid invalidates the conventional assumption of uniformly decreasing voltage magnitude along the feeder, as illustrated in Figure 9.2. The nodes behind

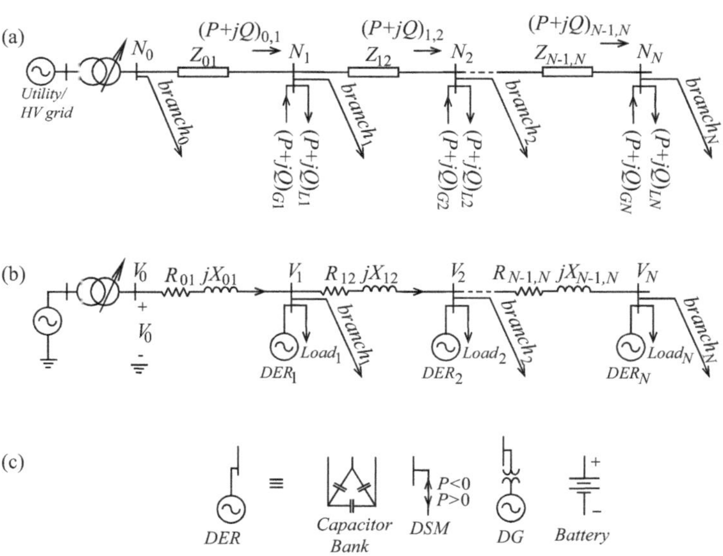

Figure 9.1 Distribution feeder as a microgrid; (a) configuration (b) single-phase representation (c) prototypes of the DERs

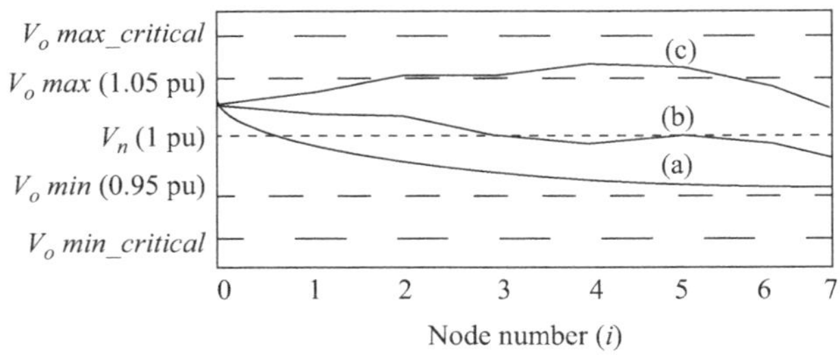

Figure 9.2 Adverse and supportive effects of DERs on the voltage profile of a distribution feeder; (a) without DER, (b) supportive effect of coordinated DERs and (c) adverse effect of uncontrolled DERs

the node with a DER will experience a decreasing voltage profile, whereas the nodes between the substation and the node with a DER will experience an increase in the voltage magnitude by receding from the substation. The unexpected change in the voltage profile leads to inappropriate operation of the voltage-regulating devices (RDs). The interference between the devices (i.e. the DER and the voltage-regulators) may lead to the violation of the voltage standard limits due to unforeseen voltage profile.

For example, if the power injection of a DER becomes more than the power consumed by the local load, the current will flow towards the substation. The upward current changes the voltage profile from a monotonically decreasing form to an alternating profile in different sections of the feeder. In a smart grid with a communication infrastructure, monitoring devices and cooperative operation of the controllable DERs with the conventional voltage RDs will resolve the problem [8]. Furthermore, optimal operation of the total system aiming at minimising various cost functions (e.g. loss minimisation in the feeder) will be attainable. Figure 9.2 illustrates the adverse effect of uncontrolled DERs and supportive influence of coordinated connection of DERs.

9.1.2 Microgrids

High penetration of the DERs is the main developing foundation of microgrids. Commonly, microgrids are combination of sets of DERs, distributed generation and storage units, loads, communication systems and the utilities [9]. Microgrids are capable of operating in the grid-connected and islanded mode. In a grid-connected operating mode, the grid frequency and the substation voltage are pre-set by the utility. In islanded mode of operation, appropriate control loops are necessary for regulation of the voltage and frequency of the system to be inside the normal operation bounds in order to preserve voltage/frequency stability.

A microgrid entails a power administration scheme (PAS) for safe and secure operation in both modes [10]. PAS could be a conventional regional transmission organisation, also named as distribution system operator, as well as a local independent management/supervision system. Comparing the microgrids with the conventional bulk power system, decisions of the PAS has significant influence on the stability and performance of the microgrids due to the following:

- Penetration of various distributed generation (DG) units with diverse generation profiles
- Absence of a slack bus
- Low inertia of the microgrids due to presence of power electronically interfaced DERs (PEI-DERs)

The PAS determines the references for the controllable RDs and the dispatchable power generation. The PEI-DERs are commanded by the PAS, whereas directly connected controllable DERs follow the results of conventional hierarchical optimal power flow method. Reference power assignment by the PAS is intended for

- load sharing between DG units as well as the utility grid
- power reference updating due to the changes in loads, demands and disturbances
- steady-state power reference update for frequency restoration
- Resynchronisation of the islanded microgrid with the main utility for the purpose of reconnection to the main grid at the point of common coupling (PCC).

The voltage stability, power quality and the quality of service are directly affected by the PAS. Analogous to the conventional power system, the control architecture of each grid-connected PEI-DER is classified as (I) the PCC voltage and active power control scheme, and (II) the injected PCC reactive and active power control scheme. The frequency of the grid is regulated by the utility in grid-connected operating mode. Similar to a slack bus, the utility compensates the difference between the demand and the DGs' provided power, in order to preserve the voltage/frequency stability. In a hierarchical control scheme, as well as the microgrid's PAS, the reference values for each DG are determined by the upper control level [11]. On the contrary, in order to preserve the voltage/frequency stability in islanded operating mode, the system frequency is regulated by the DERs. In islanded mode, the load-shedding is inevitable when the power demand and the total power loss exceed the available power of the DERs. The PAS is also responsible for

- Appropriate seamless transition between islanded and grid-connected mode of operation
- Sharing the load in regarding the available power and capacity of each DER
- Considering the environmental and operational constraints of the DERs such as type, fuel cost, dynamics and time constants of the DER.

Fundamentally, the grid interfacing systems are classified into three categories: (1) systems with no power electronics; (2) partially rated power electronics (PRPE) systems; and (3) full-scale power electronics (FSPE) systems. For instance, a conventional synchronous generator in a wind energy conversion system (WECS) may be connected directly to the power grid. A doubly fed induction generator may use a PRPE interface, whereas a PMSG may be connected to the grid through a FSPE interface in a WECS. The FSPE interface systems provide full control on output voltage and frequency that are compatible with the utility system. Although FSPE system has more losses in the power electronic devices compared to other interfacing topologies, this topology enables higher degree of freedom in different modes of operation including fast controls of the active/reactive power flows, the voltage and the frequency. The PEI-DERs have fast response compared to the conventional directly connected DG units such as diesel generators [12].

Decoupled power control in the stationary reference frame (i.e. $\alpha\beta$ coordinate) and the rotating reference frame (i.e. dq coordinate) are the foremost scheme used for the PEI control systems. Despite the decoupled power control architecture, the main shortcoming of the direct power control scheme is the absence of limits on the output current at the a.c. side of the PEI system. Decoupled control scheme has a cascaded structure consisting of an outer and an inner control loop which are designed in the $\alpha\beta$ stationary or dq rotating reference frame. The inner current regulation loop can apply the necessary constraints (i.e. maximum current for safe operation) to the interface system. In the decoupled power flow control, the output real and reactive powers are regulated using the d- and q-axis currents of the PEI at the PCC, respectively. The local inner dq control loop of the PEI-DER uses a phase-locked loop to synchronise the output voltage of the PEI with the voltage at the PCC and to calculate the voltage and the current in dq-reference frame [13]. The dq reference currents are determined by the outer control loop. The inner loop is designed to be approximately five times faster than the outer loop.

9.1.3 Motivation of cooperative decentralised control in smart grid

Conventionally, the drop of the feeder voltage profile is determined mainly by the load profile and the pre-knowledge of the feeder impedance. The presence of DERs changes the power-flow equations known to the DNO. Thus, the conventional model-free calculation of the voltage drop becomes ineffective [14]. In the smart grid integrated with DERS, the voltage profile depends on the power injection, the location, the interface type of the DERs as well as other conventional voltage RDs [15].

Changing the taps in the conventional mechanically actuated ULTCs (voltage change of 5/16% for each tap) results in a large delay and a dead-band. Generally, the time delay of the first tap change is larger than 10 s, and the time delay for the next consequent changes are 5 to 10 s. Therefore, the slow response of the ULTC represents further challenge for the sudden interruption of DERs. Disconnection of the DERs due to relaying of under/over-voltage protection system in voltage sags/swells deteriorate the voltage stability because the other protection devices of the DERs may experience further under/over-voltage and the slowly actuated ULTC fail to increase/decrease the voltage magnitude in time.

High penetration of DERs overburdens the voltage regulations using substation's ULTC transformer and voltage RDs. Moreover, the controllable and dispatchable DERs (i.e. PEI-DERs) can contribute to the voltage regulation in accordance with the DNO. Although the DERs provide fast response compared to the ULTC transformer, accurate coordination is required to avoid interference between the conventional pre-installed voltage RDs and the DERs. Otherwise, the use of the DERs for voltage regulation may deteriorate voltage stability of the grid instead of improving the quality of the service.

The conventional model-based control and protection systems require the online updating of the system's dynamic model for large changes and variations (e.g. connection and disconnection of a DER). Progressively, high penetration of DERs in smart grid environment requires plug-and-play integration of the DERs to minimise the need for updating the dynamic model, thus leading to better robustness and higher reliability of the system [16].

Concerning the coordination of the voltage regulators and DERs, a communication infrastructure connecting the PASs of the DERs with the DNO is advantageous although the security, data delay and the connectivity of the communication system impose further challenges to the stability of the power grid.

Under the assumption of secure and fast communication between DERs and DNO in the distribution grid, the problem of voltage regulation can be formulated in similar approach to the optimal power flow and the economic dispatch problems in the transmission level. Furthermore, the communication infrastructure facilitates the power dispatch of the DERs.

9.2 Decentralised and distributed control systems

Centralised, N-channel and interconnected configurations are the fundamental structures of the control systems of a smart grid as shown in Figure 9.3. In a centralised configuration, all the measurements and the control signals are analysed and calculated in a decision centre, respectively. In an interconnected structure, a subsystem interacts with other subsystems of the entire system. In an N-channel structure, the control and the output measurements are decomposed, whereas the system is still an integrated unit [17]. Figure 9.3 illustrates the schematic structures of the systems. In Figure 9.3, *Sys.*, u and y are the system, the control signals and the output measured signals. The reference control signals are defined by the controller or a decision-making block.

The difficulties of the control design which are induced by the complexity of the interconnected power grid are principally attributed to the large dimension of the power grid, the information constraints imposed by the communication infrastructure, uncertainties and the delay in the control and measured data transfers. Decentralisation of the control scheme releases the constraints and resolves the

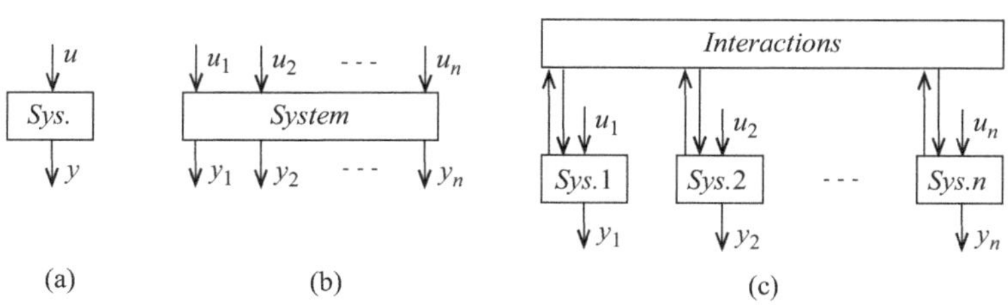

Figure 9.3 Fundamental structures of the systems; (a) centralised, (b) N-channel and (c) interconnected

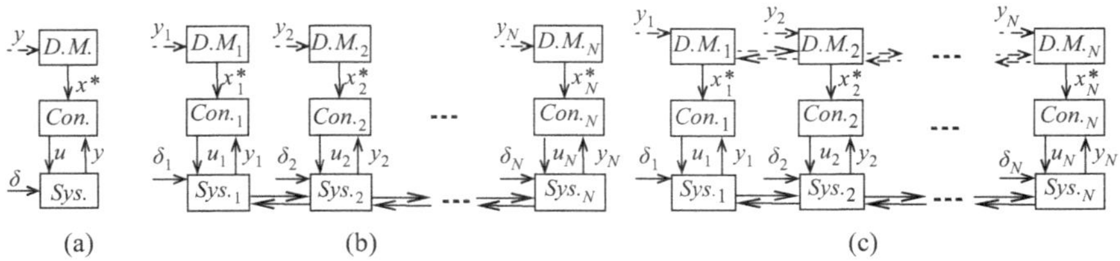

Figure 9.4 Fundamental structures of feedback control; (a) centralised, (b) decentralised and (c) distributed (dashed lines shows the information flow)

difficulties. Decentralisation, decomposition and model reduction accompanied by robust control design are the general methodologies for implementation of the controller [18].

The feedback control structures are classified as centralised, decentralised and distributed control architecture as demonstrated in Figure 9.4. The blocks *D.M.*, *Con.*, and the signals δ and $\mathbf{x}^*$ are the decision-making system block, the controller, the external disturbance signal and the reference value. The interactions between the subsystems and the intercommunications (i.e. the information/data flow between the controllers and the decision-makers) are shown using solid and dashed lines, respectively.

Decentralised control design establishes the local control centres independent of other control stations and is utilised for the systems with (1) strong coupling between the subsystems, in which the models of the subsystems are required for the control design process, and (2) weak coupling between the subsystems where considering the links among the subsystems has insignificant impact on the control design practice.

Decomposition of the design task reduces the computational effort and complication. Decomposition methods concern specific structures of the systems which are classified as (1) disjoint, (2) overlapped, (3) symmetric subsystems where the subsystems are connected with strong couplings, (4) hierarchical and (5) multiple-time-scale systems in which the coupling between the subsystems is weak.

For disjoint subsystems, the design task is accomplished through tearing the whole system into interconnected subsystems, and then reducing the strength of the interconnections (i.e. the couplings). In a system with overlapped subsystems, a number of states and outputs are shared among the subsystems; the design task is accomplished through expanding the subsystems into separate systems with larger dimensions compared to the original subsystems. The expanded subsystems are joined back after design of the independent controllers using the inclusion principle. Provided that systems are identical, and the interconnections are symmetric; the design task is simplified to a robust control scheme for a centralised system.

Decentralised networked control architecture assumes a communicative feedback control and communication between different subsystems and controllers. The network can be positioned in various locations of the control scheme. Figure 9.5

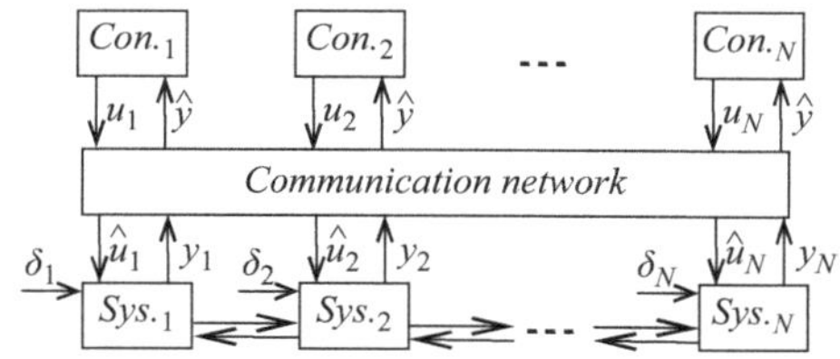

Figure 9.5 Decentralised networked control architecture

illustrates an example of a networked control prototype. In this architecture, u and y are the actual real-time control and measurement signals whereas $\hat{u}$ and $\hat{y}$ show the control and measurement signals which are affected by the characteristic of the communication network such as delays and the packet loss. The output vector $\hat{y} = f(\hat{y}_1, \hat{y}_2, \ldots, \hat{y}_N)$ is the selective combination of the measurements.

9.2.1 Contraction-based multi-agent systems

Multi-agent systems (MAS) promote the communication-based distributed control scheme using whether (1) multi-agent supervisory control [19], or (2) self-organising MAS [20].

The supervisory control requires supervision to guarantee the convergence of the optimisation solution. The necessity of the supervisory PAS arises from the complications similar to central control scheme [21]. However, self-organising MAS uses the advantage of cooperative agents and eliminates the need for supervisory action which makes it more reliable for plug-and-play practice. A plug-and-play approach also requires less information on the model of the system in comparison to the supervisory control.

Solving the voltage-regulation problem using a multi-agent-based distributed control architecture requires a communication infrastructure, coordination between the DNO and the PAS of the DERs, and the use of PEI-DERs as a supportive source for fast reactive power compensation. The communication channel is also used to send the dispatching commands to the power management control loop of the DERs as well as gathering the data from the grid.

In a MAS, the agents follow a set of tasks and goals that cannot be achieved individually. Distributed decision-making determines how a goal can be achieved by dividing the global problem into sub-problems that the agents can manage independently by sharing the problem and the knowledge.

An agent is able to propose, accept, reject, retract, disagree, counter-propose the actions and interact with other agents through a communication protocol. The communication could be active and/or passive. The agents take the information through assertions. The active agents distribute the queries and claim the assertions. The passive agents receive the queries and respond through replies. An agent can also be both active and passive (i.e. peer agents) [22].

The communication protocol identifies the type of the communication messages (i.e. an assertion or a query). The mechanism for ensuring that the activities

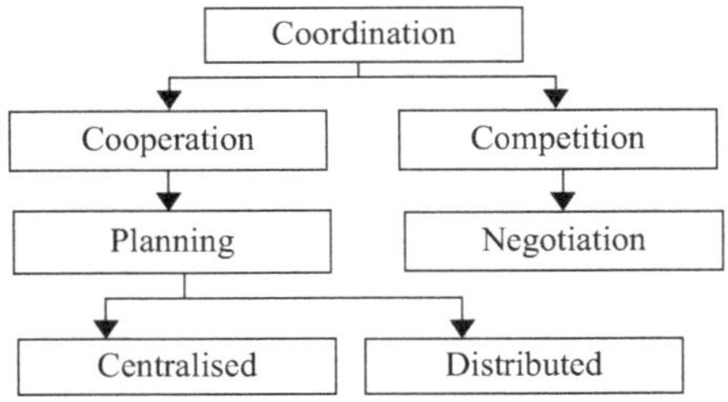

Figure 9.6 Communication coordination for MAS

of the agents hold the settled relationship (e.g. the sequence) is called *coordination*. As shown in Figure 9.6, the coordination is either cooperative (i.e. the agents cooperate entirely by planning an approach to the goals) or competitive (i.e. the agents negotiate with each other to bind the contracts). There are some motives to establish coordination between the agents as

- To attain the goals of their society
- Integrated global behaviour without a control centre
- Avoiding confliction and sharing the knowledge.

The range of the autonomy of an agent for making its next decision varies from zero autonomy (corresponding to total control) to total autonomy (corresponding to no control). The agents with some degree of autonomy (control) can cooperate with each other. Through cooperation, agents are able to complement their tasks with the assistance of other agents, share the resources and avoid detrimental interactions. There are three modes of cooperation: unintentional, unilateral and mutual. The degree of cooperation between the agents is classified as benevolent (i.e. full cooperation between the agents), antagonistic (i.e. no cooperation between the agents or blocking the goals of other agents) and partial cooperation (occasional and/or incomplete cooperation).

The fundamental strategy of the cooperation protocol is to decompose the procedure into tasks by the system designer, and then disperse the tasks to the agents. The mechanism of the distribution of the tasks are classified as (1) contract net using bids and awards, (2) market mechanism negotiating comprehensive agreements and mutual selection, (3) multi-agent planning employing task assignment for the agents and (4) organisational structure with fixed responsibility for each agent.

Definition: The cooperative compromise on common goals can be improved by reducing the contradiction and by compensating the uncertainty. *Negotiation* is defined as the process to achieve an agreement through the organised exchange of the related information. If an agent, who has a requirement, cannot manage it independently, the agent searches for the other agents with the necessary capabilities through negotiation. The negotiation consists of the following:

- Bargaining: undertaking a promise for an exchange
- Bidding: proposing a service at a particular definite price (cost)

- Contracting: undertaking the commitment of provision of a service at a particular definite price.

The process of the negotiation is summarised as

- Contact of an agent with other agents to demand/ propose a service/resource.
- Identification of the available supply and the required conditions by other agents.
- Acceptance of the conditions of supply by the *initiating agent(s)*.
- Commitment of the *supplying agent(s)* to provide the service/resource.

The interactions between the agents are governed by the set of rules defined as *negotiation protocol or strategy*. Negotiation protocols are classified on the basis of the number of the participated agents, interposition of a third party and the degree of the autonomy of the agents. The black board model by Reddy *et al.*, actor model by Hewitt, the contract net protocol by Davis and Smith and BEING by Lenat are some basic coordination paradigms which are proposed to organise the communication between the agents.

9.2.2 Contract net interaction protocol

Negotiation of a contract in contract net interaction protocol (CNIP) is the act of distributing the task(s) which consists of the following phases [23]:

- ***Recognition* of a problem by the agent**
  An agent recognises a problem which is beyond its capability to achieve the goal independently. This agent is named as the Initiator and will act resembling a manager. The Initiator may also intend to optimise a cost function related to the task.
- ***Announcement* of task by the Initiator to other agents (the participants)**
  The announcement is a call for proposal (CFP) by the Initiator and includes the required descriptions, specifications, constraints of the task, maximum time duration for the accomplishment of the task and conditions of implementation. The participants may refuse the task or grant a *proposal*. The Initiator also determines the deadline for the replies in the CFP. The late proposals after the deadline will be rejected automatically.
- ***Bidding* of the participants for the announced task**
  The recipient agents must consider their capability of completing the task and then decide to bid for the task or refuse the CFP. The participants which grant a proposal must determine the quality of the performance and inform the price (cost).
- ***Awarding* the bids**
  The Initiator (announcement-sender) agent selects the preferred proposals and communicate to fulfil the contract by sending an act of accept- or reject-proposal to the participants.

The representation of the process in CNIP is given in Figure 9.7.

After *awarding*, the participants are acquired for commitment to the accomplishment of the task. The result of the performance is sent to the Initiator by an

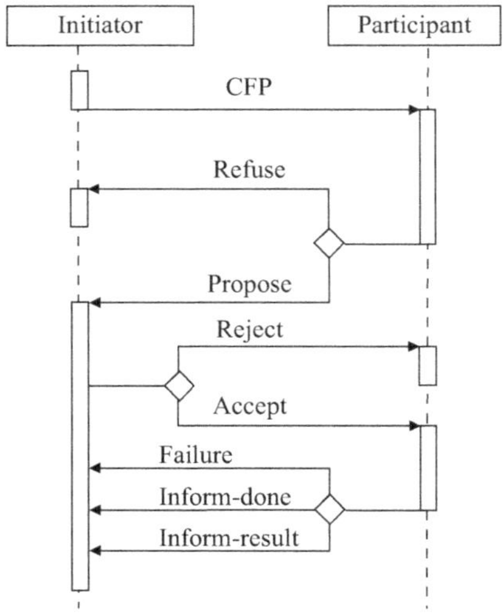

Figure 9.7 Contract Net Interaction Protocol (CNIP)

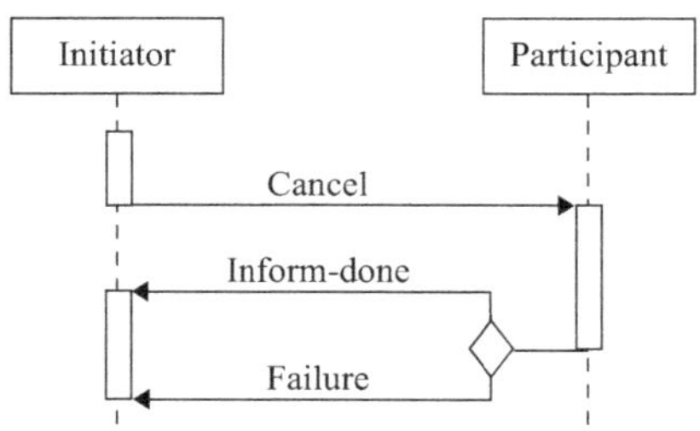

Figure 9.8 Cancel protocol

inform-result (i.e. done or failure). If an agent fails to perceive the communicated conversation, it may send a message of *not-understood*. The communication of this message may lead to the termination of the IP. Also, the Initiator can cancel the initiated IP by the cancel-meta-protocol as in Figure 9.8.

Furthermore, if it is required to further divide the current contract into subcontracts, an *expediting* stage is also established to generate more groups of agents (i.e. contract nets).

Table 9.1 summarises the advantages and shortcomings of the primary CNIP. Two supportive approaches to tackle the problem of the interactions are

1. Introduction of a mediator for coordination of the interactions by reducing the conflicts
2. Modification of the plans by negotiating the potential conflicts and sharing the plans, tasks and goals to other agents.

Table 9.1 Advantages and shortcomings of the primary CNIP

Advantages	Shortcomings
• Dynamic and self-organised tasks-allocation • Dynamic selection and elimination of the agents • Natural and self-favourite balance of the agents' burden	• Ignorance of the effects of opposing agents • Lack of the procedure to deal with incompatible agents • Absence of detection and solution for conflicts

9.3 Centralised hierarchical control of the DERs

In centralised control architecture, the power administrator as the higher control level determines the task (i.e. output references) of the DERs' controllers as the first control level of the hierarchy. Output references may include the active and reactive power injection of the DERs as well as the PCC voltage magnitude and frequency. It is assumed that the grid frequency is mainly subjected to the active power flow, whereas the voltage magnitudes at the nodes are mainly affected by the reactive power flow in the system.

9.3.1 Frequency regulation

In a grid-connected operation mode, the frequency of the microgrid is dominantly fixed through the active power support by the utility. Hence, the active power reference of the DERs in this case can be set for purposes other than the frequency stability such as maximum power point tracking (MPPT). However, in islanded operation mode, a frequency-regulation scheme (i.e. frequency-droop control) is required since the frequency of the microgrid swings spontaneously by total active power imbalances and fluctuations during the transients (e.g. balancing the demand and the generation) [13].

Accordingly, the reference output active power based on the droop control structure is defined as

$$P_{t_i}^* = P_{dr_i} + P_{re_i} \tag{9.1}$$

where $P_{t_i}^*$, P_{dr_i} and P_{re_i} are total reference power, the frequency-drooped active power and the steady-state frequency-restoration active power of the ith PEI-DER. The term P_{dr_i} reacts to local frequency deviation based on a droop characteristic, whereas P_{re_i} restores the steady-state frequency of the system in islanded mode of operation. Based on the droop control method, we have

$$\begin{cases} P_{dr_i} = -\dfrac{1}{D_{\omega\_i}}(\omega^* - \omega_i) + P_i^* \\ P_{re_i} = G_c(e_i) \end{cases} \tag{9.2}$$

where P_i^* is the primary reference power assigned by the upper control level (i.e. PAS); D_{ω_i} is the droop constant; ω^* is the reference frequency; ω_i is the measured frequency; $e_i = \omega^* - \omega_i$ the defined frequency error; and $G_c(\cdot)$ is the control function, respectively. Generally, the control function $G_c(\cdot)$ is a simple proportional–integral (PI) controller as (9.3), although advanced control schemes can also be employed for $G_c(\cdot)$ [24]

$$G_{c\_i}(e_i) = k_{P\_i}(e_i) + k_{I\_i}\int(e_i)dt \tag{9.3}$$

where k_{P_i} is the proportional, and k_{I_i} is the integral coefficient. For frequency restoration, a controller with a large time constant is preferred in order to have smooth frequency regulation without overshoot in the response. Time constant of the closed-loop system is tuned using appropriate values of k_{P_i} and k_{I_i} in a PI control architecture.

Furthermore, in order to have fair load sharing between the DERs, the droop constants are pre-set according to the rating of the power interface (i.e. S_i) as

$$D_iS_i = D_jS_j, \quad \forall\, i,j \tag{9.4}$$

The d-axis reference current is then calculated as

$$i^*_{d\_i} = Sat\Big(G_{cd\_i}\big(P^*_{t_i} - P_{\mathrm{PCC}\_i}\big)\Big) \tag{9.5}$$

where $i^*_{d\_i}$, $Sat(\cdot)$, $G_{cd\_i}(\cdot)$ and P_{PCC_i}, are d-axis reference current of the interface, the saturation function, the controller (i.e. a PI) and PCC active power. The saturation function (9.6) is used to limit the output current of the interface to ensure the safe operation.

$$Sat(x) = \begin{cases} x_{\max} & x > \epsilon \\ \dfrac{x}{\epsilon} & |x| \le \epsilon \\ x_{\min} & x < \epsilon \end{cases} \tag{9.6}$$

9.3.2 Voltage magnitude regulation

Reactive power control is the main tool for regulation of voltage in a microgrid [25]. The voltage/VAR control problem is generally formulated as the task of optimising a defined cost function, considering the operational constraints. Some of these cost functions are power loss in the feeder, load balancing, voltage magnitude deviation, the outage tariffs and the combinations of the aforementioned functions as a multiobjective optimisation. The main voltage control is categorised as predefined constant voltage control at PCC, V–Q droop control, power factor regulation control to accomplish optimised system's operation.

9.3.2.1 Predefined constant voltage control

In the predefined voltage control strategy, the reactive power injected at the PCC is varied to regulate the PCC voltage. For instance, using a simple PI controller (G_{cq_i}) in a dq reference frame, the q-axis current reference of the interface is selected as

$$i^*_{q\_i} = Sat(G_{cq\_i}(V^* - V_{rms\_i})) \tag{9.7}$$

where $i^*_{q\_i}$, V^* and V_{rms_i} are the q-axis reference current of the interface, the reference voltage of the PCC and the *rms* value of the measured voltage of the PCC.

9.3.2.2 V–Q droop control at the PCC

In the V–Q droop control strategy, the reference reactive power at the PCC of a PEI-DER is determined similar to f–P droop as

$$Q^*_{dr\_i} = -\frac{1}{D_{V\_i}}\left(V^*_{rms\_i} - V_{rms\_i}\right) + Q^*_i \tag{9.8}$$

where $Q^*_{dr\_i}$, D_{V_i}, $V^*_{rms\_i}$, V_{rms_i} and Q^*_i are drooped reference reactive power (i.e. output of the V–Q droop control strategy), the droop coefficient, the reference *rms* voltage of the PCC determined by the higher control level, the PCC *rms* voltage magnitude and the reactive power reference determined by the higher control level, respectively. The droop coefficient D_{V_i} is calculated according to the rating of the interface as (9.9) to share the required reactive power among all DERs of the microgrid.

$$D_{V\_i}S_i = D_{V\_j}S_j, \ \forall\, i,j \tag{9.9}$$

In the case that the DER is providing a specified active power at the PCC (i.e. in a frequency droop control strategy), its available reactive power is limited by the rating of the interface, as

$$Q_{i_m}(t) = \sqrt{S_i^2 - P_{io}^2(t)} \tag{9.10}$$

where Q_{i_m}, S_i and P_{io} are the available reactive power, the rating of the interface and the output active power at the operating point. In this case, the output reactive power reference of the DER must be selected according to the available reactive power and thus the droop coefficient must be adaptively updated for the next contingency. Therefore, the reference q-axis current for the inner regulator loop is determined as

$$i^*_{q\_i} = Sat\left(G_c\left(Q^*_{dr\_i} - Q_i\right)\right) \tag{9.11}$$

where $i^*_{q\_i}$, $G_c(\cdot)$ and Q_i are q-axis reference current of the interface, the controller (i.e. a PI controller) and the reactive power at the PCC.

9.3.2.3 Power factor regulation

In this control scheme, the power factor of the load at the PCC is regulated to a pre-specified value for diverse active power of the load, through injecting the required reactive power by the DER according to the consumed active power:

$$i^*_{q\_i} = \left(\frac{PF^{*2} - {PF_L}^2}{PF^{*2}\left(1 - {PF_L}^2\right)}\right) i_{qL\_i} \tag{9.12}$$

where $i^*_{q\_i}$, i_{qL_i}, PF_L and PF^* are the reference q-axis current of the interface and the load, the load power factor and the desired load power factor.

9.4 DER integration concealment

To alleviate the voltage deviation due to the integration of the DERs, DNOs necessitate the DERs to operate at constant or unity power factor. This subsection aims at calculating the reference reactive power of a controllable DER to minimise the voltage perturbation caused by the active power injection of the DERs [26].

The voltage drop through the section of the feeder from the substation to node i with power injection capability (i.e. a node with a DER) is calculated approximately as

$$\begin{aligned} \overline{V}_0 - \overline{V}_i &= \overline{I} Z_{0i} = \frac{(P_i - jQ_i)}{\overline{v}_i^*}(R_{0i} + jX_{0i}) \\ |V_0| - |V_i| &\approx -R_{0i}(P_d - P_L) - X_{0i}(Q_d - Q_L) \end{aligned} \tag{9.13}$$

where R_{0i} and X_{0i} are resistance and reactance of the feeder section between the substation and node i; V_i and V_0 are the voltage at the node with DG and the substation bus; P_d and Q_d are the injected active and reactive power; P_L and Q_L denote the load.

The injected reactive power can be used to compensate for the voltage deviation imposed by the DER active power output (P_d) at the PCC. The required reactive power in this approach is calculated as

$$\begin{aligned} 0 &\approx -R_{0i}(P_d - P_L) - X_{0i}(Q_d - Q_L) \\ Q_d^* &\approx -\frac{R_{0i}}{X_{0i}}(P_d - P_L) + Q_L \end{aligned} \tag{9.14}$$

where Q_d^* is the reference value to be injected by the controllable-DER. This approach is not practical since the load reactive power demand is often higher than the capacity of the DERs. Also, the reference value depends on the load, and any variation in the load requires updating the reference value and necessitates online measurement of the load and the power factor in the CPF mode of operation.

Reactive power control scheme overcomes the above-mentioned problems of CPF mode. It is desired to calculate the reference Q_d^* to minimise the impact of P_d on the voltage of the node of DER (v_i). Hence, one can assume that the distribution feeder contains no DER and conventional voltage drop calculations are valid. Therefore, the conventional settings for the ULTC transformer, voltage RDs and the protection system could remain unchanged for a vast region of active and reactive power flows.

To calculate the required reactive power avoiding the load measurements, consider the DER output complex power as

$$S_d = P_d + jQ_d = \overline{V}_i(\overline{I}_i)^* \tag{9.15}$$

where $(\cdot)^*$ is the complex conjugate operator. Assuming $\overline{V}_0 = |V_0|e^{j0}$ and $\overline{V}_i = |V_i|e^{j\delta}$, the voltage drop is calculated as

$$|V_i|e^{j\delta} = \frac{1}{|V_0|e^{j0}}\Big(\big(|V_i|^2 - R_{0i}P_d - X_{0i}Q_d\big) + j(-X_{0i}P_d + R_{0i}Q_d)\Big) \tag{9.16}$$

With a view to zero voltage drop (i.e. $|V_i| = |V_0|$), we have

$$R_{0i}P_d + X_{0i}Q_d = \frac{1}{2|V_i|^2}(R_{0i}^2P_d^2 + X_{0i}^2Q_d^2 + X_{0i}^2P_d^2 + R_{0i}^2Q_d^2) \tag{9.17}$$

Noticing $|Z_{0i}| = \sqrt{R_{0i}^2 + X_{0i}^2}$, we have

$$Q_d^2 - 2|V_i|^2\frac{X_{0i}}{|Z_{0i}|^2}Q_d + P_d^2 - 2|V_i|^2\frac{R_{0i}}{|Z_{0i}|^2}P_d = 0 \tag{9.18}$$

By solving (9.18) for Q_d, we have

$$Q_d^* = \frac{|V_i|^2}{|Z_{0i}|^2}X_{0i} \pm \frac{|V_i|}{|Z_{0i}|}\sqrt{\frac{X_{0i}^2}{|Z_{0i}|^2}|V_i|^2 - \frac{|Z_{0i}|^2}{|V_i|^2}P_d^2 + 2R_{0i}P_d} \tag{9.19}$$

In the presence of load, the voltage variation caused by P_d will be minimised (and will be diminished approximately) as illustrated in Figure 9.9, if the DER injects

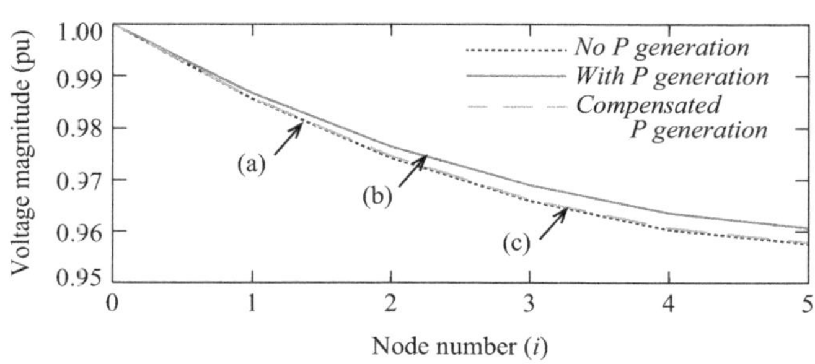

Figure 9.9 Voltage rise compensation (a) without DER, (b) voltage rise due to active power injection and (c) DER integration concealment

the reactive power calculated in (9.19). As shown in Figure 9.1(a) (the red solid line), the nodes experience voltage rise following $P_d = 1$ pu active power injection at node 3. This generation is completely concealed by injecting the reactive power $Q_d^* = 0.26$ pu into node 3 (the green dashed line) compared to the case where there is no active power generation at node 3 (the blue dotted line).

Although the voltage variation due to active power injection by the DER is minimised using this technique, it is noticed that the output of the ULTC transformer will experience more variations in comparison with CPF control scheme. The performance of the overall scheme is limited by the upper tap position. The voltage regulation may fail if ULTC reaches its highest level especially if the power system is weak. In addition, the interference between the conventional DNO-controlled voltage RDs and the proposed independent reactive power reference regulation is minimised for a vast variety of loadings. Consequently, the voltage profile with the proposed approach will be similar to the case when there is no DER.

9.5 Reactive power dispatch

Reactive power dispatch can be expressed in form of a constrained optimisation problem to minimise a particular cost function (9.20) and subject to constraints described in (9.21) including the power-flow equations, secure voltage deviation magnitude and the available reactive power.

$$CF = \sum_i f(P_i(t), Q_i(t), V_i(t)) \tag{9.20}$$

$$\begin{cases} \mathbf{I} = \mathbf{YV} \\ v_{\min} \leq v_i \leq v_{\max} \\ 0 \leq Q_i \leq Q_{i_m} \end{cases} \tag{9.21}$$

where CF is a cost function to be determined later, and $\mathbf{I} = \left[\overline{I}_i\right]^T$, $\mathbf{V} = \left[\overline{V}_i\right]^T$, $\mathbf{Y} = \left[y_{ij}\right]$ are the vectors of the current and the voltage at each bus, the admittance matrix; $v_{\min}$ and $v_{\max}$ determine the minimum and maximum allowable secure voltage range; and P_i and Q_i are the power injection of the ith DER at the operating point. The operational voltage limits (i.e. $v_{\min}$ and $v_{\max}$) are determined by the local code and the regional and national standards. The elements of the admittance matrix $\mathbf{Y}$ are calculated as

$$\begin{cases} y_{ii} = y_i + \sum_{k,k \neq i} y_{ik} \\ y_{ij} = y_{ji} = -\dfrac{1}{R_{ij} + jX_{ij}} = \left(G_{ij} + jB_{ij}\right) \end{cases} \tag{9.22}$$

The portion of the maximum available reactive power to be dispatched (i.e. $Q_i^* = \sigma Q_{i_m}, 0 \leq \sigma \leq 1$) may be further selected based on the various optimisation targets [27] such as (I) feeder's active power loss minimisation (CF_1) as in (9.23), (II) the cost of reactive energy generation (CF_2) as in (9.24), (III) variance of the voltage along the feeder (CF_3) as in (9.25) and (IV) maximum deviation in voltage magnitude (CF_4) as in (9.26).

$$\min CF_1 = \sum_i P_{G_i} - \sum_i P_{L_i} \tag{9.23}$$

where P_{G_i} and P_{L_i} are the generated and the consumed active power at node i.

$$\min CF_2 = P_{\text{loss}} c_{\text{loss}} + Q_u c_{Q_u} + \sum_{q \in RD} Q_q c_q \tag{9.24}$$

where RD is the set of voltage RDs (voltage regulator, capacitor banks, PEI-DERs and distribution-level FACTS devices); and P_{loss} and c_{loss} are total active power loss in feeders, that is CF_3 (9.25), and its relevant cost; Q_u and Q_q are the reactive power imported from the utility and injected/absorbed by RDs; and c_{Q_u} and c_q are relevant costs for Q_u and Q_q, respectively.

$$\min CF_3 = \frac{1}{N_n} \sum_{i=1}^{N_n} \left(|V_i|^* - |V_i| \right)^2 \tag{9.25}$$

where N_n is the number of the nodes.

$$\min CF_4 = \| \, |V_i|^* - |V_i| \, \|_\infty \tag{9.26}$$

where $\| \cdot \|_\infty$ is the infinity norm.

9.5.1 Power-flow equations

With respect to Figure 9.1(a) and the complex power injected to node i (i.e. $P_i + jQ_i = \overline{V}_i \overline{I}_i^*$), we have

$$\begin{cases} P_i = |V_i| \sum_k |V_k| G_{ik} \cos(\delta_i - \delta_k) + |V_k| B_{ik} \sin(\delta_i - \delta_k) \\ Q_i = |V_i| \sum_k |V_k| G_{ik} \sin(\delta_i - \delta_k) - |V_k| B_{ik} \cos(\delta_i - \delta_k) \end{cases} \tag{9.27}$$

where V_i and δ_i are voltage and angle of the ith bus, respectively (i.e. $\overline{V}_i = V_i e^{j\delta_i}$). For a section of the distribution feeder between two nodes as in Figure 9.10, the power-flow equations are simplified as

$$\begin{cases} P_m = |V_m|^2 G_{mm} + |V_m||V_{m+1}| \left(G_{m,m+1} \cos(\delta_m - \delta_{m+1}) + B_{m,m+1} \sin(\delta_m - \delta_{m+1}) \right) \\ Q_m = -|V_m|^2 B_{mm} + |V_m||V_{m+1}| \left(G_{m,m+1} \sin(\delta_m - \delta_{m+1}) - B_{m,m+1} \cos(\delta_m - \delta_{m+1}) \right) \end{cases} \tag{9.28}$$

As there is no shunt reactance connected to bus m, and the parallel capacitance for a short transmission line is ignorable, we have

$$\begin{cases} G_{m,m} = G_{i,i+1} \\ B_{m,m} = B_{i,i+1} \\ G_{m,m+1} = -G_{i,i+1} \\ B_{m,m+1} = B_{i,i+1} \end{cases} \tag{9.29}$$

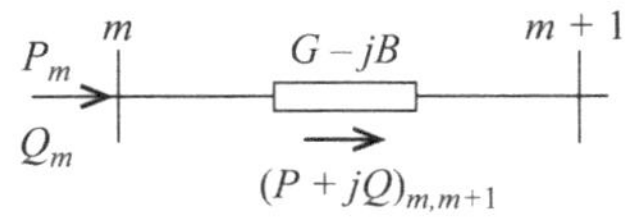

Figure 9.10 Bus m connected to only Bus m + 1

Substituting (9.29) into (9.28), we simplify the power-flow equations for a distribution feeder as

$$\begin{cases} P_i = |V_i|^2 G - |V_i||V_{i+1}|\left(G\cos(\delta_{i,i+1}) - B\sin(\delta_{i,i+1})\right) \\ Q_i = |V_i|^2 B - |V_i||V_{i+1}|\left(B\cos(\delta_{i,i+1}) + G\sin(\delta_{i,i+1})\right) \end{cases} \tag{9.30}$$

where $G = G_{i,i+1}$, $B = B_{i,i+1}$ and $\delta_{i,i+1} = \delta_i - \delta_{i+1}$.

9.5.2 Sensitivity calculations

Using the power-flow equations, we calculate the sensitivity of the power flow to the nodes' voltage magnitude (i.e. $|\mathbf{V}|$) and angle (i.e. $\boldsymbol{\delta}$) as

$$\begin{bmatrix} \Delta\mathbf{P} \\ \Delta\mathbf{Q}/|\mathbf{V}|_o \end{bmatrix} = \underbrace{\begin{bmatrix} \mathbf{J_{P\delta}} & \mathbf{J_{PV}} \\ \mathbf{J_{Q\delta}} & \mathbf{J_{QV}} \end{bmatrix}}_{\mathbf{J}} \begin{bmatrix} \Delta\boldsymbol{\delta} \\ \Delta|\mathbf{V}| \end{bmatrix} \tag{9.31}$$

where $|\mathbf{V}|_o$ is the voltage magnitude vector calculated at the operating point. The Jacobian matrix $\mathbf{J}$ contains the partial derivatives of (9.30) (e.g. $\mathbf{J_{P\delta}} = \partial\mathbf{P}/\partial\boldsymbol{\delta}|_{\mathbf{V}=\mathbf{V}_o}$) and is calculated at the operating point.

As the reactive power dispatch of the DERs is the objective, the reactive power injected by the DERs is expressed on the basis of the nodes' voltage magnitude under some assumptions such as

- The influence of the voltage magnitude on active power flow is negligible (i.e. $\mathbf{J_{PV}} \approx \mathbf{0}$).
- The effect of voltage angle on reactive power flow is insignificant (i.e. $\mathbf{J}_{Q\delta} \approx \mathbf{0}$).
- During the study period, the power consumption of the loads remains constant and consequently the generated active power by DERs are not varied (i.e. $\Delta\mathbf{P}_{\mathrm{DG}} = \Delta\mathbf{P}_L = \Delta\mathbf{Q}_L = \mathbf{0}$).

Therefore, we simplify the power-flow equations for the load and DER busses as

$$\begin{bmatrix} \begin{bmatrix} \mathbf{0} \\ \mathbf{0} \\ \Delta\mathbf{Q}_{\mathrm{DG}}/|\mathbf{V}_{\mathrm{DG}}|_o \end{bmatrix} \end{bmatrix} \approx \begin{bmatrix} \mathbf{J_{P\delta}} & \mathbf{0} \\ \mathbf{0} & [\mathbf{J_{QV}}] \end{bmatrix} \begin{bmatrix} \begin{bmatrix} \mathbf{0} \\ \Delta|\mathbf{V}_L| \\ \Delta|\mathbf{V}_{\mathrm{DG}}| \end{bmatrix} \end{bmatrix}, \quad [\mathbf{J_{QV}}] = \begin{bmatrix} \mathbf{J_{QV}}_{11} & \mathbf{J_{QV}}_{12} \\ \mathbf{J_{QV}}_{21} & \mathbf{J_{QV}}_{22} \end{bmatrix} \tag{9.32}$$

It is noted that we only need to calculate $\mathbf{J_{QV}}$. Using (9.32), the power-flow constraint at (9.21) is simplified to (9.33). Equation (9.33) also defines the relationships between the voltage deviation of the load nodes (i.e. $\Delta|\mathbf{V}_L|$) and the nodes with dispatchable DERs as well as the deviation in the injection of reactive power (i.e. $\Delta \mathbf{Q}_{\mathrm{DG}}$).

$$\begin{cases} \Delta|\mathbf{V}_L| = -\mathbf{J}_{\mathbf{QV}_{11}}^{-1}\mathbf{J}_{\mathbf{QV}_{12}}\Delta|\mathbf{V}_{\mathrm{DG}}| \\ \Delta|\mathbf{V}_{\mathrm{DG}}| = \left(-\mathbf{J}_{\mathbf{QV}_{21}}\mathbf{J}_{\mathbf{QV}_{11}}^{-1}\mathbf{J}_{\mathbf{QV}_{12}} + \mathbf{J}_{\mathbf{QV}_{22}}\right)^{-1}\Delta \mathbf{Q}_{\mathrm{DG}}/|\mathbf{V}_{\mathrm{DG}}|_o \end{cases} \quad (9.33)$$

9.5.3 Modal analysis

Considering the linearised power-flow equations with partitioned Jacobean matrix, and assuming $\mathbf{\Delta P} = \mathbf{0}$, the ith modal voltage magnitude variation is

$$\Delta V_i = \sqrt{\frac{1}{\sum_j \xi_{ji}^2}}\xi_i \quad (9.34)$$

where ξ_i is the right-column eigenvector of the Jacobean matrix $\mathbf{J}_R$ (i.e. $\mathbf{\Delta V} = \mathbf{J}_R\mathbf{\Delta Q}$) corresponding to the ith eigenvalue (i.e. λ_i) [13]. Detailed approach to calculate $\mathbf{J}_R$ is described in Section 9.5.2.

The modal variation of the voltage regarding only the variation of injected reactive power is then calculated to be $\Delta V_{mi} = \lambda_i \Delta Q_{mi}$. Small λ_i shows weak correspondence. Very high values of λ_i corresponds to instability and voltage collapse since any slight change in ΔQ_{mi} will result in large variations in ΔV_{mi}. Obviously, in this case the eigenvalues with negative real part causes instability unlike common dynamic systems. The system is voltage stable if all eigenvalues are positive and bounded. For large eigenvalues, participation factor of a bus to mode i, defined as (9.35), specify the region close to region of voltage instability.

$$pf_i = \xi_i \eta_i \quad (9.35)$$

where η_i is the left-row eigenvector corresponding to λ_i. Furthermore, participation factors of the branches and the generators show the corresponding important branches and important generators which influence the voltage stability for a specific mode. Contingency analysis for the important branches and minimum required reactive power reserve for the important generators are vital for the voltage stability analysis. Linear analysis of a sample microgrid implies that

- The modes of the PEI-DERs have faster convergence compared to the mechanical modes of the conventional DG units since the modes of the PEI-DERs have high natural frequency compared to mechanical modes
- Total convergence time constant of the microgrid is mainly affected by the mechanical modes since the natural frequency of the eigenvalues of the mechanical modes are lower than the electrical modes
- The lower level controllers of the PEI-DERs have high impact on damping the mechanical modes of conventional synchronous machine-based DERs.
- The higher level PAS has low impact on damping the mechanical modes (low-frequency oscillation from 0.1 to 3 Hz) and has high impact on the lower level control system of the PEI-DERs (frequency oscillation from 5 to 55 Hz).

9.6 Distributed voltage control schemes

9.6.1 Optimisation based on the Lagrange multipliers method

In the following, an optimal reactive power dispatch is introduced on the basis of the sensitivity calculation at Section 9.5.2 and similar to the well-known economic dispatch method.

As discussed in [16], if there exists at least one feasible solution for the constrained problem of optimal reactive power dispatch in any distribution feeder of a microgrid, then the voltage magnitude of the last node of the feeder is within the limits and has the lowest magnitude. In other words, after optimal reactive power dispatch, the last node will have the minimum magnitude. Therefore, with respect to the equipment of the last node of the feeder, there will be three cases to deal with

1. The last node of the feeder provides both measurement and reactive power control.
2. The last node is equipped with measurement, but it is not connected to a dispatchable DER.
3. The last node has neither available measurement nor control.

Case I: Assume that the last node is connected to a controllable and dispatchable DER with available measurements. Define the sensitivity of the voltage to reactive power injection as

$$\mathbf{H} = \left[h_{ij}\right] = \left(-\mathbf{J}_{\mathbf{QV}_{21}}\mathbf{J}_{\mathbf{QV}_{11}}^{-1}\mathbf{J}_{\mathbf{QV}_{12}} + \mathbf{J}_{\mathbf{QV}_{22}}\right)^{-1} \tag{9.36}$$

The last row of the equation $\Delta|\mathbf{V}_{\mathrm{DG}}| = \mathbf{H}\Delta\mathbf{Q}_{\mathrm{DG}}/|\mathbf{V}_{\mathrm{DG}}|_o$ yields the voltage variation of the last node (n) with respect to the injected reactive power as

$$\Delta|V_n| = \sum_i s_i \Delta Q_{i\mathrm{DG}} \tag{9.37}$$

where $s_i = h_{ni}/|V_{i\mathrm{DG}}|_o$ are the sensitivity coefficient in which h_{ni} is the elements of the last row of $\mathbf{H}$.

The optimisation problem (9.20) is simplified as (9.38)

$$\begin{gathered} \min CF = \sum_i \Delta Q_i \\ \text{s.t.} \begin{cases} v_{\min} - |V_n|_o \leq \Delta|V_n| \leq v_{\max} - |V_n|_o \\ 0 \leq Q_i \leq Q_{i_m} \end{cases} \end{gathered} \tag{9.38}$$

Similar to the problem of economic dispatch, using the Lagrange multipliers of the constraints, we conclude that the cost function CF is minimised with incremental cost as

$$\lambda = \min_i \left\{\frac{1}{s_i}\right\} \tag{9.39}$$

Case II: By assuming a virtual DER with $Q_{max} = 0$ connected to the last node, this case becomes identical to **case I**.

Case III: Assume that the last dispatchable DER is installed at node n which is not the last node of the feeder, and its voltage is $\overline{V}_n = |V_n|e^{j0}$ (identical to case I). The last node of the feeder is assigned a node number $d > n$ (i.e. $\overline{V}_d = |V_d|e^{j\delta_d}$). The section of the feeder between nodes n and d has the impedance $R_{nd} + jX_{nd}$. Then, the voltage of the last node is calculated as

$$\overline{V}_d = \overline{V}_n - (R_{nd} + jX_{nd})\frac{(P_n - jQ_n)}{\overline{V}_n^*} \tag{9.40}$$

where P_n and Q_n are powers flowing from the node n to d and are measured at the node d. Thus, the voltage magnitude of the node d is calculated as

$$|V_d| = \frac{1}{|V_n|}\sqrt{\left(|V_n|^2 - R_{nd}P_n - X_{nd}Q_n\right)^2 + (R_{nd}Q_n + X_{nd}P_n)^2} \tag{9.41}$$

In order to satisfy the voltage-limit constraint at the end of the feeder (i.e. $v_{min} \le v_d \le v_{max}$), we have to consider the voltage drop through the section nd of the feeder and modify the voltage-limit constraint in (9.38) as

$$\begin{cases} v_{min} \le |V_i| \le v_{max} & i < n \\ v_{min} + \Delta v_{nd} \le |V_n| \le v_{max} & i = n \\ v_{min} \le |V_i| \le |V_n| & i > n \end{cases} \tag{9.42}$$

where $\Delta v_{nd} = |V_n| - |V_d|$ is calculated at the operating point.

9.6.2 Distributed voltage control via multi-agent system

For distributed voltage support using MAS configuration, several agents are assigned to DERs and the DNO [16–22]. Each agent can play the role of the manager. The manager at each iteration is determined on the basis of a predefined scheme. One possible arrangement is to have a fixed list in which the specification numbers '$ASN = \{1, 2, = efn\}$' is assigned to each potential manager agent. Then, the agent assigned with the specification number $i \in ASN$ has the role of the manager. Then, the agent assigned with the specification number $i + 1 \in ASN$ will take the role if the ith agent is absent.

Each DER has an agent as a dispatcher which bids to the requests to proposal. An agent as a voltage monitor detects the voltage sags/swells and other voltage quality issues. This agent is also responsible for sending the voltage support request to the manager.

The bids of the agents include at least the sensitivity coefficient at the operating point, its cost and maximum available reactive power as

$$bp_i(t, S_i) := \{s_i(t_1), Q_i(t_1), \overline{Q}_i(t_1), CF_i(Q_i)\} \tag{9.43}$$

9.6.2.1 CNIP-based reactive power dispatch

Step-by-step algorithm of the reactive power dispatch using sensitivity priority scheme can be summarised as the following:

1. Determine the sensitivity coefficients online by monitoring the voltage variations.
2. Sort the priority set of the available DERs for dispatching according to their sensitivity; the highest are dispatched first
3. Dispatch the first DER in the priority set.
4. If the dispatched DER reached its maximum available reactive power, dispatch the next DER in the priority set
5. Continue dispatching new DERs according to the priority set until the voltage of the end node is justified to the limits.

In this method, the constraints on DERs' available reactive power are satisfied online during the operation.

9.6.2.2 Distributed multi-agent approach for sensitivity calculation

The fundamental of distributed dispatching architecture is the determination of the sensitivity factor. A model-based analytical approach to determine the sensitivity factors provides high performance accuracy. Nevertheless, this approach requires the exact model of the system, a management centre and laborious offline and online calculations. Also, the model-based approach restricts the plug-and-play capability of the administration framework (i.e. the PAS), because the analytical model of the system changes by connecting and disconnecting the DERs and the variations in loads.

Concerning self-sufficiency of the MAS-based PAS, it is desirable to eliminate the dependency of the multi-agent control architecture to the system's model. Model-free structure develops the scalability of the control scheme and improves the robustness of the PAS against various contingencies.

The system of the agents sets up a cooperative framework to calculate the sensitivity factors during the operation of the system. Particularly, the agents can cooperate under the supervision of the manager to intentionally alter the reactive power injection/consumption to realise the incremental increase/decrease of the voltage magnitude of the nodes. The measurements are reported by the monitoring agents and must be synchronised with the power variations. Then, the sensitivity coefficients are calculated as (9.44) and may be stored for later use. The manager must iteratively update the coefficients at specific time points of operation or after any remarkable change in the operating point (i.e. due to the change in power injection or load variations), whichever occurs earlier. Subsequently, the sensitivity coefficients (i.e. $\tilde{s}_{ij}$) are updated as

$$\tilde{s}_{ij}\big(V_j(Q_i(t_1)), Q_i(t_1)\big) \approx \frac{|V_j(t_2)| - |V_j(t_1)|}{Q_i(t_2) - Q_i(t_1)} \tag{9.44}$$

where t_1, t_2 are synchronised points of time, and $\tilde{s}_{ij}$ is the experimentally calculated sensitivity coefficient at the operating point which demonstrates the influence of the ith DER on the voltage of the jth node, at the operating point $V_j(Q_i(t_1))$.

The observation technique could be further simplified for the distribution of radial feeders under steady-state conditions. Consider a voltage variation at the PCC of the ith DER as $\Delta|V_i|$. Assume that the loads' variations (i.e. ΔP_j and ΔQ_j $j > i$) on account of the voltage variation (i.e. $\Delta|V_i|$) is negligible, and the DERs which are connected to the nodes downstream of the node i (i.e. $j = i+1, \ldots, n$) are not in constant PCC voltage mode of operation. The section of the feeder, consisting of the nodes $i+1$ to n with fixed and stable power flows in steady state, is a linear system so that the voltage variation $\Delta|V_i|$ will spread out to all the nodes $i+1$ to n. Therefore, the end node of the feeder will experience the same voltage variation in steady state. Consequently, under the mentioned assumption that the sensitivity coefficient of the voltage of the last node of the feeder ($\Delta|V_n|$) to the reactive power deviation of the each DER (ΔQ_i) is approximated as

$$\tilde{s}_i \approx \frac{\Delta|V_n|}{\Delta Q_i} \approx \frac{\Delta|V_i|}{\Delta Q_i} \tag{9.45}$$

As a result, each dispatchable agent determines its sensitivity coefficient locally by monitoring the simultaneous deviation of its PCC reactive power and voltage magnitude. The actual sensitivity of the fifth node of the feeder in Figure 9.1(a) is compared to the sensitivity of the prior nodes in Table 9.2. Comparing the values of $\Delta V_5/\Delta Q_i$, reactive power injection at node 4 has the highest impact on the voltage of node 5. Also, the impact of ΔQ_4 on ΔV_5 is approximately the same as its impact on ΔV_4 which confirms the approximation in (9.45).

Figure 9.11 compares the performance of the scheme by selecting a non-optimum node to inject reactive power. In this case, node 3 is selected for reactive power injection by the MAS after the outage of the DER at node 5 and rejecting the proposal of the DER at node 4. The MAS-based dispatching of the reactive power, based on the locally measured sensitivity in (9.45), yields higher voltage profile compared to the dispatching based on the actual sensitivity factor. Even though the dispatched power is 1.34 times the dispatch with actual sensitivity factor, both factors yield adequate voltage compensation.

Table 9.2 Sensitivity approximation by local measurements (% pu)

i	*1*	*2*	*3*	*4*	*5*
$\Delta V_5/\Delta Q_i$	*0.51*	*0.88*	*1.22*	*1.57*	*1.91*
$\Delta V_i/\Delta Q_i$	*0.27*	*0.61*	*0.91*	*1.57*	–
Relative error	*47*	*30*	*25.41*	*0*	–

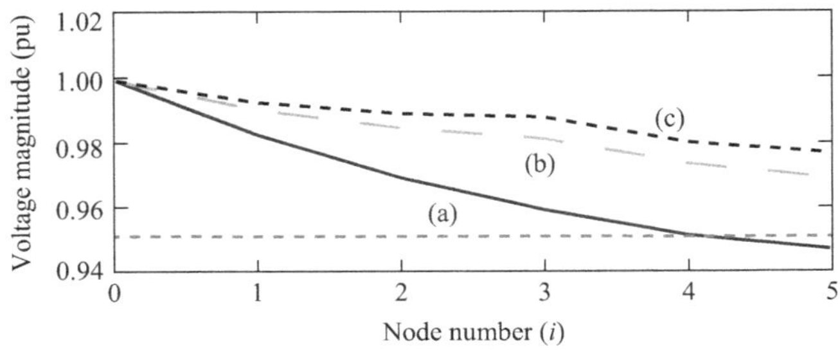

Figure 9.11 Voltage profile evolution using MAS-based control for DER at node 3 of Figure 9.1(a): (a) voltage drop after contingency; (b) reactive power dispatched with respect to actual sensitivities and (c) reactive power dispatched with respect to local sensitivities

9.6.3 Distributed voltage control with simplified model-based sensitivity calculation

9.6.3.1 Simplified model-based sensitivity calculation

For the calculation of the sensitivity coefficient, suppose the DERs and loads are modelled as constant active/reactive power sources (i.e. $P, Q > 0$ and $P, Q < 0$ corresponds to consuming loads and the generating DERs respectively); and the shunt capacitance and conductance of the feeders are negligible and the sections of the feeder between sequent nodes are modelled as constant *RL* branches. We can also assume that the ULTC at the substation regulates the bus voltage of the substation and thus the substation is modelled as a constant-voltage source. Considering the radial feeder as a tree-shaped graph, the node i (i.e. the vertices) is connected to the substation through a path (i.e. p_i) consisting the sequence of sections of the feeder (i.e. the edges; $Z_{\varepsilon,\iota} = R_{\varepsilon,\iota} + jX_{\varepsilon,\iota}, \varepsilon, \iota \in p_i$) [28]. For instance, the path between the substation and the node $i = 3$ in Figure 9.1 is $p_3 = \{Z_{01}, Z_{12}, Z_{23}\}$. Therefore, the voltage difference between the node i and the substation is calculated as

$$|V_i| = |V_0| - \sum_{\varepsilon,\iota \in p_i} |V_{\varepsilon,\iota}| \tag{9.46}$$

where $|V_0|$ and $|V_i|$ are voltage magnitudes of the substation and the node i with reference to ground. $|V_{\varepsilon,\iota}|$ is the voltage drop between the adjacent nodes ε and ι as

$$|V_{\varepsilon,\iota}| = |V_\varepsilon| - |V_\iota| = \frac{R_{\varepsilon,\iota}P_\iota + X_{\varepsilon,\iota}Q_\iota}{|V_\iota|} \tag{9.47}$$

where P_ι and Q_ι are the active and the reactive powers measured prior to and at the entrance to node ι. By neglecting the power losses through the distribution feeder, we get

$$P_\iota + jQ_\iota \approx \sum_{k \in Ch_\iota} P_k + j \sum_{k \in Ch_\iota} Q_k \tag{9.48}$$

where Ch_ι is the set of the subtrees and children connected to the root node ι in the graph of the feeder; and P_k and Q_k are the absorbed/injected active and reactive power of the kth node. Further, the voltage drop is approximated assuming that the operational voltage limits are not violated (i.e. $|V_\iota| \approx |V_r|$ and $|V_r|$ is the nominal voltage of the system).

$$|V_i| \approx |V_0| - \frac{1}{|V_r|} \sum_{\varepsilon,\iota \in p_i} \left(R_{\varepsilon,\iota} \sum_{k \in Ch_\iota} P_k + X_{\varepsilon,\iota} \sum_{k \in Ch_\iota} Q_k \right) \tag{9.49}$$

The voltage magnitude is the function of the injected powers P_k and Q_k. By linearising the voltage magnitude around the operating point (i.e. (P_k, Q_k)), we have

$$\Delta|V_i(P_k, Q_k)| = \sum_{k \in Ch_\iota} \frac{\partial|V_i|}{\partial P_k} \Delta P_k + \frac{\partial|V_i|}{\partial Q_k} \Delta Q_k \tag{9.50}$$

With constant model of the loads/DERs, the power injection/absorption at each node is independent of other nodes (i.e. $\dfrac{\partial P_{k_1}}{\partial P_{k_2}} = 0$ for $(k_1 \neq k_2) \in Ch_\iota$). Also, the reactive power consumption/dispatch variations are independent of the active power variations (i.e. $\dfrac{\partial Q_{k_1}}{\partial P_{k_2}} = 0$ for $k_1, k_2 \in Ch_\iota$) [28]. Therefore, we have

$$\begin{cases} \dfrac{\partial}{\partial P_k} \displaystyle\sum_{\varepsilon,\iota \in p_i} \left(R_{\varepsilon,\iota} \sum_{k \in Ch_\iota} P_k + X_{\varepsilon,\iota} \sum_{k \in Ch_\iota} Q_k \right) = \sum_{\varepsilon,\iota \in p_i} R_{\varepsilon,\iota} \\ \dfrac{\partial}{\partial Q_k} \displaystyle\sum_{\varepsilon,\iota \in p_i} \left(R_{\varepsilon,\iota} \sum_{k \in Ch_\iota} P_k + X_{\varepsilon,\iota} \sum_{k \in Ch_\iota} Q_k \right) = \sum_{\varepsilon,\iota \in p_i} X_{\varepsilon,\iota} \end{cases} \tag{9.51}$$

The sensitivity factors are approximated as

$$\begin{cases} s_{VP}(i,k) \approx \dfrac{\partial|V_i|}{\partial P_k} = -\dfrac{1}{|V_r|} \displaystyle\sum_{\varepsilon,\iota \in p_i} R_{\varepsilon,\iota} \\ s_{VQ}(i,k) \approx \dfrac{\partial|V_i|}{\partial Q_k} = -\dfrac{1}{|V_r|} \displaystyle\sum_{\varepsilon,\iota \in p_i} X_{\varepsilon,\iota} \end{cases} \tag{9.52}$$

In the matrix form, we have

$$\Delta|\mathbf{V}| \approx \mathbf{s_{VP}} \Delta\mathbf{P} + \mathbf{s_{VQ}} \Delta\mathbf{Q} \tag{9.53}$$

where $\Delta|\mathbf{V}| = [\Delta|V_i|]^T$, $\Delta\mathbf{P} = [\Delta P_i]^T$ and $\Delta\mathbf{Q} = [\Delta Q_i]^T, i = 1, \ldots, n$ are the vectors of voltage variation, active and the reactive power variations, respectively. The matrices $\mathbf{s_{VQ}} = [s_{VQ}(i,k)]$ and $\mathbf{s_{VP}} = [s_{VP}(i,k)]$ are the approximate sensitivity of the node's voltage magnitudes with respect to variations in the reactive and active powers, respectively. In the matrices $\mathbf{s_{VP}}$ and $\mathbf{s_{VQ}}$, the row i corresponds to the node at which the voltage must be regulated. The column k corresponds to the DERs capable of controlling the power variations (DGs and DSM).

However, the regulation of the voltage by altering the active power of DERs is refused by the DNOs, and it is desired to use only reactive power control. Presence of the minus sign in (9.52) indicates that, to increase the voltage magnitude at node i, we must reduce the absorption or increase the injection of the reactive power ($\Delta Q < 0$).

9.6.3.2 Distributed sensitivity-based reactive power dispatch

After detection and report of voltage-limit violation in node i by the operator (i.e. DNO, the manager in the MAS or an independent control centre), it selects the device m_1 (i.e. DERs, DGs, DSM, capacitors) which has the higher impact on the voltage of the node i by scanning for the maximum sensitivity factor (i.e. the maximum value at the row i of $\mathbf{s_{VQ}}$) to alter its reactive power as

$$\Delta Q_m^* = \frac{\Delta |V_i|^*}{s_{VQ}(i, m_1)} \tag{9.54}$$

where $\Delta |V_i|^*$ is the required voltage variation to return the voltage magnitude into the standard limits and $\Delta Q_{m_1}^*$ is the corresponding reactive power variation reference of the device m_1. As discussed before, if the device m_1 reaches its minimum or maximum available reactive power and the voltage limits are still violated, the next device with highest sensitivity factor is selected to be dispatched.

Following a disturbance and the consequent variation in the voltage of a node, one or more DERs will be determined to compensate the required reactive power based on a proposed scheme (i.e. sensitivity factor priority). The cooperation level of the DERs is determined on the basis of the order of the sensitivity factors. If a large disturbance results in a large voltage drop, the reactive power injection of the most-effective DER will be dispatched to its limits. This architecture has the advantage of solving the problem of voltage regulation with minimum reactive power injection and consequently the minimum increase (or a reduction) in loss of the feeder. Nonetheless, the main drawback of the method is that the most effective DER is operating at its limits and cannot be utilised for further active power support which reduces the reliability of the system. Therefore, it is recommended to avoid dispatching the total available reactive power of the DER and define a cap on maximum allowable reactive power injection of the DER (e.g. 80% of the available reactive power). The cap must be specified considering the security factors and must be updated online to release the constraints for optimality.

9.6.3.3 Droop-shaped sensitivity-based reactive power dispatch

The equation $\Delta|\mathbf{V}| \approx \mathbf{s_{VQ}}\Delta\mathbf{Q}$ implies that we can utilise all (at least more than one) available DERs for reactive power support concerning the sensitivity coefficients, rather than utilising one or two of the most effective DERs. A simple scheme to use the support of several DERs is to disperse the required voltage support

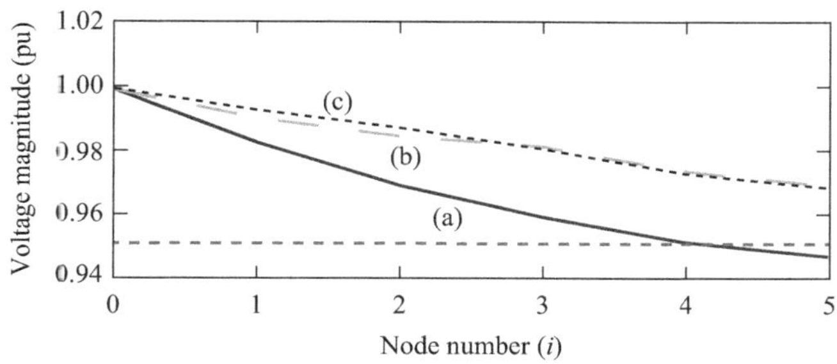

Figure 9.12 Voltage profile evolution: (a) voltage drop after contingency; (b) dispatching most effective DER and (c) cooperative sensitivity-based dispatch

according to the sensitivity factors and the portion of the available reactive power of each DER as

$$Q_i^* = \frac{s_{VQ}(i,m)}{\sum\limits_{m=1}^{m_n} S_{VQ}^2(i,m)} \Delta|V_i|^* \tag{9.55}$$

where $\overline{Q}_m$ is the maximum available capped reactive power of the mth DER; μ_m is the portion of $\overline{Q}_m$ that is desired to be dispatched; m_n is the number of the DERs participated in voltage support architecture at each time; $\Delta|V_i|^*$ is the desired voltage deviation to be applied at node i; and $s_{VQ}(i,m)$ is the sensitivity of ith node voltage to reactive power injection at mth node.

The effect of the proposed compensation schemes for sensitivity-based reactive power injection is depicted in Figure 9.12. The last node's voltage is dropped below the standard limit due to the trip of the last two DERs of the grid. Two cases are considered: (I) the voltage drop is compensated using the most effective DER which has the highest sensitivity factor, and (II) cooperation of all DERs according to the value of corresponding sensitivity factor as (9.55). The voltage compensation at the last node is identical in both methods. The cooperative approach avoids the saturation of the DER's capacity and yields higher average voltage through the feeder.

9.6.4 Decentralised cooperative optimisation using self-organised sensor network

Nonlinear programming through optimisation [29] establishes the process of decision-making in a centralised control architecture. The decision includes the states of the voltage RDs. The optimal references of the devices are set to minimise the cost function consisting of control, technical and economic objectives subject to operational and security constraints. The centralised control architecture provides fast and accurate globally optimal decisions. However, centralisation limits the scalability of the system, requires high computational capabilities, necessitates fast and reliable communication infrastructure with high bandwidth and entails large resources for data storage [27].

A decentralised optimisation approach based on a network of smart sensors and cooperative agents [27] can eliminate the drawbacks of the centralised optimisation technique, though it results in rather larger convergence time and may lead to local optima.

For implementing the decentralised voltage profile optimisation, it is assumed that each reactive power controller connected to the feeder is equipped with smart sensors (i.e. measuring the PCC's voltage, active and reactive powers), communication interface devices and a local oscillator as self-synchronised sensor agents for state estimation.

The role of the oscillator is to provide an approximation of the global cost function through limited communication with neighbour devices [30]. The prototype network of the coupled sensors provides fast synchronisation of the oscillators. The oscillators converge to the weighted average value of the communicated variables. Dynamics of the oscillator is selected as [30]

$$\dot{\mathbf{x}} = \mathbf{v} + \mathbf{\Lambda} diag\left(\mathbf{M}_{Ad}\mathbf{F}\left(\mathbf{1}_n^T \otimes \mathbf{x} - \mathbf{x}^T \otimes \mathbf{1}_n\right)\right)\mathbf{K} \tag{9.56}$$

where $\mathbf{x} = [x_i]^T = [x_1, x_2, \ldots, x_n]^T$, $i = 1, \ldots, n$, and n are the vector of state function and the number of states to be observed respectively; $\mathbf{v} = [v_i]^T$, $i = 1, \ldots, n$ is the initial condition vector; $\mathbf{K} = [k_i]^T$ is control gain vector with positive real values (i.e. $k_i > 0 \in \mathbb{R}$); $\mathbf{M}_{Ad} = [m_{ad}(i,j)]$ is the adjacency matrix of the communication links; $\mathbf{\Lambda} = \text{diag}(\lambda_i)$ is the adaptability factors matrix with positive real values (i.e. $\lambda_i > 0 \in \mathbb{R}$); $\mathbf{F}(\cdot)$ is an arbitrary differentiable odd vector-valued function; and $\mathbf{1}_n = [1, \ldots, 1]^T$ is an $n \times 1$ ones vector in which all elements are 1. Further, in order to construct the adjacency matrix, the element (i,j) is set to be $m_{ad}(i,j) = 1$ if node i is connected to node j through a communication link, otherwise $m_{ad}(i,j) = 0$. The function $\text{diag}(\mathbf{M}_{Ad}\mathbf{F})$ extracts the diagonal elements of the matrix $\mathbf{M}_{Ad}\mathbf{F}$ into a diagonal matrix of size $n \times n$.

It is proved that under the assumption of a connected communication network, each agent can observe the weighted average of the global states of the system since the states of the dynamic local system (9.56) asymptotically converge to the following [30]:

$$\dot{x}_i^* = v_i^* = \frac{1^T \mathbf{\Lambda v}}{1^T \mathbf{v} 1}, \quad i = 1, \ldots, n \tag{9.57}$$

For example, if each oscillator communicates its PCC's voltage magnitude (i.e. $\mathbf{x} = [|V_i|]$), the coupled oscillators of all n devices will converge to the average voltage across the system (i.e. $v_i^* = \frac{1}{n}\sum_{i=1}^{n} |V_i|$). Assuming $k_i = k > |v^* - v_i|$ [30], $\lambda_i = 1$, and $\mathbf{F}(\mathbf{x}) = \sin(\mathbf{x})$, (9.56) is calculated as:

$$x_i(t) = v_i(t) + k \sum_{j=1}^{n} m_{ad}(i,j) \sin\left(x_j - x_i\right) \tag{9.58}$$

A global cost function is defined as (9.59) on the basis of the observed states as $CF_g(\mathbf{v})$ which is the function of the decentrally synchronised global average states ($\mathbf{v}$). To evaluate the total loss of the active power in the grid and the cost of voltage support, one may also include the net active power injected outside the node i (i.e. $P_{G_i} - P_{L_i}$), and the cost of generation at node i (i.e. $c_{P_i}P_{G_i}$ and $c_{Q_i}Q_{G_i}$).

$$CF_g = \frac{1}{2N}\sum_{i=1}^{N}|V_i| + \frac{1}{4N}\sum_{i=1}^{N}|1 - V_i|^2 \tag{9.59}$$

Therefore, the states of the self-organised sensor network are selected as $\mathbf{x} = [|V|, |1 - V|^2]^T$. Figure 9.13 shows the convergence of the observed states at each node to the average values. In this case, k is selected to be 10, and the oscillator at each node is mutually coupled to its neighbours. Thus, the adjacent matrix has the form of

$$\mathbf{M}_{Ad} = \begin{bmatrix} 1 & 1 & 0 & 0 \\ 1 & 1 & 1 & 0 \\ 0 & 1 & 1 & 1 \\ 0 & 0 & 1 & \ddots \end{bmatrix} \tag{9.60}$$

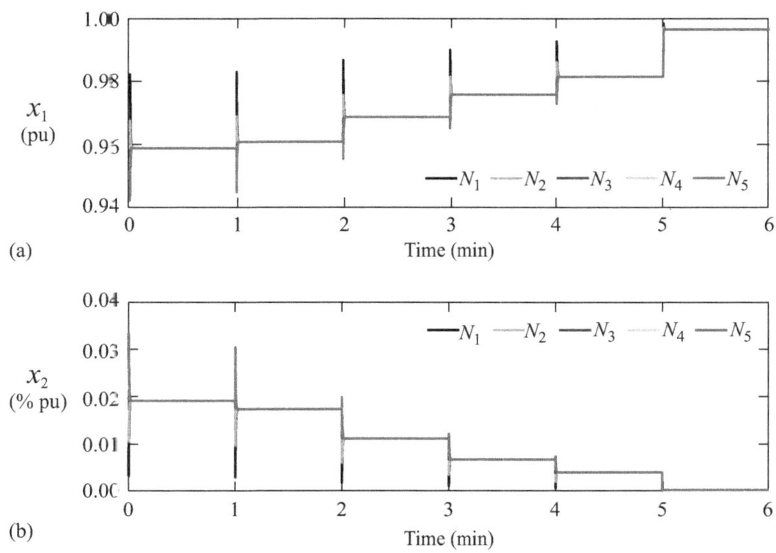

Figure 9.13 States of the self-organised sensor network for the feeder in Figure 9.1; (a) average voltage magnitude of nodes (b) average voltage deviation of the nodes

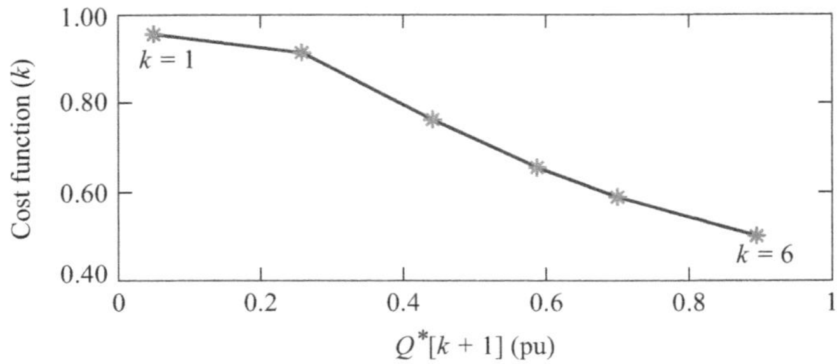

Figure 9.14 Cost function evaluation vs. reactive power reference

The controller at each node is designed and updated using the gradient decent algorithm [22] to minimise CF_g as

$$Q_i^*[k+1] = Q_i[k] - \mu \frac{CF_g[k] - CF_g[k-1]}{Q_i[k] - Q_i[k-1]} \tag{9.61}$$

where k, $Q_i^*[k]$ and $\mu \in \mathbb{R} > 0$ are the discrete updating iteration, the reference reactive power of the local controller of the device connected to the node i, and the design gain which effects the convergence rate. In order to have a feasible signal output of (9.61) in the first step, $Q_i^*[1]$ is selected arbitrarily with slight change in the current reactive power injections (i.e. $Q_i^*[1] = \acute{k} Q_i, \acute{k} \in \mathbb{R}$).

Although gradient descent optimisation method has the deficiency of converging to a local optimum, this method provides an advanced tool for reactive power dispatch due to its stability and fast convergence characteristics. A termination criterion is also necessary to stop the algorithm. For instance, the optimisation is terminated if the change in the updated reference Q_i^* for several consequent iterations becomes less than a predefined value, that is $|Q_i^*[k+1] - Q_i^*[k]| < \epsilon$, where $\epsilon \in \mathbb{R} > 0$ is a small constant. After convergence of the optimisation, the controller Q_i^* is set and the system continues with the determined optimum settings until detecting a variation in CF_g.

Figure 9.14 shows the evolution of the cost function CF_g and the reference reactive power to be injected for the first several iterations. The loads of the last two nodes are doubled at $t = 0$ which leads to 0.06-pu voltage drop at the last node. The first reference reactive power at each node is selected as 10% of its injection before the load change (i.e. $Q_i^*[1] = 0.1 Q_i(0^-)$). References are updated with 1-min time interval.

9.6.5 Distributed cooperative gradient-descent optimisation of reactive power dispatch

The main target of the voltage regulators in the system is to minimise the error of the node's voltage from its reference value (i.e. $|V_i|^* = 1$ pu). The error is minimised though optimisation approach by defining the cost function as

$$CF_6 = \sum_{i=1}^{N} (|V_i| - |V_i|^*)^2 \tag{9.62}$$

For a cooperative distributed optimisation scheme [31], the control update law of the optimisation process can be introduced similar to the gradient descent method as

$$Q_i^*[k+1] = \sum_{j=1}^{N} m_{ad}(i,j)Q_i^*[k] - \mu Q_{i_m} \frac{\partial CF_6}{\partial Q_i} \tag{9.63}$$

The available dispatchable reactive power of each DER as a RD is limited by its nominal capacity and injected active power at the operating point (i.e. $0 \le Q_i^*[k] \le Q_{i_m}(t[k])$). The first term on the right-hand side of (9.63) provides the average observation of the reactive power of the other nodes which are connected through the communication links, and the last term makes the observation Q_i^* gradually converge to the global optimal equilibrium point.

$$\frac{\partial CF_6}{\partial Q_i} = \frac{\partial CF_6}{\partial |V_i|}\frac{\partial |V_i|}{\partial Q_i} = 2(|V_i| - |V_i|^*)\frac{\partial |V_i|}{\partial Q_i} \tag{9.64}$$

The term $\frac{\partial |V_i|}{\partial Q_i}$ is calculated with respect to the power-flow equations as [32]

$$\frac{\partial |V_i|}{\partial Q_i} = \frac{1}{\frac{Q_i}{|V_i|} - B_{ii}|V_i|} \tag{9.65}$$

where B_{ii} is the imaginary part of the diagonal elements of the bus admittance matrix (i.e. $\mathbf{Y}_{bus}$). Substituting (9.64) and (9.65) into (9.63), we have

$$Q_i^*[k+1] = \sum_{j=1}^{N} m_{ad}(i,j)Q_i^*[k] + 2\mu \frac{(|V_i[k]| - |V_i|^*)|V_i[k]|}{B_{ii}|V_i[k]|^2 - Q_i[k]} \tag{9.66}$$

Equation (9.66) implies that this method requires the information of the bus admittance matrix. These values are commonly available or can be approximated by local observations and measurements.

Practically, there are many nodes incapable of reactive power control such as the passive load nodes, the saturated PEI-DERs (operating at maximum output power), conventional DERs with constant power factor control scheme and others. For these nodes, the available overhead of the reactive power is zero (i.e. $Q_{i_m} = 0$). As the target is to control the voltages of all nodes rather than some critical nodes, the uncontrollable nodes can get the support of the controllable nodes. Two approaches are proposed as

1. The average weighted overhead reactive powers of the controllable nodes are assumed as the available reactive power for the mentioned uncontrollable nodes [32]. Similar to the aforementioned control update law, the local controller of each DER is supposed to iteratively approximate the average available reactive power adopting the update law as

$$[k+1] = \sum_{j=1}^{N} \acute{m}_{ad}(i,j)O_{j\Sigma}[k] \tag{9.67}$$

where $\acute{m}_{ad}(i,j)$ are the elements of the adjacency matrix. Note that this approach applies more restrictions on the connectivity of the communication system since it is necessary to have a double-stochastic adjacency matrix. Therefore, the nodes must be connected such that (9.68) is satisfied.

$$\sum_{i=1}^{N} \acute{m}_{ad}(i,j) = \sum_{j=1}^{N} \acute{m}_{ad}(i,j) = 1 \tag{9.68}$$

2. A simpler approach is to dedicate some portions of the overhead of specific controllable nodes to each uncontrollable node. The dedicated nodes may be chosen on the basis of the sensitivity analysis or available communication links. This approach relaxes the constraint on the adjacency matrix.

Figure 9.15 compares the voltage profile evolution for cases I and II. The ULTC transformer regulates the utility voltage to 1 pu. Each controller has data exchange only with its neighbour nodes (i.e. the adjacent matrix has the form of (9.60)). In the first scenario, all nodes are equipped with dispatchable DERs (case I; solid line). In the second scenario, it is assumed that only two dispatchable DERs are connected to nodes 2 and 4 (case II; dashed line). The update of the necessary reactive power injection at nodes 1 and 2, and the nodes 4 and 5 are dedicated to DERs at nodes 2 and 4, respectively. Also the reference reactive power injection at node 3 is provided equally by nodes 2 and 4. The proposed reactive power regulation method improves the voltage profile, reduces the loss of the feeder, and lessens the reactive power injection from the utility, as presented in Table 9.3.

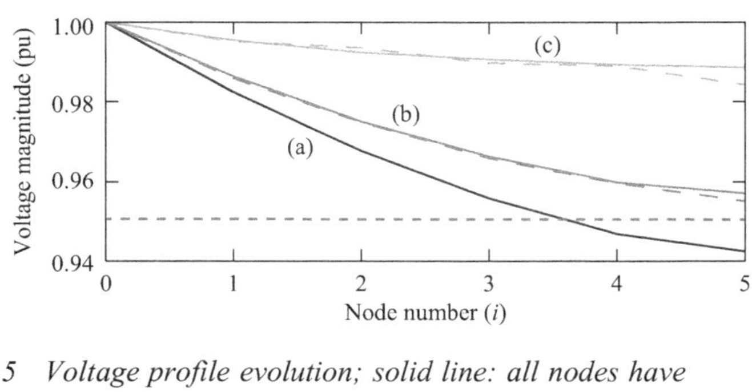

Figure 9.15 Voltage profile evolution; solid line: all nodes have dispatchable DERs; dashed line: only two nodes have dispatchable DERs; (a) heavy load, (b) first dispatch update and (c) second dispatch update

Table 9.3 Results of the optimal cooperative gradient-based reactive power dispatch

	***Utility reactive power injection* (pu)**	***Loss* (pu)**	***Average voltage* (pu)**	***Cost function***
Heavy load (1)	*0.9756*	*0.0984*	*0.9659*	*16.00e−4*
First dispatch update (2)	*0.7758*	*0.0777*	*0.9742*	*8.961e−4*
Second dispatch update (3)	*0.3412*	*0.0561*	*0.9927*	*0.681e−4*

References

[1] Momoh J. *Smart Grid: Fundamentals of Design and Analysis*. Hoboken, NJ: Wiley-IEEE Press; 2012.

[2] Majumder R. 'Aspect of voltage stability and reactive power support in active distribution'. *IET Generation, Transmission & Distribution*. 2014; **8**(3):442–50.

[3] Abapour S., Zare K., Mohammadi-Ivatloo B. 'Evaluation of technical risks in distribution network along with distributed generation based on active management'. *IET Generation, Transmission & Distribution*. 2014;**8**(4): 609–18.

[4] Abapour S., Zare K., Mohammadi-Ivatloo B. 'Dynamic planning of distributed generation units in active distribution network'. *IET Generation, Transmission & Distribution*. 2015;**9**(12):1455–63.

[5] Duman S., Sonmez Y., Guvenc U., Yorukeren N. 'Optimal reactive power dispatch using a gravitational search algorithm'. *IET Generation, Transmission & Distribution*. 2012;**6**(6):563–76.

[6] Yoshida H., Kawata K., Fukuyama Y., Takayama S., Nakanishi Y. 'A particle swarm optimization for reactive power and voltage control considering voltage security assessment'. *IEEE Transactions on Power Systems*. 2000; **15**(4):1232–9.

[7] Han B., Bae B., Kim H., Baek S., 'Combined operation of unified power-quality conditioner with distributed generation'. *IEEE Transactions on Power Delivery*. 2006;**21**(1):330–8.

[8] Muttaqi K. M., Le A. D. T., Negnevitsky M., Ledwich G. 'A coordinated voltage control approach for coordination of OLTC, voltage regulator, and DG to regulate voltage in a distribution feeder'. *IEEE Transactions on Industry Applications*. 2015;**51**(2):1239–48.

[9] Hatziargyriou N., Asano H., Iravani R., Marnay C. 'Microgrids'. *IEEE Power and Energy Magazine*. 2007;**5**(4):78–94.

[10] Li P., Wang X., Lee W. J., Xu D. 'Dynamic power conditioning method of microgrid via adaptive inverse control'. *IEEE Transactions on Power Delivery*. 2015;**30**(2):906–13.

[11] Guerrero J. M., Vasquez J. C., Matas J., de Vicuna L. G., Castilla M. 'Hierarchical control of droop-controlled ac and dc microgrids – a general approach toward standardization'. *IEEE Transactions on Industrial Electronics*. 2011;**58**(1):158–72.

[12] Zhong Q.-C., Hornik T. *Control of Power Inverters in Renewable Energy and Smart Grid Integration*. Chichester, UK: John Wiley & Sons Inc., Wiley-IEEE Press; 2013.

[13] Katiraei F., Iravani M. R. 'Power management strategies for a microgrid with multiple distributed generation units'. *IEEE Transactions on Power Systems*. 2006;**21**(4):1821–31.

[14] Jenkins N., Allan R., Crossley P., Kirschen D., Strbac G. *Embedded Generation*. London: Institute of Engineering and Technology (IET); 2000. pp. 50–93.

[15] Bokhari A., Raza A., Diaz-Aguiló M., *et al*. 'Combined effect of CVR and DG penetration in the voltage profile of low-voltage secondary distribution networks'. *IEEE Transactions on Power Delivery*. 2016;**31**(1):286–93.

[16] Baran M. E., El-Markabi I. M. 'A multiagent-based dispatching scheme for distributed generators for voltage support on distribution feeders'. *IEEE Transactions on Power Systems*. 2007;**22**(1):52–9.

[17] Lunze J. *Feedback Control of Large-Scale Systems*. London: Prentice Hall; 1992.

[18] Siljak D. D., Zecevic A. I. 'Control of large-scale systems: beyond decentralized feedback'. *Annual Reviews in Control*. 2005;**29**(2):169–79.

[19] Romanovski I., Caines P. E. 'On the supervisory control of multiagent product systems'. *IEEE Transactions on Automatic Control*. 2006;**51**(5):794–9.

[20] Loia V., Vaccaro A. 'Decentralized economic dispatch in smart grids by self-organizing dynamic agents'. *IEEE Transactions on Systems, Man, and Cybernetics: Systems*. 2014;**44**(4): 397–408.

[21] Shahidehpour M., Wang Y. Communication and Control in Electric Power Systems: Applications of Parallel and Distributed Processing. *Piscataway*, NJ: IEEE Press; 2003. pp. 332–48.

[22] Lewis F. L., Zhang H., Movric K. H. *Cooperative Control of Multi-Agent Systems: Optimal and Adaptive Design Approaches*. London: Springer; 2014.

[23] Foundation for Intelligent Physical Agents. FIPA Contract Net Interaction Protocol Specification [online]. Available from http://www.fipa.org/specs/fipa00029/SC00029H.html.

[24] Khalil H. K. *Nonlinear Systems* (3rd Edition). Englewood Cliffs, NJ: Prentice Hall; 2001.

[25] Islam S. R., Sutanto D., Muttaqi K. M. 'Coordinated decentralized emergency voltage and reactive power control to prevent long-term voltage instability in a power system'. *IEEE Transactions on Power Systems*. 2015;**30**(5):2591–603.

[26] Carvalho P. M. S., Correia P. F., Ferreira L. A. F. M. 'Distributed reactive power generation control for voltage rise mitigation in distribution networks'. *IEEE Transactions on Power Systems*. 2008;**23**(2):766–72.

[27] Vaccaro A., Velotto G., Zobaa A. F. 'A decentralized and cooperative architecture for optimal voltage regulation in smart grids'. *IEEE Transactions on Industrial Electronics*. 2011;**58**(10):4593–602.

[28] Brenna M., Berardinis E. D., Carpini L. D., *et al*. 'Automatic distributed voltage control algorithm in smart grids applications'. *IEEE Transactions on Smart Grid*. 2013;**4**(2):877–85.

[29] Augugliaro A., Dusonchet L., Favuzza S., Sanseverino E. R. 'Voltage regulation and power losses minimization in automated distribution networks by an evolutionary multiobjective approach'. *IEEE Transactions on Power Systems*. 2004;**19**(3):1516–27.

[30] Barbarossa S., Celano F. 'Self-organizing sensor networks designed as a population of mutually coupled oscillators'. *IEEE Sixth Workshop on Signal Processing Advances in Wireless Communications*; Columbia University, New York, NY, USA, June 2005. IEEE; 2005. pp. 475–79.

[31] Nedic A., Ozdaglar A. 'Distributed subgradient methods for multi-agent optimization'. *IEEE Transactions on Automatic Control*. 2009;**54**(1):48–61.

[32] Maknouninejad A., Qu Z. 'Realizing unified microgrid voltage profile and loss minimization: a cooperative distributed optimization and control approach'. *IEEE Transactions on Smart Grid*. 2014;**5**(4):1621–30.

Further Readings

[1] Katiraei F., Iravani R., Lehn P. 'Small-signal dynamics model of a micro-grid including conventional and electronically interfaced distributed resources'. *IET Generation, Transmission & Distribution*. 2007;**1**(3):369–78.

[2] Pujol G., Rodellar J., Rossell J. M., Pozo F. 'Decentralised reliable guaranteed cost control of uncertain systems: an LMI design'. *IET Control Theory & Applications*. 2007;**1**(3):779–85.

[3] Li T., Zhang J. F. 'Asymptotically optimal decentralized control for large population stochastic multiagent systems'. *IEEE Transactions on Automatic Control*. 2008;**53**(7):1643–60.

[4] Etemadi A. H., Davison E. J., Iravani R. 'A generalized decentralized robust control of islanded microgrids'. *IEEE Transactions on Power Systems*. 2014;**29**(6):3102–13.

[5] Vaccaro A., Loia V., Formato G., Wall P., Terzija V. 'A self-organizing architecture for decentralized smart microgrids synchronization, control, and monitoring'. *IEEE Transactions on Industrial Informatics*. 2015;**11**(1):289–98.

[6] Dou C., Lv M., Zhao T., Ji Y., Li H., 'Decentralised coordinated control of microgrid based on multi-agent system'. *IET Generation, Transmission & Distribution*. 2015;**9**(16):2474–84.

Chapter 10
Techno-economic issues of power quality

Jayesh Joglekar[1]

10.1 Introduction

The Electricity Act 2003 rings in changes in the Indian power sector. The distribution company was forced to adopt reformation to reduce technical and business losses, responsibility of power offer, improvement in client satisfaction, improvement in power quality, and rationalization of electricity tariff for improving service. For restructuring of this sector, there is a requirement of "smart grid" technology to extend the operational strength of the distribution network to fulfill the growing energy demand of India in line with the GDP growth of the country. The "smart grid" enhances digitally powered systems' ability wherever the employment of contemporary communications and management technologies are required. It is more sturdy, economical, and versatile compared to present power systems. In a very sensible grid, all the points have to be compelled to interconnect and share information as and wherever required. The sensible grid imagines providing selections to all clients for deciding the temporal order, and quantity consumed power supported the worth of the ability at a specific moment of your time, aside from providing selections to the patron and motivating them to participate in the operations of the grid, inflicting energy potency, and accommodating all generation and storage choices. The sensible grid additionally predicts numerous properties for the grid, like self-healing and adaptation of restoration. The sensible grid merchandise and technologies facilitate increasing system period, while additionally serving to the utility quickly to revive power to home-associated/businesses within the event of an outage. The Government of India has recently fashioned "Smart Grid Forum" and "Smart Grid Task Force" for enablement of sensible grid technology into Indian Power Distribution Utilities as a neighborhood of their sensible grid initiative to fulfill their growing energy demand in parallel with the developed countries [1].

[1]Electrical Engineering Department, Maharashtra Institute of Technology, Kothrud, Pune 411038, India

10.2 Different approaches for finding out power quality impact on tariff

Indian economy is growing at the speed of 9% during 2011–12, which within the returning years is ready to achieve inclusive number growth (10% and more). However, India suffers from serious power shortage that is probably going to worsen over the following few decades. On one hand, the square measure issues with the shortage of adequate generation capability with power cuts go several hours, which is still rife in many cities. On the other hand, there are square measure issues with the shortage of transmission infrastructure with many thousand villages that are still not connected to the national grid. In India, 15%–20% power transmitted out of total is lost within the transmission and distribution network, whereas 10%–20% is lost due to theft across the utilities. Although these losses are normalizing slowly over the recent years, there is still need to select the utilities to realize the specified state of operations. India has additionally been missing its generation infrastructure enlargement plans for the last many decades. Recently, the Government of India proclaimed the National Mission on increased energy strength that aims at rising energy power in the industrial sector as a business, whereas for the economic sector a theme the same as the carbon mercantilism, referred to as the energy potency certificate theme is in work, the case with the business and residential sector is slightly sophisticated. The industries will look out of themselves once the binding force targets are stipulated to them; however, the government must place in an additional effort for rising potency within the commercial/residential sector. Therefore, such a state of affairs requires the implementation of the sensible grid. A sensible grid would facilitate the utilities for getting information concerning the electricity used by the customers. It might doubtlessly adopt by the utility as a distribution method relevant with the time and quantum of power demand. The smart grid that uses smart meters may doubtlessly be used for investigating power crime. In addition, the data that the customers would have access to via the sensible meters may facilitate them to manage their energy use in an exceedingly higher and in a lot of economical methods. Such hypergrowth can exact unprecedented quantity of energy, which will solely be met through economical energy utilization through sensible metering and sensible grid technology [1].

10.3 Design of modules to associate disturbance and economic loss [2]

(Courtesy: APQI: Asia Power Quality Initiative India)

The power quality (PQ) is critical to the industries attributable to increasing application of electronic masses and digital controllers. This area is unit sensitive to the standard of power. The minor power disruptions that one would have noticed solely as fleeting loss of the lights might currently interrupt the entire machine-driven manufacturing assembly lines with sensitive electronic controllers. Hence,

PQ problems are not solely technical; however, those can even have a robust economic impact. If not addressed well, dangerous PQ management selections will value business uncountable rupees every year in revenue loss, method enhancements, chance loss, failure, and electrical hazards like the fireplace and scrapped product.

Within the Indian context, the huge economic loss attributable to poor PQ that would be suffered by industries continues to be not well understood or captured. The area unit of economical losses is reciprocally proportional to PQ, that is, poor power quality would increase the losses and vice versa. The power accessibility, in the Indian context has invariably been an important base for the entire commercial associates. In nurturing social unit operations since decades, its quality at an optimum level is additionally necessary. Currently as we tend to grow into a digital the world, area of policy makers is deeply engaged in addressing the demand gap within the country. PQ problems area unit still has not been given the due thought that can aggravate our crucial assets within the network. One study done by Wartsila Asian in August 2009 indicated that India suffers a staggering loss of Indian National Rupees (INR) 100,000 large integers (INR 1,000 billion and $ 15.4 billion) attributable to nationwide power disturbances (including power outages). It also indicated that Indian industries are spending INR 30,000 crores (INR 300 billion and $ 4.62 billion) annually to operate inefficient power backups using generator and inverters. At constant time, trendy industrial instrumentation is additionally sensitive to those PQ problems than ever before and hope for the higher quality of wattage.

10.3.1 Case study 1: cement plant

Problem: *The facility is one of the major cement manufacturing plants with state-of-the-art technology. It has two manufacturing lines, and each line has eight cooler fans, four ESP transformers, and raw mill classifier. The facility receives power from utility at 33-kV level. The plant team decided to install variable frequency drives (VFDs) for the cooler fans in order to maintain quality of the product and optimize energy cost. However, after installation of VFDs, it started facing losses in its key equipment like motors, conductor, transformers, and others resulting in high-economic loss for the plant. The economic loss incurred by the plant was over INR 60 lakhs per annum on account of these factors. It also faced the following problems:*

- *Nuisance tripping of circuit breakers*
- *Failure of motor winding due to high temperature*

10.3.1.1 Solutions and benefits

In view of above problems, the plant team carried out a detailed power-quality study specific for harmonics at the facility. It was observed that both current and voltage harmonics distortion were up by 32.5% and 7.4%, respectively. Both these values were above permissible limits as specified by IEEE-519 standard. As suggested by the PQ consultant, the plant team opted to install active harmonic filters (AHFs) at all locations where they have installed VFDs and made an investment of INR 2.20 crores. By installing AHFs, the plant team successfully mitigated harmonics with an annual monetary benefit of INR 60 lakhs (payback of 3.8 years).

10.3.1.2 Conclusion

The plant team reduced harmonic level by significant level and also gained following benefits:

- Reduction in losses at motors, distribution system, and transformers giving direct energy savings.
- Avoided any nuisance tripping of circuit breaker.
- Reduction in RMS value of current, failure rate of sensitive electronics, lesser production downtime, and lesser production waste.

10.3.2 Case study 2: industry

Problem: *Delphi-TVS is a JV between Delphi Corporation, USA, and T.V. Sundaram Iyengar & Sons, India, and manufactures Diesel Fuel Injection Equipment for different segments of light motor vehicles, tractors, and others. Fuel injection system is very complex and needs precision control for better combustion and efficiency of the system. Initially, the plant was receiving energy supply from 11-kV feeder that was also catering power requirement of various other loads like mix of industries, domestic, and agricultural customers. Because of this, the plant started facing various power-quality issues (external like frequent interruptions, voltage unbalance and internal like failure of equipment, its tripping, and others.). Due to poor PQ, the plant started incurring economic loss of nearly 15–20 crores. Annually, the plant had to be operated for all 7 days (instead of 6 days) to meet its production, thereby incurring additional loss of INR 520 lakhs per year and also impacting the business in the following ways:*

- *Too many electronic failures resulting in increased downtime.*
- *Purchase of costly tools resulting in increased M/C tools.*
- *Customer dispatch was seriously affected.*
- *Increase in scrap & rejection cost.*

10.3.2.1 Solutions and benefits

The plant team carried out the power-quality audit and for each identified issues took mitigation measures. Presence of noise was affecting performance of CNC machines. The team installed isolation transformer at the equipment level to mitigate the noise produced by various equipment. Due to switching of load, voltage dips were observed when plant was operating on DG sets. As a temporary measure, plant changed operating practice to avoid voltage dips due to switching of loads. Motors above 50 HP were started first, and then power was resumed to the equipment whenever power was supplied through DG set. The plant went on to invest in captive power generation to improve the supply condition. With above all and installation of PQ-mitigation equipment like capacitors, conductors, circuit breakers, and others, the plant invested a total INR 32 crores over few years. Due to this, it saved nearly annual INR 10 crores, which if not incurred, would have caused plant annual economic loss of over INR 30 crores.

10.3.2.2 Conclusion

By making considerable PQ investments, the plant had remarkable annual loss savings. It also got additional advantages as mentioned below:

- Workspace saving (removal of voltage stabilizers as DG sets were supplying critical loads).
- Saving in additional working day and its associated cost (to meet the target, the plant started operations for 7 days incurring additional cost of INR 520 lakhs per year. By saving a day and working for 6 days, it saved INR 500 lakhs per year).
- Meeting customer schedule and others.

10.3.3 Case study 3: hospital

Problem statement: *The hospital is one of the major hospitals in Hyderabad with specialization in treatments of heart disease, neurosurgery, cancer, and other life-threatening diseases. The hospital was facing problems in their distribution system like higher temperature in the transformer, humming noise in the capacitor bank, disturbance on monitor screens of machines, higher energy charges, and others, resulting in high economic loss. It noticed the presence of harmonics at point of common coupling, higher than 30% as per the limits set by standards.*

10.3.3.1 Solutions and benefits

After detailed power-quality study, the hospital management decided to mitigate these harmonics by installing 120-A AHF at an investment of INR 7.5 lakhs. This has resulted in significant reduction in energy consumption (thereby resulting in monetary benefit due to reduction in energy cost), improved services, and diagnosis capabilities due to reduction in harmonics. Post-installation study was carried out to analyze the benefits achieved by installing AHF. Table 10.1 shows the results of study before and after installation of harmonic filter.

By installing PQ-mitigation equipment at an investment of INR 7.5 lakhs, the management gained substantial monetary benefit in terms of reduced electricity charges. Energy charges were reduced by INR 7.9 lakhs, and the hospital had simple payback period of less than 1 year. The facility also got additional advantages as mentioned below:

- Reduction in total harmonic distortion (THD) being injected in the grid, thus avoiding any future penalty levied on THD limits by SEB.
- Reduction in humming noise at the capacitor bank.

Table 10.1 Effect of harmonic filter [2]

Component	Before harmonic filter	After harmonic filter	Change
Total harmonic distortion (THD)	25%	5%	↓80%
Power factor	0.97	0.995	↑2.5%
Avg. kVA	216 kVA	204 kVA	↓5.55%
Transformer temperature	80–85 °C	45 °C	↓47%

10.3.3.2 Conclusion

It is quite evident from the case studies presented above that considering PQ mitigation does save end-user's economic loss. While accurate estimation is difficult, however, to select an optimum PQ-mitigation method in the network, an economic impact analysis (EIA) is to be carried out by the involved parties to validate compliance, improve system stability, and eliminate unplanned downtime, thereby contributing to the overall turnaround of utilities and their services for end-users.

10.4 The relationship between duration of disturbance and its cost–benefit analysis index [3]

In the context of debates regarding the cost-effectiveness of good grid dependableness investments, it is necessary for utilities and regulators to own a typical framework for analytic thinking that properly accounts for the social advantages that arise from utility investments in dependableness. The key challenge in developing this framework lies in adopting sensible rules for assessing the value of service dependableness. Good grid investments will yield a large variety of advantages that represent completely different classes, like environmental and security advantages. This discussion, however, addresses the economic advantages that will result from dependableness enhancements. Economic advantages replicate potency enhancements within the use of capital, fuel, and labor within the generation, transmission, and distribution of electricity, as well as conservation impacts ensuing from providing higher data and evaluation to customers. Depending on advantages area, unit reduced prices to utilities and customers ensuing from fewer service interruptions and power-quality deterioration. In general, dependableness issues represent two distinct categories—service interruptions/outages and power-quality disturbances (i.e., momentaneous interruptions and voltage disturbances). And for many intents and functions, interruption prices and power-quality prices may be treated within the same approach once considering the value of dependableness enhancements. In addition, there are three vital dimensions of electric service reliability: the quantity of shoppers affected, the frequency with which outages or voltage disturbances occur, and also the period of service interruptions. Dependableness may be measured in a number of ways. For the needs of transmission and distribution designing, it is measured in two ways, of which one approach is to estimate the amount of unserved energy (in kilogram–watt–hours) that results, or is predicted to result, from service interruptions. This approach needs associate in nursing estimation of the number of kilowatt–hours that may be demanded if electricity had been accessible throughout those interruptions. This methodology rests on an awfully tough estimation downside—specifically, the necessity to estimate the amount of electricity that may be demanded if unreliableness had not occurred. The estimation of unserved kilowatt–hours would wish to require under consideration the time of day, the day of the week, and also the season of the year throughout which the outages occurred. These assumptions will powerfully have an effect on the estimates of expected unserved energy, adding a major supply of uncertainty. In objectively applied mathematics indicators of area dependable, additional unit used once assessing the dependability of the systems. Most utilities and regulative bodies within the United States ordinarily describe the dependableness of transmission and distribution circuits in terms of straightforward, without delay getable and clear indicators.

10.5 Improvement of power quality in the system and its expected benefits [4]

Power quality programs (PQP) became a number of the foremost recent services among distribution corporations, each for personal and state suppliers. For any distribution system to satisfy its shoppers, the utility should keep up power quality during a manner that accommodates the raised demand for electricity. The need the PQP to be enforced to begin endeavoring the difficulties faced the distribution utilities in sustaining a high customary of power quality. A PQP will facilitate in reducing the large range of complaints from finish users and, therefore, the prices diagrammatic within the harm to their instrumentality. It can even have a positive impact on the electrical distribution corporations, up to their service and saving some of the numerous resources spent. Therefore, the distribution corporations got to implement a PQ investigation program despite all the previous facts indicating a rise in PQDs within the last 20 years specially [5]. Establishing a clear vision has multiple barriers. The shortage of transparent strategy, client awareness, accommodating economic process, equipment's standards, network style, resources, employee's awareness, principal management responsibility and power-quality standards are the barriers in the entire process. In conjunction with an excessive increase of electronic devices, then any efforts to enhance power quality are the waste in time and resources. Consequently, many undeveloped countries have compelled their utilities to implement PQPs as a necessary service condition in response to the increase in consumer complaints. The poor PQ caused by the rise in refined industrial and industrial equipment, whereas customers do not demand PQ standards to run it. LDNs square measure among those systems facing poor power quality in underdeveloped countries. Sadly, applied mathematics information shows that within the last 20 years, LDNs have not enforced power-quality program. This can primarily be as a result of there being no power-quality department established, nonetheless, to influence the measuring of PQDs. This absence of an influence quality department is because of lack of awareness on a part of prime management concerning the importance of power quality. As a result, LDNs have faced terribly important difficulties in implementing PQP. In addition, lack of power-quality awareness has crystal rectifier LDNs to face 12 important difficulties through not implementing PQP.

10.6 Power-quality investment and gross domestic product in developing countries: case study [5,6]

(Courtesy: United States Agency for International Development under South Asia Regional Initiative for Energy Integration (USAID SARI / EI))

10.6.1 Case 1: Nepal

The electricity trade, which is crucial for economic and social development, has a very important role to play within the development of the Nepalese economy. The study is meant to spotlight the importance of active end-user participation within the decision-making method of the electricity offer trade. The study examined the economic impact of poor quality of electricity delivery on the commercial installations in Nepal. The study was designed to think about the economic and direct environmental impacts of power interruptions, voltage fluctuations, and alternative vital issues, such as offer harmonics to supply a basis for the assessment, existing tips on power quality were reviewed, the economic impact of power quality was analyzed for a sample of commercial customers, self-generation prices and their environmental impact were calculable, and suggestions for power-quality enhancements were developed.

The investigation was applied employing an elaborated nationwide sample survey of quite 200 installations covering the most classes of industries causative to the country's gross domestic product (GDP) growth. A structured form was administered throughout August to October 2002. The information employed in the study square measure as reported for FY 2001.

The study found that industrial sector losses in Nepal, thanks to unplanned interruptions averaged 0.49 US$/kWh; however, such losses were solely 0.14 US $/kWh for planned outages. Thus, the unplanned interruptions lead to loss 3.5 times larger than related to planned interruptions. Further, the voltage fluctuations will cause major issues in sure industrial classes, like food, beverages, and tobacco producing, and textile and animal skin producing, whereas alternative classes, such as the non-metallic and mineral industries, square measure will be utterly unaffected. Since the response on alternative offer issues, such as harmonics, was very restricted, no analysis was applied in those areas.

The study found that the frequency of unplanned interruptions throughout the year was virtually one interruption each 2 days, which lasted regarding associate hour and a 0.5. However, planned interruptions were less frequently occurring, solely double during a month; however, every interruption lasted regarding a pair of hours. Therefore, 86 of the entire unserved energy demand ensuing from interruptions is as a result of unplanned outages; however, solely 14 July, thanks to the planned outages. In total, about V-day of the commercial sector demand cannot be met by the electrical utility as a result of planned and unplanned interruptions.

Further, these outages lead to associate economic loss in Nepal's industrial sector amounting to US$ 24.7 million a year. This interprets into 4.43% of the commercial sector value or 0.47% of the national value in 2001.

On average, 76 of the commercial installations within the country have standby generation (SBG) facilities; however, they are able to offer solely regarding 74 of the total load electricity needs, assumptive that the SBGs square measure operating throughout the grid offer outages, the extra monetary burden on the commercial sector may be calculable to be regarding US$ 1.2 million a year.

Over 80 of the industries taking part reported that they overpay for the standard of power they receive; nonetheless, 51 of the samples still would be willing to pay even a lot of for higher power quality. Also, about common fraction of the sampled industries would have invested with a lot in electricity-dependent industries if the standard of power offer was higher.

The study jointly examined the Indian Electricity Regulation 2003 act related to power offer quality. Underneath this regulation, the licensee ought to make sure that the voltage at the distribution level below 33 kV does not fluctuate on the far side ±5%. (Sections 4.0–4.2). For transmission voltages that square measure 33 kV and higher than this, fluctuation has to be inside ±10%. The frequency of fluctuations has to be maintained inside ±2.5% of the quality frequency of 50 Hz alternative power-quality problems, such as frequency and period of planned and unplanned interruptions, offer harmonics, and others, do not seem to be addressed within the existing legislation.

The comparatively high value of unplanned outages as compared with planned outages needs special attention to cut back the frequency and period of unplanned outages and to convert such outages to planned outages whenever applicable. Voltage fluctuations cause a major adverse impact on sure sorts of industries, and this impact has to be seriously addressed. The terribly high level of SBG within the industrial sector is probably going to extend within the future unless the mandatory measures square measure is taken to cut back grid offer outages.

If the quantity of outages and therefore the voltage fluctuations square measure reduced, the next tariff is probably going to be acceptable to special trade classes, such as food, beverages, and tobacco; textile and animal skin products; and iron and steel. Further, the implementation of energy conservation measures within the industrial sector at the moment is comparatively low. Hence, this sector desires associate aggressive energy conservation program.

To address these problems, the subsequent recommendations may be made

- Strengthen end-user participation within the electricity offer trade and establish a daily dialog between the end-users, utilities, and policy-makers.
- Improve codes and standards on the standard of service, particularly needs regarding offer interruptions.

- Adopt measures to scale back unplanned interruptions by changing them to planned outages where doable.
- Strengthen in progress South Asia regional initiative on energy (SARI/Energy) initiatives on energy potency, such as energy service company activities and commonplace setting and labeling of end-use appliances, to confirm wider awareness and implementation of energy potency measures.
- Improve the facility system performance by timely and acceptable transmission capability additions, furthermore as distribution system enhancements.

Table 10.2 shows result of voltage variation on industrial shoppers, and Table 10.3 shows period for varied interruptions.

Based on the study results and analysis shown in Tables 10.2 and 10.3, the following conclusions will be drawn as shown in Figure 10.1:

- The average operating amount of the economic installations sampled is long each in terms of variety of operating hours on a daily basis, furthermore because the variety of operating days per week, and thus, the economic impact of offer outages within the industrial sector is important.
- Electricity is employed expeditiously in terms of important addition per unit of electricity within the following trade categories: iron and steel; chemical, petroleum, and rubber; food, beverages, and tobacco; and textile and animal skin. Therefore, the economic advantage of giving discriminatory treatment to these industries, notably at the time of planned power offer interruptions, must be seriously thought of in system operations.

Table 10.2 Cost of voltage variations per installation [5]

Industry	US$/interruption range with 90% confidence	
	From	**To**
Food, beverages, and tobacco	5.46	24.42
Chemical petroleum, rubber	0.63	3.43
Textile and leather	8.65	33.30
Iron and steel	1.38	18.54
Hotels	0.23	1.37
Nonmetallic and minerals	0.00	0.00
Miscellaneous	0.00	16.78
Industry sector average	**4.02**	**9.47**

Table 10.3 Frequency and duration of interruptions [5]

Type	Frequency/year	Period of each (h)
Unplanned outages	185.37	1.57
Planned outage	22.51	2.08
Voltage variation	111.93	2.24

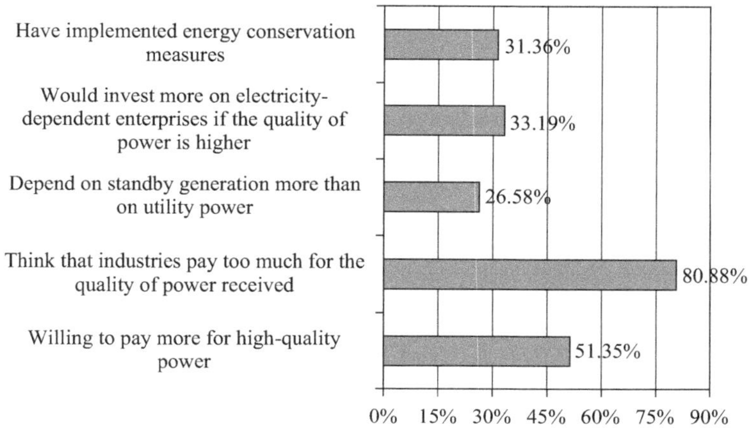

Figure 10.1 Frequency and duration of interruptions

- The operation and maintenance value of the diesel-based SBGs is more than the common value of utility delivered electricity within the industrial sector. Thus, a lot of reliable electricity offer from the utility even at a comparatively higher value would still be a lot of value economical than the present state of affairs.
- The comparatively high value of unplanned outages compared with planned outages necessitates special attention to scale back the frequency and period of unplanned outages or to create strenuous efforts to convert such outages to planned outages whenever doable. In this regard, additional stress is needed for the textile and animal skin, and iron and steel classes because of the high value of such outages.
- Voltage fluctuations cause a major adverse impact on the economic sector output, that must be seriously self-addressed by the utilities.
- The industrial sector incorporates a terribly high level of SBG, and it seems to extend within the future unless necessary measures are taken to scale back offer outages, notably unplanned outages.
- Overall economic losses because of planned and unplanned power offer outages within the industrial sector are terribly high and have a major impact on the country's economic process.
- Although the industries surveyed contemplate the current electricity tariff to be too high for the standard of power they receive, a better tariff is probably going to be acceptable to big trade classes, such as textile and animal skin, and iron and steel, if the number of outages and also the voltage fluctuations are reduced.
- Emphasis on energy conservation measures within the industrial sector at the moment is very low. Implementing AN aggressive energy conservation program within the sector would beyond question facilitate the electricity offer trade furthermore as shoppers.

10.6.2 Case 2: Sri Lanka [5]

Figure 10.2 shows rate of growth of electricity consumption against GDP. This study, as shown in Tables 10.4 and 10.5, states that associational planned power outages of 300 h annually translate into a calculable US$ 81 million reduction within the GDP of country. On similar lines, associate assumed that 100 h of unplanned outages annually causes a discount of regarding US$ 47 million in GDP.

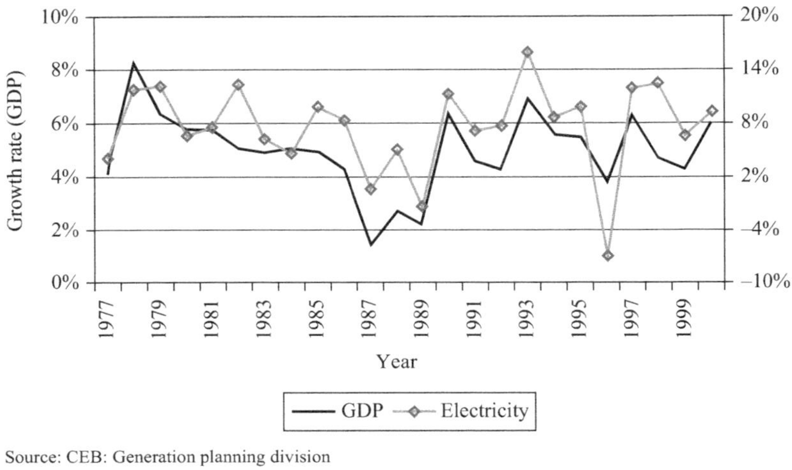

Figure 10.2 Correlation between electricity demand and economic activity in Sri Lanka

Table 10.4 GDP contribution of each industry category considered [6]

Categories	GDP US$ million	Percentages
Food, beverages, and tobacco	545	22
Textile, wearing apparel and leather products	772	32
Wood and wood products	17	1
Paper and paper products	31	1
Chemical, petroleum, rubber and plastic products	197	8
Nonmetallic mineral products	158	7
Basic metal products	11	0.4
Fabricated metal products, machinery, and transport equipment	86	4
Products not specified elsewhere	44	2
Tea	158	7
Rubber	23	1
Coconut	133	5
Hotels	250	10
Total	2,425	100

Table 10.5 Economic losses due to supply interruptions in each industry category average losses (US$) [6]

Industry category	Momentary (US$)	Planned (US$/h)	Unplanned (US$/h)
1. Food, beverage and tobacco products	301.61	153.39	363.24
1.1 Food and others	211.68	166.18	279.28
1.2 Liquor	2.60	7.41	7.41
1.3 Beverage	139.48	160.67	160.67
1.4 Tobacco products	4,362.88	15.55	4,378.43
2. Textile, wearing apparel, and leather products	121.55	107.94	186.36
2.1 Apparel	120.99	81.77	161.25
2.2 Textile	109.87	133.55	198.70
2.3 Leather	157.93	247.98	354.08
3. Chemical, petroleum, rubber, and plastic products	155.57	83.12	217.90
3.1 Chemicals, paints and fertilizers	220.98	315.81	315.81
3.2 Rubber	66.63	56.97	110.07
3.3 Plastic and PVC	187.17	110.17	288.27
3.4 Pharmaceuticals, detergent, and others	179.52	4.55	199.49
3.5 Petroleum products	0.00	65.00	65.00
4. Nonmetallic mineral products	169.32	108.08	266.51
4.1 Diamond processing	50.48	50.99	50.99
4.2 Ceramic products	100.02	98.34	171.81
4.3 Cement	782.39	581.59	1,636.31
4.4 Building materials and others	134.79	54.00	151.10
5. Fabricated metal products, machinery, and transport equipment	125.25	415.90	429.31
6. Tea industry	4.23	21.75	21.75
7. Coconut industry	0.82	75.17	131.20
8. Hotel industry	0.00	25.57	25.57

It is going to be noted that these figures are supported by domestic GDP of country of roughly US$ 12.5 billion within the year 2000.

These losses owing to planned and unplanned interruptions can also be diagrammatic within the sort of economic loss (in US$) per unit of provide loss (in kWh) known because the value of unserved energy. The study shows that for the country system, these values are US$ 0.66 per kWh (59.34 INR/kWh) for planned outages and US$ 1.06 per kWh (97.20 INR/kWh) for unplanned outages as compared with the common industrial tariff, that ranges between US$ 0.077 per kWh (7.0 INR/kWh) and US$ 0.083 per kWh (7.50 INR/kWh), the price of unserved energy owing to planned outages is 7.9–8.5 times the common value of electricity, and therefore the value of unplanned outages is from 13.0 to 13.8 times the common value of electricity to industrial customers. Table 10.6 shows result of power quality on trade.

The study unconcealed that 92% of commercial facilities have backup generation capability that takes 5–15 min to induce the backup generation method,

Table 10.6 Highest and lowest losses by category [6]

Type of outage	Highest loss	Lowest loss
Momentary outage	Food, beverage, and tobacco products	Hotel, tea
Unplanned outage	Food, beverage, and tobacco products	Hotel, tea
Planned outage	Fabricated metal products, machinery, and transport equipment	Hotel, tea

once started. In some cases, like the glass trade, backup power generation units were used for traditional operation, and power from electricity board was used below abnormal circumstances. This sort of operation clearly indicates the shortage of religion that such customers have within the quality and dependableness of utility provided power.

10.7 Conclusions

The power quality issue is as old as the electrical invention. The complexity in the system and day-by-day increasing size forces supply company to pay attention to power quality with new dimensions. With the implementation of new Electricity Act in India, consumers have become more aware of power quality and hence the power supply companies. The chapter focuses on the tariff-related issues correlation with power quality and need of revenue for improving the condition. It is observed that all developing countries have moreover same condition and policies to address power quality. It needs to change and should become more consumer-friendly as well as it should reduce burden from power supply or distribution companies so that it can encourage use of quality power.

Acknowledgment

Author would like to thank the original contributors in References who published their work in the particular area. The original articles were referred to from online/offline source available.

References

[1] A. Sinha, S. Neogi, R.N. Lahiri, S. Chowdhury, S.P. Chowdhury and N. Chakraborty "Smart Grid Initiative for Power Distribution Utility in India", IEEE Power and Energy Society General Meeting, San Diego, CA, 2011, pp. 1–8; 2011 (doi: 10.1109/PES.2011.6038943).

[2] Asia Power Quality Initiative (APQI) *India, Economic Impact and Losses due to Poor Power Quality*, dated: 13 October 2015.

[3] M.R. Haghifam and M. Daneshgar, "Developing a New Algorithm to Participate Power Quality Variations in Tariff: An Experience in Iran," 16th International Conference and Exhibition on Electricity Distribution, 2001. Part 1: Contributions. CIRED. (IEE Conf. Publ No. 482), Amsterdam, Netherlands, Vol. 2, 2001 (doi: 10.1049/cp:20010751).

[4] Saad S. Sultan and Mohamed K. Darwish, "Barriers and Benefits of Implementing a Power Quality Program: Case Study Libyan Distribution Networks," *International Journal of Scientific & Engineering Research*, vol. 3, no. 4, Apr. 2012, ISSN 2229-5518.

[5] United States Agency for International Development under South Asia Regional Initiative for Energy (USAID-SAIE) report, "Economic Impact of Poor Power Quality on Industry: Sri Lanka," January 2002.

[6] United States Agency for International Development under South Asia Regional Initiative for Energy, (USAID-SAIE) report, "United States Agency for International Development under South Asia Regional Initiative for Energy, Economic Impact of Poor Power Quality on Industry: Nepal," October 2003.

Chapter 11

An economic robust programing approach for the design of energy management systems

Felipe Valencia Arroyave[1] *and Alejandro Marquez Ruiz*[2]

Energy management systems have a key role in the operation of microgrid applications. Due to their importance, several approaches have been reported in the specialized literature. In this chapter, a novel formulation for robust optimization-based EMS for microgrid applications. The novelty recast in the introduction of transmission constraints and in the avoidance of the use of prediction models for the uncertain variables. In this formulation, a time-variant uncertainty set is used to include the uncertainty in the optimization process. Unlike in the approaches reported in the specialized literature, the sets are previously defined based on the empirical statistics of historical data available. Thus, at each sampling time, the boundaries for the uncertain variables are obtained without using a prediction model. This reduces the complexity of the EMS making possible to compute the solution of the associated optimization problem in the range of the seconds. A modified version of the IEEE 9-bus system benchmark was used as test bench. In the modified version, the hydraulic generation unit (the largest unit in the system) was changed by an equivalent wind power station. The EMS was in charge of assigning the real power and the voltage level of the remaining two generation units.

11.1 Introduction

The development of energy management systems (EMS) has received much attention in recent years due to its key role in the operation of microgrids, providing an optimal use of local energy resources, and infrastructure available to supply the demand of a predefined area (very often remote and disconnected of a main grid). An EMS is a control loop used in microgrid applications to achieve reliable and cost-effective operation through the dispatch and commitment of the available

[1]Solar Energy Research Center, SERC-Chile, Energy Center, Department of Electrical Engineering, University of Chile, Santiago, Chile
[2]Control Systems Group, Department of Electrical Engineering, Eindhoven University of Technology, 5600 MB Eindhoven, The Netherlands

generation units. Often the EMS is implemented using optimization-based control techniques that additionally provide the required coordination among energy sources for a safe and secure operation. Indeed, at this level, the most cost-effective combination of generation units to meet the predicted load and reserve requirements is defined [27]. Since microgrids have become one of the more appealing alternatives for electrification purposes, it is required the design of an EMS able to properly accomplish with the aforementioned requirement (coordination among energy resources considering a cost-effective, save and secure operation of the microgrid) [6,11,15]. However, it has been found that an adequate performance of optimization-based EMS is highly related with the degree of representativity of the model used in the optimization to represent the behavior of a microgrid. One way to enhance the performance of optimization-based EMS approaches is to incorporate an uncertainty model into the formulation of the optimization problem, so that the effects of the differences between the model and the real system (due to parameter mismatches, unmodeled phenomena, the presence of stochastic variables such as wind and solar generation) are included within the solution of the EMS problem.

Several optimization methods with application to management and decision making under uncertainty have been reported in the specialized literature (see e.g., [2,26]). In [14,24–26], these approaches have been classified as: stochastic and robust. In the former, several realizations of the uncertain variables are considered. Each realization has a probability of occurrence. Depending on this probability, a solution is obtained. In the later, it is assumed that the parameters and/or constraints in the optimization model are uncertain and prescribed into a defined set [3]. Depending on the construction of the uncertainty set, a solution is obtained. The advantage of robust over stochastic approaches is that in the robust framework (i) the use of probability density functions is not required, that is, only the range of the uncertain data is required, and (ii) the solution is robust against all realizations of the uncertain data within a deterministic uncertain set [2]. The disadvantage is that there are no guidelines available for the construction of the uncertainty set. Consequently, a solution more conservative than necessary could be achieved. Examples of both a stochastic and a robust EMS have been presented in [29,30]. In [30], probability density functions were used to generate the scenarios for each uncertain variable, while in [29] a methodology to generate time-dependent uncertainty sets was proposed.

The use of stochastic and robust optimization has not only been limited to microgrid applications. In large-scale power systems some additional approaches has also been reported. In [2,28,31], multi-stage optimization-based methodologies were proposed for the unit commitment in systems with wind generation and price responsive demand. In these approaches, the first decision stage consisted on the unit commitment and the second decision stage consisted on the power dispatch. Moreover, in [31], a third decision stage was added to include the uncertainty in the price-elastic demand curve. Furthermore, in water resources management the authors of [14,24–26] investigated the use of two-stage stochastic fuzzy programming with interval-valued membership functions. Specifically, the support of water resources management was addressed in [24], the flood-diversion planning was analyzed in [25], and the municipal solid waste management under uncertainty was studied in [14,26]. Since interval-valued membership functions provided upper and lower values for the uncertain variables, in these approaches the first decision was made considering each boundary of the fuzzy model as a deterministic scenario, then a feasibility test was performed to eliminate the unfeasible solutions, and finally the second decision was made over the feasible solutions. At the end of the process, a set of solutions was obtained. How to select an alternative from the resulting set of solutions is still a pendent issue.

Notwithstanding the broad use of stochastic and robust optimization approaches in decision making under uncertainty, particularly in energy management in power systems of different scales (microgrid applications and large-scale power systems), very often their performance depends upon the performance of prediction models used to generate representative realizations of the uncertain variables (stochastic approaches) and to define the expected range of variation of the uncertain variables (robust approaches). Moreover, the use of predictions increases the complexity of the EMS, which increases the computational burden of the solution accordingly. Indeed, several EMS has solution times in the range of the minutes (e.g., the EMS proposed in [16] and in [15]) requiring additional control loops to avoid permanent frequency and voltage deviations produced by the action of local primary controllers. Furthermore, in the context of microgrid applications little attention has been paid to the addition of transmission constraints in the EMS.

The present chapter presents a novel formulation for robust optimization-based EMS for microgrid applications. In this formulation a time-variant uncertainty set is used to include the uncertainty in the optimization process. Unlike in the approaches reported in the specialized literature, the sets are previously defined based on the empirical statistics of historical data available. Thus, at each sampling time the boundaries for the uncertain variables are obtained without using a prediction model. This reduces the complexity of the EMS making possible to compute the solution of the associated optimization problem in the range of the seconds. In this EMS transmission, constraints are also included in the formulation of the optimization problem. Since transmission constraints are nonlinear nonconvex, the formulation proposed in [7–10] was used. Such formulation allows transforming the transmission constraints in linear convex ones through a nonlinear bidirectional variable-transformation. The resulting optimization problem then remains convex (as in the case of the uninodal formulations reported in the literature) and therefore can be solved in finite steps and has a unique solution. Indeed, the EMS becomes a second-order cone programing problem (SOCP) which can be efficiently solved and whose complexity scales-up proportionally to the square root of the number of decision variables. The aforementioned features of the proposed EMS in addition to the reduction of the sample time from the minutes to the seconds allowed controlling both the frequency and the voltage profile all over the microgrid, without including a secondary control. A modified version of the IEEE 9-bus system benchmark was used as test bench. In the modified version, the hydraulic generation unit (the largest unit in the system) was changed by an equivalent wind power station. The EMS was in charge of assigning the real power and the voltage level of the remaining two generation units. The remainder of this chapter is organized as follows: Section 11.2 introduces some basic mathematical concepts of robust programing; Section 11.3 presents the proposed EMS; Section 11.4 provides the case study used to test the proposed EMS and the simulation results obtained; and Section 11.5 presents the concluding remarks.

11.2 Robust programing framework

In this section, some mathematical concepts related with robust programing based on scenarios are introduced. These concepts will be used in Section 11.3 to formulate the proposed robust EMS. For this purpose, consider an optimization problem whose objective is to minimize the difference between a mathematical model and a set of measured values. Very often an optimization problem is formulated to solve this problem. Let $J(x)$ denote the objective function of such an optimization problem, where x is a set of decision variables that determines the fulfilling of the objective of the optimization model. Then, if $J(x)$ is assumed linear such an objective function has the form $J(x) = \|Ax - b\|$. Without loss of generality, assume that A is a random variable with an empirical probability density function describing their variations. Then, finding a value of x that minimize $J(x)$ becomes a minimization of the expected value of $J(x)$, whose solution is closely related with the description of the uncertainty of A. indeed, depending of the way the uncertainty in A is addressed the resulting optimization problem could be tractable or not. For this reason, uncertainty description by means of polytopes or norm-bounded models is proposed in the literature (see [4] for details).

A particular case in which minimizing the expected value of $J(x)$ is tractable is the worst-case problem. In worst-case problem, the uncertainty is described by a set of values for A that constitutes an uncertainty set. Given this set, the uncertainty is analyzed to find the worst conditions for $J(x)$. Once these conditions are identified, the value of x minimizing $J(x)$ is computed. Thereby, the expected value becomes a min–max optimization problem. In the specialized literature, several approaches have been reported to solve min-max optimization problems [1,12,13,17,20]. But often worst-case problems became nonlinear programing problems. With currently computation resources, it is almost impossible to find the exact solution of this type of optimization problems in real time applications [5]. Therefore, transformations and/or linear approximations are used to speed up the computation of the solution of worst-case optimization problems.

One of the most used transformations is the norm-bounded error. In this case, the uncertainty of A is modeled as an ellipsoid that represents the expected variations of A. Let $A = \overline{A} + \Delta A$ represent the variations of A, with $\overline{A}$ the mean value of A and $\|\delta A\| \leq \Upsilon$ the variations of A. Then, the worst-case scenario for $J(x)$ is given by: $\sup\{\|\overline{A}x - b + \delta Ax \mid \delta A \leq \Upsilon\|\}$. Let $\|\cdot\|$ denote the Euclidean norm. Then the resulting worst-case scenario is given by $\|\overline{A}x - b\|_2 - a\|x\|_2$, with $A = auv^T$. Then, the robust approximation minimization problem is formulated as follows:

$$\min_x \|\overline{A}x - b\|_2^2 + a\|x|_2^2 \tag{11.1}$$

which is a regularized norm problem, solvable as the SOCP [4]:

$$\begin{aligned}
&\min_{x,t_1,t_2} t_1 - at_2 \\
&\textbf{s.t.:} \\
&\|\bar{A}x - b\|_2^2 \le t_1 \\
&\|x\|_2^2 \le t_2 \\
&t_1, t_2 \ge 0
\end{aligned} \tag{11.2}$$

Other alternative to represent the uncertainty of A is the use of scenarios. In this case, it is assumed that A belongs to the set $\Xi = \{A_1, \ldots, A_h\}$. In this case, the worst-case scenario is determined by the values of A that maximize the uncertainty. That is, the worst-case scenario is given by the value of $A \in \Xi$ that maximizes the deviation of Ax. Mathematically, this optimization problem can be formulated as follows:

$$\begin{aligned}
&\min_{x,t} t \\
&\textbf{s.t.:} \\
&\|A_i x - b\|_2^2 \le t, \quad i = 1, \ldots, H \\
&t \ge 0
\end{aligned} \tag{11.3}$$

where H is the number of scenarios. Note that minimization problem in (11.3) only requires the definition of the scenarios to represent the uncertain variables. These scenarios could be generated from interval models or from the empirical statistics of the available data. In this chapter, the later approach has been used. To avoid the requirement of prediction models, empirical statistics have been used. Empirical statistics have been selected because they can be easily estimated offline.

11.3 Energy management system as a robust programing problem

The current investigation involved the formulation of a robust optimization-based EMS for microgrid applications. The proposed EMS was formulated on the basis of the mathematical concepts presented in Section 11.2. For this purpose, the statistics of historical data available of the uncertain variables has been used to derive their expected range of variation. In this sense, the use of prediction models was avoided reducing the complexity of the EMS. Furthermore, transmission constraints were included to account for the coupling between real and reactive power that arises in microgrid applications, and in low-voltage distribution systems in general. Since transmission constraints are nonlinear and nonconvex, the variable transformation proposed in [7–10] was used to keep the optimization problem convex. As a consequence, the EMS became a SOCP problem that can be efficiently solved and whose complexity scales with the square root of the number of decision variables. The relationship between complexity and number of decision variables allows scaling-up the size of the microgrid (measured as number of nodes and transmission lines). This provides some degree of flexibility for the application of the proposed EMS in real systems (microgrids and even small and medium size power systems).

In real applications, an EMS is included with the objective of providing online real power set-points for each generation unit [16]. In this case, in addition to real power set-points the EMS also sends voltage set-points to each generation unit. Then, it is expected a better frequency and voltage control without using additional control loops (e.g., secondary frequency and voltage controllers). Both real power and voltage set-point are calculated so that the operational costs are minimized. These costs are often related with the use of diesel generators and/or the use of the energy stored in battery banks to supply the demand. Let C_{gi} denote the generation cost of the ith diesel generation unit; $P_{gi}(k)$ denote the real power provided by the ith diesel generation unit; C_{Bj} denote the cost of using the jth battery bank; $P_{Bj}(k)$ denote the real power provided by the jth battery bank; C_{NS} denote the cost of the unsupplied power; $P_{NS}(k)$ denote the total amount of unsupplied power; $P_{sl}(k)$, $P_{wm}(k)$, and $P_{dn}(k)$, respectively, denote the real power supplied by the lth solar power plant, by the mth wind power plant and demanded by the nth load center. Then the single-node formulation of an EMS is as follows:

$$\min_{\substack{P_{gi}(k),P_{Bj}(k),\\ P_{Lo}(k)}} \sum_{i=1}^{I} C_{gi}P_{gi}(k) + \sum_{j=1}^{J} C_{Bj}P_{Bj}(k) + C_{NS}P_{NS}(k)$$

s.t.:

$$\begin{aligned}
&\sum_{i=1}^{I} P_{gi}(k) + \sum_{j=1}^{J} P_{Bj}(k) + P_{NS}(k) = \sum_{n=1}^{N} P_{dn}(k) - P_{Lo}(k) - \sum_{l=1}^{L} P_{sl}(k) - \sum_{m=1}^{M} P_{wm}(k)\\
&V_{gi}^{\min} \leq V_{gi}(k) \leq V_{gi}^{\max}\\
&E_{Bi}^{\min} \leq E_{Bi}(k) \leq E_{Bi}^{\max}\\
&P_{gi}^{\min} \leq P_{gi}(k) \leq P_{gi}^{\max}\\
&P_{Bi}^{\min} \leq P_{Bi}(k) \leq P_{Bi}^{\max}\\
&P_{NS}(k) \geq 0\\
&P_{Lo}(k) \leq 0
\end{aligned} \tag{11.4}$$

where $P_{Lo}(k)$ is the unused power; $V_{gi}^{\min}$ and $V_{gi}^{\max}$ are the allowed minimum and maximum values of diesel volume for the ith generation unit; $E_{Bj}^{\min}$ and $E_{Bj}^{\max}$ are the allowed minimum and maximum values of stored energy in the jth battery bank; $P_{gi}^{\min}$ and $P_{gi}^{\max}$ are the allowed minimum and maximum values of real power generation for the ith diesel generator; $P_{Bi}^{\min}$ and $P_{Bi}^{\max}$ are the allowed minimum and maximum values of real power generation of the jth battery bank; and I, J, L, M, and N the number of diesel generators, battery banks, solar power plants, wind power plants, and load centers. Note that $P_{Lo}(k)$ is a slack variable used to relax the power balance equality constraint. Physically, this variable indicates the energy to be dissipated for maintaining the balance in the microgrid.

In (11.4), only constraints over the real power are considered. Moreover, since a single-node transmission/distribution grid is assumed, it is not possible to guarantee that the scheduled powers $P_{gi}(k)$ and $P_{Bj}(k)$ can be transferred from the sources (diesel generators and battery banks) towards the load centers. Indeed, since the transmission capacity of the lines is not included in the formulation, it is possible to obtain values of $P_{gi}(k)$ and $P_{Bj}(k)$ that could exceed their maximum transmission capacity. Additionally, the real power set-points computed using (11.4) could lead to operating conditions highly sensitive to the variations in $P_{si}(k)$, $P_{wj}(k)$, and $P_{dj}(k)$. To overcome these issues, transmission constraints were added in the formulation of the EMS problem in (11.4).[1] Let $Y_{\text{bus}} = G + jB$ denote the admittance matrix of a microgrid. Then, the power-balance equations for a node r of the system are the following:

$$\begin{aligned}P_r(k) &= v_r{}^2(k)G_{r,r} \\ &+ \sum_{s=1,s\neq r}^{S} \left[v_r(k)v_s(k)G_{r,s}\cos(\delta_r(k)-\delta_s(k)) + v_r(k)v_s(k)B_{r,s}\sin(\delta_r(k)-\delta_s)(k))\right] \\ Q_r(k) &= -v_r{}^2(k)B_{r,r} \\ &- \sum_{s=1,s\neq r}^{S} \left[v_r(k)v_s(k)G_{r,s}\cos(\delta_r(k)-\delta_s(k)) - v_r(k)v_s(k)B_{r,s}\sin(\delta_r(k)-\delta_s(k))\right]\end{aligned}$$

with S the number of nodes in the grid, $P_r(k)$ and $Q_r(k)$ the difference between the generated and demanded power at the rth node, $v_r(k)$ the voltage at the rth node, and $\delta_r(k)$ the voltage angle at the rth node, $r, s = 1, \ldots, S$. Note that a balance between generated, demanded, and transmitted real and reactive powers could be achieved with several values of $P_r(k)$, $Q_r(k)$, $v_r(k)$, and $\delta_r(k)$. This makes including transmission constraints in an EMS a challenging task.

Let $R_{r,s}(k) = v_r(k)v_s(k)\cos(\delta_r(k)-\delta_s(k))$; $T_{r,s}(k) = v_r(k)v_s(k)\sin(\delta_r(k)-\delta_s(k))$; and $u_r = \frac{v_r^2(k)}{\sqrt{2}}$. Then, from the properties of $\sin(x)$ and $\cos(x)$ the following relations hold: $R_{r,s}(k) = R_{s,r}(k)$, $T_{r,s}(k) = -T_{s,r}(k)$ and $2u_r(k)u_s(k) = R^2_{r,s}(k) + T^2_{r,s}(k)$. Furthermore, the power balance equations become

$$\begin{aligned}P_r(k) &= \sqrt{2}G_{r,r}u_r(k) + \sum_{s=1,s\neq r}^{S}\left[G_{r,s}R_{r,s}(k) + B_{r,s}T_{r,s}(k)\right] \\ Q_r(k) &= -\sqrt{2}B_{i,i}u_r(k) - \sum_{s=1,s\neq r}^{S}\left[B_{r,s}R_{r,s}(k) - G_{r,s}T_{r,s}(k)\right]\end{aligned} \tag{11.5}$$

In (11.5), the balance between generated, demanded, and transmitted real and reactive powers depends upon the values of $P_r(k)$, $Q_r(k)$, $u_r(k)$, $R_{r,s}(k)$, and $T_{r,s}(k)$. The values of these variables can be obtained by solving a system of linear

[1] For simplicity in the formulation, in the sequel only electric constraints will be considered in the EMS. In [22,23] how to relate nonelectric variables, such as the diesel volume, and the energy stored in a battery bank with the decision variables of (11.4) is presented.

equations, which has a unique solution. Accordingly, transmission constraints can be included in the formulation of the EMS without significantly increasing its complexity. Let define $P_r(k)$ and $Q_r(k)$ in terms of the diesel generation, battery generation, solar power, wind power, and the demand as $P_r(k) = P_{gr}(k) + P_{Br}(k) + P_{sr}(k) + P_{wr}(k) - P_{dr}(k)$ and $Q_r(k) = Q_{gr}(k) + Q_{Br}(k) - P_{dr}(k)$. Then, the minimization problem in (11.4) can be reformulated as follows:

$$\min_{\substack{P_{gi}(k),P_{Bj}(k),\\ b(k),b_p(k),b_q(k)\\ u_r(k),R_{r,s}(k),T_{r,s}(k)}} \sum_{i=1}^{I} C_{gi}P_{gi}(k) + \sum_{j=1}^{J} C_{Bj}P_{Bj}(k) + \alpha b(k) + \beta b_p(k) + \gamma b_q(k)$$

s.t.:

$$\sum_{i=1}^{I} P_{gi}(k) + \sum_{j=1}^{J} P_{Bj}(k) = \sum_{n=1}^{N} P_{dn}(k) - \sum_{l=1}^{L} P_{sl}(k) - \sum_{m=1}^{M} P_{wm}(k) + b_p(k)$$

$$\sum_{i=1}^{I} Q_{gi}(k) + \sum_{j=1}^{J} Q_{Bj}(k) = \sum_{n=1}^{N} Q_{dn}(k) + b_q(k)$$

$$P_{gr}(k) + P_{Br}(k) + P_{sr}(k) + P_{wr}(k) - P_{dr}(k) = \sqrt{2}G_{r,r}u_r(k) + \sum_{s=1,s\neq r}^{S} \left[G_{r,s}R_{r,s}(k) + B_{r,s}T_{r,s}(k)\right] + b(k)$$

$$Q_{gr}(k) + Q_{Br}(k) - Q_{dr}(k) = -\sqrt{2}B_{i,i}u_r(k) - \sum_{s=1,s\neq r}^{S} \left[B_{r,s}R_{r,s}(k) - G_{r,s}T_{r,s}(k)\right] + b(k)$$

$$\left|(Y_{r,s}(u_r(k)) - u_s(k))\right| \le I_{r,s}^{\max}$$

$$2u_r(k)u_s(k) = R_{r,s}^2(k) + T_{r,s}^2(k)$$

$$P_{gi}^{\min} \le P_{gi}(k) \le P_{gi}^{\max}$$

$$P_{Bi}^{\min} \le P_{Bi}(k) \le P_{Bi}^{\max}$$

$$Q_{gi}^{\min} \le Q_{gi}(k) \le Q_{gi}^{\max}$$

$$Q_{Bi}^{\min} \le Q_{Bi}(k) \le Q_{Bi}^{\max}$$

$$b(k), b_p(k), b_q(k) \ge 0 \tag{11.6}$$

where $I_{r,s}^{\max}$ is the maximum allowed current flowing through the line connecting the nodes r and s; and $b(k)$, $b_p(k)$, and $b_q(k)$ are slack variables used to relax the equality constraints associated with the power balances. These slack variables can be interpreted as active and reactive loses in each case. Moreover, since currently solar and wind power plants are not being forced to supply reactive power, their contribution to the reactive power balance has not been considered.

Minimization problem in (11.6) represents the more general case of energy management in a microgrid. In this case, a node r could have all kind of generation resources (here diesel generation, battery bank, solar generation, and wind generation) connected at the same time. Since real microgrid projects attempt to take advantage of the distributed local generation resources available very often only one type of generation is connected to a particular node. Assuming that a node only has one generation resource the minimization problem in (11.6) becomes the following optimization problem:

$$\min_{\substack{b(k),b_{\mathrm{p}}(k),b_{\mathrm{q}}(k)\\ u_r(k),R_{r,s}(k),T_{r,s}(k)}} \sum_{i=1}^{I} C_{\mathrm{g}i}P_{\mathrm{g}i}(k) + \sum_{j=1}^{J} C_{\mathrm{B}j}P_{\mathrm{B}j}(k) + \alpha b(k) + \beta b_{\mathrm{p}}(k) + \gamma b_{\mathrm{q}}(k)$$

s.t.:

$$\sum_{i=1}^{I} P_{\mathrm{g}i}(k) + \sum_{j=1}^{J} P_{\mathrm{B}j}(k) = \sum_{n=1}^{N} P_{\mathrm{d}n}(k) - \sum_{l=1}^{L} P_{\mathrm{s}l}(k) - \sum_{m=1}^{M} P_{\mathrm{w}m}(k) + b_{\mathrm{p}}(k)$$

$$\sum_{i=1}^{I} Q_{\mathrm{g}i}(k) + \sum_{j=1}^{J} Q_{\mathrm{B}j}(k) = \sum_{n=1}^{N} Q_{\mathrm{d}n}(k) + b_{\mathrm{q}}(k)$$

$$P_{\mathrm{g}i}(k) - P_{\mathrm{d}i}(k) = \sqrt{2}G_{i,i}u_i(k) + \sum_{s=1,s\neq i}^{S} \left[G_{i,s}R_{i,s}(k) + B_{i,s}T_{i,s}(k)\right] + b(k), \quad i = 1,\ldots,I$$

$$P_{\mathrm{B}j}(k) - P_{\mathrm{d}j}(k) = \sqrt{2}G_{j,j}u_j(k) + \sum_{s=1,s\neq i}^{S} \left[G_{j,s}R_{j,s}(k) + B_{j,s}T_{j,s}(k)\right] + b(k), \quad j = 1,\ldots,J$$

$$P_{\mathrm{s}m}(k) - P_{\mathrm{d}m}(k) = \sqrt{2}G_{m,m}u_m(k) + \sum_{s=1,s\neq i}^{S} \left[G_{m,s}R_{m,s}(k) + B_{m,s}T_{m,s}(k)\right] + b(k), \quad m = 1,\ldots,M$$

$$P_{\mathrm{w}n}(k) - P_{\mathrm{d}n}(k) = \sqrt{2}G_{n,n}u_m(k) + \sum_{s=1,s\neq i}^{S} \left[G_{n,s}R_{n,s}(k) + B_{n,s}T_{n,s}(k)\right] + b(k), \quad n = 1,\ldots,N$$

$$-P_{\mathrm{d}r}(k) = \sqrt{2}G_{r,r}u_r(k) + \sum_{s=1,s\neq i}^{S} \left[G_{r,s}R_{r,s}(k) + B_{r,s}T_{r,s}(k)\right] + b(k), \quad r = 1,\ldots,S-N-M-J-I$$

$$Q_{\mathrm{g}i}(k) - Q_{\mathrm{d}i}(k) = -\sqrt{2}B_{i,i}u_i(k) - \sum_{s=1,s\neq r}^{S} \left[B_{i,s}R_{i,s}(k) - G_{i,s}T_{i,s}(k)\right] + b(k), \quad i = 1,\ldots,I$$

$$
\begin{aligned}
&Q_{\mathrm{B}j}(k) - Q_{\mathrm{d}j}(k) = -\sqrt{2}B_{j,j}u_j(k) - \sum_{s=1,s\neq r}^{S}\left[B_{j,s}R_{j,s}(k) - G_{j,s}T_{j,s}(k)\right]\\
&\qquad + b(k), \quad j = 1,\ldots,J\\
&-Q_{\mathrm{d}r}(k) = -\sqrt{2}B_{r,r}u_r(k) - \sum_{s=1,s\neq r}^{S}\left[B_{r,s}R_{r,s}(k) - G_{r,s}T_{r,s}(k)\right]\\
&\qquad + b(k), \quad r = 1,\ldots,S-N-M-J-I\\
&\left|(Y_{r,s}(u_r(k)) - u_s(k))\right| \leq I_{r,s}^{\max}, \quad r,s = 1,\ldots,S\\
&2u_r(k)u_s(k) = R_{r,s}^2(k) + T_{r,s}^2(k), \quad r,s = 1,\ldots,S\\
&P_{gi}^{\min} \leq P_{\mathrm{g}i}(k) \leq P_{gi}^{\max}, \quad i = 1,\ldots,I\\
&P_{\mathrm{B}i}^{\min} \leq P_{\mathrm{B}i}(k) \leq P_{\mathrm{B}i}^{\max}, \quad j = 1,\ldots,J\\
&Q_{gi}^{\min} \leq Q_{\mathrm{g}i}(k) \leq Q_{\mathrm{g}i}^{\max}, \quad i = 1,\ldots,I\\
&Q_{\mathrm{B}i}^{\min} \leq Q_{\mathrm{B}i}(k) \leq Q_{\mathrm{B}i}^{\max}, \quad j = 1,\ldots,J\\
&b(k), b_{\mathrm{p}}(k), b_{\mathrm{q}}(k) \geq 0
\end{aligned}
\tag{11.7}
$$

It is important to remark that in (11.7) unlike in (11.6) the real and reactive power injected by each generation unit is determined by the values of $u_r(k), R_{r,s}(k), T_{r,s}(k)$, $r, s = 1,\ldots,S$. Thus, these are the decision variables in such optimization problem. Accordingly, the constraints and the cost function in (11.7) can be expressed as a linear function of them. Furthermore, both (11.6) and (11.7) provide an additional flexibility to the EMS. The additional flexibility consists on the possibility of sending voltage set-points to any connected device able to regulate the voltage in the connection point.

Minimization problem in (11.6) (and as a special case minimization problem (11.7)), corresponds to the deterministic formulation of the proposed optimization-based EMS. The implementation of this EMS in a real microgrid requires a perfect knowledge of $P_{\mathrm{s}m}(k)$, $P_{\mathrm{w}m}(k)$, $P_{\mathrm{d}r}(k)$, and $Q_{\mathrm{d}r}(k)$. In practice, these variables are highly uncertain and a mismatch in their estimation could lead the operation of a microgrid to undesirable points. To include the uncertainty in (11.6) (and therefore in (11.7)) the mathematical framework introduced in Section 11.2 has been used. In this framework, a set of possible scenarios is constructed using the combinations of the extreme values of the uncertain variables. In this research, unlike in [22,23] and the references therein, these extreme values are determined using the statistics of historical data available. Therefore, the use of prediction models or estimators is avoided. Based on the statistics of the historical data, and given the desired sample time for the EMS, the mean value and the standard deviation are computed per sample time. For instance, if the EMS is designed to be solved every minute, then for each of the 1,440 min of the day the mean $\mu(k)$ and the standard deviation $\sigma(k)$ are computed. With the values of $\mu(k)$ and $\sigma(k)$, the extreme values are computed as: $x^{\min}(k) = \mu(k) - \xi_1\sigma(k)$ and $x^{\max}(k) = \mu(k) + \xi_2\sigma(k)$ with ξ_1 and ξ_2 tunable parameters determined by the desired confidence level.

Let $P_{sm}^{\min}(k)$, $P_{wm}^{\min}(k)$, $P_{dr}^{\min}(k)$, $Q_{dr}^{\min}(k)$ denote the minimum expected value for the real power supplied by the solar and wind power plants, and for the real and reactive power demanded by the load centers. Let $P_{sm}^{\max}(k)$, $P_{wm}^{\max}(k)$, $P_{dr}^{\max}(k)$, $Q_{dr}^{\max}(k)$ denote the maximum expected value for the real power supplied by the solar and wind power plants, and for the real and reactive power demanded by the load centers. Note that in the minimization problem in (11.6), both $P_{sm}(k)$ and $P_{wm}(k)$ are not being controlled. Then, an equivalent demand could be formulated to integrate all the sources of uncertainty into a single variable. Let $P_{Lr}(k)$ denote the equivalent real power demand at node r. Then the extreme values of $P_{Lr}(k)$ are given by: $P_{Lr}^{\max}(k) = P_{sr}^{\max}(k) + P_{wr}^{\max}(k) - P_{dr}^{\max}(k)$ and $P_{Lr}^{\min}(k) = P_{sr}^{\min}(k) + P_{wr}^{\min}(k) - P_{dr}^{\min}(k)$. Given these definitions, a robust EMS with transmission constraints can be formulated (from (11.6)) as follows:

$$\min_{\substack{P_{gi}(k),P_{Bj}(k),\\ b(k),b_p(k),b_q(k)\\ u_r(k),R_{r,s}(k),T_{r,s}(k)}} \sum_{i=1}^{I} C_{gi}P_{gi}(k) + \sum_{j=1}^{J} C_{Bj}P_{Bj}(k) + \alpha b(k) + \beta b_p(k) + \gamma b_q(k)$$

s.t.:

$$\sum_{i=1}^{I} P_{gi}(k) + \sum_{j=1}^{J} P_{Bj}(k) + \sum_{r=1}^{S} P_{Lr}^{\max}(k) - b_p(k) = 0$$

$$\sum_{i=1}^{I} P_{gi}(k) + \sum_{j=1}^{J} P_{Bj}(k) + \sum_{r=1}^{S} P_{Lr}^{\min}(k) - b_p(k) = 0$$

$$\sum_{i=1}^{I} Q_{gi}(k) + \sum_{j=1}^{J} Q_{Bj}(k) - \sum_{n=1}^{N} Q_{dn}^{\max}(k) - b_q(k) = 0$$

$$\sum_{i=1}^{I} Q_{gi}(k) + \sum_{j=1}^{J} Q_{Bj}(k) - \sum_{n=1}^{N} Q_{dn}^{\min}(k) - b_q(k) = 0$$

$$P_{gr}(k) + P_{Br}(k) + P_{Lr}^{\max}(k) = \sqrt{2}G_{r,r}u_r(k) + \sum_{s=1,s\neq r}^{S} \left[G_{r,s}R_{r,s}(k) + B_{r,s}T_{r,s}(k)\right] + b(k)$$

$$P_{gr}(k) + P_{Br}(k) + P_{Lr}^{\min}(k) = \sqrt{2}G_{r,r}u_r(k) + \sum_{s=1,s\neq r}^{S} \left[G_{r,s}R_{r,s}(k) + B_{r,s}T_{r,s}(k)\right] + b(k)$$

$$Q_{gr}(k) + Q_{Br}(k) - Q_{dr}^{\max}(k) = -\sqrt{2}B_{i,i}u_r(k) - \sum_{s=1,s\neq r}^{S} \left[B_{r,s}R_{r,s}(k) - G_{r,s}T_{r,s}(k)\right] + b(k)$$

$$Q_{gr}(k) + Q_{Br}(k) - Q_{dr}^{\min}(k) = -\sqrt{2}B_{i,i}u_r(k) - \sum_{s=1,s\neq r}^{S} \left[B_{r,s}R_{r,s}(k) - G_{r,s}T_{r,s}(k)\right] + b(k)$$

$$\begin{aligned}
&\left|\left(Y_{r,s}(u_r(k))-u_s(k)\right)\right| \leq I_{r,s}^{\max}\\
&2u_r(k)u_s(k)=R_{r,s}^2(k)+T_{r,s}^2(k)\\
&P_{gi}^{\min} \leq P_{gi}(k) \leq P_{gi}^{\max}\\
&P_{Bi}^{\min} \leq P_{Bi}(k) \leq P_{Bi}^{\max}\\
&Q_{gi}^{\min} \leq Q_{gi}(k) \leq Q_{gi}^{\max}\\
&Q_{Bi}^{\min} \leq Q_{Bi}(k) \leq Q_{Bi}^{\max}\\
&b(k), b_p(k), b_q(k) \geq 0
\end{aligned} \tag{11.8}$$

Until here, this section has been focused on the formulation of electrical constraints. But, as presented in [16], some other features could be added into the proposed EMS, for example, demand-side management signals and programmable loads. Since both of them are deterministic and the associated constraints are linear (or at least convex see [16] for some details), adding these elements would not significantly increase the complexity of the EMS. The main expected change is that in addition to real and reactive power set-points and voltage set-points, the EMS should send consumption signals to the customers and ON/OFF signals for the programmable loads. In this sense, although the EMS problem formulated as the minimization problem in (11.6), and even as the minimization problem in (11.7), seems to be more complex than those presented in, for example, [15,16,22,23], the formulation in (11.6) provides more flexibility to the grid from the control viewpoint. Indeed, using (11.6), it is possible to define real and reactive power set-points but also voltage set-points. Then a better control of the behavior of the grid is expected. Furthermore, since transmission constraints are considered in the formulation, it is possible to guarantee that the power injections computed by the EMS can be transferred from the sources to the load centers. Next the case study is described and the simulation results are analyzed.

11.4 Case study and simulation results

Results obtained in previous studies [22,23,29,30] using stochastic and robust programing frameworks to implement an EMS in microgrid applications indicated that the use of EMS approaches in which only predictions to deal with the uncertainty [16] resulted in a lower grid performance from the security viewpoint. According to the results reported in [22], robust EMS assigned larger power reserves to mitigate the negative effects of the intermittency of wind generation. The present study proposed a novel optimization-based robust EMS with transmission constraints. The novelty of the proposed EMS recast in both the addition of transmission constraints and the avoidance prediction models to determine the time-variant uncertainty set used to represent the uncertain variables. The performance of the proposed EMS was evaluated using a modified version of the IEEE

9-nodes power system benchmark. In the modified version of the IEEE 9-nodes power system benchmark, the generation unit originally connected to the node 1 (which is the slack machine) was changed by an equivalent wind generator (here equivalent means same terminal voltage and maximum capacity). Figure 11.1 presents a single-line diagram of the power system used as test bench.

The system shown in Figure 11.1 was implemented in MATLAB®/Simulink™ software. The parameters used in [18,19] were considered to implement the simulation model. To implement the simulation model, the following considerations were done: the generation units were modeled through subtransient models (seventh-order model); the transmission lines were modeled by means of the π lumped-parameters model; the load centers were modeled as constant impedance with nominal powers of: $125 + j50$ MVA for the load center connected to node 5, $90 + j30$ MVA for the load center connected to node 6 and $100 + j35$ MVA for the load center connected to node 8; and the wind power generator was modeled as a voltage source converter, using an average model to represent the converter dynamics (this model includes the dynamic response of both the voltage and the current inner control loops). A 1-month length wind power signal measured in Atacama Desert (north of Chile), with a resolution of 1 min, was scaled up to match the capacity equivalence between the original and the wind power station. This signal was used to generate the current reference signal required by the inner current loop of the converter. The simulation model was implemented with a fixed integration step of 50 μs (numerical method ODE4 of MATLAB/Simulink) to solve the differential equations. Since the EMS is discrete by nature, it was executed every 10 s.

Due to the differences between the execution times, rate-transition blocks were added to emulate the communication delay that appears in real applications. Furthermore, since the load centers were modeled as constant impedance only uncertainty in the wind generation was considered in the EMS. The deterministic EMS in (11.7) was implemented for comparison purposes. Figure 11.2 presents the wind power signal used during the simulation, and its extreme values derived from the statistics of the signal. A base-power of 255.47 MW was used for the normalization of the signal. As can be seen, the power supplied by the wind generator (solid line) is always bounded by the extreme variables derived from the statistics of the signal, which is consistent with the procedure used to derive the interval. In this case, $\xi_1 = 0.3$ and $\xi_2 = 0.3$. In the deterministic, the measured wind power signal was used to define the power injected by the wind power generator. In the robust EMS, both extreme values were used to represent the power injected by the wind power generator. Since these values were obtained off-line prediction models were not required to implement the robust EMS. Note that the robust EMS never exactly knows the real value of the wind power injected to the system.

Figure 11.3 shows the results obtained with both the deterministic and the robust EMS. In this case, equal generation costs were assigned to generation units 2 and 3. These values were: $C_{q2} = \frac{1.5}{1.92}$ and $C_{q3} = \frac{1}{1.28}$, since in a base of 100 MW their maximum capacities are 1.92 and 1.28, respectively. As it can be seen in the

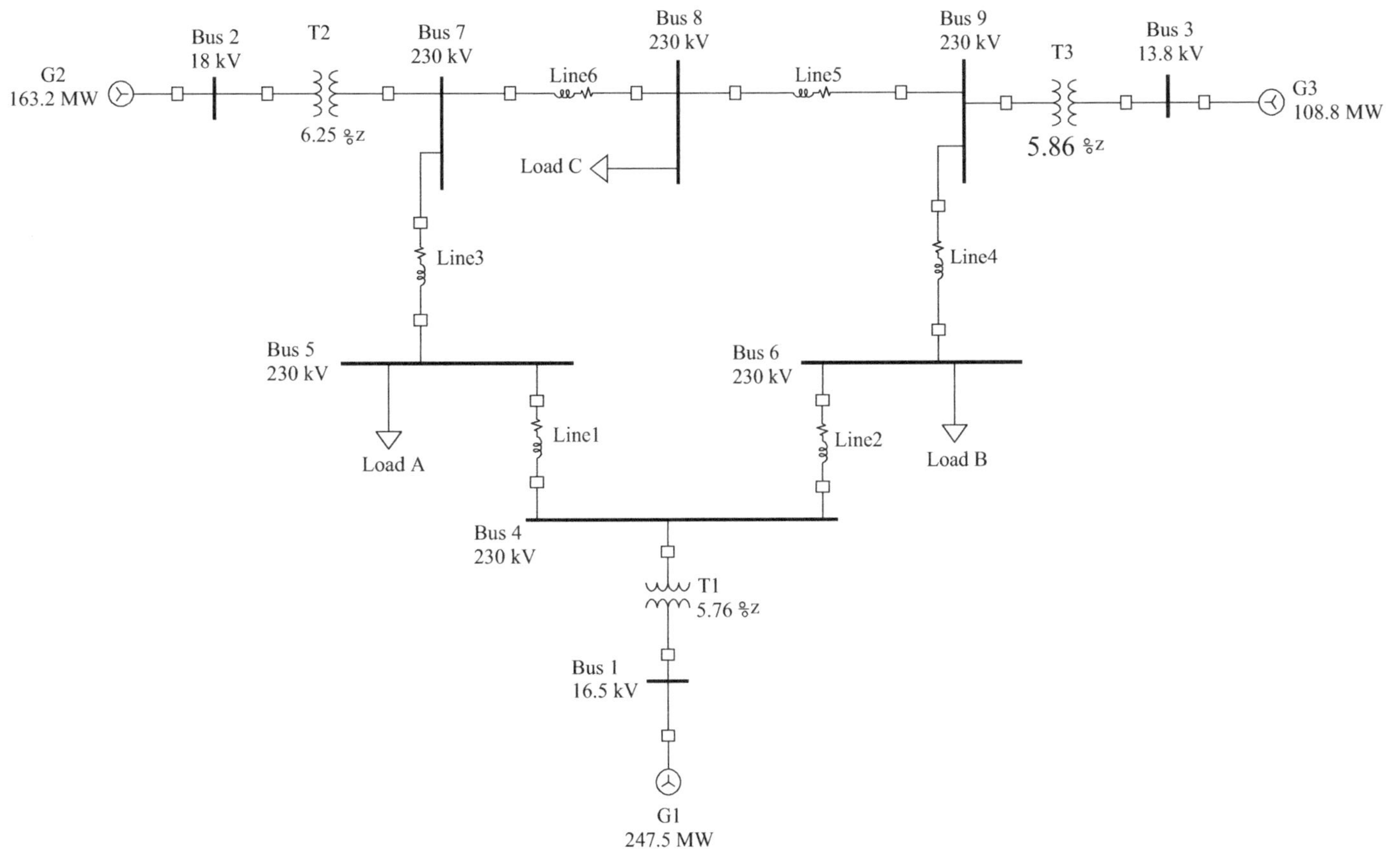

Figure 11.1 *Single-line diagram of the EEE 9-nodes power system benchmark (taken from: http://www.al-roomi.org/power-flow/9-bus-system)*

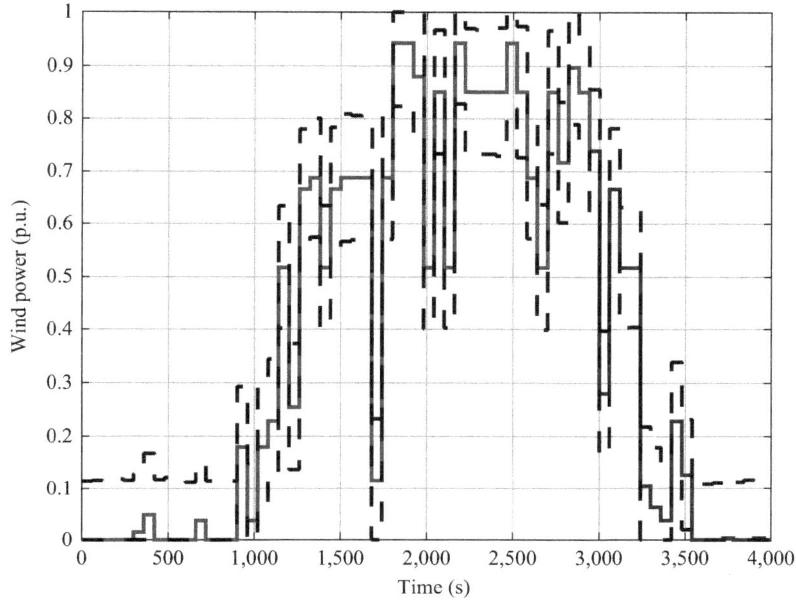

Figure 11.2 Wind power time-series used in the simulation (solid line), and extreme values derived from the statistics of the available data (dashed lines)

top and bottom left-hand-side panels, the deterministic EMS makes an underestimation of the real power that generation units 2 and 3 should supply. Consequently, a difference between the scheduled (black dashed line) and the injected (solid line) real power by each generation appears throughout the entire simulation. By contrast, the robust EMS adequately estimated the real power that generation units 2 and 3 should supply (top and bottom right-hand side panels). Accordingly, the difference between scheduled and supplied real power does not appear. Note that in both cases, there exists a delay in the scheduling of the generation units. The delay is about 10 s and is a consequence of the discrete nature of both EMS. In the literature, several alternatives to compensate such a delay have been proposed (e.g., the approach reported in [21] and the references therein). Any of these alternatives could be combined with the EMS proposed in this chapter to enhance the performance of the microgrid as a whole.

A clear difference between the behavior of the system with the deterministic and the proposed EMS is evident from Figure 11.3. Such a difference is reinforced by the behavior of the frequency of the system throughout the simulation. Figure 11.4 presents the frequency measured in all generation nodes (top panels). It can be observed that during the entire simulation the frequency measured in the generation nodes was the same (for the three generators) despite the changes in the wind power. Moreover, it also can be observed that, in the case of the

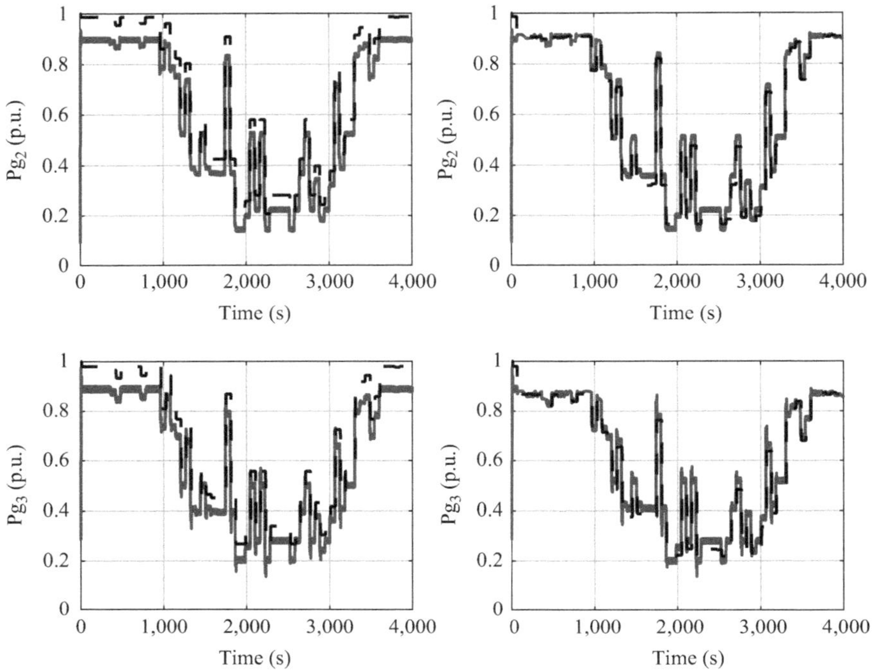

Figure 11.3 Scheduled (dashed lines) and supplied (solid line) real power for generation units 2 and 3 throughout the simulations. The left-hand-side panels correspond to the results obtained with equal generation costs for both generation units. The right-hand-side panels correspond to the results obtained with different generation costs

deterministic EMS, the difference between the scheduled and the supplied real power resulted in a frequency deviation in steady-state throughout the simulation (left-hand-side top panel). Indeed, after the transients produced by the fast changes in the wind power the frequency always stabilized over 1 p.u. This phenomenon did not appear when the proposed robust EMS was used (see right-hand-side top panel). It is important to remark that voltage and reactive power results have not been included in this section since significative differences were not evidenced. Instead of these variables, the execution time of each EMS was included in Figure 11.4. Throughout this chapter and special emphasis has been given to the possibility of efficiently solving the minimization problem in (11.7). The execution time was computed using the TIC-TOC function of MATLAB/Simulink software. As it can be noticed, in both cases, the execution time was always less than the sample time of 10 s, being lower for the deterministic EMS. Also, in both cases, the execution time was about 0.8 s in average. This fact puts forward the low computational burden of the solution of the EMS analyzed here.

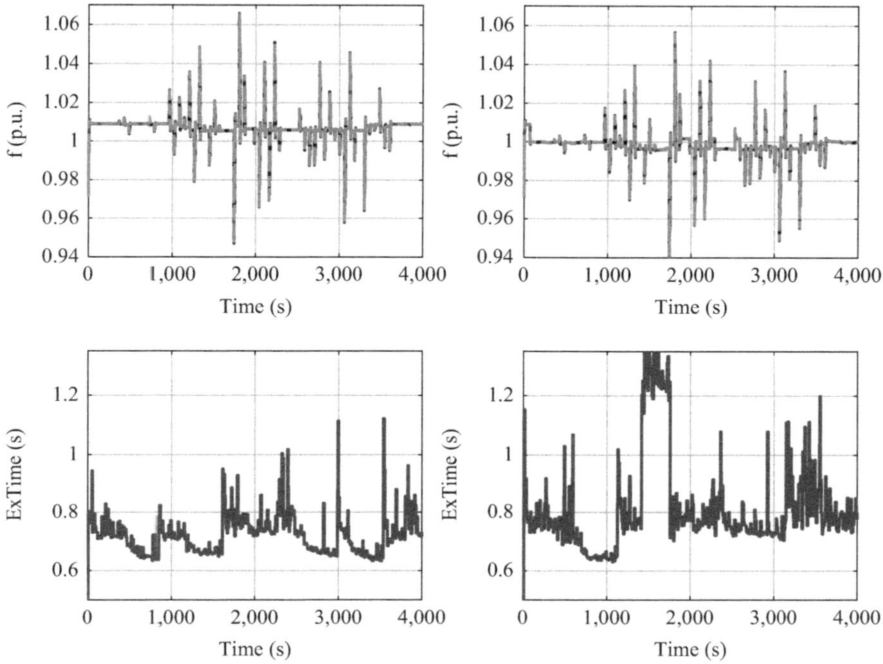

Figure 11.4 System frequency (top panels) and execution time (bottom panels) throughout the simulation. The frequency was measured in the generation nodes, namely, in nodes 1, 2, and 3. As it can be seen all frequencies are overlapped showing that the frequency along the system is the same

The results presented in Figures 11.3 and 11.4 were obtained considering the same generation cost for both generation units. However, these results could change in these costs are assumed different. Indeed, one of the main issues in EMS approaches is how to select the generation costs of each technology so that the resulting optimization model adequately represents the objectives of the EMS. To illustrate this statement, the generation costs were changed for the following values: $C_{q2} = \frac{1}{1.92}$ and $C_{q3} = \frac{1}{1.28}$. With this selection, the cost of using the generation unit 2 to supply the demand is lower than the cost of using the generation unit 3 for the same purpose. Figure 11.5 presents the results obtained with equal generation costs for both generation units (left-hand-side panels) and the results obtained with different generation costs (right-hand-side panels). As it can be noticed, in left-hand-side panels both generation units are scheduled so that they evenly share the demand. By contrast, in the right-hand-side panels the EMS tends to use generation unit 2 more than generation unit 3 to supply the demand. It is important to notice that in the time range (1,000 s, 3,300 s) the EMS assigned almost 0 MW of real power to generation unit 2. This range corresponded to the time range with the maximum wind

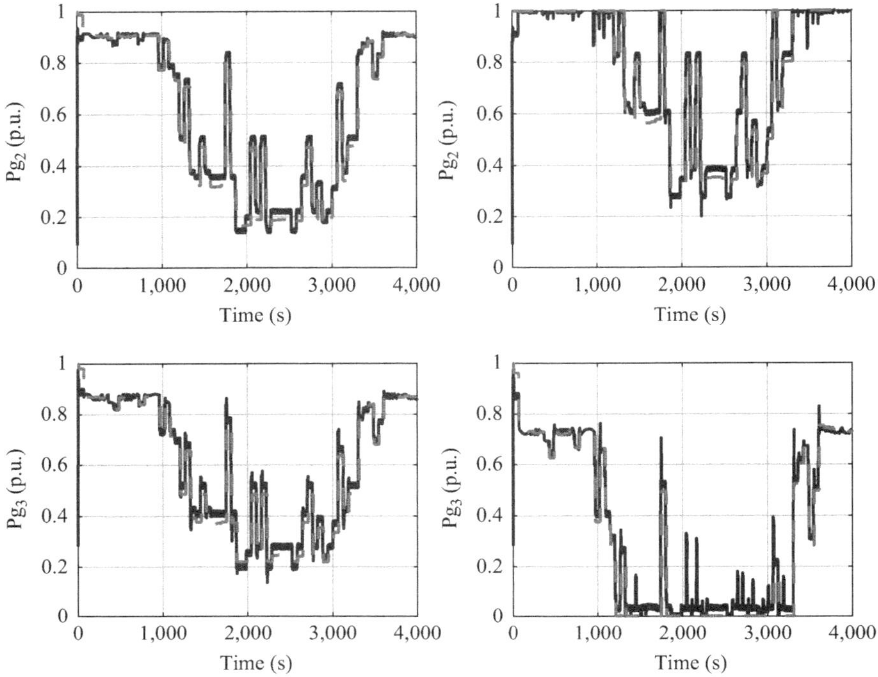

Figure 11.5 System frequency (top panels) and execution time (bottom panels) throughout the simulation. The frequency was measured in the generation nodes, namely, in nodes 1, 2, and 3. As it can be seen, all frequencies are overlapped showing that the frequency along the system is the same

power supply. This condition might correspond in real applications to turn off the diesel generator and use only the battery bank to supply the demand.

The results presented throughout this section suggest that the proposed framework for the development of robust EMS is a suitable alternative to include transmission constraints into the formulation of the EMS problem, without significantly increasing its complexity, without compromising its application in real microgrids, and avoiding the use of prediction models. Furthermore, the use of the proposed robust EMS allows having a better control of the variables of the grid, namely, real and reactive powers but also voltages and system frequency.

11.5 Concluding remarks

Prior work has documented the importance of the robustness of EMS in microgrid applications in guaranteeing the security in the operation of the grid and in maintaining an optimal economic operation; the authors in [29,30], for example, report the improvements in the operation of a microgrid achieved when using different alternatives to deal with the uncertainty in EMS formulations. However, these studies have either been affected by the performance of prediction models and have not included transmission constraints. Then it is not possible to guarantee that the scheduled powers can be transferred from the sources to the demand centers in a proper way. In this study, it has been proposed a novel formulation for robust optimization-based EMS for microgrid applications, in which prediction models are avoided, and where transmission constraints are included in the formulation. To represent the uncertainty, the statistics of the available data of the uncertain variables is used to derive a time-variant uncertainty set. It has been found that in all cases analyzed the proposed robust EMS provides a better scheduling of the generation units. For comparison purposes, a deterministic version of the proposed EMS has been used. Moreover, it has also found that including transmission constraints into the EMS improves the performance of the microgrid and adds more flexibility to the EMS without significantly increasing its complexity. These results extend those of [16,23], confirming that improving the microgrid models used in the formulation of EMS approaches tend to increase the type of elements to be able to be controlled by the EMS and to extend the variables that can be directly/indirectly regulated by the EMS. In addition, the formulation presented in this chapter is independent of the topology of the microgrid and the type of generation units. This study therefore indicated that the benefits gained with the implementation of the proposed EMS could be extended to a broad range type of microgrids. Future work should therefore include implementing the proposed EMS in different types of microgrids.

Acknowledgment

Felipe Valencia thanks to the FONDAP/CONICYT Project: Solar Energy Research Center, SERC-Chile, grant number 15110019.

References

[1] W. Al-Gherwi, H. Budman, and A. Elkamel. A robust distributed model predictive control algorithm. *Journal of Process Control*, 21(8):1127–1137, 2011.

[2] D. Bertsimas, E. Litvinov, X.A. Sun, J. Zhao, and T. Zheng. Adaptive robust optimization for the security constrained unit commitment problem. *IEEE Transactions on Power Systems*, 28(1):52–63, 2013.

[3] S. Boyd, L. Vandenberghe, and M. Grant. Efficient convex optimization for engineering design. In *IFAC Symposium on Robust Control Design*, Rio de Janeiro, Brazil, 1994.

[4] S. Boyd and L. Vandenberghe. *Convex Optimization*. Cambridge, UK: Cambridge University Press, 2004.

[5] J. Espinosa, J. Vandewalle, and V. Wertz. *Fuzzy Logic, Identification, and Predictive Control*. London: Springer-Verlag London Limited, 2005.

[6] N. Hatziargyriou, G. Contaxis, M. Matos, *et al.* Energy management and control of island power systems with increased penetration from renewable sources. In *Power Engineering Society Winter Meeting, 2002. IEEE*, volume 1, pages 335–339, 2002.

[7] R.A. Jabr and B.C. Pal. Conic programming approach for static voltage stability analysis in radial networks. *IET Generation, Transmission & Distribution*, 1(2):203–208, 2007.

[8] R.A. Jabr. Radial distribution load flow using conic programming. *IEEE Transactions on Power Systems*, 21(3):1458–1459, 2006.

[9] R.A. Jabr. A conic quadratic format for the load flow equations of meshed networks. *IEEE Transactions on Power Systems*, 22(4):2285–2286, 2007.

[10] R.A. Jabr. Optimal power flow using an extended conic quadratic formulation. *IEEE Transactions on Power Systems*, 23(3):1000–1008, 2008.

[11] J.D. Kueck, R.H. Staunton, S.D. Labinov, and B.J. Kirby. *Microgrid Energy Management System*. Oak Ridge, TN: Oak Ridge National Laboratory Report ORNL/TM-2002/242, 2003.

[12] J.H. Lee and Z. Yu. Worst-case formulations of model predictive control for systems with bounded parameters. *Automatica*, 33(5):763–781, 1997.

[13] D. Limon, I. Alvarado, T. Alamo, and E.F. Camacho. Robust tube-based MPC for tracking of constrained linear systems with additive disturbances. *Journal of Process Control*, 20(3):248–260, 2010.

[14] X.H. Nie, G.H. Huang, Y.P. Li, and L. Liu. IFRP: A hybrid interval-parameter fuzzy robust programming approach for waste management planning under uncertainty. *Journal of Environmental Management*, 84(1): 1–11, 2007.

[15] D.E. Olivares, C.A. Canizares, and M. Kazerani. A centralized optimal energy management system for microgrids. In *Power and Energy Society General Meeting, 2011 IEEE*, pages 1–6, July 2011.

[16] R. Palma-Behnke, C. Benavides, F. Lanas, *et al.* A microgrid energy management system based on the rolling horizon strategy. *IEEE Transactions on Smart Grid*, 4(2):996–1006, 2013.

[17] G. Pannocchia. Robust model predictive control with guaranteed setpoint tracking. *Journal of Process Control*, 14(8):927–937, 2004.

[18] A. Pizano-Martínez, C. Fuerte-Esquivel, E.A. Zamora-Cárdenas, and J. Segundo Ramírez. Conventional optimal power flow analysis using the MATLAB optimization toolbox. *ROPEC International*, pp. 189–194, XII Reunión de Otoño de Potencia, Electrónica y Computación, 2010.

[19] P.W. Sauer and M.A. Pai. Power system dynamics and stability. *Urbana*, 51:61801, 1997.

[20] P.O.M. Scokaert, J.B. Rawlings, and E.S. Meadows. Discrete-time stability with perturbations: Application to model predictive control. *Automatica*, 33(3): 463–470, 1997.

[21] P.L. Tang and C.W. de Silva. Compensation for transmission delays in an ethernet-based control network using variable-horizon predictive control. *IEEE Transactions on Control Systems Technology*, 14(4):707–718, 2006.

[22] F. Valencia, J. Collado, D. Sáez, and L.G. Marín. Robust energy management system for a microgrid based on a fuzzy prediction interval model. *IEEE Transactions on Smart Grid*, 7(3):1486–1494, 2016.

[23] F. Valencia, D. Sáez, J. Collado, F. Ávila, A. Marquez, and J.J. Espinosa. Robust energy management system based on interval fuzzy models. *IEEE Transactions on Control Systems Technology*, 24(1):140–157, 2016.

[24] S. Wang and G.H. Huang. An interval-parameter two-stage stochastic fuzzy program with type-2 membership functions: An application to water resources management. *Stochastic Environmental Research and Risk Assessment*, 27(6):1493–1506, 2013.

[25] S. Wang and G.H. Huang. A two-stage mixed-integer fuzzy programming with interval-valued membership functions approach for flood-diversion planning. *Journal of Environmental Management*, 117:208–218, 2013.

[26] S. Wang, G.H. Huang, and B.T. Yang. An interval-valued fuzzy-stochastic programming approach and its application to municipal solid waste management. *Environmental Modelling & Software*, 29(1):24–36, 2012.

[27] B. Wright. *A Review of Unit Commitment*, ELENE4511, 28 May 2013 [online]. Available from http://www.ee.columbia.edu/lavaei/Projects/Brittany_Wright.pdf.

[28] Z. Yu, L. McLaughlin, L. Jia, M.C. Murphy-Hoye, A. Pratt, and L. Tong. Modeling and stochastic control for home energy management. In *Power and Energy Society General Meeting*, 2012 IEEE, pages 1–9, July 2012.

[29] Y. Zhang, N. Gatsis, and G.B. Giannakis. Robust energy management for microgrids with high-penetration renewables. *IEEE Transactions on Sustainable Energy*, 4(4):944–953, 2013.

[30] B. Zhao, Y. Shi, X. Dong, W. Luan, and J. Bornemann. Short-term operation scheduling in renewable-powered microgrids: A duality-based approach. *IEEE Transactions on Sustainable Energy*, 5(1):209–217, 2014.

[31] C. Zhao, J. Wang, J.-P. Watson, and Y. Guan. Multi-stage robust unit commitment considering wind and demand response uncertainties. *IEEE Transactions on Power Systems*, 28(3):2708–2717, 2013.

Chapter 12

Future trends in power quality

A.M. Sharaf[1], *Abdelazeem A. Abdelsalam*[2], *Hossam A. Gabbar*[3] *and Ahmed Othman*[3]

12.1 Introduction

Power quality issues and problems are of paramount concern to electric utility industry and industrial, commercial and residential users of electricity.

The rising concerns stem from increased use of non-linear loads and new large-scale integration of renewable energy sources (PV; wind; tidal/wave; battery storage; micro-, hydro and hybrid multisource microgrid grids).

This chapter addresses the growing needs and issues associated with power quality monitoring, measurements, future trends, complexities and challenging issues associated with measurements, assessment tools as well as mitigation and use of FACTS switched compensators, dynamic voltage regulators, active power filters and STATCOM devices.

The chapter covers power quality, monitoring, measurement, performance indices, equipment requirements, emerging issues and PQ impacts as well as power quality trends and future interface issues with smart grid.

The chapter concludes by a case study presenting three novel low-cost FACTS-based modulated/switched filter-compensation devices and intelligent dynamic control strategies using dynamic fast controlled PWM switching strategies.

Future problems and challenges in measurements, monitoring and analysis tools are also presented.

12.2 Contracts of PQ in a reconfigured electric power industry

The vertically integrated old structure of the electric utility industry is a multitier inter-coupled, fully regulated enterprise comprising generation, transmission, distribution and customer services. The service models and tariffs are set by fixed and regulated business models that are controlled by government regulatory board/

[1]Sharaf Energy Systems, Inc., Fredericton-NB, Canada
[2]Electrical Engineering Department, Suez Canal University, 41522 Ismailia, Egypt
[3]University of Ontario Institute of Technology (UOIT), Oshawa, Canada

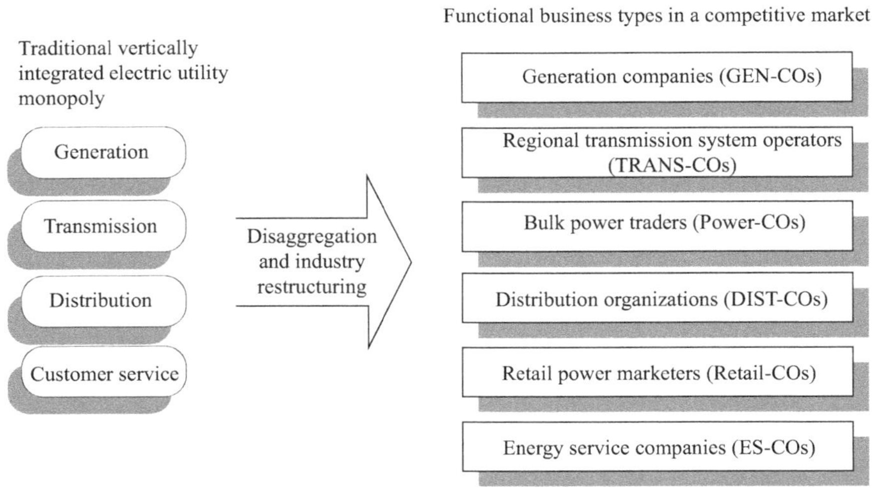

Figure 12.1 The utility industry configuration changes due to deregulation and competition

commission. The company is allowed to seek special rate structures or increases based on projected rate of return on the capital investments and debt services to ensure reliable, secure, modern generation, transmission and distribution systems and expected power quality electric service [1].

Figure 12.1 shows the new model and change in utility configuration based on the deregulation and separation of business entities to ensure open competition and new enterprising structure. Customers will have the choice to buy/sell electric energy and secure electric power from competitive low-cost markets and a number of key independent energy suppliers. In this model, the price and service contracts will be based on transactions and short- and medium-term competitive contracts and marketplace that are scheduled and coordinated to match supplied energy to meet customer demands and contractual requirements. Generation, transmission and distribution are treated as separate business entities with full decoupled business models, structures, mandates and investment strategies.

The retail marketer is only one of the different elements that is the outcome of the restructuring. Generation companies or distributed generation owners (DG-Os) will offer power in contracts to bulk power traders. Regional transmission system operators (TRANS-COs) will give transmission access to get the ability to the distribution networks supplying the customers. At last, distribution organizations (DIST-COs) will give the last conveyance of the power to individual end-use customers.

These distribution or transmission companies will probably still be the sole supplier of conveyance services, and the entrance costs and some service parts will most likely still be directed.

One intriguing consequence of the new configuration is the chance for unbundled customer services as new income-producing opportunities. Utilities are making unregulated backups at a gigantic pace with a specific end goal to tap this potential business sector for customer services. These business could be a piece of the retail energy marketing business or they can be separate energy service companies (ES-COs) [2].

Assessment of power quality concerns and execution of poor power quality mitigation technologies are clear opportunities for these businesses. Note that these businesses do not have any geographic limits. Whenever advancements and ability are created to offer a scope of services, the services can be offered around the world, not simply in a conventional service domain. Power quality improvement is one of these services.

This new model for the electric utility industry makes numerous requirements for contracts between the distinctive elements included. The contracts will need to address issues of reliability and power quality, in addition to the conspicuous issues of costs and conveyance necessities.

The prerequisites of a specific power quality contract and the worries that should be tended shall rely on the parties included and the specifications of the system. The power quality contracts could include the following areas:

- Reliability/power quality terms to be evaluated
- Achievement indicators to be applied
- Prospective level of achievement (base case)
- Penalization for achievement outside the prospective level and/or remunerations for achievement superior to the prospective level
- Measurement/estimation techniques to validate the achievement
- Duties of every party to fulfil the required achievement
- Duties of all parties to solve the issue

In this part, the important interests that must be tended are collected for every kind of the contract. Under every interest, imperative variables that must be incorporated into the contract are depicted. Figure 12.2 shows a portion of the essential connections in which power quality contemplations will be incorporated into the contracts [3].

Power quality contracts will require few methods for assessing consistence. This will typically include a mix of observation and analysing instruments. The observation devices are desired to precisely evaluate the system achievement; however, this is only one of the significant goals of observing the system.

1. *Observing to identify the system status*
   First, the system status should be evaluated to identify the baseline of the power quality. System status identification is a *pro-active* way for observing the power quality. Once baseline levels are specified, outstanding situation measurements are applied to characterize problem region and confirm that sufficient performance is preserved.
2. *Observing to identify a certain problem*
   Short-term observation devices are placed at an end-user of difficult customer to identify a specific cause of the mal-function or faults at these customers to suggest the mitigation tool. This process is considered as a re-active PQ observation.
3. *Observation to enhance PQ problems*
   Enhancement of PQ problems is a part of services offered by RETAIL-COs and ES-COs. The required PQ enhancement and situation levels are stated in the contracts that are signed between customers and DIST-COs or TRANS-COs.

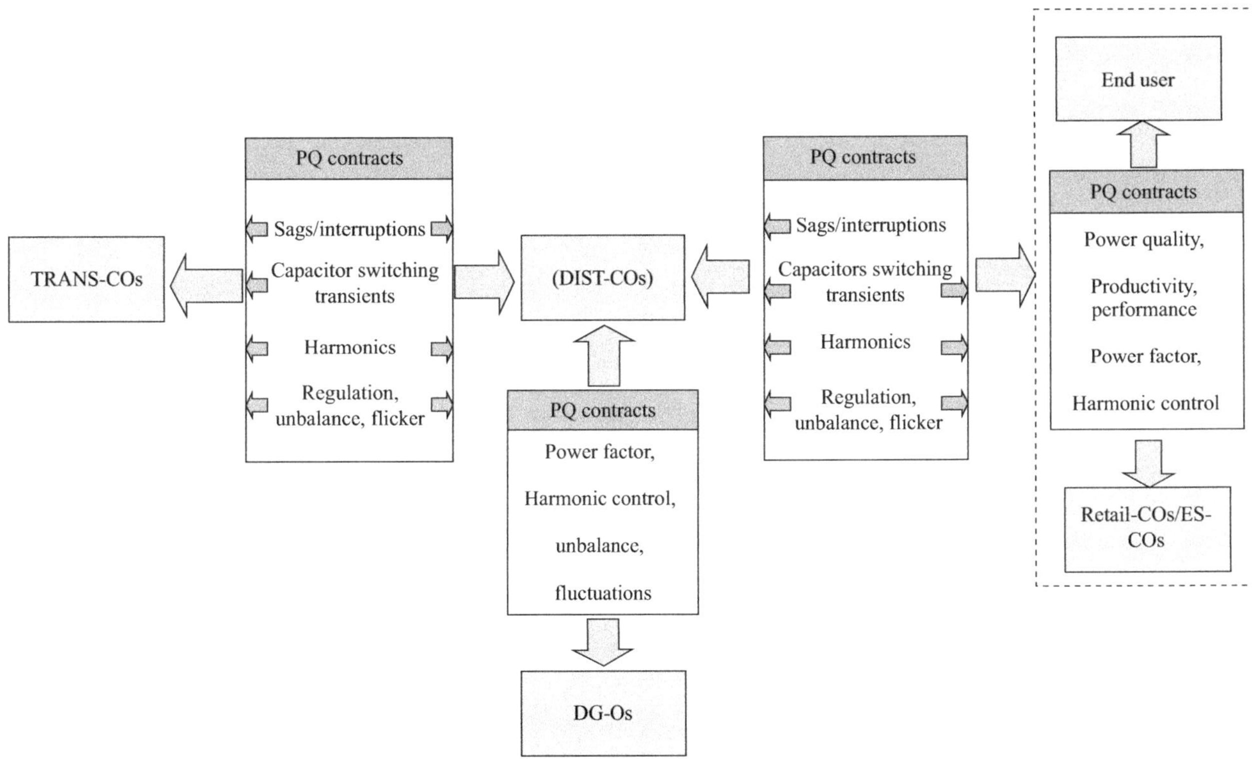

Figure 12.2 Contracts of PQ in the new configuration of electricity industry

Table 12.1 Required monitoring/measuring devices and analysis schemes for different categories of PQ events

PQ category	Monitoring/measuring devices	Analysis/display schemes
Harmonics distortion	• Three-phase voltages and currents • Single phase acceptable for balanced three-phase load • Sampling frequency • Sinusoidal form • Synchronized sampling	• Fast Fourier transform capability • Frequency spectra and waveforms plots
Long-duration variations	• Three-phase voltages • Root mean square • Sinusoidal form	• Magnitude/duration display
Short-duration variations and interruption	• Three phases voltages • Root mean square • Threshold level • One-cycle root mean square resolution	• Magnitude/duration display
Low-frequency transient	• Three-phase voltages and currents • Sampling frequency • Frequency responses ≥5 kHz • Threshold level	• Waveforms display prevent and recovery
High-frequency transient	• Three-phase voltages and current • Frequency responses ≥1 MHz • Impulse max and width disclosure • Threshold level	• Waveforms display location of impulse imposed on fundamental frequency signal

These services include an offering of different levels of PQ to verify the requirements of end-users. The supplier and end-user can together achieve this objective by reconstructing power systems or by inserting instruments within the end-user's premises. In either case, observation process is very important to set the standard for the different services and to validate that providers verify PQ levels in the contract [4].

The observing or monitoring instruments' needs shall rely on the particular PQ events that should be identified to evaluate the system situation. Some essential prerequisites of equipment are collected in Table 12.1 for different PQ categories.

With the degree of the monitoring prerequisites constantly expanding, it is getting to be basic to computerize the data gathering capacities and give simple-to-use data investigation and reporting highlights as a component of the general system. The system ought to have programmed downloading of data and describing of the data into a database that can be utilized for data investigation purposes. Data should be accessible to a wide assortment of people – the World Wide Web gives a helpful mechanism to this reason.

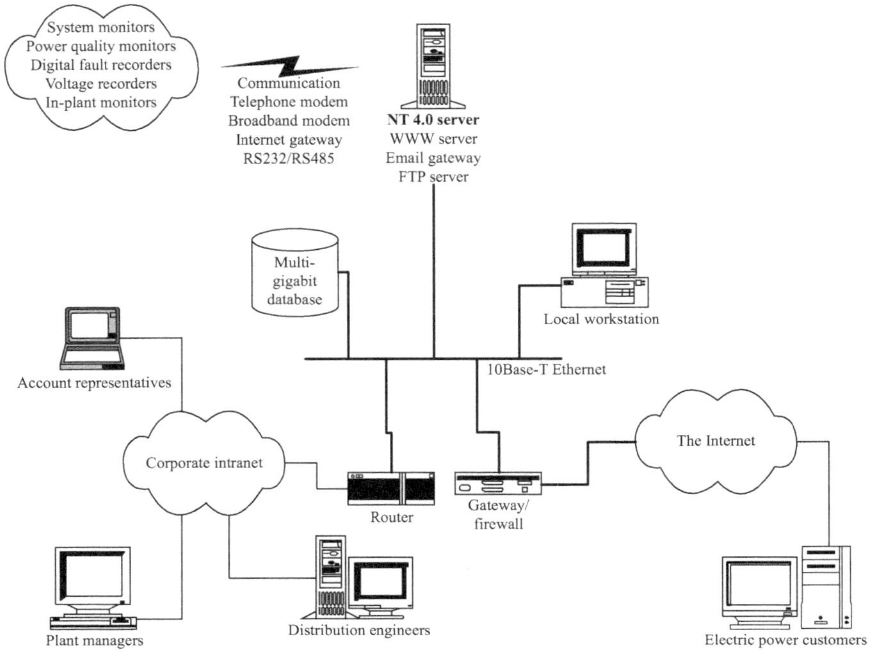

Figure 12.3 The power quality measurement/monitoring system structure

Figure 12.3 shows a monitoring system structure that exploits the World Wide Web for data access (either intranet for inside access or web for external access). At the heart of the monitoring system is a server PC advanced for database administration and examination. The Microsoft NT 4.0 operating system is demonstrating more than sufficient for keeping up World Wide Web and FTP administrations, email and paging notice and taking care of limit of several gigabytes of observed information.

Once the information has been gathered, this data should be analysed and summaries in a form to assess the contract compliance. To make the system hardware compatible, the system must be able to identify information from various measuring instruments [5].

12.3 Emerging power quality measurements

On-site measurement of PQ in electricity networks is imperative to system administrators to guarantee that an associated plant does not bring about undue disturbance to their system. Whilst PQ surveys can be utilized to decide the nearby impact of a specific disturbance impact, its impact on the more extensive system stays obscure. For instance, the generation of local harmonics by a large installation such as a HVDC station could be noteworthy, setting off the requirement for costly moderation hardware. Be that as it may, if that unsettling influence had been 'normally' lessened with separation at the wider grid level, the requirement for this moderation can be relaxed. On the other hand, the resonances in the system could offer ascent to an expansion in the disturbances to a wide level of the system [6].

It is suggested that the power quality spread is broken down by the synchronous measurements of PQ events utilizing various GPS time-stamped A/D converters disseminated around the system of concern. The way of the PQ spread will rely on the particular system structure and impedances. To make any general determinations about spread instruments, it is important to decide the spreading specifications by measuring in a choice of distribution and transmission systems with different important power quality events.

Smart grid networks are emerging as electric utilities that improve automation and security as well as grid efficient utilization by incorporating telecommunications, smart metering and fast fault-clearance technologies. Topology and components modelling tools are needed to evaluate the PQ issues, harmonic flow, disturbances, faulty events to assess required grid management and enhancements as well as future planning to adhere to specified future-regulating standards and regulations. In case of severe PQ problems and serious issues and/or fault events, there is a need to distinguish the three key indicators of source, coupling and effect in order of mitigating the correct solution and action. Locating faults and anomaly requires a harmonic power flow tools as well as event transient recorders and fault locators based on power quality collected data. Using various GPS and system impedance profiles as well as synchronized power quality voltage, current and power measurements, it is possible to locate serious noteworthy events [7].

Phasor measurement units (PMUs) are the next-generation RTU-remote terminal units for measuring future power system. PMUs are based on GPS technique and are widely utilized in monitoring operational states and power system signals. These efficient and accurate devices are capable of directly measuring frequency, voltage and current waveforms along with bus phase angle differences at very high sampling rate and excellent accuracy. Modern PMUs are equipped with added features, including frequency deviations, power components, power quality indices and harmonics and switchgear status of substation apparatus. The basic operation needed for PMUs is described in [8].

PMUs are emerging as a required fast acting instrumentation tool for the new generation of smart grids to allow system operators to monitor status and grid conditions. Using PMU data, key choices are made to ensure grid security, stability and supply integrity. The outcome of a missing PMU is loss of the grid including brownouts, blackouts or costly failure-event requiring load shedding or running extra backup generation with added costs.

It is fundamental that PMU-synchronized measurements are accurate, reliable and exact. Metrology base for traceable static estimations on PMUs has been created at NIST and EU smart grid venture [9]. PMU new developments are based on standardization of key functions and communications technology with new metrology. Future utilization of PMUs in distribution grid systems may require complex units with higher phase accuracy devices capable of operating in a noisy distorted PQ environment.

Knowledge of the system impedance is important for system operators, for instance (i) to use new load-flow models that can utilize PMU data directly to monitor system security and stability and (ii) to mitigate the PQ problems and severe disturbances due to new generations of non-linear loads or renewable energy sources. System impedance data is additionally needed to determine PQ-related location-specific problems and issues. Real-system impedance measurements are difficult to determine or measure and require different techniques to utilize numerous PMUs together [10]. It is crucial to estimate fundamental and harmonic impedances under different switching and operating conditions.

Sensors and transducers are fundamental components of the functionality of a robust PMU for additional PQ measurement; however, little attention is focused on the impact of their limited amplitude and phase spectral frequency response on measured signals and waveforms that are detected by the PMU/PQ calculations. The objective is to enhance the utilization of non-intrusive current transformers to enhance the measurement accuracy and exactness. Test laboratories are developed to characterize current and voltage transformers (VT) frequency response under severely distorted voltage and current waveforms. Digital real-time filtering and pre-filtering compensation techniques are also used and modified to account for sensor and transducer-variable spectral frequency response. Sensor/transducer fidelity and propagation of transducer uncertainties through complex PMU measurements are also addressed and analysed.

12.4 Power quality: impacts, harmonics estimation and mitigation

Harmonics are complex and problematic issues produced by multitude of sources and interactions between system components and control systems. Harmonics are manifested in different forms of dynamic, quasi-dynamic, static, quasi-steady state integral, fractional, sub, super, inter and modulating types. All leads to increased system losses, equipment damage, reduced life spans, inefficient operation, safety issues and electric fires [11].

Disturbances and excursions in voltage and current waveforms can be triggered by electric utilities as well as customer equipment. Short-duration and long-duration disturbances ranging from microseconds to milliseconds include spikes, notching, surges, voltage sags and swells, flicker, voltage and current imbalances and phase angle imbalances are produced by non-linear, inrush, arc type, cyclical and temporal type loads such as motors, drives, arc furnaces and LED lighting. Disturbance severity is classified by IEC Std according to short and long duration. Frequency disturbances and excursions causing frequency deviations can cause loss of synchronism and modulating type inter-harmonics at AC–DC interface. Series and shunt resonance conditions can magnify harmonic distortion due to AC system and customer voltage regulating and power factor correction capacitor banks [12].

Power quality monitoring and metering are required to record voltages, currents, powers, frequency, short-term and long-term harmonics and disturbances and associated harmonic spectra of integral type quasi-static harmonics as well as total harmonic distortion (THD) of voltage and current waveforms in all phases. Special DSP-signal processing metering devices with multi-channels (seven and above) with sample and hold metering units are used with special transducers sensors and pre-filtering conditioning units [13].

Mitigation and corrective measures for persistent and severe PQ problems violating key adopted standards covers

- Robust designs of devices and systems that is less sensitive to PQ problems
- Limit PQ disturbance sources and system conditions that affect sources, coupling and effective grounding, isolation systems.
- Use of conditioning, filtering and FACTS devices and systems in addition to shielding and solid grounding schemes.

IEEE-Std. 1100-1992 is intended to facilitate the selection of best conditioning unit suitable for a specific PQ problem [14].

The injection and harmonic power flow in the host electric utility can cause key performance degradation and severe problems including

1. Increased voltage drops and voltage regulation problems
2. Increased THD levels
3. Hot neutrals and ground potential rise grid potential rise (GPR)
4. Low-order harmonics exceeding recommended IEEE Std. 519-1992.
5. Increased resonance conditions and damage of capacitor banks and current transformers as well as overheating of electric machines
6. False tripping of overcurrent relays as well as negative sequence increased levels on feeders and transmission lines
7. Vibration and bearing, shaft damages due to shaft torque ripples and SSR oscillations of generators and large motor drives.

All PQ problems and issues are costly and problematic as they result in shut down, blackouts, brownouts, equipment damage, loss of revenue, increased losses, personal and equipment safety, increased electric fires, lower power factor and reduced energy-efficient utilization.

12.5 Power quality indices and standards

Dynamic PQ indices, assessment and measurement tools are required to deal with the PQ problems. Steady-state-based PQ indices include waveform THDs for voltage and current, harmonic magnitudes and phase angles, THD, PF, TIF, C-MSG Index, IT, VT products, K-Factor, crest-factor, flicker factor and unbalanced factor.

Power factor and PQ indices are not yet defined for transient quasi-dynamic harmonics and distorted voltage and current waveforms [15].

The true powers and power factor definitions are also lacking for highly distorted voltage and current waveforms. This applies to both balanced THD as well as residual THD indices. These definitions can be guides to calculating losses and losses as well as transformer and equipment de-rating.

Power quality standards are used to help electric utility and customers assess the severity and need for corrective mitigation and conditioning units [16,17].

PQ standards and interface units are produced by organizations such as IEEE, CENELEC, IEC, ANSI, UL SEMI, NFPA Associations and NEMA.

The following are key standards used around the world:

IEEE Std, 141-1993, IEEE Std. 142-1991, IEEE Std. 242-2001, IEEE Std. 446-1995, IEEE Std 519-1992, IEEE Std. 493-199, IEEE Std. 493-199, IEEE Std. 929-2000, IEEE-Std. 1100-2005, IEEE Std. 1159.3-2003, IEEE Std. 1250-1995, IEEE Std. 1346-1998, IEEE Std. 18-2002. IEEE Std. 1453-2004, IEEE Std. 1159-2009, SEMIE E-10-1999, SEMI F-42 1993, UL 1449, ANSI C 84.1-1995, EMA MG-1-1998, NEMA-LSI-1992, IEC 61000-4-15 and EN 50160.

Current PQ standards are only limited to quasi-steady state periodic harmonics and include IEEE, PES, IAS, ANSI, UL, NEMA, NFPA, NIST and IEC publications.

Harmonic control standards adopted by IEEE includes IEEE 519 voltage and current harmonic levels tables as well as IEEE 1159 monitoring power quality.

Transient standards includes ANSI C62.41 and C.62.41.

All PQ standards assess key indices and their effects and impact on cost, lifespan, EMC, power losses, safety, supply quality and potential for noise interference and malfunctioning of telecommunications, telecontrol and expected failures, faults and downtimes.

Future key indices, regulations and proposed standards addressing dynamic power quality phenomena and online assessment, estimation, measurements and monitoring tools should help make alleviate persistent power quality concerns and problems and provide better evaluation tools and suggested mitigation that are based on

- *Customer-Equipment:* Equipment should be designed with full immunity to fast decaying notches and transients as well as less sensitive to acceptable standard levels of THD of voltages and currents. The full immunity is a high-cost target but key design modifications and use of grounding, shielding and electronic components with high specifications to rate of rise of voltage and current can help reduce damage, failure and shutdown at reasonable cost.
- *Consumers:* Electricity consumers realizing power quality issues and concerns and educated on possible damage and cost of shutdowns, brownouts and blackouts will be keen in ensuring proper mitigation using isolation, grounding and shielding and use of robust interfaces and isolation transformers as well as active and passive power filters in their equipment designs. The layouts of electrical systems will take into account power conditioners based on FACTS to verify requirements of sensitive/critical equipment. Also power conditioners

options will be taken into account during the design stage. Power factor improvements and harmonics mitigation will be considered together.

- *Electric Utilities:* Electric utilities with the help of rigid and robust interface guidelines and well protected, grounded and isolated host smart grid systems will be able to provide more shielding and pose less power-quality problems to the customers and assurances of the expected system performance. The electric utility future investments in smart metering, automation, isolation, SCADA and demand control and fast isolation and switching will ensure an added higher level of performance with the added use if power conditioning systems and FACTS devices and fast acting control, relaying and protection systems.

Power quality is a complex issue requiring full coordination of all stakeholders: the electric utility, customers, equipment manufacturers, government health and safety organizations. Power quality can be ensured by active early coordination and participation of all parties based on established guidelines, enforced standards and use of successful effective mitigation.

The estimated cost of PQ problems can range from 3 to 6 billion dollars annually as result of voltages sags, swells, transients, flicker, voltage and current imbalances, glitches, equipment damage, telecontrol malfunction, solid state failures and converter and capacitor bank damages in addition to EMC and telecommunications interference and additional power losses.

Harmonic power quality measurements are only limited to integral quasi-steady state periodic voltage and current waveforms: THD, V and I harmonics, power factor using industrial synchronized sample and hold DFT/FFT-based meters and analysers including Drantz, Fluke and VIP meters, AEMC, LISN and other manufacturers. PQ indices are determined for site and system separately [18].

The 95 per cent probability of daily 3-s RMS values and harmonic components as well as the 99 per cent probability of weekly 10-min RMS values and harmonic components.

Transients and glitches and transient voltage are current waveforms measurements that are both tricky and un-reliable given the limitations of standardized test equipment; however, new DSP-based wavelets and meta-heuristic search tools can be utilized.

Harmonic power flow and online estimation techniques are not mature enough but promises the use of wavelets and DSP-probabilistic and estimation search tools using S-transform, Morley and discrete wavelets as well as DSP and Kalman-based filtering algorithms.

PQ will emerge as the new challenge for electric utilities as the focused mission of smart grid systems.

12.6 Power quality and smart grid

Key mitigation technologies for the smart grid, power quality are selected with full understanding of common sources, coupling and effects of and how economically to mitigate various problems. The power quality measurement tools are emerging using DSP-signal processing, digital technology and smart metering or intelligent electronic devices (IEDs).

Power quality analysis functions and measurement procedures require established electric utility protocols addressing sampling, location, duration and test conditions and variations from the norm for the smart grid objectives stated below [19–21].

- *Smart meter or IEDs with advanced key indices and power quality indexing functions:*
  The smart meters and IEDs will perform all voltage, current and power as well as harmonic measurement. Measured electric quantities are recorded and transmitted for further complex analysis and classifications of harmonics, inter-harmonics, flickering, modulation, distortion, notching and waveform time displays. Substation automation standards based on IEC 61850 can provide more developments and advanced power quality analysis.
- *Wide-area measurement:*
  The collected power quality recorded signals and indices are utilized in the wide-area global synchronized measurements with key time-synchronization and online information. Smart-grid and global-network-based power quality monitoring requirements can enhance data interpretation and supply quality information for disturbances and faults.
- *Mechanisms for power quality tracking and identification using reliable communication network:*
  Real-time efficient identification and tracking of faults, disturbance and anomaly conditions require full duplex and reliable two-way communications for fast detection and online decision-making allowing remedial and clearance strategies for power quality problems.
- *Real-time power quality signal analysis methods:*
  Applications of DSP digital signal processing analysis techniques for synchronized and filtered measurements can provide fast detection and diagnosis of power quality issues, faults, disturbance and failure/fault/anomaly types and suggest effective mitigation solutions and tools. Advanced signal processing DSP techniques including wavelets, FFT, DFT and probabilistic algorithms are required for monitoring power system operation and severe power quality disturbances.
- *Power quality data management:*
  Efficient data analysis and mining tools for large recorded series of power quality data in real time are required for identifying system abnormalities and power quality issues.
- *Power quality indices and guidelines/standards:*
  The use of FACTS devices and other energy-efficiency technologies can have negative impacts on power quality. The increased use of fast switching and automation tools in smart grids can also cause additional power quality problems. Power quality guidelines and standards are required to limit use of

correction and FACTS-based equipment that may have negative impact with the system. Smart grid will impact power quality in different ways. Will the power quality be improved? This will depend on the new standards and regulatory framework.

Smart grid is aiming at achieving grid reliability, security, energy efficiency and fast isolation using newly enhanced communications and supervisory control systems for energy delivery and fault isolation as part of the fully automated grid.

The goal for the smart grid highest power supply quality can be achieved with reduced waveform distortion, harmonics, transient and quasi-steady state conditions with controlled THD, limited harmonics sub, super and inter-harmonics.

The mushrooming use of non-linear loads and solid state switching converters, industrial control systems, UPS, FACTS devices and switched modulated, temporal, inrush and cyclical type loads coupled with emerging high penetration of renewable energy, PV, wired, wave, tidal, micro-hydro, battery storage and battery charging (V2H and V2G charging stations) devices and interfacing to the smart grid will pose a new challenge and cause deteriorating effect and severe on the host electric grid.

This new challenge will impact power quality and increase host grid-renewable energy DC–AC interface problems with required voltage stabilization decoupling and transient voltage recovery using the modified active power filters, STATCOM and switched/modulated LC filter – capacitive compensation devices. The impact on power quality, modelling tools, measurement requirements and test procedures is also massive.

12.7 Power quality trends and future requirements

The rising concern by electric utilities of power quality issues problems is a by-product of the mushrooming use of non-linear loads and new demands for DC–AC and AC–DC–AC interfacing with renewable energy schemes including solar PV, wind, wave/tidal, small hydro and fuel cell-battery storage systems and devices. The use of FACTS technology, active power filters and STATCOM devices for voltage stabilization, inrush current limitations and efficient energy utilization created new challenges and problems for dynamic and quasi-dynamic power quality issues and supply quality deterioration.

The problem is complicated by the new inter-harmonics-sub-harmonics, super-harmonics and fractional-order harmonics created by inter-modulation and system–control interaction as well as solid state switching and use of cascaded multi-stage MV and HV converter topologies for interfacing renewable energy schemes to distribution, sub-transmission and transmission systems.

The future needs and requirements include the following, Figure 12.4:

1. New established measurement procedures, tools, devices and synchronized measurements using robust DSP algorithms and signal-filtering tools.
2. New robust models of non-linear loads, renewable energy DC–AC interface scheme and FACTS devices for enhanced accurate harmonic flow studies.

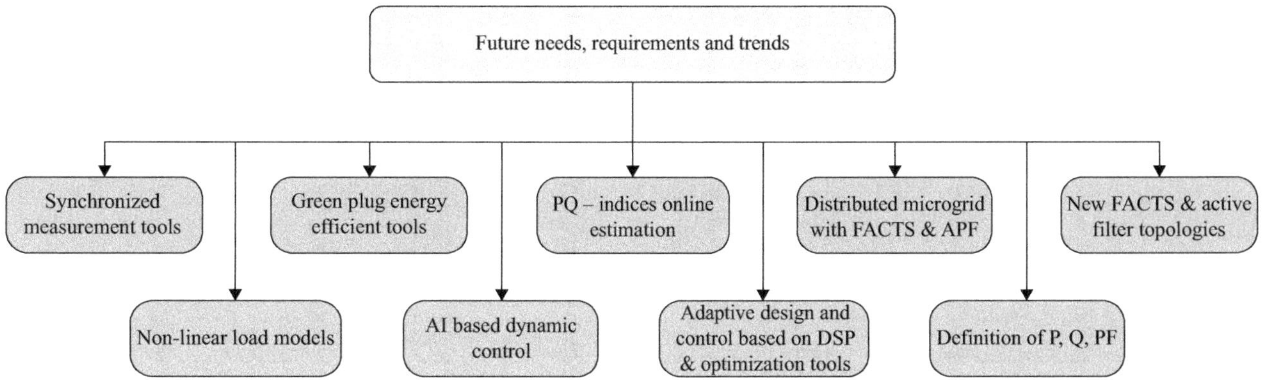

Figure 12.4 Future trends of power quality

3. New active filter topologies based on the voltage source converters (VSCs) with possibility of multilevel cascaded converters for MV and HV interface to distribution and transmission grid systems.
4. New dynamic control strategies using heuristic and meta-heuristic search and optimization algorithms (PSO, Harmony, GA, Ant search, weed invasion, Binary search and Bacteria foraging algorithms).
5. Hybrid VSC with six pulse, 12 pulse and higher pulses for modified active power filter topologies to be used with battery storage, V2H, V2G battery charging DC–AC interface converters.
6. New definitions for the active power, reactive power and power factor in case of heavily distorted voltage and current.
7. New digital metering tools based on active power, reactive power and power factor modified definitions.
8. Definitions of power quality steady state and dynamic indices with online smart grid estimation tools using new wavelets (S-transform, Morley and discrete wavelets algorithms) and meta-heuristic search and estimation tools.
9. Development of new FACTS compensation strategies and dynamic control strategies for D-STATCOM and hybrid power filters integration with smart grid and non-linear load centres.
10. Development of efficient green plug, switched filters, switched/modulated capacitive compensators for energy-efficient operation and limited transient over-voltages and inrush current conditions.
11. Applications of APF and FACTS in distributed micro-grid systems.

12.8 Case studies

It has become necessary to develop a new kind of custom power devices based on series/shunt compensator which can mitigate a wider range of power quality problems. In this section, some of power quality mitigation techniques are demonstrated and tested as a new trend for solving power quality disturbances.

Figure 12.5 depicts the single-line diagram of the tested system that is used to check the validation of the new techniques. The new techniques are based on design and work on the intermittent complementary switching between filters and compensation circuits [22–30]. The tested system includes the new mitigation technique beside a hybrid load that comprises a linear load, a converter-type non-linear load and an induction motor load.

A multi-loop error-driven dynamic controller is used to modulate PWM switching, Figure 12.6. Where the output pulses of the PWM are the complementary pulses that are used for the intermittent switching between filters and compensation circuits. The global error (e_g) is the summation of the three-loop individual errors including voltage stability, current limiting and synthesis dynamic

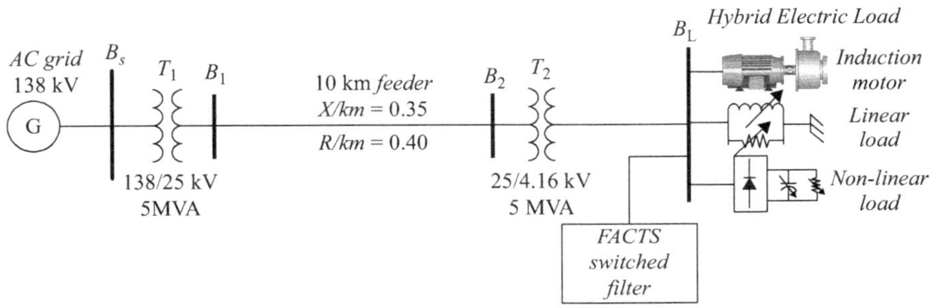

Figure 12.5 The single-line diagram of the tested distribution system with the FACTS switched filter compensation

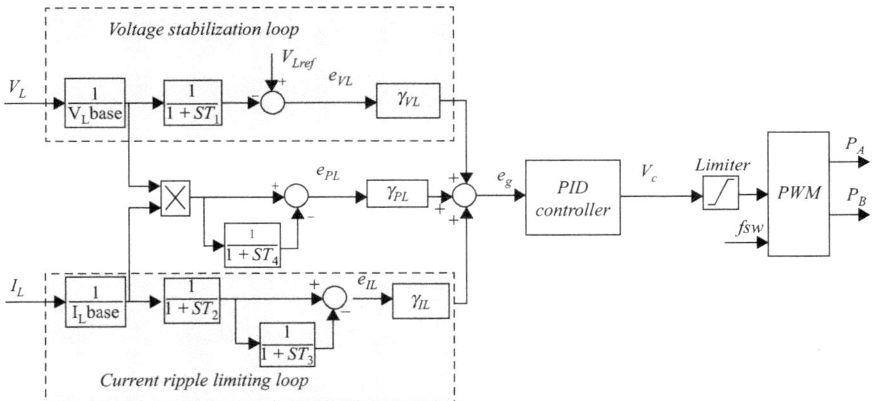

Figure 12.6 The block diagram of the intermittent switching process controller

power loops. The per-unit three-dimensional error vector (e_{VL}, e_{IL}, e_{PL}) is governed by (12.1)–(12.5):

$$e_{V_L} = \frac{V_{L\text{ref}} - V_L(1/(1+ST_1))}{V_{L\text{base}}} \tag{12.1}$$

$$e_{I_L} = \frac{I_L(1/(1+ST_2))}{I_{L\text{base}}}\left(1 - \frac{1}{1+ST_3}\right) \tag{12.2}$$

$$e_{P_L} = \left(\frac{V_L}{V_{L\text{base}}} \times \frac{I_L}{I_{L\text{base}}}\right)\left(1 - \frac{1}{1+ST_4}\right) \tag{12.3}$$

The global error e_g:

$$e_g = \gamma_{V_L}|e_{V_L}| + \gamma_{I_L}|e_{I_L}| + \gamma_{P_L}|e_{P_L}| \tag{12.4}$$

The solid-state switches (S_A, S_B) are usually (GTO, IGBT/bridge or MOSFET/ bridge) turns 'ON' when a gating pulse P is applied by the activation switching circuit. Removing the pulse will turn the solid-state switch 'OFF'.

$$T_{sw} = \frac{1}{f_{sw}} = T_{\text{on}} + T_{\text{off}} \tag{12.5}$$

where f_{sw} is the switching frequency and $0 < T_{\text{on}} < T_{sw}$.

The equivalent impedance for the hybrid capacitors and the tuned arm filter (TAF) is modulated by the pulse width modulation (PWM) index as function of the duty cycle *m* and switching frequency f-switching, both are dynamically regulated and controlled by the error-driven decoupled multi-loop PID controller. The output of the PWM block is a train of pulses with modified duty cycle at constant switching frequency. The equivalent of reactive compensation is dependent on the duty cycle *m* modulating the apparent Thevenin's impedance at interface bus.

All detailed parameters of the new mitigation schemes, multi-loop error and gains pf PID controller are given in the appendix section.

12.8.1 The hybrid FACTS SPFC-filter compensator

A novel mitigation scheme called the hybrid filter compensation scheme is presented, Figure 12.7. This scheme comprises two power filters in parallel, a TAF and

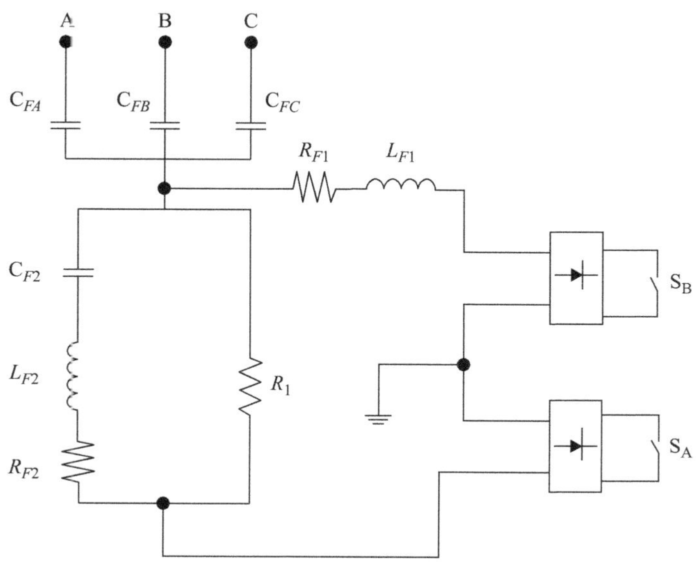

Figure 12.7 The hybrid FACTS SPFC-filter compensation scheme

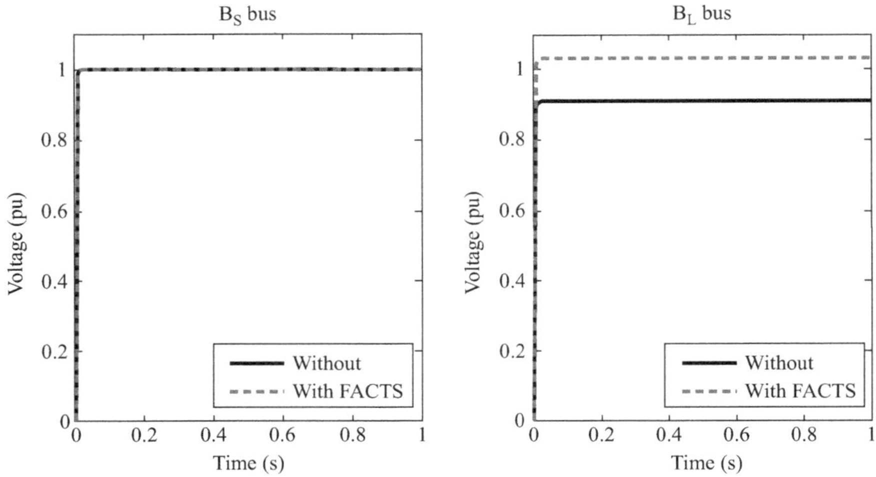

Figure 12.8 The rms voltage at the source bus, B_s, and the load bus, B_L

C-type filter (CTF), connected in series with three-phase shunt capacitor bank. The two power filters are switched on/off intermittently using two switches (S_A and S_B). These switches are controlled by two complementary switching pulses (P_A and P_B) that are generated from PWM driven by PID controller. The hybrid filter compensator has two topologies based on the switching process of the arm filters using the complementary switching pulses as follows:

Topology one: when the S_A is switched on and S_B is switched off, the TAF will be disconnected and the CTF will provide the required shunt capacitive compensation to the AC system to improve the power factor and eliminate the harmonic distortion of the typical harmonic orders that adjusted using CTF.

Topology two: here S_A is switched off and S_B is switched on, the CTF filter will be disconnected and TAF filter will be connected into the circuit to mitigate typical orders of harmonics and enhance the system power factor.

The simulation result shows that with inserting the hybrid filter compensator, the voltage at the load bus is enhanced to 1.04 pu, Figure 12.8, and the power factor at the system buses is improved, where the power factor at the source and load buses is enhanced to 0.95 and 0.76, respectively, Figure 12.9.

The fast Fourier transform is used to show the harmonic spectra of the voltage and current waveforms of system buses. The harmonic spectrum of voltage and current waveforms at the load bus are depicted in Figures 12.10 and 12.11. Also, the voltage and current harmonic analysis in term of the THD at all the system buses are calculated and are summarized in Table 12.2. It is obvious that the voltage and current harmonics at the system buses are significantly reduced to a level within the limit set by the IEEE Std. 519-1992 (IEEE, 1992) regarding the THD of bus voltage at low-voltage system (less than 69 kV) [31].

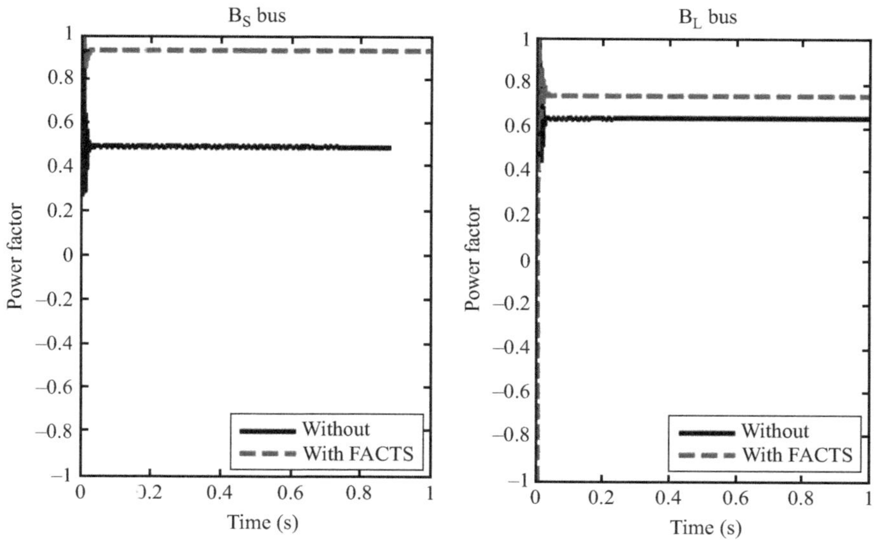

Figure 12.9 The power factor at the source bus, B_s, and the load bus, B_L

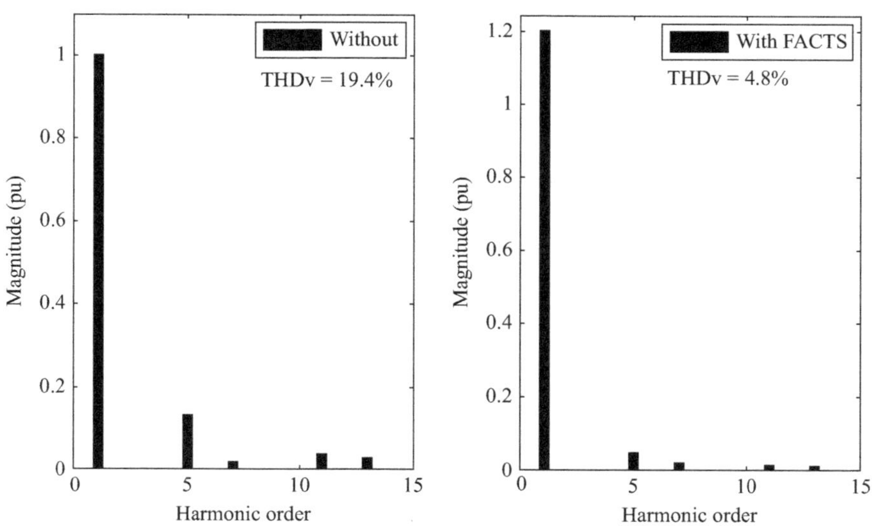

Figure 12.10 The frequency spectrum of voltage waveform at the load bus, B_L, without and with the FACTS filter compensator

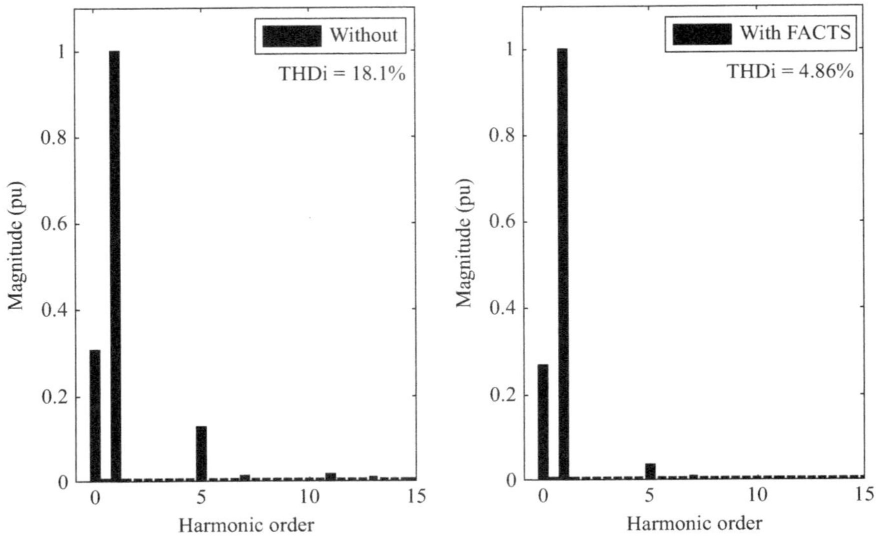

Figure 12.11 The frequency spectrum of current waveform at the load bus, B_L, without and with the FACTS filter compensator

Table 12.2 THD values at system buses

	% THD of voltage waveforms		**% THD of current waveforms**	
	Without	**With FACTS**	**Without**	**With FACTS**
B_s bus	0.071	0.05	17.9	4.85
B_1 bus	8.4	2.8	18.0	4.85
B_2 bus	10.1	3.3	15.9	4.86
B_L bus	19.4	4.8	18.1	4.86

12.8.2 FACTS-MPFC Modulated power filter compensator scheme I

The structure of the modulated power filter compensator (MPFC) scheme I is a shunt three-phase capacitor bank in series with an uncontrolled bridge rectifier. The DC side of the uncontrolled rectifier is short circuit or connected to an arm filter based on an intermittent control signals.

The two switches (S_A and S_B) on the DC side of the uncontrolled rectifier are controlled by two complementary switching pulses (P_A and P_B) that are generated

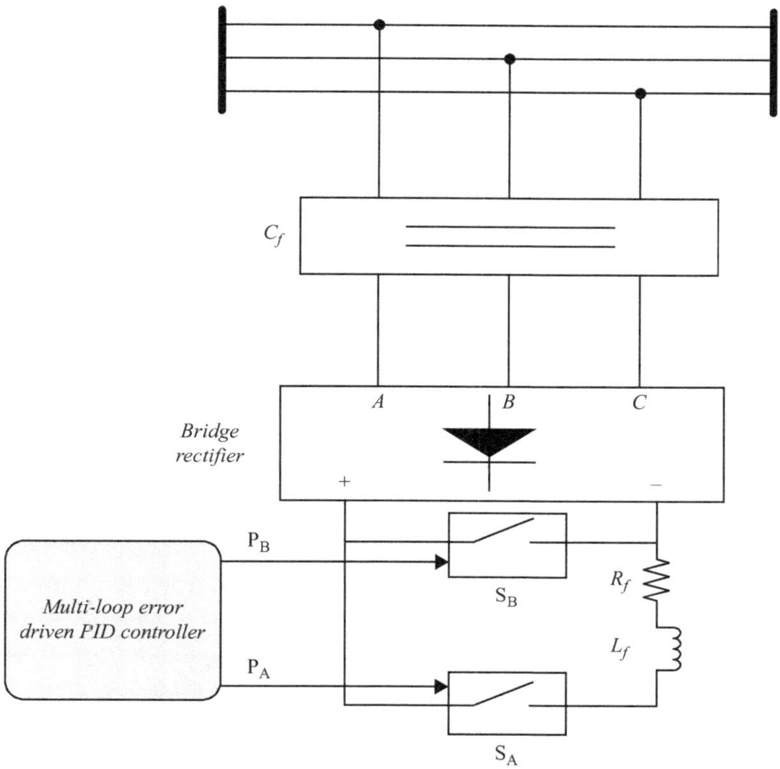

Figure 12.12 The FACTS-MPFC switched power filter compensator scheme I

by PWM, Figure 12.12. The variable MPFC scheme I topology can be structured by complementary switching PWM square signals as follows:

Case 1: If P_A is low and P_B is high, the arm filter will be disconnected and the shunt capacitor will supply the reactive power compensation to the tested system.

Case 2: If P_A is high and then P_B is low, the arm filter will be linked to the tested system to eliminate the harmonic distortion.

The simulation results of inserting the scheme I of MPFC show that the voltage is enhanced from 0.9 to 1.0 pu at the load bus. In Figure 12.13 also, the power factor is improved at the system buses, where the power factor is enhanced to 0.9 and 0.78 at the source and load buses, respectively, Figure 12.14.

The voltage and current waveforms at the system buses are captured and processed using Fourier transform to show the harmonic contents presented in Table 12.3. The results in Table 12.3 show that the THD is enhanced using the MPFC to be within IEEE standard limits.

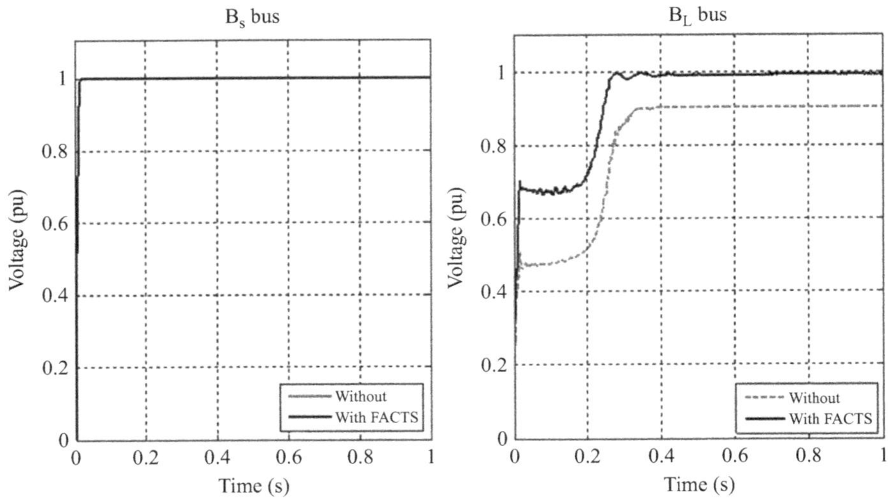

Figure 12.13 The rms value of the voltage at B_s and B_L buses, with and without MPFC I

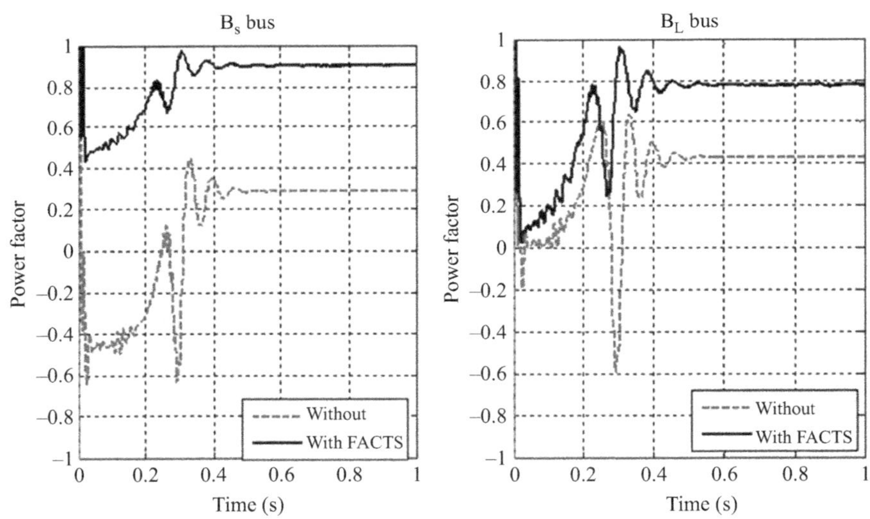

Figure 12.14 The power factor at B_s and B_L buses, with and without MPFC I

12.8.3 FACTS–MPFC switched power filter compensator scheme II

In this scheme, a low-cost MPFC, switched type filter, is used to provide measured filtering in addition to reactive compensation. The MPFC is controlled by the on–off timing sequence of the PWM switching pulses that are generated by PID controller.

Table 12.3 The percentage values of THD of voltage and current waveforms at system buses

	% THD of voltage waveforms		% THD of current waveforms	
	Without	**With FACTS**	**Without**	**With FACTS**
B_s	0.071	0.1	17.9	3.98
B_1	8.4	4.93	18.0	3.67
B_2	10.1	4.61	15.9	4.12
B_L	19.4	4.23	18.1	4.69

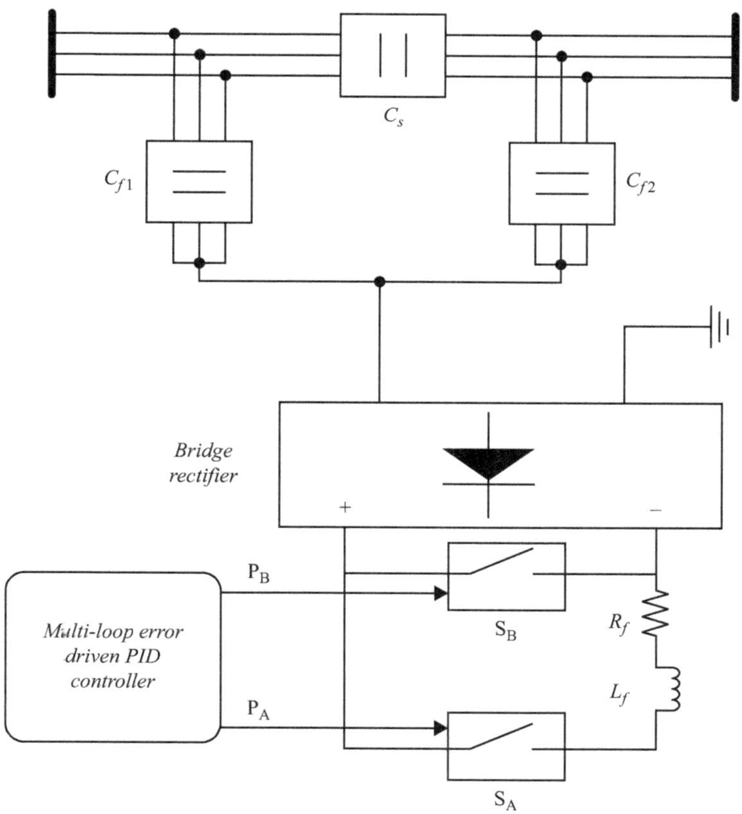

Figure 12.15 The FACTS–MPFC switched power filter compensator scheme II

The MPFC scheme II structure comprises a series fixed capacitor bank and two shunt fixed capacitor banks are connected to a modulated PWM switched TAF through two pulses uncontrolled rectifier.

The used SPFC scheme II, Figure 12.15, is a combination of one series capacitor bank and two parallel capacitor banks with an arm power filter.

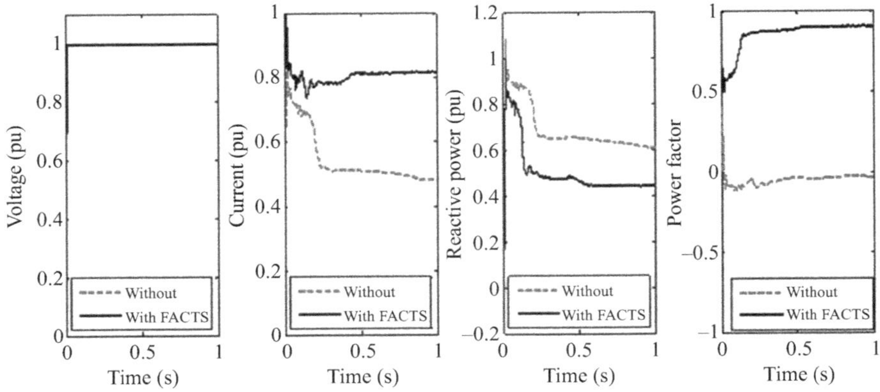

Figure 12.16 The voltage, current, reactive power and power factor at the B_s without and with MPFC II

The discharged energy of the two parallel capacitor banks is through a single-phase uncontrolled rectifier in series with a resistance (R_f) and an inductance (L_f) branch that shapes a filter at the DC side of the uncontrolled rectifier. The IGBT switches (S_A and S_B) are switched on/off by complementary pulses P_A and P_B that are produced by the PID controller. The variable MPFC topology can be configured by complementary switching square signals generated from PWM as follows:

Topology 1: If P_A is in lower form and P_B is in higher form, the arm filter will have disconnected and the parallel and series capacitor banks will inject the needed shunt/series reactive power compensation to the tested system.

Topology 2: If P_A is in higher form and then P_B is in lower form, the arm filter will have inserted into the tested system as a low pass filter.

The results show that the scheme II of MPFC has the capability to compensate the reactive power at the system buses to stabilize the voltage profile and enhance the power factor. When the power factor at B_s bus is improved to 0.9 lag from 0.07 lead, this means that the reactive power is transferred from the system to the utility grid after using the MPFC, Figure 12.16. Also with using the MPFC, the voltage profile is stabilized at 1 pu and the power factor is improved at B_L bus, Figure 12.17.

The THD of voltage and current waveforms are calculated and are summarized in Table 12.4. The results in Table 12.4 show that the harmonic distortion is improved to be less than that of the IEEE standard limits.

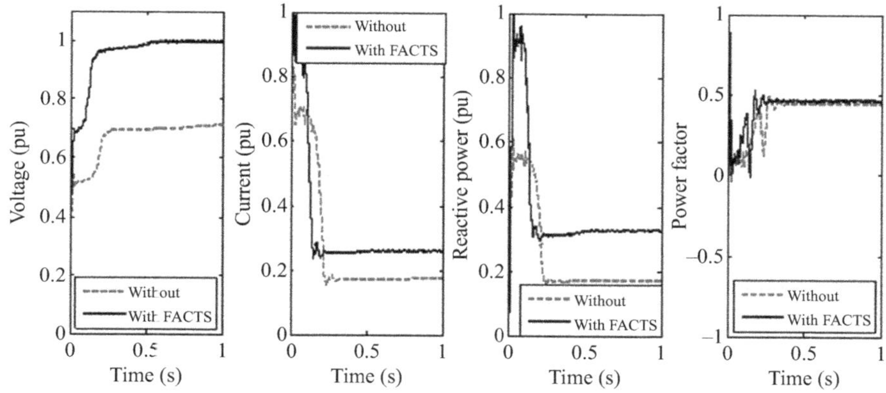

Figure 12.17 The voltage, current, reactive power and power factor at the B_L without and with MPFC II

Table 12.4 The percentage THD of voltage and current at system buses

	THD of voltage waveform		THD of current waveform	
	Without	**With FACTS**	**Without**	**With FACTS**
B_S	0.071	0.1	17.9	4.55
B_1	8.4	4.92	18.0	4.56
B_2	10.1	4.82	15.9	4.12
B_L	19.4	4.71	18.1	4.3

12.8.4 Modulated/switched series-shunt power filter compensator scheme III

The proposed MPFC scheme III, Figure 12.18, is a combination of one series capacitor bank and two parallel capacitor banks with an arm power filter. The series capacitor is inserted in series with the line feeders to suppress a part of the line reactance to enhance the load flow and decrease the reactive losses of the line. Also, the series capacitor bank works as a voltage-booster and inrush-current reduction instruments. The parallel capacitor banks inject the required reactive compensation and enhance the regulation of the system feeders. The discharged energy of the two parallel capacitor banks is through a full bridge three-phase uncontrolled rectifier in series with a resistance (R_f) and an inductance (L_f) branch that shapes a filter at the DC side of the uncontrolled rectifier. The IGBT switches S_A and S_B are switched on/off by complementary pulses PA and PB that are produced by the PID controller. The lower pulse P_A switches S_A, while the upper pulse

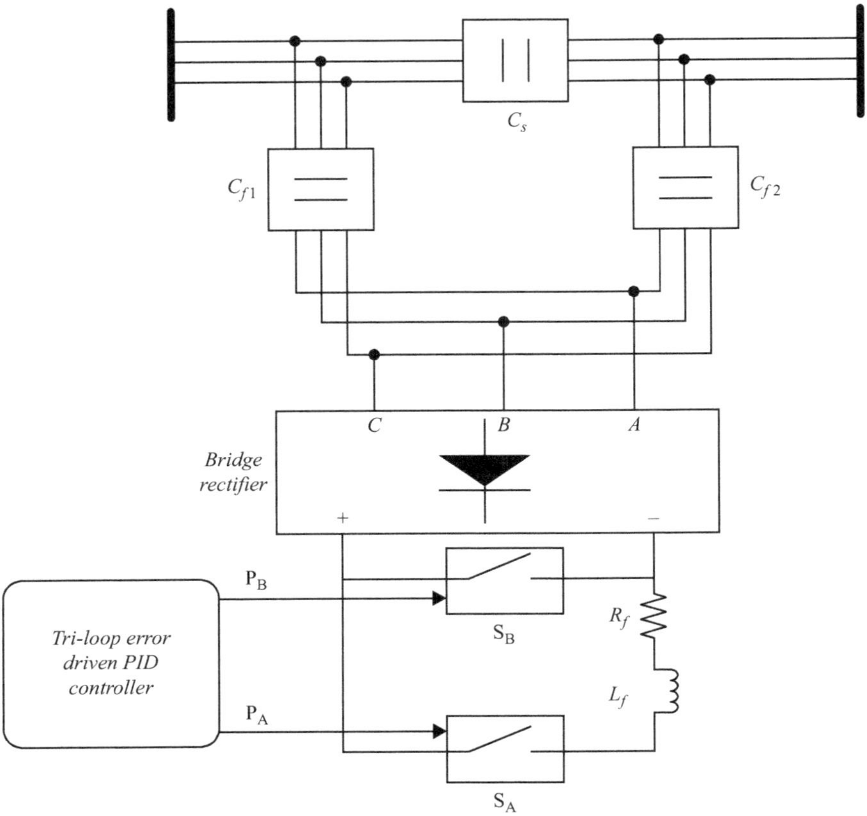

Figure 12.18 The FACTS switched series-shunt power filter compensator scheme III

P_B controls S_B. There are variable topologies of the MPFC based on the intermittent switching of the switches S_A and S_B.

The voltage and power factor are enhanced using the proposed MPFC as shown in Figures 12.19 and 12.20, respectively.

The captured voltage and current waveforms at the system buses are analysed using Fourier transform, and their frequency spectra are shown in Figures 12.21 and 12.22, respectively. A comparative analysis based on calculating the THD of the captured quantities of the system without and with inserting scheme III of MPFC are done and collected in Table 12.5. The results show that the THDs of the captured quantities are enhanced and reduced to a level within the limits stated by the IEEE Std. 519-1992.

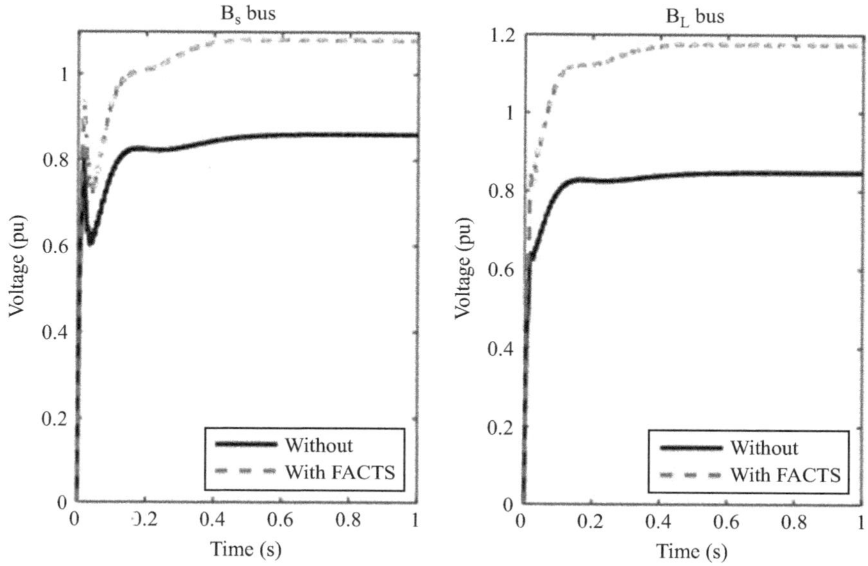

Figure 12.19 The rms values of the voltage at the source, B_s, and the load, B_L, buses

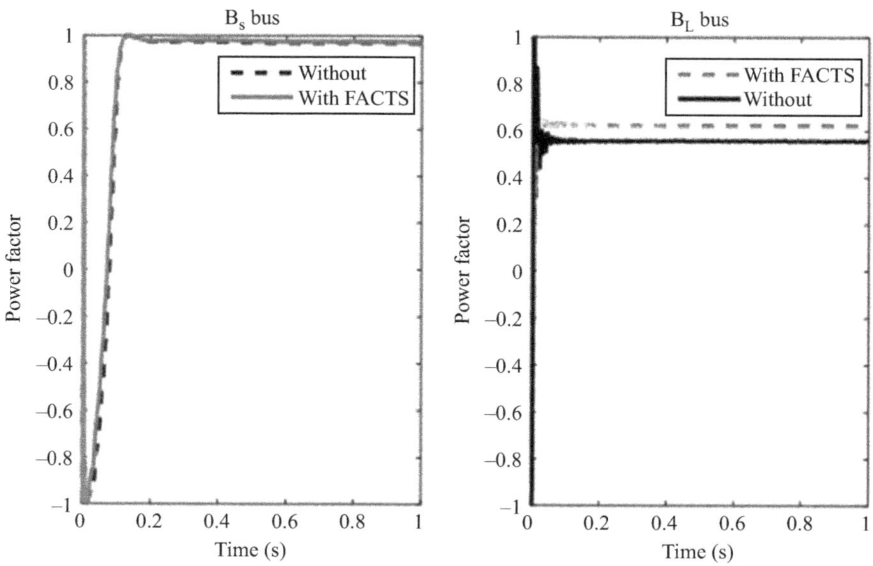

Figure 12.20 The power factor values at the source, B_s, and the load, B_L, buses

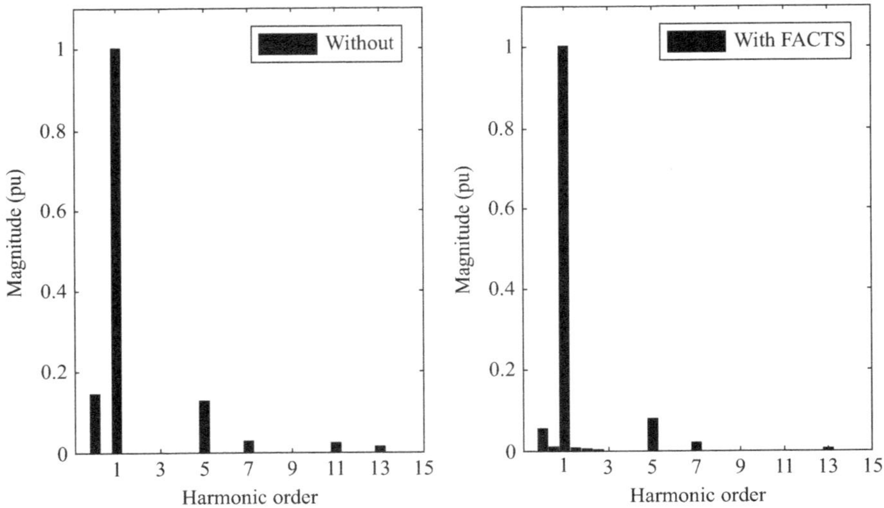

Figure 12.21 The frequency spectrum of voltage waveform at the load bus, B_L

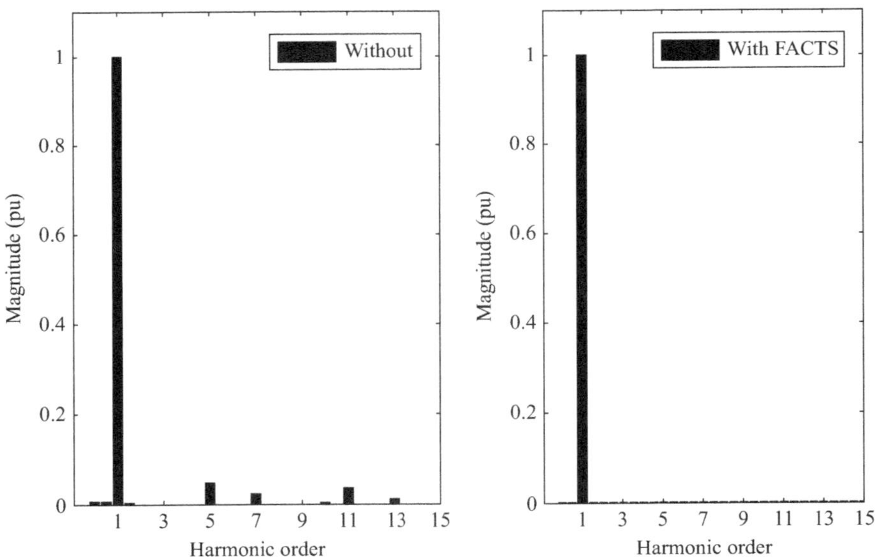

Figure 12.22 The frequency spectrum of current waveform at the load bus, B_L

Table 12.5 THD values of voltage and current waveforms

	% THD of voltage waveforms		% THD of current waveforms	
	Without	**With FACTS**	**Without**	**With FACTS**
$\mathbf{B_s}$	0.071	0.6	17.9	3.5
$\mathbf{B_1}$	8.4	2.2	18.0	4.6
$\mathbf{B_2}$	10.1	4.61	15.9	4.12
$\mathbf{B_L}$	19.4	4.8	18.1	2.64

12.9 Conclusions

The chapter presents a review of power quality issues, measurements, monitoring, performance indices, impacts and standards associated with established PQ indices, impacts, emerging limitations and PQ future trends and challenges.

The case study presented at the chapter-end presents three new designs of FACTS-based low-cost switched filter compensation devices acting as dynamic Modulated/PWM switched filters to reduce power quality THD of voltages and currents at point of common coupling.

The three FACTS devices include hybrid switched power filter, switched/ MPFC and series-shunt FACTS switched power filter.

The chapter is intended as a short guide and a review of power quality issues, limitations and emerging challenges due to the increasing use of renewable Energy AC-DC FACTS devices and compensating systems, mushrooming use of non-linear loads and new smart metering and interface issues with the smart grid.

Appendix

Hybrid FACTS filter compensator parameters:

$$C_f = 375\ \mu F,\ R_{f1} = R_{f2} = 0.05\ \Omega,\ L_{f1} = L_{f2} = 1\ mH,\ \text{and}\ C_{f2} = 20\ \mu F$$

MPFC scheme I parameters:

$$C_f = 100\ \mu F,\ R_f = 0.2\ \Omega,\ L_f = 3\ mH$$

MPFC scheme II parameters:

$$C_s = 1{,}000\ \mu F,\ C_{f1} = C_{f2} = 50\ \mu F,\ R_f = 0.25\ \Omega,\ L_f = 3\ mH$$

MPFC scheme III parameters:

$$C_s = 150\ \mu F,\ C_{f1} = C_{f24} = 40\ \mu F,\ R_f = 0.1\ \Omega,\ L_f = 1\ mH$$

Gains of multi-loop error and PID controller:

Multi-loop error gains: $\gamma_{vL} = 1$, $\gamma_{iL} = 0.75$, $\gamma_{pL} = 0.5$
PID controller gains: $K_p = 1$, $K_i = 0.5$, $K_d = 0.2$
PWM frequency: $f_s = 1{,}750$ Hz.

References

[1] McGranaghan M., Kennedy B. W., Samotyj M. 'Power quality contracts in a competitive electric utility industry'. *Eighth International Conference on Harmonics and Quality of Power*; Athens, Greece, 1998, pp. 170–175.

[2] Degand C., Fauquembergue P., Regnier M., Levillain C., Serres E. 'EDF customized approach of power quality, an opportunity to meet customer's expectations'. *Eighth International Conference on Harmonics and Quality of Power*; Athens, Greece, 1998, pp. 190–196.

[3] Chandler T. 'Quality of power contracts'. *2008 International Conference on Electricity Distribution*; Guangzhou, China, 2008, pp. 1–5.

[4] Dettloff A. 'Power quality performance component of the special manufacturing contracts between energy provider and customer'. *IEEE Power Engineering Society Summer Meeting*; Seattle, WA, USA, 2000.

[5] Milanović J. V. 'Economic impact of power quality disturbances on customers and utilities'. *IET Conference on Power in Unity: A Whole System Approach*; London, 2013, pp. 1–37.

[6] Tse N. C. F., Chan J. Y. C., Lau W. H., Poon J. T. Y., Lai L. L. 'Real-time power-quality monitoring with hybrid sinusoidal and lifting wavelet compression algorithm'. *IEEE Transactions on Power Delivery*. October 2012, vol. 27(4), pp. 1718–1726.

[7] Muscas C. 'Power quality monitoring in modern electric distribution systems'. *IEEE Instrumentation & Measurement Magazine*. October 2010, vol. 13(5), pp. 19–27.

[8] Auler L. F., d'Amore R. 'Power quality monitoring controlled through low-cost modules'. *IEEE Transactions on Instrumentation & Measurement*. March 2009, vol. 58(3), pp. 557–562.

[9] IEEE Std 1159-2009 (Revision of IEEE Std 1159-1995): 'IEEE Recommended Practice for Monitoring Electric Power Quality'. June 26, 2009.

[10] Parle J. A., Madrigal M., Acha E. 'Trends in power quality monitoring'. *IEEE Power Engineering Review*. 200, vol. 21(10), pp. 3–21.

[11] Bongiorno M., Svensson J., Sannino A. 'Effect of sampling frequency and harmonics on delay-based phase-sequence estimation method'. *IEEE Transactions on Power Delivery*. July 2008, vol. 23(3), pp. 1664–1672.

[12] Tiwari V. K., Jain S. K. 'Hardware implementation of polyphase-decomposition-based wavelet filters for power system harmonics estimation'. *IEEE Transactions on Instrumentation & Measurement*. July 2016, vol. 65(7), pp. 1585–1595.

[13] Jain S. K. 'Algorithm for dealing with time-varying signal within sliding-window for harmonics estimation'. *IET Science, Measurement & Technology*. 2015, vol. 9(8), pp. 1023–1031.

[14] IEEE Std 1100-1992. '*IEEE Recommended Practice for Powering and Grounding Sensitive Electronic Equipment*'. December 31, 1992, pp. 1–256.

[15] Lin T., Domijan A. 'On power quality indices and real time measurement'. *IEEE Transactions on Power Delivery*. October 2005, vol. 20(4), pp. 2552–2562.
[16] Kandil M. S., Farghal S. A., Elmitwally A. 'Refined power quality indices'. *IEE Proceedings – Generation, Transmission and Distribution*. November 2001, vol. 148(6), pp. 590–596.
[17] Andreotti A., Bracale A., Caramia P., Carpinelli G. 'Adaptive Prony method for the calculation of power-quality indices in the presence of nonstationary disturbance waveforms'. *IEEE Transactions on Power Delivery*. April 2009, vol. 24(2), pp. 874–883.
[18] Morsi W. G., El-Hawary M. E. 'Wavelet packet transform-based power quality indices for balanced and unbalanced three-phase systems under stationary or nonstationary operating conditions'. *IEEE Transactions on Power Delivery*. October 2009, vol. 24(4), pp. 2300–2310.
[19] Wright P. S., Rietveld G., van den Brom H. E., Crotti G., Braun J. P. 'Smart grid power quality and stability measurements in Europe'. *2016 Conference on Precision Electromagnetic Measurements (CPEM 2016)*; Ottawa, ON, Canada, 2016, pp. 1–2.
[20] Wright P. S., Christensen A. E., Davis P. N., Larson B., Lippert T., Patel P. 'Power quality propagation measurements in smart grids'. *2016 Conference on Precision Electromagnetic Measurements (CPEM 2016)*, Ottawa, ON, Canada, 2016, pp. 1–2.
[21] Bollen M. H. J., Das R., Djokic S., *et al.* 'Power quality concerns in implementing smart distribution-grid applications'. *IEEE Transactions on Smart Grid*. Jan. 2017, vol. 8(1), pp. 391–399.
[22] Sharaf A. M., Abdelsalam A. A. 'A novel FACTS based dynamic voltage compensation scheme for smart electric grid stabilization and efficient utilization'. *The 24th EEE Canadian Conference on Electrical and Computer Engineering CCECE'11*; Niagara Falls, Ontario, Canada, 2011, pp. 42–47.
[23] Sharaf A. M., Abdelsalam A. A. 'A novel switched filter compensation scheme for power quality enhancement and loss reduction'. *International Symposium on Innovations in Intelligent Systems and Applications INISTA 2011*, Istanbul, Turkey, 2011, pp. 398–403.
[24] Sharaf A. M., Abdelsalam A. A. 'Power quality enhancement in wind grid interface based on switched filter compensator'. *International Symposium on Innovations in Intelligent Systems and Applications INISTA 2011*, Istanbul, Turkey, 2011, pp. 404–409.
[25] Sharaf A. M., Abdelsalam A. A., Eldesouky A. A., Sallam A. A. 'A novel FACTS based modulated power filter compensation scheme for smart electric grid stabilization and efficient utilization'. *International Journal of Distributed Energy Resources*. 2011, vol. 7(4), pp. 311–328.
[26] Abdelsalam A. A., Sharaf A. M. 'A novel FACTS compensation scheme for power quality improvement in wind smart grid'. *The 25th IEEE Canadian Conference on Electrical and Computer Engineering CCECE'12*; Montréal, Canada, 2012.

[27] Sharaf A. M., Abdelsalam A. A. 'A FACTS based stabilization compensation scheme for smart distribution grid'. *Australian Journal of Electrical & Electronics Engineering*. 2013, vol. 10(1), pp. 65–73.

[28] Abdelsalam A. A., Sharaf A. M., Eldesouky A. A. 'Power quality enhancement in wind-grid interface based on dynamic filter compensator'. *International Journal of Distributed Energy Resources and Smart Grids*. October 2013, vol. 9(4), pp. 367–381.

[29] Sharaf A. M., Abdelsalam A. A. 'A FACTS switched filter compensation scheme for smart grid-distribution nonlinear loads'. *International Journal of Distributed Energy Resources*. 2012, vol. 8(3), pp. 169–183.

[30] Abdelsalam A. A., Gabbar H. A., Sharaf A. M. 'Performance enhancement of hybrid AC/DC microgrid based D-FACTS'. *International Journal of Electrical Power & Energy Systems*. 2014, vol. 63(1), pp. 382–393.

[31] IEEE Std 519-1992. 'IEEE Recommended Practices and Requirements for Harmonic Control in Electrical Power Systems'. 1993.

Index